Socioecology and Biodiversity Conservation—2nd Edition

Socioecology and Biodiversity Conservation—2nd Edition

Guest Editor

Alfredo Ortega-Rubio

MDPI

Basel • Beijing • Wuhan • Barcelona • Belgrade • Novi Sad • Cluj • Manchester

Guest Editor
Alfredo Ortega-Rubio
Centro de Investigaciones
Biológicas del Noroeste
Planeación Ambiental y
Conservación
La Paz
Mexico

Editorial Office
MDPI AG
Grosspeteranlage 5
4052 Basel, Switzerland

This is a reprint of the Special Issue, published open access by the journal *Diversity* (ISSN 1424-2818), freely accessible at: https://www.mdpi.com/journal/diversity/special_issues/E2A93QURUJ.

For citation purposes, cite each article independently as indicated on the article page online and as indicated below:

Lastname, A.A.; Lastname, B.B. Article Title. *Journal Name* **Year**, *Volume Number*, Page Range.

ISBN 978-3-7258-8167-3 (Hbk)
ISBN 978-3-7258-8168-0 (PDF)
https://doi.org/10.3390/books978-3-7258-8168-0

Contents

About the Editor vii

Preface ix

Muhammad Syazwan Omar, Rona Dennis, Emily Mae Meijaard, Syafiie Sueif, Syahmi Zaini, Muiz Mohamdih, et al.
Centering Communities in Biodiversity Monitoring and Conservation: Preliminary Insights from a Citizen Science Initiative in Kalimantan, Indonesia
Reprinted from: *Diversity* **2025**, *17*, 679, https://doi.org/10.3390/d17100679 1

Sérgio Murilo de Souza Filho, Marco Tadeu Grassi, Mayara Padovan dos Santos, Juliano Morimoto, Marcelo Soeth and Luís Fernando Fávaro
Differential Accumulation of Particulate Pollutants in Gills and Gastrointestinal Tracts in *Sphoeroides* Fish from Tropical and Subtropical Estuaries in Brazil
Reprinted from: *Diversity* **2025**, *17*, 300, https://doi.org/10.3390/d17040300 17

M. Shafiqul Islam, Hossain Zamal, Muhammad Shah Alam, Md. Maheen Mahmud Bappy, Abu Hena Mustafa Kamal, M. Belal Hossain and Takaomi Arai
Biodiversity and Seasonal Succession of Macrobenthos in Saltmarsh Habitat Adjacent to a Ship-Breaking Area
Reprinted from: *Diversity* **2025**, *17*, 231, https://doi.org/10.3390/d17040231 30

Miriam Sosa and Antonina Ivanova
Assessment of Financing for Biodiversity Conservation in Mexico: Links Between Biodiversity and Climate Change Adaptation Funds
Reprinted from: *Diversity* **2025**, *17*, 185, https://doi.org/10.3390/d17030185 46

Ana Cano-Ortiz, Carmelo Maria Musarella and Eusebio Cano
Phytosociological Research and Meaningful Learning: Constructivist Approaches for Teaching Vegetation
Reprinted from: *Diversity* **2025**, *17*, 172, https://doi.org/10.3390/d17030172 70

Jonathan Pérez-Flores and Jorge Omar López-Martínez
Understanding Wildlife Biodiversity Awareness: Rural Children's Knowledge, Attitudes and Perceptions of Conservation in the Selva Maya
Reprinted from: *Diversity* **2025**, *17*, 152, https://doi.org/10.3390/d17030152 87

Ian D. Rotherham
How Ancient Trees and Botanical Indicators Evidence Both Change and Continuity
Reprinted from: *Diversity* **2025**, *17*, 118, https://doi.org/10.3390/d17020118 105

Pavol Prokop, Simona Todáková and Jana Fančovičová
Beauty Bias? Exploring the Influence of Attractiveness on Conservation Intentions for Plants and Their Pollinators
Reprinted from: *Diversity* **2025**, *17*, 71, https://doi.org/10.3390/d17010071 128

Andrey A. Chemagin, Elena I. Popova and Martin Schletterer
Effects of Flow Velocity on the Dynamics of Juvenile Fish Habitats in River Meanders of the Irtysh River
Reprinted from: *Diversity* **2025**, *17*, 68, https://doi.org/10.3390/d17010068 144

Cleo M. Gaganis and Andreas Y. Troumbis
Bayesian Inference of Human-Made Hazards in Networks of Island Wetlands: The Case of the Aegean Archipelago
Reprinted from: *Diversity* **2025**, *17*, 54, https://doi.org/10.3390/d17010054 . . . . **163**

Hejie Wei, Ke Wang, Yu Ma, Qingxiang Meng, Yi Yang and Mengxue Liu
Exploring the Relationship between Ecosystem Services and Sustainable Development Goals for Ecological Conservation: A Case Study in the Hehuang Valley of Qinghai-Tibet Plateau
Reprinted from: *Diversity* **2024**, *16*, 553, https://doi.org/10.3390/d16090553 . . . . **183**

Elizabeth Olmos-Martínez, Heidi Leticia Romero-Schmidt, María del Carmen Blázquez, Camilo Arias-González and Alfredo Ortega-Rubio
Human Communities in Protected Natural Areas and Biodiversity Conservation
Reprinted from: *Diversity* **2022**, *14*, 441, https://doi.org/10.3390/d14060441 . . . . **209**

Immaculate K. Constantine, Richard A. Giliba and Angela G. Mkindi
Ethnobotanical Knowledge and the Distribution of Pesticidal Plants in Northern Tanzania: A Multi-Ethnic Perspective
Reprinted from: *Diversity* **2026**, *18*, 253, https://doi.org/10.3390/d18050253 . . . . **230**

Gisela Kaplan
A Step Too Far: Culling a Native Australian Honeyeater, the Noisy Miner (*Manorina melanocephalis*), for 'Conservation': Biases, Contradictions, and Myth-Making
Reprinted from: *Diversity* **2026**, *18*, 99, https://doi.org/10.3390/d18020099 . . . . **250**

About the Editor

Alfredo Ortega-Rubio

Alfredo Ortega Rubio, Ph.D. in Ecology, has been granted 123 awards, including three national awards presented by the President of Mexico, for his career in scientific research applied to biodiversity conservation and local communities' improvement in Natural Protected Areas. He is the author of 242 original research articles, has published 29 books and 125 book chapters, and has led 171 strategic projects of resource management, ecological restoration, and environmental impact assessment, including those required to create the Sierra de La Laguna, the Vizcaíno Desert, and the Revillagigedo Archipelago Biosphere Reserves. He has directed 86 theses, which include 48 Ph.D. and 32 Master of Sciences. Since 2004 he has been a Researcher "E", the top Researcher level at the CONACYT Research Center System, and has been ratified by the fourth consecutive time as National Researcher Level III, the top category. He has now been promoted to National Emeritus Researcher.

Preface

Due to worldwide co-evolutive reasons, hot spots of biodiversity coincide with hot spots of cultural, and even linguistic, diversity. Therefore, to preserve global biodiversity, socio-ecological studies on local communities must be conducted. These studies must take into account the social, economic and cultural needs and aspirations of the local communities in relation to biodiversity conservation policies. This contribution will help us to understand social phenomena in light of ecological knowledge, and how cultural and economic aspects influence the use, management and conservation of their ecological environment. Also, it provides a theoretical basis, methodological tools and key case studies, which contribute to the advancement of the socio-ecological perspectives for biodiversity conservation. I hope this contribution will have a dramatic impact on the next generations of researchers and professionals contributing to biodiversity conservation all over the world.

Alfredo Ortega-Rubio
Guest Editor

diversity

MDPI

·le

·ntering Communities in Biodiversity Monitoring and ·nservation: Preliminary Insights from a Citizen Science ·itiative in Kalimantan, Indonesia

·ammad Syazwan Omar *, Rona Dennis, Emily Mae Meijaard, Syafiie Sueif, Syahmi Zaini, Muiz Mohamdih, ·i Erman and Erik Meijaard

Borneo Futures, Bandar Seri Begawan BS 8811, Brunei; rona.dennis@borneofutures.org (R.D.); emilymeijaard@borneofutures.org (E.M.M.); syafiie.sueif@borneofutures.org (S.S.); syahmi.zaini@borneofutures.org (S.Z.); muiz.wafiy@borneofutures.org (M.M.); andiiptkk8009@gmail.com (A.E.); emeijaard@borneofutures.org (E.M.)

* Correspondence: syazwan.omar@borneofutures.org

Abstract: This paper presents preliminary findings on the effectiveness of a citizen science initiative that engages local communities in rural Kalimantan in collecting wildlife observations within their village forests. By leveraging the power of community participation, the initiative aims to build on local knowledge, promote sustainable management practices, and collect valuable data on species distribution. Through a combination of focus group discussions, training workshops, field surveys, and mobile app-based data collection from 2023 to 2025, the initiative successfully mobilized community members, particularly those with limited technological experience, to actively participate in biodiversity monitoring. We recently introduced a small 'payment for wildlife observations' system that significantly boosted observations. The initial results highlight the potential for citizen science to generate valuable species trend data and foster a sense of pride, ownership, and stewardship among community members. While the current manuscript does not provide statistical analyses of the wildlife data, we describe how we plan to overcome data biases that are inherent to opportunistic, unstructured survey efforts. The project continues, but the lessons learned thus far can inform future citizen science initiatives and contribute to the development of sustainable, long-term, low-cost and effective community-based conservation strategies in the region.

Keywords: citizen science; community-based conservation; biodiversity monitoring; indigenous knowledge; sustainable resource management

1. Introduction

Global biodiversity loss is a pressing issue, and effective and inclusive monitoring is crucial to assess environmental changes, evaluate the impact of policies and management practices, and understand the benefits that biodiversity provides to people. Citizen science (CS), which involves the public in data collection, often as part of scientific research, offers a promising approach to address this challenge, particularly in the environmental sciences [1]. By engaging local communities with unique ecological knowledge but limited scientific training, CS can expand the reach and frequency of biodiversity monitoring, especially in biodiversity-rich but remote regions.

CS initiatives have been applied across a wide range of disciplines, including biodiversity monitoring, climate change studies, and informing global and public health policies [2–4]. These projects typically rely on the participation of volunteers to gather data that would otherwise be difficult or impossible to collect due to resource constraints. In many cases, CS initiatives have successfully produced data that are comparable in volume and quality to that collected by professional scientists, given that appropriate training and oversight are in place [5].

As environmental changes directly impact communities, their involvement in conservation is not just beneficial but essential. Integrating their knowledge and experiences through CS offers a socially inclusive, context-specific approach to biodiversity monitoring and protection. By empowering local communities to take ownership of the research process, CS can bridge the gap between scientific inquiry and traditional ecological knowledge, fostering a deeper understanding of local ecosystems and promoting management practices that maintain wildlife populations.

In this study, we explore the implementation of a CS initiative in Kalimantan (Indonesian Borneo), a region renowned for its rich biodiversity [6,7] and part of Sundaland, one of the top five global biodiversity hotspots [7]. By leveraging the power of community participation, we aim to demonstrate the effectiveness of CS in promoting biodiversity conservation and empowering local communities to be an integral part of the solution.

Despite the growing adoption of CS in rural and remote areas, there are substantial gaps in our understanding of its effectiveness in these contexts [8]. Concerns persist regarding the accuracy and reliability of data collected by community members, particularly in cases where participants may have limited formal education or scientific training. Furthermore, because CS surveys are often unstructured, resulting in opportunistic wildlife encounters, standard statistical models are challenging to apply [9]. Additionally, there is a lack of research examining the factors that motivate participation and sustain engagement over time, especially if communities do not recognize the direct benefits. The long-term sustainability and opportunity for impact of CS initiatives within these unique socio-cultural and environmental landscapes are not well understood. Addressing these gaps is essential to ensuring that our CS initiative not only generates robust data but also contributes meaningfully to local conservation efforts and community empowerment.

This study aims to:

1. Evaluate the quality and reliability of data generated through CS initiatives in select rural communities in Indonesian Borneo, with an emphasis on identifying the variables that influence data accuracy and exploring methods to enhance data integrity within these contexts;
2. Examine the motivations for participation, levels of engagement, and demographic diversity of CS participants in these communities, with a particular focus on the roles of different demographic groups, including women, youth, and elders, and the integration of traditional knowledge into the data collection process;
3. Analyze the long-term impacts of CS involvement on both the participants and the scientific outcomes, including how participation influences local attitudes toward environmental conservation and the propensity for continued engagement in scientific and conservation activities;
4. Develop evidence-based recommendations for the design and implementation of CS projects in rural communities, ensuring that initiatives are culturally sensitive, sustainable, and capable of producing high-quality scientific data while promoting local community engagement and empowerment.

2. Materials and Methods

2.1. Study Area

This initiative was conducted across four village forest areas (hutan desa) in Kalimantan, Indonesia (referred to here as Village A, Village B, Village C, and Village D). These communities are situated in a tropical rainforest landscape in Kalimantan, characterized by rich biodiversity and varying degrees of human disturbance. The exact village locations remain anonymous to respect the confidentiality of these villages whilst the project is in place. The participating villages all contain village forests, a social forestry scheme in which state forest areas are managed by the village to improve village welfare [10].

Participants in the monitoring program are generally from the Iban Dayak and Malay ethnic groups. Their daily livelihoods are often shaped by customary gender roles. Iban and Malay men typically engage in fishing, farming, hunting, and the harvesting of forest products, often spending extended periods in forested or riverine areas [11]. Iban women commonly participate in agriculture and the collection of forest resources, while Malay women are often involved in fishing, fish processing, gardening, and gathering non-timber forest products [12]. These gendered livelihood practices and daily movements contributed to broad spatial and temporal coverage in biodiversity observations.

2.2. Local Engagement and Contextual Knowledge

The monitoring program engaged over 300 unique local observers, who contributed wildlife observations on both a voluntary and an incentivized basis. Participation was open to all community members. Before participation, all observers were introduced to the project's goals and methods through a First, Prior, and Informed Consent (FPIC) process, with which they had to agree to participate.

To gain a clearer understanding of the communities involved in this program and to complement the collected observation data, we conducted focus group discussions (FGDs) and semi-structured interviews in all participating villages to determine the social baseline. This process aimed to capture local perceptions, priorities, logistical challenges, and the community's understanding and experience of the program. Such information was essential for interpreting ecological data within its social context and for ensuring that the monitoring approach remained grounded in and responsive to local realities.

Semi-structured interview data were primarily collected through our local facilitator, who lives near the participating villages and maintains regular informal contact with community members, both in person and through WhatsApp (version 2.23–2.25 since 2023). The local facilitator is uniquely qualified for this role, not only as he originates from the area, but he has also worked on numerous community livelihood and conservation projects in this area over the past 25 years. Through open-ended conversations, the facilitator collected insights on wildlife population conditions, village priorities, forest use, and views on the program. Discussions also addressed practical aspects, including access to banking services, mobile phones, and digital payment systems. WhatsApp groups created for each village, initially intended for coordination, became valuable platforms for ongoing communication. Observers used them to share wildlife photos, discuss species identification, raise questions, and report issues. Program staff also used these channels to respond directly, offer clarifications, and gather feedback. After each village visit, the facilitator documented key observations and community feedback in short internal reports, capturing both positive and critical views to support program refinement.

Further observational data were obtained during field visits by a core team member who assessed how monitoring tools were used and engaged in informal discussions with villagers. These visits provided insight into participation patterns, barriers to technology use, and broader community engagement, which informed the development of our program. When presented with satellite imagery on project tablets, villagers demonstrated detailed spatial knowledge of their environment, quickly identifying landmarks, orangutan nest sites, old fruit gardens, rivers, and forest features. Targeted inquiries, such as a participatory orangutan (*Pongo pygmaeus*) population survey using the Pooled Local Expert Opinion (PLEO) method [13], enriched our understanding of wildlife perceptions, in this case, specifically about perceived orangutan population trends. The PLEO method works by systematically pooling estimates from multiple local sources and combining them through statistical weighting to derive more reliable mammal population estimates in tropical rainforests. Through a combination of local facilitation, digital communication, field-based observation, and focused qualitative approaches, community perspectives were consistently incorporated into the monitoring process. This integration informed not only the program's outcomes but also its structure, communication methods, and adaptive management strategies.

2.3. *Wildlife Data Collection and Verification*

Data collection through our program has been ongoing since May 2023 and was designed to emphasize local participation, ownership, and practicality over conventional scientific rigor. A combination of qualitative and digital tools was used, including ESRI's Survey123 [14], Field Maps, and traditional methods such as participatory mapping. Participatory mapping was conducted monthly through interviews with local residents, as well as through Focus Group Discussions (FGDs) held in each village. Since March 2024, community members have been using mobile applications, including Survey123 and Field Maps, on smartphones and tablets to record biodiversity observations. These observations include direct sightings, vocalizations, and indirect signs such as nests, footprints, claw marks, and feeding traces. Each submission typically includes the observer's names, date and time, species name, location name with GPS coordinates, a description or notes about the observation, and supporting evidence in the form of a photo/audio recording, and/or video.

All observations submitted by community members are reviewed by a trained verification team of general wildlife experts, including specific taxonomic experts for difficult species. This team applies rule-based criteria and uses field guide references to assess the validity of each submission. Verification involves cross-checking species identification, evaluating the quality of photographic or audio evidence, and assessing spatial and temporal consistency.

2.4. *Ethical Considerations*

The FPIC process ensured that all participants understood and agreed to the goals, risks, and benefits of their involvement. We regularly re-check the FPIC process to see if any changes to the approach are necessary. Data privacy was maintained by anonymizing observer identities and generalizing sensitive location data in reports and publications.

2.5. *Data Collection Methods*

The community-based observation program was implemented in three distinct phases since project commencement: voluntary, target-based incentive, and payment rewards for observations.

2.5.1. Phase 1: Voluntary Data Collection (May 2023–August 2024)

During the initial voluntary phase, data were gathered through participatory mapping involving sketch maps and semi-structured interviews, including Pooled Local Expert Opinion surveys to determine perceived orangutan trends (Text S1, Supplementary Materials). Designated village-based collectors conducted ground-level data collection by traveling across hamlets (dusun) to engage with local residents. Observations were initially marked with physical stickers on printed maps and subsequently digitized in ArcGIS for spatial analysis and storage.

2.5.2. Phase 2: Incentive-Based Tablet Mapping (September–December 2024)

To enhance participation, an incentive-based approach was introduced in September 2024. Each participating village received a digital tablet equipped with the ArcGIS Field Maps application, alongside a goal of submitting 600 observations per village within four months. Successful completion of this target entitled the community to retain the tablet permanently.

2.5.3. Phase 3: Payment for Observations (January 2025–Ongoing)

In late January 2025, a pilot monetary incentive system was introduced in one village. Verified orangutan sightings were compensated at USD 3 and orangutan nest observations at USD 0.90. This pilot yielded a notable increase in reporting rates, averaging 300–400 submissions per month. Given the success, the scheme was expanded to all four participating villages starting on 1 May 2025, along with an expanded list of reward-eligible species. This list primarily includes fauna of cultural significance to the community, as well as species of national and global conservation concern.

3. Results

3.1. Participant Engagement and Data Quality

3.1.1. Participant Engagement

Most participants are knowledgeable about wildlife and have positive attitudes toward wildlife, and some noted that it now feels more financially rewarding to record bird calls than to sell birds in the market. In terms of feedback, the project shares information on the global conservation status of species, the scientific names, and where certain bird species appear the same but are scientifically recognized as two distinct species. Nonetheless, traditional hunting of specific species such as pigs and deer for protein [15] and song birds, snakes. and turtles for external markets prevails. The community is generally aware of which species are protected by law, although there is little on-the-ground enforcement by government authorities.

Participants have also demonstrated a high level of engagement and enthusiasm, with many reporting frequent observations of wildlife around their homes and farmland. Common sightings include orangutan nests, a variety of bird species, squirrels, and macaques. This underscores the accessibility of biodiversity data collection in these rural settings, where wildlife encounters are a regular part of daily life.

3.1.2. PLEO Interviews

From May to July 2024, a total of 92 respondents across four villages participated in the PLEO interviews regarding orangutan population trends. Respondents ranged in age from their 20 s to over 80, with the majority aged 41–60. The pooled orangutan population estimates were 2135 orangutans in 2000, 1067 in 2019 and 870 in 2024, indicating a perceived decrease of 59.3% across 24 years (Table 1). We note that this is in line with other population estimates from the same area and a similar sampling frame [16].

Table 1. Orangutan population estimates generated with the PLEO method for each of the four villages in the years 2000, 2019, and 2024.

Villages	Year		
	2000	2019	2024
A	419.1 (2.0)	164.2 (2.5)	61.8 (4.0)
B	1322.4 (1.3)	774 (1.2)	750 (1.0)
C	273.7 (1.7)	89.6 (1.9)	30.2 (1.8)
D	119.8 (2.2)	39.1 (1.8)	25.9 (1.6)
Total	2135.0	1066.9	867.9

3.1.3. Observation Data

Since program initiation, over 300 community observers from diverse demographic backgrounds have actively contributed to data collection efforts. As of September 2025, they have collectively recorded a total of 19,069 observations of 165 species using multiple data collection methods since the program began in May 2023. Of these, 15,891 observations (83%) were recorded using Survey123, while 2645 observations (14%) were gathered through Field Maps. The remaining 533 observations (3%) were obtained through participatory mapping using stickers and focus group discussions (FGDs), incorporating traditional knowledge and community-driven spatial documentation. Among these observers, 49 observers have recorded more than 100 observations, and the top 5 contributors across the four villages have individually recorded between 1023 and 47 observations, underscoring the dedication of key individuals in biodiversity monitoring (Figure 1).

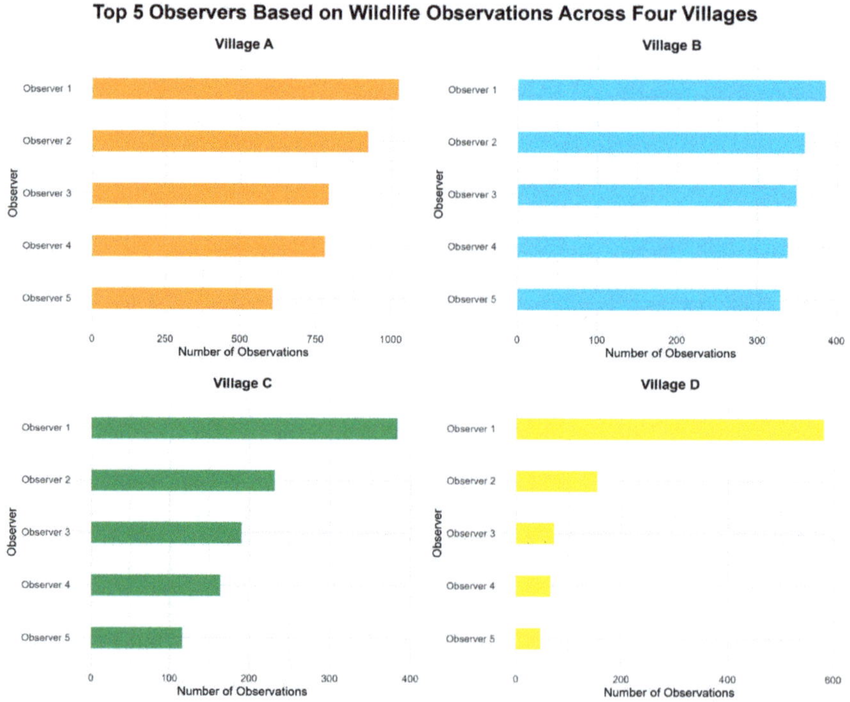

Figure 1. List of the top 5 observers in each village.

Observation submission varied significantly across the three phases, reflecting the changes in community motivation driven by incentive structures introduced in the different phases (Figure 2). During phase 1, participation remained relatively low and steady, averaging 17 observations per village per month. Observations were collected voluntarily through participatory mapping without any direct incentive, resulting in a low submission rate that served as a baseline for subsequent phases.

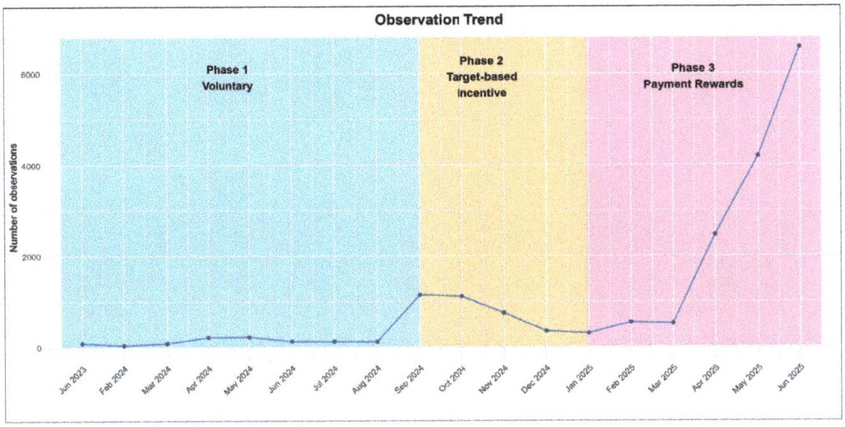

Figure 2. Observation trends since the initiation of the program in May 2023.

A sharp increase in observations occurred during phase 2. With the introduction of digital tablets and a community-level reward (permanent tablet ownership upon reaching 600 observations), engagement surged. This approach resulted in high engagement, with participants collectively submitting 2645 observations, exceeding the initial target. This phase demonstrated the motivating power of collective, tangible incentives. Phase 3 further amplified the participation rate, initially piloted in one village in late January and expanded to all four villages by May 2025. As shown in Figure 2, this approach resulted in a sustained increase in observation activity. Initial reporting averaged 300–400 observations per month, but the number of observations surged in the subsequent months to 2000, 4000, and eventually over 6000 observations per month. By June 2025, the cumulative total had reached 14,085 observations in phase 3. This increase was partly attributed to the expansion of the list of reward-eligible species to include taxa of cultural, national, and global conservation significance, but also due to increased familiarity with the smartphone application, and understanding of the program objectives. These trends indicate a positive correlation between the incentives and community engagement. The progression from voluntary to monetary rewards resulted in measurable increases in participation, showing the effectiveness of phased, adaptive motivation strategies in CS within rural communities. We have built in a dynamic and flexible financial reward structure, which means it can change to suit both requests from communities and our data requirements.

Across the three monitoring phases, a total of 104 unique species were recorded during phase 1, increasing to 133 species in phase 2, and then slightly decreasing to 120 species in phase 3 (Figure 3). On average, villages A, B, C, and D recorded 64, 72, 68, and 60 unique species across the three phases, respectively. The higher species count in phase 2 likely reflects participants recording any wildlife sightings, as they aimed to reach the target of 600 observations, while the slight decline in phase 3 coincides with the introduction of a paid species list, which guided community observations toward specific species eligible for rewards.

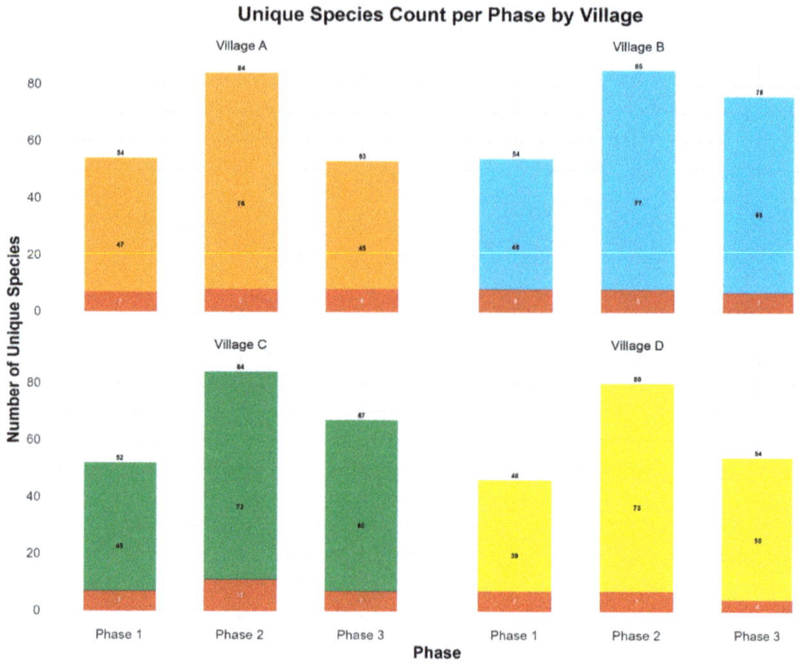

Figure 3. The number of unique species and rare or endangered species identified during each phase across the four villages (red indicates the number of rare or endangered species).

3.1.4. Data Quality

Among the recorded biodiversity observations, several species have been documented over 100 times. The distribution of these community wildlife observations across the study villages is shown in Figure 4. The most frequently reported are orangutan nests (9766 observations), white-rumped shamas (*Copsychus malabaricus*, 1814 observations), signs of sun bears (*Helarctos malayanus*, 467 observations), stingless bees (*Trigona* spp., 370 observations), Bornean orangutans (*Pongo pygmaeus*, 316 direct observations), long-tailed macaques (*Macaca fascicularis*, 305 observations), and Oriental magpie-robin (*Copsychus saularis*, 247 observations). These records include both direct and indirect observations, such as photographs or videos of orangutans, audio recordings of gibbon and bird calls, as well as traces such as footprints and claw marks from sun bears.

3.2. Motivation

The introduction of incentives led to a noticeable increase in submitted observations (Figure 2). For most participants, the idea of being rewarded for daily wildlife monitoring was novel. Previously, compensation was limited to short-term contracts or daily labor assisting non-governmental organizations (NGOs) with camera traps or survey work. A total of 88 community observers received financial rewards through the wildlife monitoring incentive scheme. The average monthly payments to each participating village are summarized in Table 2. Village A began participating earlier than others, as the reward system was first trialed there with payments focused on orangutan and orangutan nest observations. The program was subsequently expanded to village B, followed by full implementation in the remaining two villages, C and D.

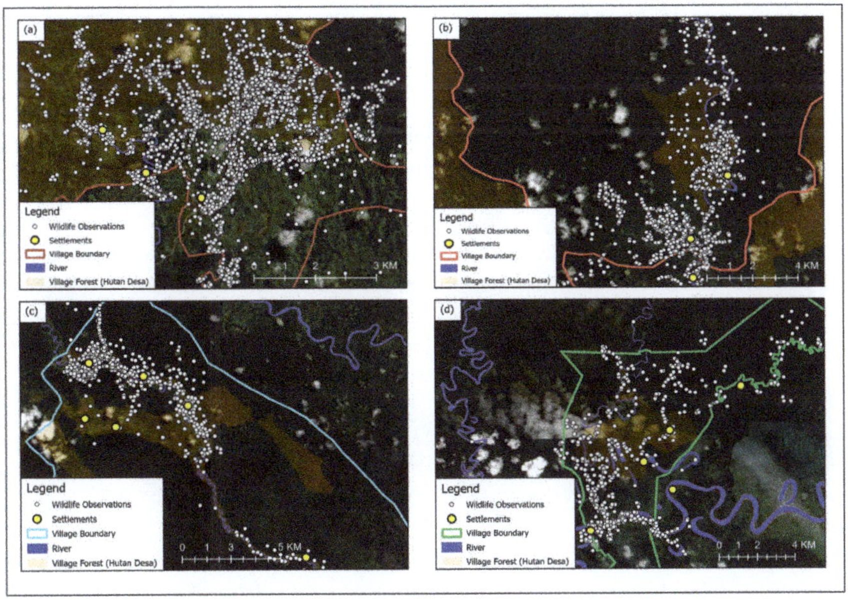

Figure 4. The distribution of community wildlife observations across four villages in rural Kalimantan: (**a**) Village A, (**b**) Village B, (**c**) Village C, and (**d**) Village D. The full administrative boundaries of the villages extend beyond the areas regularly used by communities and are therefore not shown.

Table 2. The average monthly payments distributed across the four participating villages.

Village	Average Monthly Payment (USD)
A	986
B	613
C	61
D	162

Across all observers, the mean monthly total payment per individual was approximately USD 80, with a 95% confidence interval ranging from USD 57 to USD 106. Variation in the total amount received per observer was considerable, reflecting differences in observation frequency, consistency, and individual participation levels. Around 30% of observers received payments above the average, while approximately 20% received relatively low amounts. This model of rewarding routine monitoring is both novel and empowering. Some participants even noted that recording wildlife observations was more financially rewarding than a day of manual labor, which typically pays around USD 9.

3.2.1. Observation Trends and Cost Comparison

The impact of introducing incentives is reflected in the volume of observations submitted over time. There were 1014 observations collected during the voluntary phase, using a combination of informal interviews and participatory mapping using stickers on paper maps. Observations were regularly recorded opportunistically during daily activities like fishing, farming, and going into the forest.

The target-based incentive introduced from September to December 2024 (Figure 2) resulted in 2645 observations in the four months. With the rollout of payment rewards for valid wildlife observations in 2025, expanding participation using Survey123, yielding 14,085 observations by June 2025.

The shift not only reflects increased volume but also a shift in behavior. Many observations are now made intentionally, with participants going into the forest specifically to document wildlife. To ensure data accuracy, it is required that each observation includes photos and audio recordings as supporting evidence. This has improved data quality and verification, making the dataset more robust. The incentive model also includes an incentive value based on species prioritized by the community, with higher rewards for species considered important in terms of conservation and cultural significance.

Cost analysis of this approach resulted in an average cost of USD 0.85 per observation, and aside from travel expenses incurred by the local facilitator in distributing payments across the four villages, this approach has proven to be cost-effective. While traditional biodiversity surveys conducted by trained researchers typically involve substantially higher costs, direct comparisons are not straightforward due to differences in methodology and scale. For example, scientific field teams conducting transects typically incur costs estimated at approximately USD 95 per transect [17], while camera trap monitoring requires a substantial upfront investment of around USD 401 per unit, covering hardware, memory cards, and batteries [18]. Deployment and maintenance costs are significant per recorded observation, in addition to field labor and data processing costs. A similarly run CS-based initiative in oil palm plantations with company staff, contractors, and external visitors equates biodiversity monitoring to approximately USD 0.14 per hectare [19], illustrating that CS-based approaches can collect large volumes of biodiversity data at relatively low cost compared with traditional methods.

3.2.2. Trust and Credibility

During the program, there were incidents of data falsification in which observers submitted orangutan records later identified as sourced from the internet. The verification team detected inconsistencies between the submitted media and the reported GPS location, prompting further investigation. Swift coordination between the local program coordinator and the village head led to direct clarification with the individual involved, who admitted to the falsification. Consequently, the observer was removed from the program. The community regarded this breach as a serious matter that threatened their integrity and the trust established between the community and program staff, and they supported the decision to impose sanctions.

Overall, participants tend to track their submissions and estimate their expected payments. When discrepancies arise, such as rejected entries, they are generally accepting of decisions by the verification team, especially when rejections are accompanied by clear explanations (e.g., due to unclear photos, audio, or descriptions, or duplications). Trust in the program depends heavily on the consistency and transparency of the verification and payment system. While occasional setbacks like system downtime can temporarily reduce confidence in the program, timely payments and effective communication, alongside responsive technical support, have been critical in restoring and maintaining long-term credibility.

3.2.3. Technology Adaptation

Most community members own smartphones but have limited prior experience with using mobile apps for recording observations. However, participants quickly adapted to using Survey123 and Field Maps, particularly when the benefits were tangible. In addition to individual adaptation, many participants have demonstrated strong peer-to-peer support. Those without smartphones have been supported by others willing to lend their devices. Some have even suggested entering the forest together to observe orangutan nests and other species, then sharing the benefits equally. Nevertheless, we recognize that access to mobile phones might be a barrier to participating in the program, potentially resulting in elite capture of the financial benefits. As this could undermine the community-level perception changes that we aim to seek, we are looking into ways of increasing mobile phone access for those who currently lack such access.

3.3. Data Biases

While our manuscript presents preliminary findings regarding the uptake of citizen science-based wildlife monitoring, and not yet statistical analyses of temporal changes in wildlife populations, we are aware of concerns about data biases. Citizen science data are inherently prone to spatial, temporal, and observer-related biases [9], and our pay-for-observation model introduces additional challenges that require explicit attention. Preliminary data analyses reveal that observations cluster near villages, rivers, and roads, with many areas receiving little or no survey effort, and the absence of "zero-visit" locations currently limits inference about true absences. To address this, we are testing estimates of survey effort (e.g., time spent in grid cells), and locations without observations so that future wildlife occupancy models can incorporate explicit measures of effort and detection probability.

Repeated visits to grids with previous sightings, particularly of orangutans and gibbons, risk inflating encounter rates, but the temporal sequence of detections and non-detections across many grids provides the replication structure needed for robust Bayesian occupancy models. This is something we have tested in a different citizen science setting, resulting in robust modeling outputs [19,20]. Those studies, however, were based on voluntary wildlife observation. The payment system in our current program creates potential additional behavioral biases, such as artificial attractors or an overemphasis on rewarded species. We are mitigating these risks through strict photo and audio verification, metadata checks, random audits, and by periodically adjusting incentives to reward both survey effort and species detections. Variation in observer skill remains another source of noise, but training workshops, regular feedback, and an active verification team are helping improve data quality over time.

Although direct sightings of rare species, such as orangutans and gibbons, remain sparse, the combination of improved data collection protocols, strict verification, and occupancy modeling provides a clear path toward reducing bias while maintaining the advantages of broad community participation in wildlife monitoring. Future statistical analysis of citizen science data and cross-checking with independently derived population estimates from drone surveys should clarify the extent to which opportunistic, unstructured citizen science monitoring can result in statistically robust occupancy estimates.

4. Discussion

4.1. Lessons Learned

Local communities have proven to be invaluable contributors to biodiversity monitoring. Their deep understanding of the environment and sharp observational skills, detecting species from tiny frogs to secretive, nocturnal tarsiers, have significantly improved both the quality and quantity of data collected. By integrating traditional ecological knowledge with modern tools such as ESRI's Survey123 and Field Maps, the project has been able to monitor species distribution. While data analysis is still in its early stages, preliminary occupancy modeling has been conducted, providing initial insights into species distribution and habitat use. In a future study, we aim to more rigorously test the statistical robustness of the species occupancy models derived from citizen science and compare them to other data sources.

The progressive changes in unique species counts across the three phases highlight how the program design influenced community-based data collection. The increase in phase 2 suggests that allowing participants to record all wildlife encouraged diverse observations, whereas the slight decline in phase 3 reflects the introduction of a paid species list, which focused community effort on specific species. Rare and endangered species were detected throughout the study, demonstrating that guided community participation can generate valuable biodiversity data (Figure 3). We have found that observers know exactly where to find certain species, such as orangutans (Figure 5). They have a deep knowledge of seasonality in species presence and are quick to identify tracks and signs.

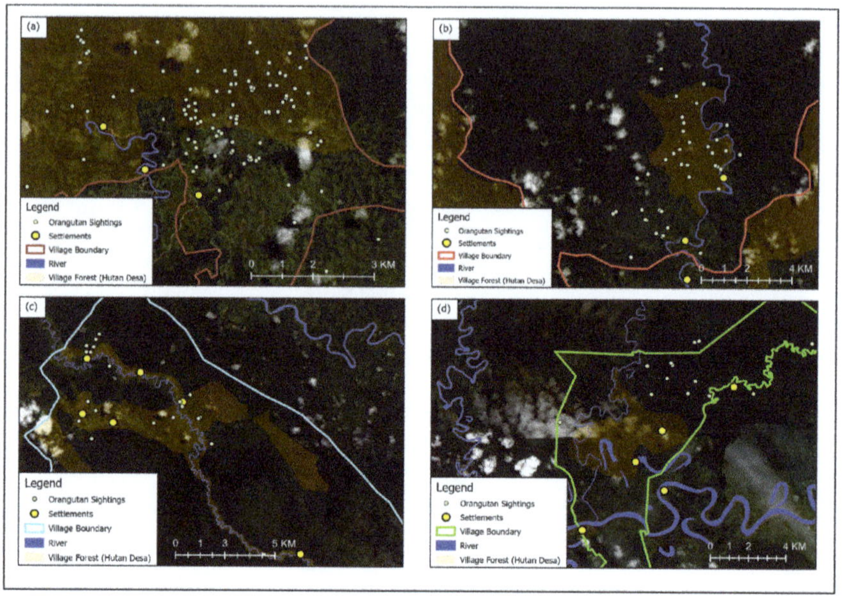

Figure 5. The distribution of orangutan sightings across the four villages: (**a**) Village A, (**b**) Village B, (**c**) Village C, and (**d**) Village D.

Preliminary findings suggest that payments for wildlife observation may encourage conservation-relevant behavior in local communities, consistent with other studies exploring whether financial incentives translate into conservation outcomes [21–27]. Research on conditional payments for ecosystem services and cash transfer programs shows that financial incentives can support conservation even when programs are not directly aimed at environmental goals [21,23,26]. However, some evidence suggests that unconditional payments may increase land clearance in certain contexts, highlighting the importance of careful program design to ensure incentives support, rather than replace, local motivations to protect wildlife [26,28,29]. While payments may be a useful tool for promoting positive conservation behavior, their long-term effects and how they interact with local knowledge, engagement, and voluntary stewardship should be closely monitored.

Although the program is still in its early stages, anecdotal information from interviews suggests that continued engagement may foster greater interest in biodiversity, management of threats to wildlife, and increased attentiveness to ecological patterns. The initiative has also provided valuable insights into the dynamics of community engagement, including motivating factors, the influence of incentives, and participants' adaptability to new technologies.

4.2. Challenges and Limitations

While this project has shown promising results, engaging communities in wildlife observation across rural Kalimantan presents several challenges. Sustaining motivation requires continuous engagement, clear incentives, and meaningful feedback to ensure that participants feel their contributions are valued. Technical difficulties, such as system downtime and limited internet access, can disrupt data collection efforts. Language barriers, including the use of different local names for species across villages, further complicate efforts to standardize and validate observations. We also recognize that citizen science can present a risk of bias due to the unstructured and opportunistic nature of sampling and observation, but as described in Section 3.3, we believe that such biases can be effectively managed.

The initial reliance on self-reported data, the unstructured nature of sampling and observation, and varying quality in observers' observations introduce further potential biases and inaccuracies [30]. Risks such as double reporting and exploitation of the payment system can affect data quality if not actively managed. Additionally, limited access to appropriate tools for recording observations restricts broader participation, often leaving only a few individuals actively contributing. However, it is worth noting that participants often demonstrate a strong spirit of cooperation, with many willing to help others by sharing devices or recording observations on their behalf. Overcoming these challenges is essential for building a reliable, inclusive, and sustainable community-driven monitoring program. Implementing quality control measures, such as regular data validation and feedback sessions, will be key to ensuring long-term success and credibility.

4.3. Future Directions

This project is in the early stages of implementation and, as such, is concentrating on encouraging participation, quality of observations, and tailoring the smartphone application to the local situation. The next stage we are planning to implement is to use the citizen science data to generate spatial occupancy estimates. The plan is to use Bayesian occupancy analyses similar to what we used in our 6-year citizen science pilot study in oil palm [19,20]. Furthermore, we aim to ground-truth the species occurrence patterns derived from the citizen science methods with traditional structured surveys, in this case, drone surveys using thermal and optical cameras for orangutans and gibbons. These are planned in year 2 of our study (we are currently in year 1), allowing for sufficient accumulation of citizen science data and resulting spatial distribution patterns of target species, before we ground-truth these.

Equally important is the need to give back to the communities that contribute this data. While financial incentives play a role in motivating participation, long-term sustainability depends on building a deeper sense of ownership and value. This includes regularly sharing the results of their observations, such as visual maps of species sightings, trends in biodiversity, and early warning signs of ecological change, so that communities can see the real-world value of their efforts. In speaking with the communities, it was made clear to us that they take pride in their contributions, and this can stand as motivation in and of itself. To support this, we aim to recognize local observation champions and promote further engagement with the project through regular community bulletins highlighting observations and leading observers.

The end-game is that we secure long-term financial support to pay for observations, which are used to calculate an annual metric for biodiversity that is a measure of how well the community is maintaining its biodiversity values (based on species occupancy against a baseline). We are in the process of seeking private investment or Indonesian government funds (Village Funds) that will provide communities with capital for developing economic activities, as well as financing the ongoing wildlife monitoring and management. We believe that by demonstrating both the environmental and socioeconomic benefits of our program, through presenting change from social baseline assessments, we will appeal to audiences with financing capabilities and interests for our initiative.

By translating raw data into accessible, meaningful information, we not only validate local contributions but also empower communities to make informed decisions about the future of their natural resources. Empowering local communities to lead the initiative ensures long-term viability while fostering trust and reinforcing the value of their traditional knowledge. Collaboration with researchers, conservation organizations, and policymakers can further amplify the project's impact and facilitate broader knowledge exchange. A feedback-rich, participatory approach can contribute to more inclusive and effective biodiversity conservation strategies.

5. Conclusions

The CS initiative demonstrates the value of leveraging indigenous ecological knowledge and redefining the power balance in community-based conservation from the data collection phase to forging more positive local perspectives on human–wildlife interactions. By empowering communities to monitor and manage 'their own' biodiversity, this project has fostered a deeper understanding of the value of nature beyond immediate economic or agricultural concerns. Previously, species like the orangutan were often viewed as a threat to livelihoods in certain circumstances. However, with incentives such as rewards for recording or photographing wildlife, community members are beginning to see these species as assets rather than adversaries. Instead of instinctively driving orangutans away to protect their fruit gardens, they now recognize an alternative value, where documenting these animals provides an economic return that may even exceed their crop losses. This development in mindset is a crucial step toward sustainable coexistence, ensuring that conservation efforts are not just about preserving biodiversity but also about strengthening the positive perceptions and interactions that communities have with their natural environment. Future initiatives should continue to build on these lessons, exploring innovative ways to engage local people in conservation while providing tangible benefits that align with their livelihoods.

Supplementary Materials: The following supporting information can be downloaded at: https://www.mdpi.com/article/10.3390/d17100679/s1. Text S1. Interview guide and control questions for the orangutan population baseline; Figure S1. Observations verified as valid or rejected due to unidentifiable species in the photographs; Figure S2. Field observations of *Egretta* sp. (a) Rejected record due to poor visibility and insufficient detail. (b) Accepted record based on clear evidence confirmed by verifiers; Table S1. Unique species recorded in Village A; Table S2. Unique species recorded in Village B; Table S3: Unique species recorded in Village C; Table S4. Unique species recorded in Village D.

Author Contributions: Conceptualization, E.M. and R.D.; Methodology, E.M., R.D. and A.E.; Software, S.Z.; Validation, M.S.O., S.S., S.Z., M.M., E.M. and R.D.; Formal Analysis, M.S.O., S.Z. and S.S.; Investigation, M.S.O. and A.E.; Data Curation, M.S.O., S.Z. and S.S.; Writing—Original Draft Preparation, M.S.O.; Writing—Review and Editing, R.D., E.M. and E.M.M.; Visualization, M.S.O., M.M. and S.S.; Supervision, E.M. and R.D.; Project Administration, M.S.O., E.M. and R.D.; Funding Acquisition, E.M. and R.D. All authors have read and agreed to the published version of the manuscript.

Funding: This work was supported by the Binnacle Fund and Wildlife Futures.

Institutional Review Board Statement: Ethical review and approval were waived for this study due to the nature of the research. The study was conducted following community-level Free, Prior, and Informed Consent (FPIC) processes, and individual informed consent was obtained from all participants before collecting any data.

Informed Consent Statement: Written informed consent has been obtained from the communities to publish this paper.

Data Availability Statement: The data presented in this study are available on request from the corresponding author. The data are not publicly available due to privacy and confidentiality agreements with local community participants.

Acknowledgments: The authors acknowledge the financial support received from the Binnacle Fund and Wildlife Futures and thank the funders for their trust in supporting this novel project. Deep appreciation is extended to the village leaders and community participants for their enthusiasm and dedication in adopting this new approach and contributing valuable observations. The authors also thank the coordinators and collectors whose commitment ensured smooth project implementation, as well as all individuals involved in training sessions, focus group discussions, and logistical support. Their contributions were instrumental to the success of this phase of the work. During the preparation of this manuscript, the authors used ChatGPT (OpenAI, GPT-4, July 2025 version) for linguistic refinement. The authors have thoroughly reviewed and edited the output and take full responsibility for the content of this publication.

Conflicts of Interest: Authors were employed by the company Borneo Futures. The authors declare that the research was conducted in the absence of any commercial or financial relationships that could be construed as a potential conflict of interest.

Abbreviations

The following abbreviations are used in this manuscript:

CS	Citizen Science
FPIC	First, Prior, and Informed Consent

ferences

Gooding, L.; Pateman, R.M.; West, S.E. Citizen Science and Its Potential for Aiding Low Carbon Energy Transitions. *Energy Res. Soc. Sci.* **2024**, *117*, 103702. [CrossRef]
Katapally, T.R. A Global Digital Citizen Science Policy to Tackle Pandemics Like COVID-19. *J. Med. Internet Res.* **2020**, *22*, e19357. [CrossRef]
Jansen, M.; Beukes, M.; Weiland, C.; Blumer, M.; Rudolfi, M.; Poerting, J.; Meißner, R.; Weiß, M.; Condori, Y.; Aramayo-Ledezma, G.; et al. Engaging Citizen Scientists in Biodiversity Monitoring: Insights from the WildLIVE! Project. *Citiz. Sci. Theory Pract.* **2024**, *9*, 6. [CrossRef]
Albagli, S.; Iwama, A.Y. Citizen Science and the Right to Research: Building Local Knowledge of Climate Change Impacts. *Humanit. Soc. Sci. Commun.* **2022**, *9*, 39. [CrossRef]
Albert, A.; Balázs, B.; Butkevičienė, E.; Mayer, K.; Perelló, J. Citizen Social Science: New and Established Approaches to Participation in Social Research. In *The Science of Citizen Science*; Vohland, K., Land-Zandstra, A., Ceccaroni, L., Lemmens, R., Perelló, J., Ponti, M., Samson, R., Wagenknecht, K., Eds.; Springer International Publishing: Cham, Switzerland, 2021; pp. 119–138, ISBN 978-3-030-58277-7.
MacKinnon, K.; Hatta, G.; Halim, H.; Mangalik, A. *The Ecology of Kalimantan: Indonesian Borneo*; Tuttle Publishing: New York, NY, USA, 2013; ISBN 978-1-4629-0505-8.
Myers, N.; Mittermeier, R.A.; Mittermeier, C.G.; Da Fonseca, G.A.B.; Kent, J. Biodiversity Hotspots for Conservation Priorities. *Nature* **2000**, *403*, 853–858. [CrossRef]
Benyei, P.; Skarlatidou, A.; Argyriou, D.; Hall, R.; Theilade, I.; Turreira-García, N.; Latreche, D.; Albert, A.; Berger, D.; Cartró-Sabaté, M.; et al. Challenges, Strategies, and Impacts of Doing Citizen Science with Marginalised and Indigenous Communities: Reflections from Project Coordinators. *Citiz. Sci. Theory Pract.* **2023**, *8*, 21. [CrossRef]
Van Strien, A.J.; Van Swaay, C.A.M.; Termaat, T. Opportunistic Citizen Science Data of Animal Species Produce Reliable Estimates of Distribution Trends If Analysed with Occupancy Models. *J. Appl. Ecol.* **2013**, *50*, 1450–1458. [CrossRef]

10. Moeliono, M.; Mulyana, A.; Adnan, H.; Manalu, P.; Yuliani, L. *Balang Village Forests (Hutan Desa): Empowerment, Business Burden?*; Agroforestry and Forestry in Sulawesi No. 51; World Agroforestry Centre—ICRAF Southeast Asia Regional Off Bogor, Indonesia, 2015.
11. Colfer, C.J.P.; White, R.L.; Howard, E.; Prabhu, R. *Inter-Generational Access to Resources: Developing Criteria and Indicators*; Cen for International Forestry Research (CIFOR): Bogor, Indonesia, 1997. [CrossRef]
12. Dudley, R.G. The Fishery of Danau Sentarum. *Borneo Res. Bull.* **2000**, *31*, 261–306.
13. Van Der Hoeven, C.A.; De Boer, W.F.; Prins, H.H.T. Pooling Local Expert Opinions for Estimating Mammal Densities in Tropi Rainforests. *J. Nat. Conserv.* **2004**, *12*, 193–204. [CrossRef]
14. Esri. *Survey123 for ArcGIS*, Version 3.23.36. Mobile Application Software. Environmental Systems Research Institute: Redlar CA, USA, 2024. Available online: https://survey123.arcgis.com (accessed on 29 July 2025).
15. Wadley, R.L.; Colfer, C.J.P. Sacred Forest, Hunting, and Conservation in West Kalimantan, Indonesia. *Hum. Ecol.* **2004**, *32*, 313– [CrossRef]
16. Yuliani, E.L.; Bakara, D.O.; Ilyas, M.; Russon, A.E.; Salim, A.; Sammy, J.; Sunderland-Groves, J.L.; Sunderland, T.C.H. Borne Orangutan *Pongo pygmaeus pygmaeus* Population Estimate within and around Danau Sentarum National Park, Kapuas Hu West Kalimantan. *Conserv. Sci. Pract.* **2023**, *5*, e12916. [CrossRef]
17. Franco, A.M.A.; Palmeirim, J.M.; Sutherland, W.J. A Method for Comparing Effectiveness of Research Techniques in Conservati and Applied Ecology. *Biol. Conserv.* **2007**, *134*, 96–105. [CrossRef]
18. Haysom, J.K.; Deere, N.J.; Wearn, O.R.; Mahyudin, A.; Jami, J.B.; Reynolds, G.; Struebig, M.J. Life in the Canopy: Usi Camera-Traps to Inventory Arboreal Rainforest Mammals in Borneo. *Front. For. Glob. Change* **2021**, *4*, 673071. [CrossRef]
19. Maharani, N.; Nardiyono; Munthe, C.R.; Swayanuar, P.; Ni'Matullah, S.; Sueif, S.; Zaini, S.; Supriatna, J.; Kusrini, M.; D nis, R.; et al. A Novel Citizen Science-Based Wildlife Monitoring and Management Tool for Oil Palm Plantations. *Prepr* **2025**, 2025011201. [CrossRef]
20. Van Strien, A.J.; Meijaard, E.; Nardiyono; Suief, S.; Zaini, S. Occupancy Estimation of Wild Species in a Palm Oil Plantation Us Unstructured Data. *Preprint* **2025**. [CrossRef]
21. Ferraro, P.J.; Simorangkir, R. Conditional Cash Transfers to Alleviate Poverty Also Reduced Deforestation in Indonesia. *Sci. A* **2020**, *6*, eaaz1298. [CrossRef] [PubMed]
22. Fletcher, R.; Büscher, B. Conservation Basic Income: A Non-Market Mechanism to Support Convivial Conservation. *Biol. Cons* **2020**, *244*, 108520. [CrossRef]
23. Jayachandran, S.; de Laat, J.; Lambin, E.F.; Stanton, C.Y.; Audy, R.; Thomas, N.E. Cash for Carbon: A Randomized Trial Payments for Ecosystem Services to Reduce Deforestation. *Science* **2017**, *357*, 267–273. [CrossRef] [PubMed]
24. Milne, S.; Niesten, E. Direct Payments for Biodiversity Conservation in Developing Countries: Practical Insights for Design a Implementation. *Oryx* **2009**, *43*, 530–541. [CrossRef]
25. Sheehan, C.; Martin-Ortega, J. Is Conservation Basic Income a Good Idea? A Scoping Study of the Views of Conservati Professionals on Cash Giving Programmes. *Biol. Conserv.* **2023**, *279*, 109914. [CrossRef]
26. Wilebore, B.; Voors, M.; Bulte, E.H.; Coomes, D.; Kontoleon, A. Unconditional Transfers and Tropical Forest Conservati Evidence from a Randomized Control Trial in Sierra Leone. *Am. J. Agric. Econ.* **2019**, *101*, 894–918. [CrossRef]
27. WWF (WWF International). *Main Report: Assessing the Application of a Conservation Performance Payment (CPP) for Communities the Greater Virunga Landscape (GVL)*; WWF: Gland, Switzerland, 2024; 66p.
28. Deci, E.L. Effects of Externally Mediated Rewards on Intrinsic Motivation. *J. Pers. Soc. Psychol.* **1971**, *18*, 105–115. [CrossRef]
29. Deci, E.L.; Koestner, R.; Ryan, R.M. A Meta-Analytic Review of Experiments Examining the Effects of Extrinsic Rewards Intrinsic Motivation. *Psychol. Bull.* **1999**, *125*, 627–668. [CrossRef]
30. Dickinson, J.L.; Zuckerberg, B.; Bonter, D.N. Citizen Science as an Ecological Research Tool: Challenges and Benefits. *Annu. R Ecol. Evol. Syst.* **2010**, *41*, 149–172. [CrossRef]

diversity

MDPI

nunication

fferential Accumulation of Particulate Pollutants in Gills and ıstrointestinal Tracts in *Sphoeroides* Fish from Tropical and ıbtropical Estuaries in Brazil

io Murilo de Souza Filho [1,2], Marco Tadeu Grassi [3,4], Mayara Padovan dos Santos [3,4], Juliano Morimoto [2,5], celo Soeth [6,7] and Luís Fernando Fávaro [1,2,*]

1 Laboratory of Reproduction and Fish Community, Federal University of Paraná, P.O. Box 19031, Curitiba 81531-980, PR, Brazil; sergio.filho@ufpr.br
2 Postgraduate Program in Ecology and Conservation, Federal University of Paraná, Curitiba 82590-300, PR, Brazil; juliano.morimoto@abdn.ac.uk
3 Department of Chemistry, Federal University of Paraná, P.O. Box 19032, Curitiba 81531-980, PR, Brazil; mtgrassi@ufpr.br (M.T.G.); mayara.dossantos@ufpr.br (M.P.d.S.)
4 Postgraduate Program in Chemistry, Department of Chemistry, Federal University of Paraná, P.O. Box 19032, Curitiba 81531-980, PR, Brazil
5 Institute of Mathematics, University of Aberdeen, King's College, Aberdeen AB24 3FX, UK
6 Fisheries Department, Falkland Islands Government, Bypass Road, Stanley FIQQ 1ZZ, Falkland Islands; marcelosoeth@yahoo.com.br
7 Postgraduate Program in Coastal and Oceanic Systems, Federal University of Paraná, P.O. Box 61, Pontal do Paraná 83255-976, PR, Brazil
* Correspondence: lufavaro@ufpr.br

Abstract: The widespread use of recyclable materials in contemporary society has led to the accumulation of pollutants in estuaries and marine ecosystems, with potential impacts on biodiversity. This study assessed the abundance and types of particulate pollutants in *Sphoeroides* fish across two Brazilian estuaries (tropical and subtropical). Our findings showed that 70 biological samples from fish (92.11%) contained debris, with the tropical estuary exhibiting the highest abundance ($n = 499$ particles—67% of the total), dominated by laminar (film) particulate pollutants (76.75%). In this estuary, the gill exhibited the highest contamination index, with most particulate pollutants (<1 mm and 1–3 mm) found in 63.6% and 54.5% of samples, respectively. In the subtropical estuary, 246 debris particles (33% of the total) were detected in the biological samples, with 58.5% of particles being of the film type. The gastrointestinal tract had the highest contamination index in this region, with 70.6% of particles in the 1–3 mm size range. These results highlight the pervasive presence of particulate pollutants in estuarine ecosystems and the organ-specific contamination patterns in tropical and subtropical regions, underscoring the ecological risks posed by plastic waste to estuarine biodiversity in hotspot regions.

Keywords: pollution; aquatic ecosystem; environmental risk; particulate debris

1. Introduction

Estuarine environments are highly productive and provide essential ecosystem services, supporting feeding, reproduction, and nursery functions for diverse animal groups, including both vertebrates and invertebrates [1–4]. However, human activities in these regions, such as urban expansion and industrial development, have led to significant ecological degradation [5,6]. These activities introduce various contaminants, including metals, organic pollutants, pesticides, plastics, and particulate pollutants, into estuarine ecosystems [7–10], which is the topic of investigation in this study.

Particulate pollutants are small matter particles suspended in a fluid medium, such as air or water. In aquatic environments, these particles can have negative effects on species health and the entire ecosystem (see review by [11]). Particulate pollutants have been detected in aquatic environments across the globe, with reports from all continents [12–14]. Among particulate waste, plastics deserve special attention. Approximately 12% of global waste consists of plastics [15]. In total, around 9 billion tons of plastic have been produced, with annual production increasing from 0.5 million tons in the 1950s to 413.8 million tons in 2023 [16]. This situation has quickly become one of the most pressing environmental issues [15]. These small plastic particles are easily integrated into the food chain, contaminating various animals—from lower trophic levels, such as zooplankton, copepods, crustaceans, and echinoderms, to vertebrates, such as fish, birds, and mammals [17]. They have been detected in aquatic environments worldwide, with reports from all continents [12–14]. Several studies have investigated particulate pollutants related to plastic, such as microplastic, in Brazilian estuaries [18–24], as well as in other countries, such as Canada [25,26], Mexico [27], Portugal [28], China [13,29] and India [30,31], and regions, such as the North Sea [32], Mediterranean Sea [33] and Yellow Sea [34]. The ecological impacts of plastic waste in estuaries have been well documented, demonstrating harm to aquatic organisms and ecosystems [7,9,10]. For instance, in Malaysia, an average of ~2.5 particles of microplastic per liter of water was found in the Klang River estuary, originating from fishing gear and synthetic clothes [35,36]. This has implications for the biodiversity of the estuary, which also accumulate microplastics in their bodies, for example, gastropods [36]. Similar results were found in bivalves in an urbanized estuary in Brazil (Santos-Sao Vicente), for which bioavailability of microplastics were linked to industrial activities and negatively correlated with population density, bioaccumulated in decreasing concentrations from clams, mussels and oysters, respectively [37].

Particulate plastic pollutants can act as vectors, facilitating the entry of secondary pollutants into the environment [38,39]. Due to their relatively large surface area and hydrophobic characteristics, these particles can adsorb contaminants, such as heavy metals present in the environment [40] and persistent organic pollutants (POPs) [41], including polycyclic aromatic hydrocarbons (PAHs) [42]. Particulate pollutants can also accumulate in microorganisms [42–44]. The ingestion of particulate pollutants by marine organisms can impair their mobility, reproduction, and development, and they may disrupt the endocrine systems in fish and invertebrates [45]. The presence of micro- and nanoplastics in atmospheric air was analyzed by Sheraz et al. (20230) [46]. In their study, the authors reported that harmful compounds, such as dichlorodiphenyltrichloroethane (DDT), can adhere to plastic particles. Additionally, these particles may act as vectors for microbial or viral diseases [47]. However, the role of acting as vectors may not be exclusive to plastic pollutants. Other particulate waste with similar physical characteristics may also behave in the same way.

This study examines particulate pollutant contamination in two distinct Brazilian estuaries—one in the tropical region (Porto Seguro) and the other in the subtropical region (Babitonga Bay). These estuaries experience different levels of anthropogenic pressure, providing an opportunity to compare particulate pollutant contamination under varying environmental conditions. According to data from the Brazilian Institute of Geography and Statistics [48], the largest urban centers within these estuaries—Porto Seguro (tropical) and Joinville (subtropical)—differ significantly in various aspects. Porto Seguro covers an area of 2285.74 km^2, with a population of 168,326 and a density of 73.64 inhabitants per km^2. In contrast, Joinville spans 1,127.94 km^2, has a population of 616,317, and a much higher density of 546.41 inhabitants per km^2. Socioeconomic indicators, such as formal employment salaries, education levels, and sanitation, are generally higher in Joinville, situated in Brazil's subtropical region. Land use patterns also differ between these regions. In Porto Seguro, 44.94% of the total area faces a moderate to very high risk of environmental degradation, driven by agricultural and pasture expansion, particularly over forested areas [37]. However, tourism remains the dominant economic activity. Meanwhile, Babitonga Bay is where the largest industrial hub of Santa Catarina is concentrated, alongside agriculture and pasturelands. Urban expansion into mangrove and forest areas has also been documented in this region [49]. This study focuses on fish of the genus *Sphoeroides*, as these species are estuarine residents and are found in both locations. We hypothesize that, despite analyzing the same genus of fish, particulate pollutant contamination levels will differ between the estuaries due to the varying anthropogenic activities in each region.

2. Material and Methods

The study was conducted in two estuarine systems in Brazil: the Porto Seguro Estuary and the Babitonga Bay Estuary (Figure 1). The Porto Seguro Estuary (tropical) is located in the tropical region of northeastern Brazil, in the state of Bahia (16°26′44.06″ S, 39°05′5.22″ W), and is influenced by human activities, such as tourism and urban development. The surrounding area is characterized by large mangrove forests and an extensive coral reef system near the estuary's mouth. The Babitonga Bay Estuary (subtropical) is situated in the subtropical region of southern Brazil, in the state of Santa Catarina (26°02′26.28″ S, 48°28′48.50″ W), and is the largest estuarine complex along the coast of the state. It covers an area of 130 km^2, and the mangrove forests in the bay account for 75% of the state's total mangrove area [50].

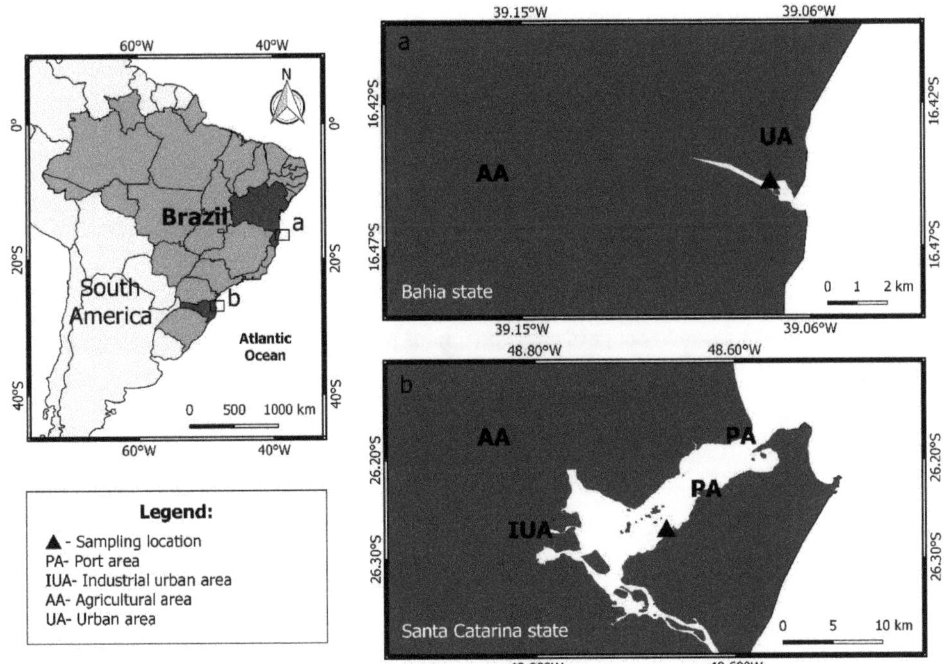

Figure 1. Map of the study areas: (**a**) Porto Seguro Estuary, tropical region of northeastern Brazil; (**b**) Babitonga Bay Estuary, subtropical region of southern Brazil.

This study involved fish of the genus *Sphoeroides* (*S. testudineus* and *S. greeleyi*), both of which are benthic, carnivorous species commonly found in tropical and subtropical estuaries [51]. *S. testudineus* is also commercially fished and consumed in some regions of Brazil. The fish were collected by local fishermen in both the Porto Seguro (May 2020) and Babitonga Bay (March 2020) estuaries. In total, 39 fish were sampled, with 22 specimens from Porto Seguro and 17 from Babitonga Bay. Upon collection, fish were transported to the laboratory on ice. The total length (tL, cm) and weight (tW, g) of each specimen were recorded. For particulate pollutant analysis, the gastrointestinal tracts and gills of each fish were removed after ventral sectioning.

Since microplastics are considered the most widely distributed particulate waste in the oceans [52], and given the size range of the particulate pollutants detected in this study, we chose to use the same analytical methodology applied for microplastics, with modifications to suit our data. The methodology used was based on and adapted from the protocols of Karami et al. (2017) and Pegado et al. (2018) [21,53]. Samples were first rinsed twice with distilled water, then preserved in sterile glass vials containing 10% formalin for 48 h before being transferred to 70% ethanol. For the detection of particulate pollutants, gills were analyzed whole under a stereomicroscope, while the gastrointestinal contents were rinsed in 70% ethanol and spread onto glass Petri dishes for stereomicroscopic analysis. The 10% formalin was prepared using 37% pure formaldehyde from the brand Êxodo Científica, a Brazilian manufacturer, located in Sumaré, São Paulo – Brazil, while the 70% ethanol was diluted from 95% P.A. ethyl alcohol from the brand ACS Científica, also a Brazilian manufacturer and located in Sumaré, São Paulo – Brazil. Particulate pollutants were photographed using a Zeiss Axiocam 503 color camera attached to a Zeiss Discovery.V12 stereomicroscope, the equipment was manufactured in Jena, Germany.

Particulate pollutants were classified based on their size and shape, with fragments (irregular shape and surface), fibers (thin, elongated), and films (thin, flat) as the primary categories [53]. Particulate pollutants were further categorized by size: ≤1 mm, 1.1–3 mm, and 3.1–5 mm, following Bessa et al. (2018) [28]. Plastic particles larger than 5 mm were excluded from the analysis. To minimize contamination, all procedures were carried out with strict controls, including the use of latex gloves, glassware, metal instruments, and dedicated cotton lab clothing. The total number of particulate pollutants in each sample (gastrointestinal tract and gills) was counted separately for each estuary. The mean number of particulate pollutants per sample was calculated using Formula (1):

$$\frac{n_i}{n} \tag{1}$$

where n_i is the total number of particulate pollutants in the samples, and n is the number of samples analyzed. To account for the fish's weight, the total particulate pollutant count was normalized by dividing the total particulate pollutant count by the log-transformed weight of each fish.

We used a two-way analysis of variance (ANOVA) to assess the difference in particulate pollutant content between organs (gastrointestinal tract and gills) and between estuaries, as well as the interaction between these factors. The particulate pollutant content was square-root transformed to improve model fit. The frequency of occurrence (FO%) of each particulate pollutant type (fragments, fibers, and films) and size class (≤1 mm, 1–3 mm, and 3.1–5 mm) was determined for each biological sample, following Formula (2):

$$FO\% = \frac{N_i}{N} \times 100 \tag{2}$$

where N_i is the number of samples containing particulate pollutants, and N is the total number of samples analyzed. These values were used to examine the relative abundance of particulate pollutants by type and size in each estuary. A two-factor Permutational

Analysis of Variance (PERMANOVA [54]) with 5000 randomizations was performed to assess differences in particulate pollutant abundance, as the data did not meet normality (Kolmogorov–Smirnov: $p < 0.05$) or homoscedasticity (Levene's test: $p < 0.05$) assumptions. The two factors in the model were: (1) particulate pollutant type (fixed), with two levels (biological samples and site), and (2) the interaction between particulate pollutant type, biological samples, and estuaries. Pairwise comparisons were used to identify significant differences between the means of each factor. All statistical analyses were conducted in R v.4.2.1 [55].

3. Results

We analyzed a total of 39 *Sphoeroides* specimens—22 from Porto Seguro Estuary and 17 from Babitonga Bay Estuary—yielding 76 biological samples (39 gastrointestinal tracts and 37 gills). A total of 757 plastic particles were detected across all samples, of which 745 were particulate pollutants (particles $\leq$ 5 mm in size). The remaining 12 particles, larger than 5 mm, were excluded from the analysis. Particulate pollutant contamination was observed in 92.11% of the samples, with only one specimen from Porto Seguro lacking detectable particulate pollutants. Some particulate pollutants detected under the stereomicroscope are shown in Figure 2.

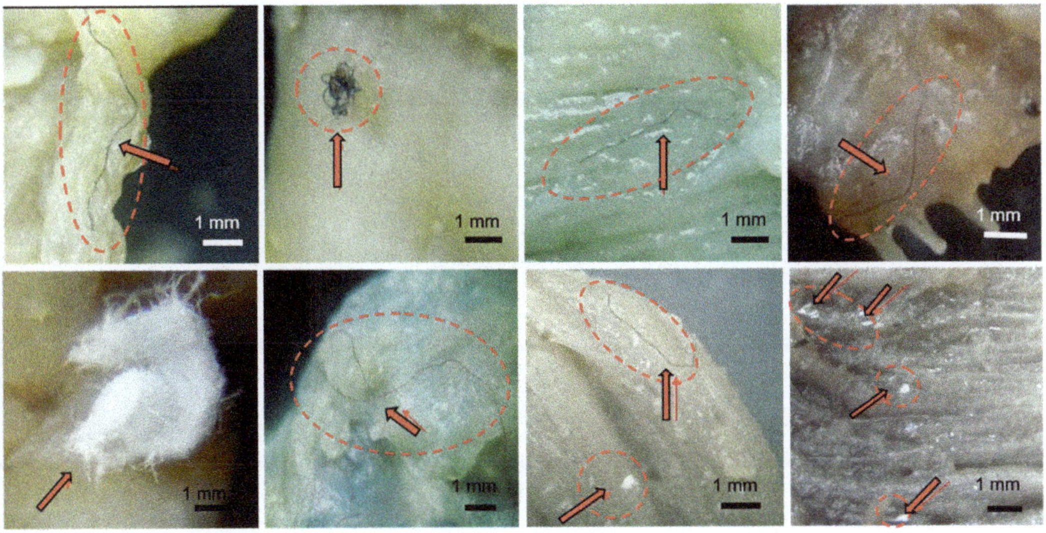

Figure 2. Examples of particulate pollutants found adhered to the gastrointestinal tracts (**top row**) and gills (**bottom row**) of *Sphoeroides* from both estuaries. Red circles and arrows showing the region and the particulate pollutant, respectively.

The analysis of the amount of particulate material in each biological sample, for each estuary, revealed that the debris content was significantly higher in the gastrointestinal tracts of fish from Babitonga Bay compared to those from Porto Seguro, while the opposite was observed for the gills, where Porto Seguro showed a higher number of particles compared to Babitonga Bay (Table 1). This interaction was statistically significant ($F_{1,72} = 16.510$, $p < 0.001$). However, no significant main effects were found for estuary ($F_{1,72} = 0.077$, $p = 0.782$) or organ ($F_{1,72} = 2.801$, $p = 0.098$, Figure 3).

Table 1. Number of particulate pollutants (PP) (n) per biological sample, in each estuary analyzed. The number in parentheses represents the average number of particulate pollutants, considering the number of samples analyzed.

Estuaries	PP Number GI	PP Number Gills	Total PP Number
Porto Seguro	80 (3.6)	419 (19.0)	499 (11.3)
Babitonga Bay	170 (10.0)	76 (5.1)	246 (7.3)

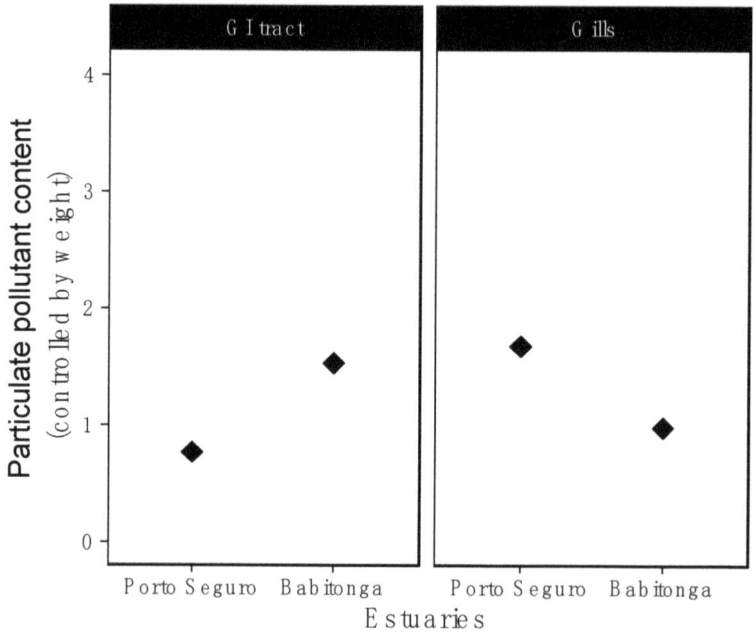

Figure 3. Standardized particulate pollutant content found in each organ (GI tract and gills) between estuaries.

The analysis of particulate pollutant types revealed that fibers were the most abundant type in the gastrointestinal tracts of both estuaries, while the gills showed varying patterns between estuaries. In Porto Seguro, films were the most frequent, whereas fibers dominated in Babitonga Bay (Tables 2 and 3). The analysis of particulate pollutant size classes revealed distinct patterns between estuaries. In the gastrointestinal tracts, fibers of all sizes were most common in both Porto Seguro and Babitonga Bay. However, for the gills, Porto Seguro showed a higher occurrence of film across all size categories, while Babitonga Bay had a higher frequency of fibers (Tables 4 and 5).

PERMANOVA revealed significant differences for the fragment ($p = 0.004$) and film ($p < 0.001$) types across biological samples (gastrointestinal tract and gills) and between estuaries (fragment $p < 0.001$. film $p = 0.006$). No significant differences were found for fiber types across sites or sample types. Pairwise comparisons showed significant differences for fragments and films between specific sample types and estuaries. In summary, for fragments: GI (Babitonga Bay) vs. gills (Babitonga Bay) ($p = 0.0088$); GI (Porto Seguro) vs. GI (Babitonga Bay) ($p < 0.001$); gills (Porto Seguro) vs. GI (Porto Seguro) ($p < 0.001$). For films: gills (Porto Seguro) vs. gills (Babitonga Bay) ($p < 0.001$); gills (Porto Seguro) vs. GI (Babitonga Bay) ($p < 0.001$); GI (Porto Seguro) vs. gills (Porto Seguro) ($p < 0.001$).

Table 2. Frequency of occurrence (FO%) of each particulate pollutant (PP) type (fragment, fiber, film) detected in the gastrointestinal tract and gill biological samples of *Sphoeroides* from each estuary.

	GI FO% of PP		
Estuaries	Fragment (n)	Fiber (n)	Film (n)
Porto Seguro	13.6%	90.9%	13.6%
Babitonga Bay	64.7%	94.1%	35.3%
	Gill FO% of PP		
Estuaries	Fragment (n)	Fiber (n)	Film (n)
Porto Seguro	4.5%	63.6%	72.8%
Babitonga Bay	33.3%	86.7%	13.3%

Table 3. Number of each particulate pollutant (PP) type detected in the gastrointestinal tract and gill biological samples of *Sphoeroides* from each estuary. The number in parentheses represents the percentage of particles relative to the total detected.

GI—Types of PP				
Estuaries	Fragment (n)	Fiber (n)	Film (n)	Total (n)
Porto Seguro	4 (5%)	70 (87.5%)	6 (7.5%)	80
Babitonga Bay	47 (27.7%)	99 (58.2%)	24 (14.1%)	170
Gill—Types of PP				
Estuaries	Fragment (n)	Fiber (n)	Film (n)	Total
Porto Seguro	1 (0.2%)	41 (9.8%)	377 (90%)	419
Babitonga Bay	10 (13.2%)	45 (59.2%)	21 (27.6%)	76

Table 4. Frequency of occurrence (FO%) of particulate pollutants (PP) by type and size class (≤1 mm, 1.1–3 mm, 3.1–5 mm) detected in the gastrointestinal tract and gill biological samples of *Sphoeroides* from each estuary.

	GI FO% of the Sizes (mm) of PP Types								
	Fragment			**Fiber**			**Film**		
tuary	≤1	1.1–3	3.1–5	≤1	1.1–3	3.1–5	≤1	1.1–3	3.1–5
rto Seguro	9%	0	4.5%	59.1%	59.1%	54.5%	13.6%	4.5%	0
bitonga Bay	41.2%	29.4%	11.8%	70.6%	70.6%	64.7%	0	35.3%	11.8%
	Gill FO% of the Sizes (mm) of PP Types								
	Fragment			**Fiber**			**Film**		
tuary	≤1	1.1–3	3.1–5	≤1	1.1–3	3.1–5	≤1	1.1–3	3.1–5
rto Seguro	0	4.5%	0	27.3%	45.5%	50%	63.6%	54.5%	40.9%
bitonga Bay	33.3%	6.7%	6.7%	73.3%	33.3%	53.3%	0	13.3%	6.7%

Table 5. Number of particulate pollutants (PP) by type and size class (≤1 mm, 1.1–3 mm, 3.1–5 mm) detected in the gastrointestinal tract and gill biological samples of *Sphoeroides* from each estuary. The number in parentheses represents the percentage of each size class in relation to the total for each pollutant type.

	GI—Sizes of the PP Types (mm)								
	Fragment			**Fiber**			**Film**		
uary	≤1	1–3	3.1–5	≤1	1–3	3.1–5	≤1	1–3	3.1–5
rto Seguro	3 (75%)	0	1 (25%)	29 (41.4%)	25 (35.8%)	16 (22.8%)	5 (83.3%)	1 (16.7%)	0
bitonga Bay	22 (46.8%)	14 (29.8%)	11 (23.4%)	47 (47.5%)	23 (23.2%)	29 (29.3%)	0	20 (83.3%)	4 (16.7%)

Table 5. *Cont.*

	Gill—Sizes of the PP types (mm)								
	Fragment			Fiber			Film		
Estuary	≤1	1–3	3.1–5	≤1	1–3	3.1–5	≤1	1–3	3.1–5
Porto Seguro	0	1 (100%)	0	11 (26.8%)	18 (43.9%)	12 (29.3%)	194 (51.4%)	162 (43%)	21 (5.6%)
Babitonga Bay	6 (60%)	3 (30%)	1 (10%)	19 (42.2%)	17 (37.8%)	9 (20%)	0	15 (71.4%)	6 (28.6%)

4. Discussion

This study presents the first comparative analysis of particulate pollutant contamination in the gastrointestinal tracts and gills of *Sphoeroides* species from tropical and subtropical estuaries in the South Atlantic, with varying levels of anthropic influence. The findings provide new insights into particulate pollutant pollution in these regions and reveal contrasts with previous studies, contributing to our understanding of how different environmental conditions affect particulate pollutant contamination in marine life. The use of fish from the *Sphoeroides* genus is justified by their status as true estuarine species with a wide distribution along the Brazilian coast, occupying both tropical and subtropical estuaries. Due to the biological characteristics of this genus, *S. testudineus* was also used in an ecotoxicological study [56] to analyze the gastrointestinal tract. The authors justified this choice by stating that the gastrointestinal tract is the primary absorption route for many pollutants, has connections to other organs, and represents a complex interface between the organism and its environment. For this reason, we also included the gastrointestinal tract in our analyses.

Plastic waste is the most common type of debris found in the ocean. Brazil is the fourth largest producer of plastic waste in the world, generating an average of 11 million tons per year [52]. Based on this, we infer that many of the particles detected in our study are likely microplastics. However, we cannot confirm this, as no polymer detection analysis was conducted. According to Araujo et al. (2018) [57], among the various techniques used for polymer detection in plastic production, Raman spectroscopy is widely applied, although it can also have limitations and disadvantages. Despite this, our overall results challenge the conclusions of Jabeen et al. (2016) [58] who suggested that marine environments typically experience higher concentrations of particulate pollutants compared to freshwater and estuarine environments. When comparing the two estuaries in our study, the Porto Seguro Estuary (tropical region) exhibited a higher number of particulate pollutants (n = 499) and a higher percentage of contaminated fish (95.4%) compared to other tropical studies, such as those by Pegado et al. (2018) and Vendel et al. (2017) [21,24]. However, the Babitonga Bay Estuary (subtropical region) also demonstrated significant contamination, with 100% of the specimens affected and 246 particulate pollutants recorded. These results are comparable to those observed by Arias et al. (2019) [59] in Blanca Bay, Argentina, where 100% of the fish specimens examined contained particulate pollutants (n = 241). The high levels of contamination in *Sphoeroides* may be attributed to their benthic lifestyle and carnivorous diet [51]. Benthic species, which forage in sediment-dominated environments, are more likely to ingest particulate pollutants along with their prey [60]. Given that 99% of particulate pollutants settle in sediments, benthic organisms, like *Sphoeroides*, are particularly susceptible to accidental ingestion. Furthermore, this study suggests that particulate pollutants are not only confined to benthic organisms but may also be transferred up the food web, making this an important issue for non-benthic species as well.

This is a notably higher prevalence than reported in other marine studies, such as those by Rummel et al. (2016) in the Baltic Sea (5.5%) [61], Davison and Ach (2011) in the North Pacific Gyre (9.2%) [62], and Lusher et al. (2016) in the Northeast Atlantic (11%) [10]. These results are more in line with the study conducted by Tanaka and Tanaka (2016) [63] in Tokyo Bay, which found particulate pollutants in 77% of analyzed specimens. Logistical constraints prevented us to chemically characterize polymer types. As a result, plastic particle detection was conducted solely through stereomicroscope analysis. This prevents us from gaining a nuanced understanding of the origins and potential long-term consequences of these particles. Nevertheless, it is important to emphasize that no methodology is 100% efficient. According to Araujo et al. (2018) [57], among the various techniques used for particulate pollutant detection, Raman spectroscopy is widely applied; however, it also has limitations and disadvantages. Raman spectroscopy is prone to fluorescence interference, has an inherently low signal-to-noise ratio, and can cause sample heating due to the use of a laser as a light source, sometimes leading to background emission and subsequent polymer degradation. Consequently, some methodologies may overestimate results, while others may underestimate them. In our study, fish specimens collected from both estuaries were carefully analyzed using the same standardized approach at each location.

The predominance of fibers in marine organisms has been well documented [10,53,64], with possible sources linked to sewage sludge [65], fishing activities, port operations, and the widespread use of fibers in industrial applications [53]. Our results were consistent with these reports in the Babitonga Bay Estuary, where fibers accounted for 58.5% of particulate pollutants. However, in Porto Seguro, a predominance of film-type particulate pollutants was observed (76.8%), which diverges from the general trend. This difference can be attributed to regional variations in the sources of particulate pollutants.

The Babitonga Bay Estuary is located near a densely industrialized area, with significant activity in different sectors, such as metallurgy, textiles, and manufacturing [66]. The surrounding population exceeds one million residents. In addition to the region's industrial activity, other anthropogenic pressures come from port operations, large-scale fishing, and urbanization [48,67]. Port and fishing activities likely contribute to the high prevalence of fibers in the estuarine system, as these areas are known to be critical points for particulate pollutant contamination. Furthermore, ships discharge ballast water, which may also contribute to the contamination of the bay with particulate waste. In contrast, Porto Seguro is a major tourist destination in Brazil, with a population of 168.326 [48], which increases 6 to 7 times annually due to tourism [68]. The increased population puts significant pressure on local waste management systems, and much of the city's sewage is released into the estuary without treatment [67]. This inadequate sanitation infrastructure, coupled with high tourist traffic, may explain the high prevalence of film-type particulate pollutants in the Porto Seguro samples. Films are commonly found in plastic packaging materials and food wrappers, which are widely used and discarded by consumers, particularly in tourist-heavy areas.

While many studies have focused on gastrointestinal tract contamination, only a few have examined particulate pollutant accumulation in fish gills. The gills are crucial for respiration and are directly exposed to environmental pollutants. Karami et al. (2017) [53] suggested that particulate pollutants are more likely to adhere to gill filaments, especially when they are smaller in size. Our study found notable particulate pollutant contamination in the gills of *Sphoeroides*, particularly of the film type, and smaller to intermediate sizes (up to 3 mm) in the Porto Seguro Estuary. This is consistent with findings that suggest particulate pollutants of smaller sizes have a higher likelihood of adhering to gill structures due to their increased surface area and ability to penetrate finer gill filaments [69]. Particulate pollutants in gills may pose a direct threat to fish health, as these particles could impair respiration and possibly cause physical damage or chemical contamination. Our results emphasize the importance of considering gills as a critical organ for particulate pollutant accumulation, especially in environments where small, film-type particulate pollutants are abundant.

Our results indicate varying levels of particulate pollutant contamination within and between tropical and subtropical estuaries, which can affect how the surrounding community interacts with, contributes to, and are affected by current land use and other anthropogenic processes. Addressing how waste is disposed within local communities should be a focus of immediate action to develop tailored approaches to each community, which takes into account local sociodemographic infrastructure, culture and economic activities. Regional studies like this are crucial for initial mapping and for informing the development of localized conservation strategies. By understanding the specific sources and types of particulate pollutants in different estuarine environments, we can design targeted interventions to mitigate pollution in those regions. However, given the global nature of the particulate pollutant crisis, a more coordinated international approach is necessary. A network of researchers focused on aquatic pollution can facilitate broader collaboration, knowledge sharing, and more effective solutions.

5. Conclusions

This study highlights the widespread contamination of estuarine ecosystems by particulate pollutants, with significant differences between tropical and subtropical estuaries. The high levels of contamination found in *Sphoeroides* species, particularly in the gastrointestinal tracts and gills, underscore the need for increased attention to particulate pollutants in estuarine environments. Further research is needed to understand the long-term ecological consequences of particulate pollutant contamination, especially in areas with high anthropic pressure. Moreover, the study draws attention to the need for improved waste management systems and more sustainable tourism practices, especially in regions with high tourist visitation, to mitigate the flow of plastic pollution into these sensitive ecosystems. Solving the particulate pollution problem requires more than just scientific research. It involves addressing the complex relationship between environmental and socioeconomic issues. To make meaningful progress, action must be taken across all sectors; researchers, local communities, industry, the third sector, and governments must collaborate. Only through joint efforts can we hope to reduce particulate pollutant contamination and safeguard aquatic ecosystems for future generations.

Author Contributions: All authors contributed to the study. Conceptualization, S.M.d.S.F. and L.F.F.; methodology, S.M.d.S.F., L.F.F., M.T.G., M.P.d.S. and M.S.; formal analysis, S.M.d.S.F., L.F.F. and J.M.; investigation, L.F.F. and S.M.d.S.F.; resources, L.F.F. and J.M.; writing—original draft preparation, L.F.F., J.M. and S.M.d.S.F.; writing—review and editing, L.F.F. and J.M.; supervision, M.T.G., M.P.d.S. and M.S.; project administration, L.F.F.; funding acquisition, L.F.F. and J.M. All authors have read and agreed to the published version of the manuscript.

Funding: This work was supported by the 'Vice-chancellor's funds for support of research activities' of the Federal University of Paraná (PRPPG 04/2018 e PRPPG 02/2020). SMSF were financed by and acknowledge the Coordenação de Aperfeiçoamento de Pessoal de Nível Superior (CAPES) (Grant 02).

Institutional Review Board Statement: Ethical review and approval were waived for this study because when samples were collected, approval from an Ethics Committee for Research was not yet a requirement. Nevertheless, we conducted our work with utmost care and respect for the animals, using a minimal number of specimens and ensuring that they were sacrificed in accordance with established ethical standards and guidelines.

Data Availability Statement: The datasets generated during and/or analysed during the current study are available from the corresponding author on reasonable request.

Acknowledgments: The authors would also like to thank researcher Larissa Ajala Batista for her contribution to the production of the maps.

Conflicts of Interest: The authors have no conflicts of interest to declare.

ferences

Einoder, L.D.; MacLeod, C.K.; Coughanowr, C. Metal and isotope analysis of bird feathers in a contaminated estuary reveals bioaccumulation, biomagnification, and potential toxic effects. *Arch. Environ. Contam. Toxicol.* **2018**, *75*, 96–110. [CrossRef] [PubMed]
Lana, P.C.; Bernardino, A.F. (Eds.) *Brazilian Estuaries: A Benthic Perspective*; Springer: Cham, Switzerland, 2018.
Paitach, P.L.; Simões-Lopes, P.C.; Cremer, M.J. Tidal and seasonal influences in dolphin habitat use in a southern Brazilian estuary. *Sci. Mar.* **2017**, *81*, 49–56. [CrossRef]
Vasconcelos, R.P.; Reis-Santos, P.; Costa, M.J.; Cabral, H.N. Connectivity between estuaries and marine environment: Integrating metrics to assess estuarine nursery function. *Ecol. Indic.* **2011**, *11*, 1123–1133. [CrossRef]
Santos, T.A.; Bonfim, T.M.; Silva, F.S.; Silva, A.G.; Bandeira, M.L.S.F.; Jesus, R.M.; Nascimento, L.D. Determination of metals in an estuarine system: *Laguncularia racemosa* as a potential indicator of contamination. *Braz. J. Environ. Sci.* **2018**, *49*, 51–65. [CrossRef]
Trevizani, T.H.; Domit, C.; Santos, M.C.O.; Figueira, R.C.L. Bioaccumulation of heavy metals in estuaries in the southwest Atlantic Ocean. *Environ. Sci. Pollut. Res.* **2023**, *30*, 26703–26717. [CrossRef]
Andrady, A.L. Microplastics in the marine environment. *Mar. Pollut. Bull.* **2011**, *62*, 1596–1605. [CrossRef]
Law, K.L.; Thompson, R.C. Microplastics in the seas. *Science* **2014**, *345*, 144–145. [CrossRef]
Lebreton, L.C.M.; van der Zwet, J.; Damsteeg, J.; Slat, B.; Andrady, A.; Reisser, J. River plastic emissions to the world's oceans. *Nat. Commun.* **2017**, *8*, 15611. [CrossRef]
Lusher, A.L.; O'Donnell, C.; Officer, R.; O'Connor, I. Microplastic interactions with North Atlantic mesopelagic fish. *ICES J. Mar. Sci.* **2016**, *73*, 1214–1225. [CrossRef]
Gokul, T.; Ramesh Kumar, K.; Prema, P.; Arun, A.; Balaji, P.; Faggio, C. Particulate pollution and its toxicity to fish: An overview. *Comp. Biochem. Physiol. Part C Toxicol. Pharmacol.* **2023**, *270*, 109646. [CrossRef]
Haward, M. Plastic pollution of the world's seas and oceans as a contemporary challenge in ocean governance. *Nat. Commun.* **2018**, *9*, 667. [CrossRef]
Li, H.X.; Ma, L.S.; Lin, L.; Ni, Z.X.; Xu, X.R.; Shi, H.H.; Yan, Y.; Zheng, G.M.; Rittschof, D. Microplastics in oysters Saccostrea cucullata along the Pearl River Estuary, China. *Environ. Pollut.* **2018**, *236*, 619–625. [CrossRef] [PubMed]
van Sebille, E.; Wilcox, C.; Lebreton, L.; Maximenko, N.; Hardesty, B.D.; van Franeker, J.A.; Eriksen, M.; Siegel, D.; Galgani, F.; Law, K.L. A global inventory of small floating plastic debris. *Environ. Res. Lett.* **2015**, *10*, 124006. [CrossRef]
Bouças da Silva, D.L.; Gil, J.; Nascimento, E.P.; Costa, H.A.; Paixão, R. Poluição plástica no litoral brasileiro: Percepções de gestores de meios de hospedagem sobre consumo de descartáveis. *Rev. Bras. Pesqui. Em Tur.* **2022**, *16*, e2481. [CrossRef]
Global Plastic Production. Available online: https://www.statista.com/statistics/282732/global-production-of-plastics-since-1950/ (accessed on 31 March 2025).
Hollman, P.C.H.; Bouwmeester, H.; Peters, R.J.B. *Microplastics in Aquatic Food Chain: Sources, Measurement, Occurrence and Potential Health Risks*; RIKILT Report; Wageningen UR: Wageningen, The Netherlands, 2013; 28p.
Krelling, A.P.; Possatto, F.E.; Maia, G.R.A.; Araki, K.; Olivieri, G.B.R.; Nogueira, H.P. First Evidence of High Microplastic Concentrations in Estuarine Litter Windrows of a World Heritage Site: The Paranaguá Estuarine Complex, Brazil. *Water Air Soil Pollut.* **2025**, *236*, 266. [CrossRef]
Menezes, K.M.; da Costa, M.B.; Schuab, J.M.; Dalbó, G.Z.; Alves, M.M.; Ocarisd, E.R.Y.; Rodrigues, F.L. Microplastic contamination in the mangroves of Piraquê-Açu and Piraquê-Mirim rivers, Aracruz (Brazil): An analysis in sediment, water, and biota. *Mar. Pollut. Bull.* **2025**, *214*, 117696. [CrossRef]
Parra, D.F.; Gimiliani, G.T.; Dos Santos, J.L.; Wetter, N.U.; Schepis, W.R.; Bereczki, A.; Cotrim, M.E.B. Microplastics in Santos São Vicente estuarine–Hotspot in sediments caused by low energy hydrodynamic events in strongly populated areas. *Mar. Pollut. Bull.* **2025**, *210*, 117286. [CrossRef]
Pegado, T.S.S.; Schmid, K.; Winemiller, K.O.; Chelazzi, D.; Cincinelli, A.; Dei, L.; Giarrizzo, T. First evidence of microplastic ingestion by fishes from the Amazon River estuary. *Mar. Pollut. Bull.* **2018**, *133*, 814–821. [CrossRef]
Possatto, F.E.; Barletta, M.; Costa, M.F.; Ivar do Sul, J.A.; Dantas, D.V. Plastic debris ingestion by marine catfish: An unexpected fisheries impact. *Mar. Pollut. Bull.* **2011**, *62*, 1098–1102. [CrossRef]
Ribeiro, V.V.; Casado-Coy, N.; Rangel, D.F.; Sanz-Lazaro, C.; Castro, Í.B. Microplastic in bivalves of an urbanized Brazilian estuary: Human modification, population density and vegetation influence. *J. Hazard. Mater.* **2025**, *482*, 136546. [CrossRef]
Vendel, A.L.; Bessa, F.; Alves, V.E.N.; Amorim, A.L.A.; Patrício, J.; Palma, A.R.T. Widespread microplastic ingestion by fish assemblages in tropical estuaries subjected to anthropogenic pressures. *Mar. Pollut. Bull.* **2017**, *117*, 448–455. [CrossRef] [PubMed]
Anderson, J.C.; Park, B.J.; Palace, V.P. Microplastics in aquatic environments: Implications for Canadian ecosystems. *Environ. Pollut.* **2016**, *218*, 269–280. [CrossRef]

26. Kabir, A.E.; Michon, E.; Mingelbier, M.; Robert, D.; Soubaneh, Y.D.; Xie, H.; Lu, Z. Microplastics in the benthic fish from Canadian St. Lawrence River and Estuary: Occurrence, spatial distribution and ecological risk assessment. *Mar. Pollut. Bull.* **2(** *212*, 117509. [CrossRef] [PubMed]
27. Sánchez-Paz, A.; Encinas-García, T.; Ávila-Félix, L.; Mendoza-Cano, F. Microplastics in surface waters surrounding a touris beach at the Gulf of California, Mexico. *Int. J. Environ. Health Res.* **2025**, 1–13. [CrossRef]
28. Bessa, F.; Barría, P.; Neto, J.M.; Frias, J.P.G.L.; Otero, V.; Sobral, P.; Marques, J.C. Occurrence of microplastics in commercial from a natural estuarine environment. *Mar. Pollut. Bull.* **2018**, *128*, 575–584. [CrossRef]
29. Wang, W.; Guo, S.; Li, B.; Xu, D.; Gao, B. Microplastics in Pearl River Estuary and Yangtze River Estuary, China: Occurre fragmentation and pollution risk. *J. Hazard. Mater. Adv.* **2025**, *17*, 100529. [CrossRef]
30. Goswami, P.; Vinithkumar, N.V.; Dharani, G. First evidence of microplastics bioaccumulation by marine organisms in the P Blair Bay, Andaman Islands. *Mar. Pollut. Bull.* **2020**, *155*, 111163. [CrossRef]
31. Kalangutkar, N.; Mhapsekar, S.; Rajagopal, P. Comparative assessment of microplastic pollution in Terekhol and Sal estuar Goa, India. *Environ. Earth Sci.* **2025**, *84*, 48. [CrossRef]
32. Foekema, E.M.; de Gruijter, C.; Mergia, M.T.; van Franeker, J.A.; Murk, A.T.J.; Koelmans, A.A. Plastic in North Sea fish. *Envi Sci. Technol.* **2013**, *47*, 8818–8824. [CrossRef]
33. Romeo, T.; Battaglia, P.; Pedá, C.; Consoli, P.; Andaloro, F.; Fossi, M.C. First evidence of presence of plastic debris in stomach large pelagic fish in the Mediterranean Sea. *Mar. Pollut. Bull.* **2015**, *95*, 358–361. [CrossRef]
34. Sun, X.; Li, Q.; Shi, Y.; Zhao, Y.; Zheng, S.; Liang, J.; Liu, T.; Tian, Z. Characteristics and retention of microplastics in the digest tracts of fish from the Yellow Sea. *Environ. Pollut.* **2019**, *249*, 878–885. [CrossRef] [PubMed]
35. Zaki, M.R.M.; Ying, P.X.; Zainuddin, A.H.; Razak, M.R.; Aris, A.Z. Occurrence, abundance, and distribution of microplas pollution: An evidence in surface tropical water of Klang River estuary, Malaysia. *Environ. Geochem. Health* **2021**, *43*, 3733–37 [CrossRef] [PubMed]
36. Zaki, M.R.M.; Zaid, S.H.M.; Zainuddin, A.H.; Aris, A.Z. Microplastic pollution in tropical estuary gastropods: Abundar distribution and potential sources of Klang River estuary, Malaysia. *Mar. Pollut. Bull.* **2021**, *162*, 111866. [CrossRef] [PubMed
37. Ribeiro de Sena, S.; de Souza Veiga, R.; De Amorim Silva, V. Análise de áreas de risco à degradação ambiental no município Porto Seguro, Bahia, Brasil. *Braz. J. Phys. Geogr.* **2023**, *16*, 3059–3072. [CrossRef]
38. Akdogan, Z.; Guven, B. Microplastics in the environment: A critical review of current understanding and identification of fut research needs. *Environ. Pollut.* **2019**, *254*, 113011. [CrossRef]
39. Sinha, P.; Saini, V.; Varshney, N.; Pandey, R.K.; Jha, H.C. The Infiltration of Microplastics in Human Systems: Gastrointesti Accumulation and Pathogenic Impacts. *Heliyon* **2025**, *11*, e42606. [CrossRef]
40. Wang, J.; Peng, J.; Tian, Z.; Gao, Y.; Zhan, Z.; Chen, Q.; Cai, L. Microplastics in the surface sediments from the Beijiang Riv littoral zone: Composition, abundance, surface textures and interaction with heavy metals. *Chemosphere* **2017**, *171*, 248–2 [CrossRef]
41. Wang, L.C.; Lin, J.C.; Dong, C.; Chen, C.; Liu, T. The sorption of persistent organic pollutants in microplastics from the coas environment. *J. Hazard. Mater.* **2021**, *420*, 126658. [CrossRef] [PubMed]
42. Cverenkárová, K.; Valachovicová, M.; Mackul'ak, T.; Žemlicka, L.; Bírošová, L. Microplastics in the food chain. *Life* **2021**, *11*, 13 [CrossRef]
43. Carpenter, E.J.; Smith, K.L., Jr. Plastics on the Sargasso Sea surface. *Science* **1972**, *175*, 1240–1241. [CrossRef]
44. Rummel, C.D.; Jahnke, A.; Gorokhova, E.; Kühnel, D.; Schmitt-Jansen, M. Impacts of biofilm formation on the fate and poten effects of microplastic in the aquatic environment. *Environ. Sci. Technol. Lett.* **2017**, *4*, 258–267. [CrossRef]
45. Cole, M.; Lindeque, P.; Halsband, C.; Galloway, T.S. Microplastics as contaminants in the marine environment: A review. *M Pollut. Bull.* **2011**, *62*, 2588–2597. [CrossRef] [PubMed]
46. Sheraz, M.; Kim, J.; Kim, J. Nano/microplastics in indoor air: A critical review of synthesis routes for toxicity testing a preventative measure strategies. *Process Saf. Environ. Prot.* **2023**, *180*, 274–304. [CrossRef]
47. VishnuRadhan, R.; Thresyamma, D.D.; Eldho, T.I.; Bhagat, J. Atmospheric plastics- a potential airborne fomite with an emerg climate signature. *J. Clim. Change Health* **2021**, *3*, 100037. [CrossRef]
48. IBGE—Brazilian Institute of Geography and Statistics. *Demographic Census 2022*; IBGE: Rio de Janeiro, Brazil, 2022.
49. Filgueras, A.S.; Silva, T.S.; Corrêa, I.C.S. Impacts of sea level rise and land use on coastal wetlands: Methodology applied to B da Babitonga (SC). *Soc. Nat.* **2023**, *36*, e69403. [CrossRef]
50. Mazzer, A.M.; Gonçalves, M.L. Aspectos geomorfológicos da Baía da Babitonga, Santa Catarina, Brazil: Caracterização m fométrica. *Rev. Bras. Geomorfol.* **2011**, *12*, 115–120. [CrossRef]
51. Figueiredo, J.L.; Menezes, N.A. *Manual de Peixes Marinhos do Sudeste do Brasil. IV. Teleostei (5)*; Museu de Zoologia-USP: São Pau Brazil, 2000.
52. Vargas, J.G.M.; Silva, V.B.; Oliveira, L.K.; Molina, E.F. Microplásticos: Uso na indústria cosmética e impactos no ambiente aquát *Química Nova* **2022**, *45*, 705–711. [CrossRef]

Karami, A.; Golieskardi, A.; Ho, B.Y.; Larat, V.; Salamatinia, B. Microplastics in eviscerated flesh and excised organs of dried fish. *Sci. Rep.* **2017**, *7*, 5473. [CrossRef]
Anderson, M.; Gorley, R.N.; Robert, K.C. *Permanova+ for Primer: Guide to Software and Statistical Methods*; Primer-E: Plymouth, UK, 2008.
R Core Team. *R: A Language and Environment for Statistical Computing*; R Foundation for Statistical Computing: Vienna, Austria, 2025. Available online: https://www.R-project.org/ (accessed on 31 March 2025).
Marinsek, G.P.; Gusso-Choueri, P.K.; Choueri, R.B.; Abessa, D.M.S.; Gonçalves, A.R.N.; Bortolotto, L.B.; Mari, R.B. Integrated analysis of fish intestine biomarkers: Complementary tools for pollution assessment. *Mar. Pollut. Bull.* **2022**, *178*, 113590. [CrossRef]
Araujo, C.F.; Nolasco, M.M.; Ribeiro, A.M.P.; Ribeiro-Claro, P.J.A. Identification of microplastics using Raman spectroscopy: Latest developments and future prospects. *Water Res.* **2018**, *142*, 426–440. [CrossRef]
Jabeen, K.; Su, L.; Li, J.; Yang, D.; Tong, C.; Mu, J.; Shi, H. Microplastics and mesoplastics in fish from coastal and fresh waters of China. *Environ. Pollut.* **2017**, *221*, 141–149. [CrossRef]
Arias, A.H.; Ronda, A.C.; Oliva, A.L. Evidence of microplastic ingestion by fish from the Bahía Blanca Estuary in Argentina, South America. *Bull. Environ. Contam. Toxicol.* **2019**, *102*, 750–756. [CrossRef]
Kane, I.A.; Clare, M.A.; Miramontes, E.; Wogelius, R.; Rothwell, J.J.; Garreau, P.; Poh, F. Seafloor microplastic hotspots controlled by deep-sea circulation. *Science* **2020**, *368*, 1140–1145. [CrossRef] [PubMed]
Rummel, C.D.; Löder, M.G.J.; Fricke, N.F.; Lang, T.; Griebeler, E.M.; Janke, M.; Gerdts, G. Plastic ingestion by pelagic and demersal fish from the North Sea and Baltic Sea. *Mar. Pollut. Bull.* **2016**, *102*, 134–141. [CrossRef] [PubMed]
Davison, P.; Asch, R.G. Plastic ingestion by mesopelagic fishes in the North Pacific Subtropical Gyre. *Mar. Ecol. Prog. Ser.* **2011**, *432*, 173–180. [CrossRef]
Tanaka, K.; Takada, H. Microplastic fragments and microbeads in digestive tracts of planktivorous fish from urban coastal waters. *Sci. Rep.* **2016**, *6*, 34351. [CrossRef]
Cozar, A.; Echevarría, F.; Gonzalez-Gordillo, J.I.; Irigoien, X.; Ubeda, B.; Hernandez-León, S.; Palma, A.T.; Navarro, S.; Garcia-de-Lomas, J.; Ruiz, A.; et al. Plastic debris in the open ocean. *Proc. Natl. Acad. Sci. USA* **2014**, *111*, 10239–10244. [CrossRef]
Browne, M.A.; Crump, P.; Niven, S.J.; Teuten, E.L.; Tonkin, A.; Galloway, T.; Thompson, R.C. Accumulations of microplastic on shorelines worldwide: Sources and sinks. *Environ. Sci. Technol.* **2011**, *45*, 9175–9179. [CrossRef]
IPPUJ. *Joinville—City in Data*; Research and Planning Institute for the Sustainable Development of Joinville: Joinville, Brazil, 2011.
IBGE—Brazilian Institute of Geography and Statistics. *Demographic Census 2010*; IBGE: Rio de Janeiro, Brazil, 2010.
Silva, A.C.R.S.; Bernardes, M.E.C.; Assireu, A.T.; Siegle, E.; Sousa, P.H.G.O.; Brown, D. Hydrodynamics of a tropical estuary: Buranhém River, Porto Seguro, Brazil. *Braz. J. Water Resour.* **2018**, *23*, 1–9. [CrossRef]
Batel, A.; Borchert, F.; Reinwald, H.; Erdinger, L.; Braunbeck, T. Microplastic accumulation patterns and transfer of benzo[a]pyrene to adult zebrafish (*Danio rerio*) gills and zebrafish embryos. *Environ. Pollut.* **2018**, *235*, 918–930. [CrossRef]

diversity

Article

Biodiversity and Seasonal Succession of Macrobenthos in Saltmarsh Habitat Adjacent to a Ship-Breaking Area

M. Shafiqul Islam [1,*], Hossain Zamal [1], Muhammad Shah Alam [2], Md. Maheen Mahmud Bappy [3], Abu Hena Mustafa Kamal [4], M. Belal Hossain [3,5,*] and Takaomi Arai [6]

1 Institute of Marine Sciences and Fisheries, University of Chittagong, Chittagong 4331, Bangladesh; hzcu@yahoo.com
2 Department of Zoology, University of Chittagong, Chittagong 4331, Bangladesh; shah-alom@gmali.com
3 Department of Fisheries and Marine Science, Noakhali Science and Technology University, Noakhali 3814, Bangladesh; bappy0213@student.nstu.edu.bd
4 Faculty of Fisheries and Food Science, Universiti Malaysia Terengganu, Kuala Nerus 21030, Terengganu, Malaysia; a.hena@umt.edu.my
5 School of Engineering and Built Environment, Griffith University, Brisbane, QLD 4111, Australia
6 Environmental and Life Sciences Programme, Universiti Brunei Darussalam, Gadong BE1410, Brunei; takaomi.arai@ubd.edu.bn
* Correspondence: prof.shafiqims@cu.ac.bd (M.S.I.); belal.hossain@nstu.edu.bd (M.B.H.)

Abstract: The Fauzderhat coast of Chattogram (Bangladesh) is increasingly affected anthropogenic pressures, necessitating an understanding of its ecological conditions to inform effective ecosystem management. Despite this urgency, the local succession patterns and environmental impacts on macrobenthic communities remain poorly understood. This study examines the saltmarsh bed macrobenthos in Fauzderhat, documenting 81,724 individuals from 54 species. These include ten families and twenty-two species of annelids, ten and twelve species of arthropods, and ten and eleven species of mollusks, as well as six and nine species from different phyla. Seasonality showed significantly different patterns of changes, with the number of species and abundance peaking during the monsoon (53 species) and post-monsoon (21,969) conditions, respectively, and being lowest in the post-monsoon condition (39 species) and winter (18,265 individuals). Species richness, diversity, and evenness were significantly higher in monsoon and lower in post-monsoon conditions, with the differences being only significant in the former. Cluster analysis and line graphs indicated that average species abundance was lowest post-monsoon, increased through the winter and pre-monsoon conditions, then declined again during monsoon conditions. SIMPER analysis revealed the highest dissimilarity between pre-monsoon and post-monsoon conditions while winter and post-monsoon conditions showed the lowest dissimilarity of microbenthic assemblages. Correlation coefficients showed the macrobenthos were positively correlated with soil salinity, dissolved oxygen, and pH, while they were negatively correlated with sand, Inundation Period, and nutrients. CCA showed that monsoon conditions (higher water temperature, inundation period, and tidal height) created unfavorable environments for most species, except for several species. Conversely, winter favored species like *M. oligobranchia*. Post-monsoon nutrient levels increased stress, reducing species presence, while pre-monsoon conditions supported balanced diversity.

Keywords: macrobenthos; abundance; diversity; seasonal variation; succession; environmental impact

1. Introduction

Macroorganisms are essential components of benthic ecosystems, contributing to energy flow, nutrient cycling, and enhanced productivity [1,2]. They inhabit a wide array of environments, including mangroves [3], seagrass beds [4], coral reefs [3], saltmarshes [5], and unvegetated mud flats [3]. Their direct dependence on environmental conditions, coupled with varying tolerance to pollutants among species, makes them key indicators of ecological health [6]. As ecosystem engineers, certain macrobenthic species shape their surroundings through activities like burrowing, which aerates sediments and influences its composition, affecting habitat structure and sympatric benthic organisms [7], while other species serve primarily as food for fish and invertebrate predators [8]. However, burrowing can also increase turbidity and resuspend pollutants, degrading water quality and impacting other organisms [9].

Saltmarshes—dynamic ecosystems bridging terrestrial and marine environments—harbor diverse microbenthic communities adapted to harsh environmental conditions [10,11], which play a crucial role in nutrient cycling. Using sediment resuspension, they enhance nutrient and dissolved gas availability in the water column, helping to maintain productivity [5]. Using aerating sediments, they support plant growth and enhance habitat complexity [12]. In turn, changes in saltmarsh vegetation can alter macrobenthic communities [13].

The Fauzderhat coast, which includes saltmarshes dominated by *Porteresia coarctata* [14], is highly vulnerable to flooding, storm surges [15], and pollution from the shipbreaking industry, which operates across roughly 40 shipyards in the region. This industry releases heavy metals and other toxic chemicals that severely impact water quality and fisheries, as well as the surrounding forests [16]. Pollution disrupts food chains through biomagnification, while habitat degradation threatens the ecological balance and marine biodiversity [17,18].

Being geographically and administratively part of the Sitakunda coast (Chattogram district), Fauzderhat faces pressing environmental challenges [17]. Thus, assessing its ecological status, species diversity, and seasonal patterns are crucial for informed conservation and restoration initiatives. Studying its succession patterns can provide insights into its macrobenthic biodiversity, community resilience, and recovery potential in degraded saltmarshes. Understanding the influence of environmental variables on macrobenthic communities is also critical for predicting responses to climate change and in guiding restoration efforts.

While studies on macrobenthic abundance and diversity in saltmarshes are extensive [19–22], these are limited in Bangladesh [12,23]. As biodiversity declines worldwide [24], exploring local ecosystems is increasingly crucial for conservation. Investigating macrobenthic succession and its associated environmental impacts can reveal tolerance limits, adding tailored restoration strategies to enhance biodiversity and ecosystem functionality. Therefore, this study aimed to (i) quantify the macrobenthic abundance in a coastal saltmarsh along the Chattogram coast, (ii) analyze the community structure, (iii) examine seasonal succession patterns, and (iv) evaluate the environmental impact on community diversity and succession.

2. Materials and Methods

2.1. Study Area

The Fauzderhat coast (22°23.89′–22°24.02′ N, 91°44.59′–91°44.84′ E) belongs to the Sitakunda upazila, Chattogram district (Figure 1), and suffers from high tidal ranges and frequent cyclonic activity. Water temperature ranges from 23 to 33 °C, pH ranges from 7.1 to 7.9, and salinity ranges from 1.6 to 21 ppt [25,26], and the coastline is characterized by a mix of wave-dominated sand and estuarine mud regions, with tidal pumping and fluvial fluxes significantly influencing sediment transportation processes, leading to significant erosion and accretion due to both natural processes and human activities [27,28].

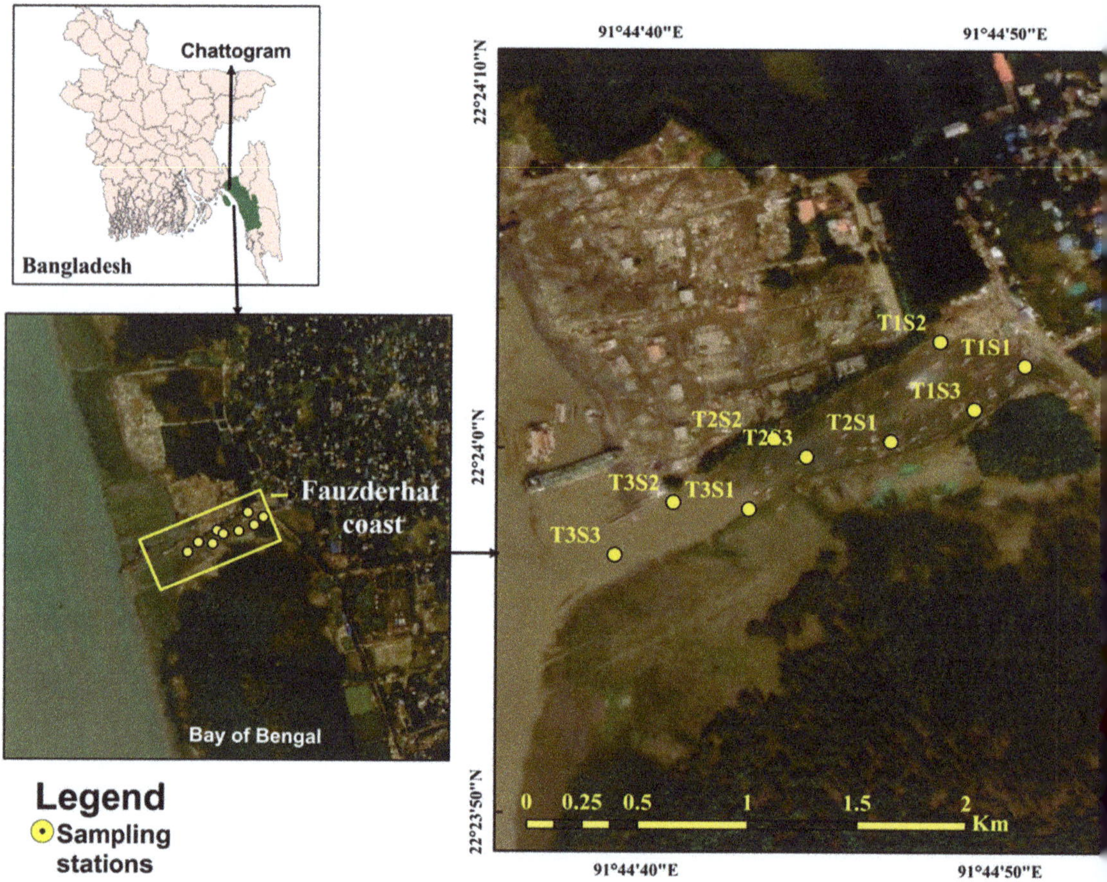

Figure 1. The location of the study area shows nine sampling stations.

2.2. Sampling Design and Collection

Samples were collected monthly during extremely low spring tides over two years divided into four seasons: monsoon (June–September), post-monsoon (October–November), winter (December–February), and pre-monsoon (March–May). Sampling was conducted across nine stations. Within (50×350 m^2) plots, three 50 m landward-to-seaward transects were drawn along the coast (Figure 2). All stations were equidistant within the transects, with their positions and depths being estimated using a portable GPS and Topographic Abney level. Samples for macrobenthos were collected using a hand-held cylindrical core (11.5×15 cm^2), with the corresponding labels being designated by, respectively, combining the station codes T1, T2, and T3 with S1, S2, and S3.

Triplicate samples that were 50 m long were collected from nine stations inside a (50×350 m^2) sized plot. The replicas were located parallel to the coast from landward to seaward, being, respectively, coded as S1–S3, plus the station code (e.g., for Station 1, labels were T1S1 and T1S3) (Figure 2). The stations were set equidistant in each transect. During the fieldwork, a portable GPS and Topographic Abney level were utilized for measuring the position fixing and elevation of the sampling stations.

Figure 2. Sampling design showing the studied stations and spots.

2.3. Sample Analysis

Five sediment replicas were collected at each station with a core (20 × 20 × 20 cm). Collected sediments were washed using a 500 µm mesh sieve [29,30], preserved in a 10% formalin solution, and stained with Rose Bengal to enhance visualization [31,32]. Preserved samples were rinsed with distilled water prior to sorting the macrobenthos into major taxonomic groups [33–35], which were transferred to labeled vials with 70% alcohol mixed with glycerol [29], identified to the lowest possible taxonomic level using relevant guides and manuals [33–35], and counted.

2.4. Data Analysis

The following univariate descriptors were used to characterize the macrobenthic assemblages: number of species, abundance (expressed as number of individuals in one square meter, ind./m^2), Shannon species diversity (H′), Margalef species richness (D′), and Pielou evenness (J′) [36–39]. Data were normalized using fourth-root transformation prior to being analyzed using an analysis of variance (ANOVA) to assess for seasonal differences. Tukey's honest significant difference test (Tukey HSD) was used to assess the responsibility for the significance of the observed differences. These analyses were performed with the SPSS (v.11.5) software. A one-way analysis of similarity (one-way ANOSIM) was conducted using PRIMER (V.7) software to assess the differences between descriptors. The total number of species and total number of individuals were counted for each season (monsoon, pre-monsoon, post-monsoon, and winter).

The similarity percentages (SIMPER) routine in PRIMER (SIMPER) was applied to determine the level of dissimilarity between the faunal assemblages of different seasons and to identify the species contributing to this dissimilarity based on the percentage contribution of individual species to average dissimilarity. Analysis was performed based on the contribution of numerically important species that cumulatively contributed 60% of total abundance. Canonical Correspondence Analysis (CCA) using the XLSTAT 2024 package was used to explore the relationships between benthic community composition and environmental variables. In total, 9 environmental factors (soil moisture, field water capacity, bulk density, particle density, porosity, (%) sand, (%) silt (%) clay, and total suspended solids) were excluded during the analysis due to avoid redundancy and ensure a more accurate representation.

Twenty-seven environmental variables (measured in triplicate; Supplementary Table S1) were analyzed for their correlation with abundance using the correlation coefficient (r) in Origin Pro v. 2024. Among the studied environmental factors, 6 environmental factors (soil moisture, soil temperature, soil pH, water transparency, field water capacity, bulk density, particle density, porosity soil organic carbon, soil organic matter, exchangeable ca, exchangeable Mg, and total suspended solids) were excluded during the analysis to avoid redundancy and ensure a more accurate representation. The season providing the most favorable environmental conditions was identified using a CCA also using the XLSTAT-2024 package.

3. Results

3.1. Abundance

The 1080 samples yielded 81,724 specimens from 48 taxa, grouped in 8 phyla (Table 1). The abundance showed significantly differed seasonally (one-way ANOSIM, global r = 0.773, $p < 0.1\%$). *Nematodes, polychaetes* (*Sigambra* sp. and *Perinereis* sp.), bivalves (*Mytella guyanensis, Tagelus plebeius, Macoma constricta,* and *Sphenia antillensis*), and decapods (*Uca* sp.) were dominant. The most abundant species, the bivalve *Tellina philippinarum,* ranged from 380 ± 247 to 950 ± 789 ind./m^2. *T. philippinarum* showed its highest abundance during winter followed by monsoon, pre-monsoon, and post-monsoon conditions. Twenty-one species exhibited higher abundance in winter, with *N. indica* and *T. philippinarum* comprising 44.69% and 35.39% of the total, respectively. Several species were absent in certain seasons. For example, *Moringua raitaborua* was not found during the pre-monsoon season, while *S. cumanense* was also absent in winter. Additionally, *Phascolosoma* sp. and *Neanthes* sp. were not observed in the post-monsoon season.

Table 1. Abundance (ind./m^2) of macrobenthos in the saltmarsh during the monsoon, post-monsoon, winter, and pre-monsoon seasons.

Taxon ID No.	Taxa	Monsoon	Post-Monsoon	Winter	Pre-Monsoon
01	*Nemertea* spp. Max Johann Sigismund Schultze, 1851	85 ± 51	31 ± 6 [b]	63 ± 10	102 ± 40 [a]
02	*Scopimera investigatoris* Alcock, 1900	34 ± 20 [a]	5 ± 2 [b]	23 ± 9	23 ± 6
03	*Tellina philippinarum* Hanley, 1844	687 ± 440	380 ± 247	950 ± 789	667 ± 399
04	*Micronephthys oligobranchia* Southern, 1921	80 ± 46 [a]	4 ± 4 [a]	278 ± 81 [b]	208 ± 95 [b]
05	*Edwardsia jonesii* Seshaiya & Cuttress, 1969	125 ± 46 [a]	16 ± 4 [b]	112 ± 75	155 ± 87 [a]
06	*Edwardsia tintrix* Annandale, 1915	93 ± 40 [b]	8 ± 4 [b]	147 ± 98 [a]	231 ± 33 [a]
07	*Sipunculus* sp. Linnaeus, 1766	128 ± 64 [a]	36 ± 16 [b]	77 ± 45	73 ± 28
08	*Heteromastides* sp. Augener, 1914	316 ± 149	90 ± 61 [b]	229 ± 174 [b]	604 ± 380 [a]
09	*Littoraria melanostoma* Gray, 1839	26 ± 8 [a]	10 ± 8 [b]	10 ± 9 [b]	14 ± 9

Table 1. *Cont.*

xon No.	Taxa	Monsoon	Post-Monsoon	Winter	Pre-Monsoon
	Nerita violacea Gmelin, 1791	6 ± 3^{a}	2 ± 1^{b}	2 ± 2^{b}	4 ± 3
	Composetia sp. Hartmann-Schröder, 1985	163 ± 73^{b}	183 ± 100^{b}	416 ± 95^{a}	80 ± 22^{b}
	Diopatra sp. Audouin & Milne Edwards, 1833	6 ± 6^{b}	8 ± 3^{b}	34 ± 24^{a}	9 ± 4^{b}
	Ocypode ceratophthalma Pallas, 1772	5 ± 5	1 ± 2	2 ± 2	4 ± 3
	Prionchulus sp. Nathan Augustus Cobb in 1916	41 ± 11	10 ± 2^{b}	43 ± 19	114 ± 122^{a}
	Alicella sp. Chevreux, 1899	8 ± 8^{a}	7 ± 1^{a}	7 ± 6^{a}	1 ± 1^{b}
	Diptera larvae	4 ± 4	1 ± 2^{b}	10 ± 8^{a}	4 ± 2
	Dendronereis sp. Peters, 1854	3 ± 3^{b}	2 ± 2^{b}	4 ± 4^{b}	12 ± 8^{a}
	Belba sculpta Mihelčič, 1957	4 ± 4^{b}	15 ± 6^{a}	7 ± 3^{b}	2 ± 2^{b}
	Moringua raitaborua Hamilton, 1822	1 ± 1	0 ± 0	1 ± 1	1 ± 1
	Siphonosoma cumanense Keferstein, 1867	16 ± 12^{a}	0 ± 0^{b}	0 ± 0^{b}	2 ± 2^{b}
	Glauconome sculpta G. B. Sowerby III, 1894	55 ± 22^{b}	48 ± 20^{b}	48 ± 12^{b}	128 ± 72^{a}
	Loimia sp. Malmgren, 1866	128 ± 74^{a}	81 ± 12^{b}	120 ± 65^{a}	66 ± 57^{b}
	Littorina undulata J. E. Gray, 1839	4 ± 3	5 ± 4	6 ± 5	4 ± 2
	Capitella sp. Blainville, 1828	31 ± 42^{b}	11 ± 5^{b}	69 ± 53^{b}	594 ± 832^{a}
	Terebellides sp. Sars, 1835	66 ± 67	13 ± 11^{b}	143 ± 82^{a}	149 ± 61^{a}
	Venitus dentipes Lucas in Guérin, 1836	15 ± 18^{a}	1 ± 1^{b}	8 ± 8^{a}	12 ± 9^{a}
	Corophium sp. Latreille, 1806	24 ± 21^{a}	2 ± 2^{b}	20 ± 6^{a}	7 ± 7^{b}
	Namalycastis indica Southern, 1921	310 ± 323	61 ± 45^{b}	1039 ± 663^{a}	915 ± 690
	Namalycastis fauveli Nageswara Rao, 1981	7 ± 7	2 ± 2^{b}	15 ± 9^{a}	12 ± 12^{a}
	Lumbrineris pseudobifilaris Fauvel, 1932	1 ± 2^{b}	4 ± 3	4 ± 5	11 ± 15^{a}
	Polydora sp. Oken, 1815	4 ± 2^{b}	4 ± 2^{b}	17 ± 1^{a}	9 ± 8
	Glycinde oligodon Southern, 1921	1 ± 1	1 ± 1	3 ± 2	3 ± 4
	Glycera lancadivae Schmarda, 1861	0 ± 0^{b}	1 ± 1^{b}	27 ± 25^{a}	2 ± 4^{b}
	Phascolosoma sp. Leuckart, 1828	11 ± 13^{b}	0 ± 10^{b}	481 ± 305^{a}	110 ± 40^{b}
	Dasybranchus sp. Grube, 1850	1 ± 1^{b}	1 ± 1^{b}	4 ± 4^{a}	0 ± 0^{b}
	Nemalycastis indica Southern, 1921	1 ± 1^{b}	0 ± 0^{b}	2 ± 1^{a}	0 ± 0^{b}
	Culicoides sp. Latreille, 1809	2 ± 3	0 ± 0	1 ± 1	1 ± 2
	Sabellidae sp. Latreille, 1825	1 ± 1^{b}	0 ± 0^{b}	7 ± 1^{a}	1 ± 11^{a}
	Sphaeroma sp. Bosc, 1801	6 ± 7^{a}	0 ± 0^{b}	3 ± 3^{a}	1 ± 1^{b}
	Gammarus sp. Fabricius, 1775	5 ± 4^{a}	1 ± 2	1 ± 1^{b}	1 ± 1^{b}
	Hesionidae sp. Grube, 1850	4 ± 4^{b}	2 ± 3^{b}	35 ± 39^{a}	15 ± 6
	Pterynotus elongatus [Lightfoot], 1786	3 ± 2	0 ± 0	1 ± 1	2 ± 1
	Procellio spinicornis Thomas Say, 1818	5 ± 4^{a}	0 ± 0^{b}	1 ± 1^{b}	1 ± 1^{b}
	Portunus sp. Weber, 1795	9 ± 6^{a}	0 ± 0^{b}	11 ± 17^{a}	0 ± 0^{b}
	Urothoe lacteal Dana, 1852	1 ± 1^{b}	0 ± 0^{b}	14 ± 3^{a}	2 ± 1^{b}
	Stenothyra deltae W. H. Benson, 1837	4 ± 3^{b}	13 ± 7^{a}	5 ± 5^{b}	3 ± 3^{b}
	Tanysyphon rivalis Benson, 1858	17 ± 7^{b}	22 ± 8^{b}	174 ± 149^{a}	16 ± 4^{b}
	Laevicaulis alte André Étienne d'Audebard de Ferussac, 1822	2 ± 2	2 ± 2	1 ± 1	2 ± 4

Note: Different superscripts in the same row denote statistically significant different subsets according to Tukey's HSD range test ($p < 0.05$).

3.2. Seasonal Diversity of Macrobenthos Communities

The total number of species significantly varied seasonally, with some species thriving in specific seasons while others declined (Figure 3), being significantly lower ($p < 0.05$) during post-monsoon compared to monsoon conditions, winter, and pre-monsoon conditions along with the total number of individuals. *Nemertea* spp. (40.13%), *Edwardsia jonesii* (37.99%), *Edwardsia tintrix* (48.23%), *Heteromastides* sp. (48.75%), and *Glauconome sculpta* (19.71%), among others, were more abundant during pre-monsoon conditions. Fourteen species concentrated the highest dominances during monsoon and pre-monsoon conditions, while only *B. sculpta* and *T. stroemi* dominated during post-monsoon conditions.

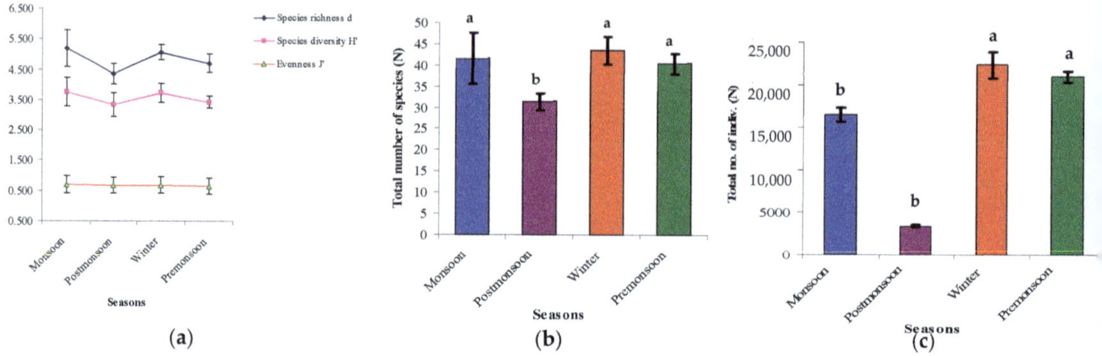

Figure 3. (**a**) Seasonal variation in species richness (Margalef, d), species diversity (Shannon, H′) and evenness (Pielou, J′), (**b**) seasonal variation in macrobenthic species number, and (**c**) seasonal variation in macrobenthic individuals observed in the Fauzderhat area of Chattogram. Different letters on the top of the bars indicate significant differences among the seasons.

Species richness ranged from 4.359 ± 0.343 to 5.189 ± 0.593 across seasons, being significantly lower in the post-monsoon season and higher in monsoon season ($p < 0.05$), followed by winter and pre-monsoon season. Diversity ranged from 3.342 ± 0.394 to 3.758 ± 0.467 bits, mirroring species richness, with monsoon conditions having the most diverse and relatively evenly distributed assemblages. Evenness ranged from 0.642 ± 0.041 to 3.758 ± 0.467, showing the most balanced distribution during monsoon conditions and the most imbalanced during post-monsoon conditions. While the variations in species diversity and evenness across seasons were not statistically significant, higher values were consistently recorded during monsoon conditions and lower during post-monsoon conditions.

3.3. Seasonal Succession of Macrobenthos

Seasonality strongly influenced community, with monsoon and post-monsoon conditions assemblages clustering together and becoming well distinguished for the winter/pre-monsoon. This suggests similar within-cluster similar communities and a substantial shift between them in terms of macrobenthic abundance and diversity. This strong seasonal influence could be related with differences in the degree of humidity (i.e., wetter vs. drier).

The average density was lower (1113 ± 58 ind./m^2) in post-monsoon conditions and increased towards winter (4813 ± 205 ind./m^2) and in pre-monsoon conditions (4519 ± 187 ind./m^2), before decreasing again in monsoon and post-monsoon conditions (Figure 4). Most likely, this reflects cyclical trends in community structure in response to fluctuating environmental conditions. The average number of individuals per species was moderate (367) during monsoon conditions, with *Heteromastides* sp. (316) and *N. indica* (310), followed by *Composetia* sp. (163), being dominant. Likely, these are the best adapted to the freshwater inputs characterizing monsoon conditions. *Heteromastides* sp. showed a similarly high abundance during pre-monsoon conditions, while *N. indica* increased its abundance.

Table 2. Results of the SIMPER analysis showing the average dissimilarity and the key discriminating species responsible for seasonal community.

Comparing Groups (1)	Comparing Groups (2)	Comparing Groups (3)
Monsoon and post-monsoon 26.98 Discriminator species *Neanthes* sp. *Siphonosoma cumanense* *Portunus* sp. *Porcellio spinicornis* *Sphaeroma* sp. *Pseudapocryptes elongatus*	Monsoon and winter 25.07 Discriminator species *Siphonosoma cumanense* *Glycera lancadivae*	Post-monsoon and winter 28.33 Discriminator species *Phascolosoma* sp. *Urothoe lactea* *Neanthes* sp. *Lycastonereis indica* *Sphaeroma* sp.
Comparing Groups (4)	**Comparing Groups (5)**	**Comparing Groups (6)**
Monsoon and pre-monsoon Average dissimilarity = 22.46 *Portunus* sp.	Post-monsoon and pre-monsoon Average dissimilarity = 29.67 *Phascolosoma* sp. *Neanthes* sp. *Urothoe lactea* *Pseudapocryptes elongatus*	Winter and pre-monsoon Average dissimilarity = 21.78 *Lycastonereis indica* *Dasybranchus* sp. *Siphonosoma cumanense*

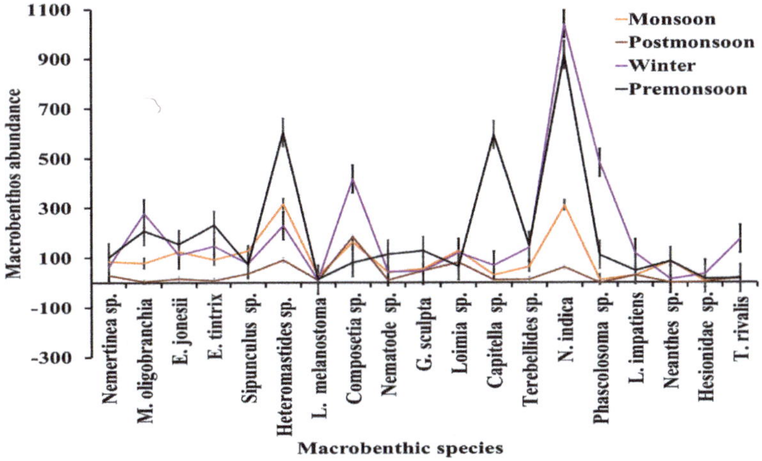

Figure 4. Distinct patterns of macrobenthic abundance in the four different seasons. Community composition significantly differed between monsoon and post-monsoon conditions (Table 2).

Numerous species experienced a reduction in abundance during post-monsoon conditions, when the average number of individuals was approximately of 338. *Composetia* sp. showed the highest abundance, followed by *Heteromastides* sp. (90 individuals) and *Loimia* sp. (81 individuals). Notably, *Corophium* sp. and *S. investigatories* showed significant declines during this season.

3.4. Relationships Between Environmental Variables and Macrobenthic Distribution

The abundance was positively correlated with soil salinity (r = 0.78), silt and clay (r = 0.054 and r = 0.33, respectively), dissolved oxygen (r = 0.47), and pH (r = 0.65) (Figure 5), while it was negatively correlated with sand (r = −0.31), tidal height (r = −0.52), inundation period (r = −0.40), and nutrients (r = 0.60) (Figure 5).

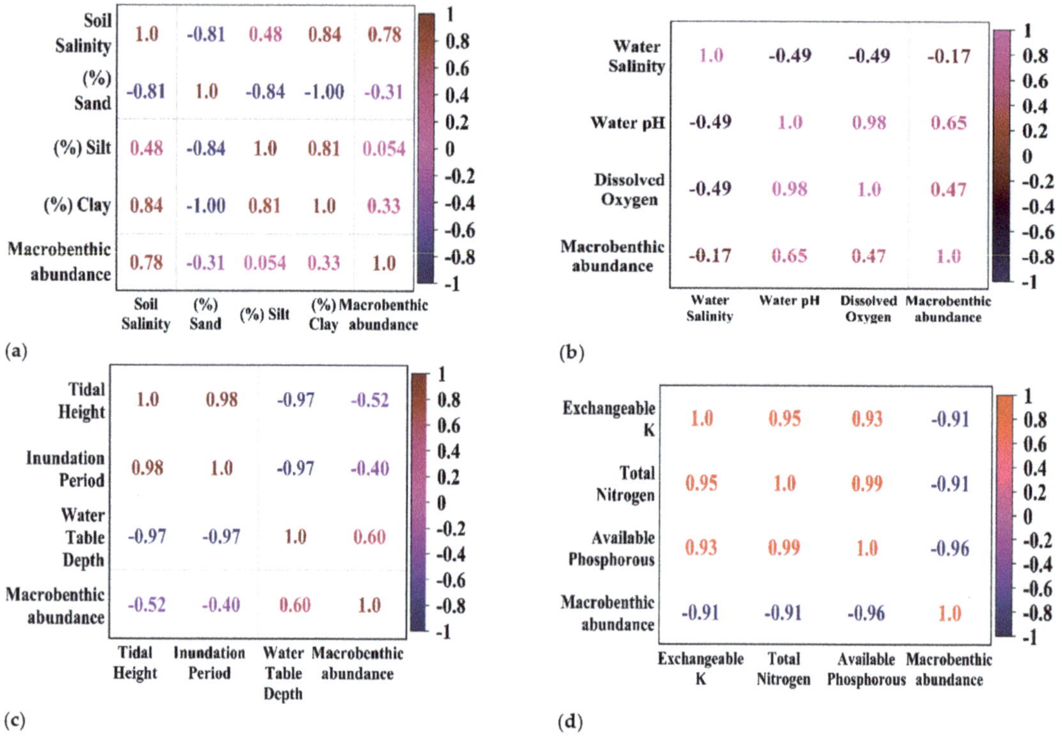

Figure 5. Correlation coefficients between macrobenthic abundance and environment: (**a**) sediment physio-chemical factors; (**b**) water factors; (**c**) tidal changes; (**d**) nutrients.

The first two axes of the CCA (Figure 6) explain 89.73% (F1 = 49.66%; F2 = 40.06%) of the total variance in macrobenthic distribution across seasons. During monsoon conditions (upper right quadrant), higher water temperature, inundation period, and tidal height, together with moderate sediment and water salinities and organic carbon were observed. These conditions led to increased water influx and mixing that were unfavorable for most species. However, species such as *G. sculpta*, *Neanthes* sp., *Nemertea* spp., *Heteromastides* sp., *E. jonesii, and Prionchulus* sp. showed strong associations with this season. In contrast, winter (lower left quadrant, Figure 6) showed a distinct community structure. Lower water temperatures and depths provided favorable conditions for many species, including *M. oligobranchia*, *T. rivalis*, *Phascolosoma* sp., and *Hesionidae* sp.

Post-monsoon conditions (upper left quadrant, Figure 6) were characterized by high nutrient levels (exchangeable K, available Phosphorus, total Nitrogen) and high dissolved oxygen. However, the increased bottom temperature, and bottom and water pH give rise to more stressful conditions, resulting in the lower presence of species such as *L. melanostoma*, *Loimia* sp., *Sipunculus* sp., and *Composetia* sp. Pre-monsoon conditions (lower right quadrant, Figure 6) supported species like *Capitella* sp., *Prionchulus* sp., and *Heteromastides* sp., with the fewer closely associated environmental factors suggesting balanced conditions. This appears to be a transition season, placed between the more favorable winter and the less favorable post-monsoon and monsoon season, with the moderate environmental conditions overall favoring species diversity.

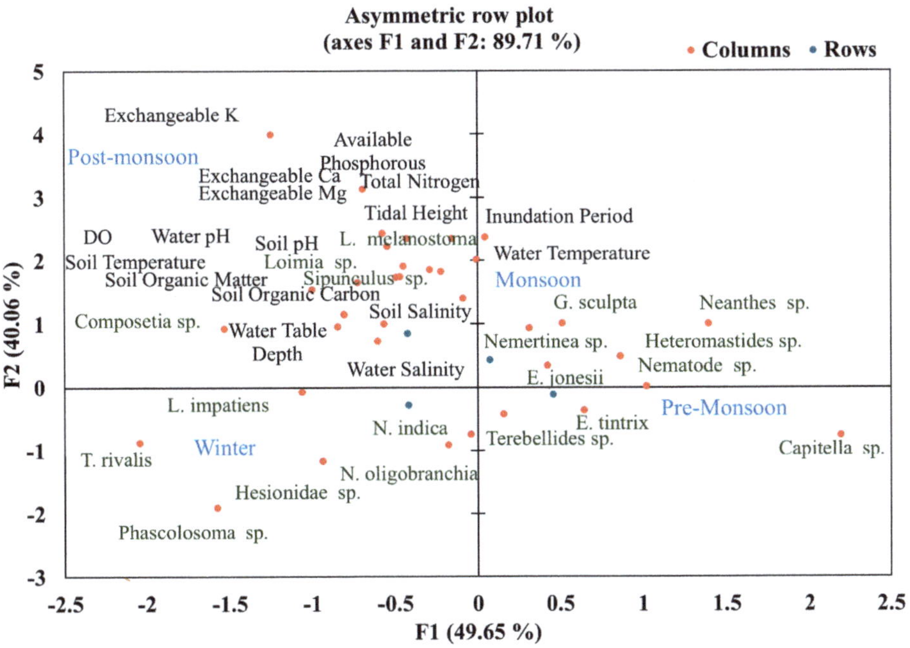

Figure 6. Canonical correspondence analysis between macrobenthic abundance and environmental factors. Seasons (monsoon, pre-monsoon, post-monsoon, and winter) are indicated in blue color.

The transition from less abundant post-monsoon conditions to more abundant winter and pre-monsoon conditions strongly suggest that macrobenthic dynamics are driven by seasonal environmental factors. Species like *Capitella* sp. showed strict seasonal preferences (i.e., pre-monsoon conditions) while species like *Sipunculus* sp.—located near the plot center (Figure 6)—showed broader tolerance.

4. Discussion

4.1. Abundance of Macrobenthos

The taxa diversity in this study was lower than in the Bohai Sea, northwest India (14 phyla), the northern Yellow Sea (7 phyla, 153 species, 301–598 specimens), and the Ulleung Basin (262 species) [2,40,41]. This lower diversity may be attributed to the high levels of pollutants and habitat degradation along the Fauzderhat coast. In contrast, the diversity appears to be higher than along the Cox's Bazar coast, were only six major groups (gastropoda, bivalvia, polychaeta, oligochaeta, Sipuncula, and crustacea) were reported [41]. Most likely, this could be due to differences in human impact, sediment types, and water quality. However, different methodologies—such as sampling area size, sampling frequency, and techniques (e.g., grabs vs. cores)—could also contribute to the abundance and diversity discrepancies. The dominance of nematodes, polychaetes, and bivalves in our study aligns with the previously reported prevalence of polychaetes and bivalves found in other studies [2,41,42]. However, crustaceans and mollusks dominated the Malaysian coastal waters [43], while polychaetes, isopods, and gastropods dominate the Brazilian saltmarshes [44]. These differences could be linked to local environmental factors and variations in saltmarsh vegetation, where polychaetes, mollusks, and crustaceans often tend to be dominant [2,41,43,44]. The adaptability of polychaetes to varying environmental conditions, such as salinity and temperature [45,46], likely explain their prevalence, underscoring their value as bioindicators for environmental health assessments. Similarly, the favorable salinity and temperature conditions could also support the prevalence of mollusk and crustacean microbenthic species [12].

4.2. Seasonal Diversity

The two species (*N. indica* and *T. philippinarum*) constituted a substantial proportion (80.08%) of the overall abundance among the 21 species that were more abundant in winter. This indicated that these two species are especially dominant or well-suited to winter conditions in the examined area. This also suggests that these species may thrive under colder temperatures or seasonal nutrient conditions, potentially benefiting from reduced interspecific competition in contrasts with the data from Minicoy Island, where a lower diversity was recorded during the monsoon season [47]. The monsoon season also favored species such as *Sipunculus* sp. and *Loimia* sp., while polychaetes may represent up to 64% of the total benthos during the monsoon season [48], when an increasing sedimentation and available organic matter may create favorable feeding and breeding conditions. We have found several taxa to be more dominant during the pre-monsoon season (i.e., *Nemertea* spp., *E. jonesii*, *E. tintrix*, *Heteromastides* sp., and *G. sculpta*). In contrast, only polychaetes were dominant during the pre-monsoon season in the Naf River estuary, where the macrobenthic abundances were overall higher (2972 ± 1994 ind./m^2) [48] compared to Fauzderhat. Therefore, warming waters and increased food availability during the pre-monsoon season seem to create optimal growth conditions favoring the multiple taxa dominance observed in Fauzderhat. Similarly, a study in a tropical estuary linked post-monsoon season declines to increased freshwater inflow reducing salinity [49]. The maximum number of species and individuals during the monsoon season and winter, respectively, in contrast with the higher diversity during the post-monsoon season in a subtropical mangrove estuary [50]. Accordingly, the dominance of a particular species during particular seasons appears to be influenced by environmental factors such as temperature, salinity, and food availability, which shape habitat suitability and competition dynamics.

Species richness was lower in the post-monsoon season and higher in the monsoon season, in contrast with previous reports of peaking in summer and lowering in winter [51]. The monsoon season had the most diverse community, in contrast with other studies showing a diversity of 2.1 bits, significant seasonal changes, and higher values in shallow waters, particularly in summer [2], or ranging from 2.55 to 2.92 bits [51]. The observed seasonal changes in evenness contrast with the uniform diversities and minor evenness fluctuations across seasons found across the western Indian coasts [52]. These differences likely arise from habitat type (shallow vs. deep water) and varying environmental conditions, such as temperature and salinity, influencing species distribution.

4.3. Seasonal Succession

The marked seasonality found in the macrobenthic community in the Sitakundu coastal saltmarsh could be caused by variations in hydrodynamic conditions, salinity, and organic matter supply between the wet (monsoon) and dry seasons. The wet season has more freshwater inputs and increasing sedimentation rates, which altered habitat conditions, leading to lower diversity and density, whereas the dry season ensured more stable salinity and oxygen levels, facilitating recruitment and growth. Additionally, monsoonal disruptions and intermittent flooding likely influence species composition by reshaping substrate topography and food sources. In return, the communities were less diverse in November and October and more diverse in February at Merchang lagoon [53], while summer conditions significantly affected species diversity and abundance in the northern Gulf of Mexico [54], leading to community structure shifts similar to those observed in Fauzderhat. Environmental factors, such as nutrient availability and hydrodynamic conditions, strongly influenced community succession patterns [55] as they occurred in Fauzderhat, but also in the Naf River estuary, where the abundance was higher during the pre-monsoon season compared to the monsoon season [48]. The reduced abundance of *Corophium* sp. and *S. investigatories* during the post-monsoon season was likely caused by the habitat being altered by increased sedimentation rates [56]. Seasonal variations in precipitation, which directly influence salinity levels and hence impact water density, seem to be a primary factor in the observed patterns of species abundance. *N. indica* demonstrated peak abundance in winter, with 1039 individuals documented, followed by a small reduced count of 915 individuals in the pre-monsoon period. This periodic fluctuation in abundance presumably indicates the species' sensitivity to salinity variations, ultimately leading to a shift in species' dominance across the seasons.

Winter's lower temperature and freshwater inflow gives rise to more stable conditions for macrobenthos [57], while water temperature, total phosphorus, and total nitrogen critically influence the abundance [1], and stable environments and higher food availability positively influence diversity [50]. The strong seasonal differences shown by the macrobenthos supports the Fauzderhat saltmarsh as being a dynamic ecosystem, where the cyclical domination of different species at different times of the year emerges as a key indicator of ecosystem succession. Different environmental conditions likely affected species presence and abundance, which resulted in significant differences in community composition and structure, as previously reported in a river-connected floodplain lake [41] and other macrobenthic communities [55,58]. Shifts in community structure across seasons affected abundance, dominance, and diversity, likely being driven by environmental changes and species interactions.

4.4. Environmental Variables and Their Effects

The correlation analysis indicated an overall good adaptation of macrobenthos to more saline conditions, as shown in previous studies in which species richness was also benefited [1,12]. Increasing sand contents has been suggested to lead to decreasing habitat complexity and resource availability, thus negatively affecting macrobenthic diversity and abundance compared to the richer organic matter and nutrient content provided by muddy or silty substrates, which tend to support more diverse communities [4,59,60]. Macrobenthos showed a preference for well-oxygenated neutral-to-slightly alkaline conditions, as found in the Heihe River Basin [1]. In contrast, the tidal height and inundation periods apparently had a detrimental effect on organism performances. Excessive increases in nutrient loads may lead to eutrophication, algal blooms, and subsequent hypoxia, as it occurs during the monsoon season [61,62]. This may adversely affect the most sensitive species, leading to altered community structure and succession, likely explaining the strong negative correlation between nutrients and abundance and favoring pollution-tolerant species over those less adapted to such environments [60].

The results of the CCA align with previous studies, in which elevated turbidity and fluctuating pH led to reduced diversity [61], while winter's colder temperatures and increased habitat stability fostered higher species richness, likely by lowering metabolic rates [61,63]. Despite nutrient enrichment can initially help in supporting diverse communities, the pre-monsoon season's extreme soil temperatures and pH fluctuations often created unfavorable conditions for macrobenthic organisms [61,62]. The post-monsoon season showed extreme environmental fluctuations with lower presence of species and the pre-monsoon season was a transition period favoring species diversity. Stable temperatures and nutrient levels associated with the transitional periods often foster more diverse communities compared to the periods with more extreme fluctuations [62,63]. Seasonal assemblage structure shifts appear to be driven by nutrient availability and physical parameters, combined with species-specific adaptations to the different seasonal conditions within the saltmarsh ecosystem.

5. Conclusions

This study investigates the abundance, diversity, and seasonal dynamics of macrobenthic communities in an anthropically influenced coastal ecosystem. Of the 53 taxa recorded, nematodes, polychaetes, decapods, and bivalves dominated at different seasons. All of our different analytical approaches showed distinct macrobenthic successional patterns reflecting ecological process dynamics, with some species declining while others becoming dominant. Bottom salinity and water depth, dissolved oxygen, and pH proved to be the key variables shaping macrobenthic abundance. The evident seasonal succession pattern at Fauzderhat saltmarsh underscores the resilience of this particular ecosystem, prompting its ability to recover from anthropic disturbances—such as habitat modification—and offering valuable insights for the conservation and sustainable management of coastal biodiversity.

Supplementary Materials: The following supporting information can be downloaded at: https://www.mdpi.com/article/10.3390/d17040231/s1, Supplementary Table S1. Average (±SD) values of different environmental factors of the Fauzderhat area in the Chittagong coast in different seasons of two years of sampling sessions.

Author Contributions: Conceptualization, M.S.I. and H.Z.; Methodology, M.S.A., A.H.M.K. and M.B.H.; Software, M.M.M.B.; Validation, A.H.M.K.; Formal analysis, T.A.; Investigation, A.H.M.K.; Resources, M.M.M.B.; Data curation, H.Z.; Writing — original draft, M.S.I. and M.M.M.B.; Writing — review & editing, M.B.H. and T.A.; Supervision, H.Z. and M.S.A. All authors have read and agreed to the published version of the manuscript.

Funding: This research was funded by Universiti Brunei Darussalam research grant number UBD/RSCH/1.4/FICBF(b)/2023/057.

Institutional Review Board Statement: Not applicable for studies not involving humans.

Data Availability Statement: Data is contained within the article and Supplementary Materials.

Conflicts of Interest: The authors declare no conflict of interest.

References

1. Wang, Y.; Liu, J.J.; Li, B.L.; Liu, W.; Zuo, Y.F.; Kong, D.X.; Zhu, J.L. Relationships between characteristics of macrobent assemblages and environmental variables in the Heihe River Basin, China. *AQUA—Water Infrastruct. Ecosyst. Soc.* **2021**, 710–730. [CrossRef]
2. Kim, S.L.; Kang, S.M.; Lee, H.G.; Han, G.H.; Yu, O.H. Species Diversity and Community Structure of Macrobenthos in the Ulleu Basin, East Sea, Republic of Korea. *J. Mar. Sci. Eng.* **2023**, *11*, 92. [CrossRef]
3. Armenteros, M.; Saladrigas, D.; González-Casuso, L.; Estevez, E.D.; Kowalewski, M. The role of habitat selection on the divers of macrobenthic communities in three gulfs of the Cuban Archipelago. *Bull. Mar. Sci.* **2018**, *94*, 249–268.
4. Ji, Y.; Song, B.; Xu, J.; Jiang, S.; Chen, L.; Li, B. Distribution and influencing factors of macrobenthos on three seagrass beds in intertidal zone of Shandong province, China. *Front. Mar. Sci.* **2024**, *11*, 1349131. [CrossRef]
5. Ullah, M.A.; Hossain, M.S.; Hossain, M.B.; Rahman, M. Belal Hossain, and Mahbubur Rahman. Intertidal variation of macrob thos in a saltmarsh habitat, Noakhali coast, Bangladesh. *Egypt. J. Aquat. Biol. Fish.* **2020**, *24*, 377–390.
6. Pawhestri, S.W.; Hidayat, J.W.; Putro, S.P. Assessment of water quality using macrobenthos as bioindicator and its application Abundance-Biomass Comparison (ABC) curves. *Int. J. Sci. Eng.* **2014**, *8*, 84–87.
7. Gage, J.D. *Macrobenthos en Encyclopedia of Ocean Science*; Scottish Association for Marine Science: Oban, UK, 2001; Volum pp. 1505–1515.
8. Jia, K.; Tang, Y.; Liu, Q.; Zhang, R.; Liao, Y.; Shou, L.; Zeng, J.; Chen, Q.; Chen, J. Spatial and temporal distribution of macrobent communities and their relationship with secondary front in Hangzhou Bay. *Front. Mar. Sci.* **2022**, *9*, 1037287. [CrossRef]
9. Muskananfola, M.R.; Purnomo, P.W.; Sulardiono, B. Impact of environmental factors on macrobenthos distribution and abunda in mangrove ecosystems on the Northern Coast of Java. *AACL Bioflux* **2020**, *13*, 2745–2756.

Touchette, B.W.; Schmitt, S.R.; Moody, J.W. Enhanced thermotolerance of photosystem II by elevated pore-water salinity in the coastal marsh graminoid Sporobolus pumilus. *Aquat. Biol.* **2020**, *29*, 111–122. [CrossRef]
Vahsen, M.L.; Kleiner, H.S.; Kodak, H.; Summers, J.L.; Vahsen, W.L.; Blum, M.J.; Megonigal, J.P.; McLachlan, J.S. Complex eco-evolutionary responses of a foundational coastal marsh plant to global change. *New Phytol.* **2023**, *240*, 2121–2136. [CrossRef]
Mely, S.S.; Hossain, M.B.; Rahman, M.; Albeshr, M.F.; Arai, T. Changes of Macrobenthic Diversity and Functional Groups in Saltmarsh Habitat under Different Seasons and Climatic Variables from a Subtropical Coast. *Sustainability* **2023**, *15*, 7075. [CrossRef]
Gandini, L.J.R.; Funk, F.A.; Pratolongo, P.D. Pratolongo. Change in the community structure and organic carbon content of meio-and macrobenthos between tidal flat and salt marsh areas colonized by Spartina alterniflora in the Bahía Blanca estuary (SW Atlantic). *Sci. Mar.* **2023**, *87*, 3.
Islam, M.S.; Imran, M.A.; Mandol, S.; Bhuyan, M.S. Restoration of saltmarsh as mitigative measure of climate change impact at Chittagong coastal area of Bangladesh. *Clim. Change* **2016**, 2, 313–329.
Miah, J.; Hossain, K.T.; Hossain, M.A.; Najia, S.I. Assessing coastal vulnerability of Chittagong District, Bangladesh using geospatial techniques. *J. Coast. Conserv.* **2020**, *24*, 66.
Fakhruddin, A.N.M.; Talukdar, M.I.; Hossain, M.A. Environmental impacts of ship breaking and recycling industry of Sitakunda, Chittagong, Bangladesh. *Adv. Nat. Sci.* **2015**, *8*, 51–58.
Hossain, M.S.; Fakhruddin, A.N.M.; Chowdhury, M.A.Z.; Gan, S.H. Impact of ship-breaking activities on the coastal environment of Bangladesh and a management system for its sustainability. *Environ. Sci. Policy* **2016**, *60*, 84–94.
Kutub, J.R.; Falgunee, N.; Nawfee, S.M.; Rabby, Y.W. Ship breaking industries and their impacts on the local people and environment of coastal areas of Bangladesh. *Hum. Soc. Stud.* **2017**, *6*, 35–58.
Li, S.; Cui, B.; Xie, T.; Shao, X.; Zhang, M. Consequences and implications of anthropogenic desalination of salt marshes on macrobenthos. *CLEAN–Soil Air Water* **2016**, *44*, 8–15.
Lee, H.G.; Yoon, K.T.; Park, H.S.; Hong, J.S.; Lee, J.H. The influence of environmental variables on distribution of macrobenthic community in salt marsh vegetation in Donggeomdo, Ganghwa on the west coast of Korea. *Ocean. Polar Res.* **2016**, *38*, 115–128. [CrossRef]
Qiu, D.; Yan, J.; Ma, X.; Gao, F.; Wang, F.; Wen, L.; Bai, J.; Cui, B. How vegetation influence the macrobenthos distribution in different saltmarsh zones along coastal topographic gradients. *Mar. Environ. Res.* **2019**, *151*, 104767.
Santos, T.M.T.; Rabelo, D.M.L.; Beasley, C.R.; Braga, C.F. Vertical distribution of macrobenthic community of tropical saltmarshes on the Amazon coast. *Reg. Stud. Mar. Sci.* **2020**, *40*, 101536. [CrossRef]
Islam, M.S.; Pervez, A.; Asseri, A.H.; Al-Mutair, M.; Sumon, M.A.A.; Taleb, M.A.; Ashik, A.A.; Rahman, M.A.; Molla, M.H.R. Diversity and seasonal succession of resident and migratory macrobenthic fauna of saltmarsh restoration site at Sonadia Island, Cox's Bazar, Bangladesh. *Reg. Stud. Mar. Sci.* **2022**, *53*, 102460. [CrossRef]
Haque, M.M.; Sharif, A.S.M.; Ahmed, M.K.; Anwar, F.; Rani, S.; Molla, M.H.R.; Khan, M.I. Distribution and diversity of macrobenthos in Sangu River, Bangladesh. *Dhaka Univ. J. Earth Environ. Sci.* **2020**, *9*, 27–34. [CrossRef]
Talukder, A.; Mallick, D.; Hasin, T.; Anka, I.Z.; Hasan, M.M. Spatio-temporal variability in hydro-chemical characteristics of coastal waters of Salimpur, Chittagong along the Bay of Bengal. *J. Fish.* **2016**, *4*, 335–344.
Islam, S.S.; Tanim, A.H.; Mullick, M.R.A. Vulnerability assessment of Bangladesh coastline using Gornitz method. In *Water, Flood Management and Water Security Under a Changing Climate: Proceedings from the 7th International Conference on Water and Flood Management*; Springer International Publishing: New York, NY, USA, 2020; pp. 301–313.
Barua, D.K. The coastline of Bangladesh—An overview of processes and forms. In *Coastal Zone'91*; ASCE: Preston, VA, USA, 1991; pp. 2284–2301.
Khan, M.S.; Das, J.; Kabir, M.H. Monitoring coastline changes in the coastal belt of chittagong in Bangladesh Using GIS and remote sensing. *Asian J. Geogr. Res.* **2019**, *2*, 1–13.
Tagliapietra, D.; Sigovini, M. Benthic fauna: Collection and identification of macrobenthic invertebrates. *Terre Et Environ.* **2010**, *88*, 253–261.
Enge, A.J.; Wanek, W.; Heinz, P. Preservation effects on isotopic signatures in benthic foraminiferal biomass. *Mar. Micropaleontol.* **2018**, *144*, 50–59. [CrossRef]
Shiozaki, T.; Itoh, F.; Hirose, Y.; Onodera, J.; Kuwata, A.; Harada, N. A DNA metabarcoding approach for recovering plankton communities from archived samples fixed in formalin. *PLoS ONE* **2021**, *16*, e0245936.
Rahman, M.K.; Hossain, M.B.; Majumdar, P.R.; Mustafa, M.G.; Abu Noman, M.; Albeshr, M.F.; Bhat, E.A.; Arai, T. Macrobenthic assemblages, distribution and functional guilds from a freshwater-dominated tropical estuary. *Diversity* **2022**, *14*, 473. [CrossRef]
Carpenter, K.E.; De Angelis, N. *The Living Marine Resources of the Eastern Central Atlantic. Volume 1: Introduction, Crustaceans, Chitons, and Cephalopods*; FAO: Rome, Italy, 2014.
Carpenter, K.E.; Niem, V.H. (Eds.) *The Living Marine Resources of the Western Central Pacific. Volume 1. Seaweeds, Corals, Bivalves and Gastropods*; FAO: Rome, Italy, 1998; p. xiv+-686.

35. Carpenter, K.E. *Introduction, Molluscs, Crustaceans, Hagfishes, Sharks, Batoid Fishes and Chimaeras*; Food and Agriculture Organizat of the United Nations: Rome, Italy, 2002.
36. Welch, P.S. *Limnological Methods*; McGraw-Hill: New York, NY, USA, 1948.
37. Shannon, C.E. A mathematical theory of communication. *Bell Syst. Tech. J.* **1948**, *27*, 379–423. [CrossRef]
38. Margalef, R. Information theory in ecology. *Real Acad. De Cienc. Y Artes De Barc. Traslation Ser.* **1973**, *2609*, 1–140.
39. Pielou, E.C. The measurement of diversity in different types of biological collections. *J. Theor. Biol.* **1966**, *13*, 131–144. [CrossR
40. Sukumaran, S.; Vijapure, T.; Mulik, J.; Ridha, H. Marine macrobenthos of NorthWest India-reviewing the known and unkno *Front. Mar. Sci.* **2021**, *8*, 671245. [CrossRef]
41. Li, F.; Ma, Y.; Song, X.; Li, S.; Zhang, X.; Wang, X.; Wang, T.; Sun, Z. Community structure and ecological quality assessmen macrobenthos in the coastal sea areas of Northern Yantai, China. *Front. Mar. Sci.* **2022**, *9*, 989034. [CrossRef]
42. Abu Hena, M.K.; Akhtar, A.; Khan, M.A.A.; Uddin, S.A.; Rezowan, M. Community structure of macro benthos in an estuar salt marsh habitat at Cox's Bazar. *Bangladesh J. Mar. Sci.* **2009**, *1*, 119–132.
43. Chowdhury, A.J.K.; John, A.; Aqilah, N.S.; Abdullah, R.; Salihah, N.T.; Basir, K.H.; Marsal, C. Macrobenthic community towa sustainable aquatic ecosystem: A systematic review along the coastal waters of Malaysia. *Geol. Ecol. Landsc.* **2024**, *8*, 57– [CrossRef]
44. Braga, C.F.; Beasley, C.R.; Isaac, V.J. Effects of plant cover on the macrofauna of Spartina marshes in northern Brazil. *Braz. A Biol. Technol.* **2009**, *52*, 1409–1420. [CrossRef]
45. Musale, A.S.; Desai, D.V. Distribution and abundance of macrobenthic polychaetes along the South Indian coast. *Environ. Mo Assess.* **2011**, *178*, 423–436. [CrossRef]
46. Kim, S.L.; Lee, H.G.; Yu, O.H. Correlation between rocky reefs and surrounding benthic habitats: Distribution and divers patterns of polychaetes in the macrobenthic community in the East Sea of South Korea. *J. Sea Res.* **2021**, *174*, 102083. [CrossR
47. Dalia Susan, V.; Satheesh Kumar, P.; Pillai, N.G.K. Biodiversity and seasonal variation of benthic macrofauna in Minicoy Isla Lakshadweep, India. *Acta Oceanol. Sin.* **2014**, *33*, 58–73. [CrossRef]
48. Noman, A.; Mamunur, R.; Islam, M.S.; Hossain, M.B. Spatial and seasonal distribution of intertidal macrobenthos with th biomass and functional feeding guilds in the Naf River estuary, Bangladesh. *J. Oceanol. Limnol.* **2019**, *37*, 1010–1023.
49. Lowe, V.; Frid, C.L.; Venarsky, M.; Burford, M.A. Responses of a macrobenthic community to seasonal freshwater flow i wet-dry tropical estuary. *Estuar. Coast. Shelf Sci.* **2022**, *265*, 107736.
50. Khatun, B.; Jewel, A.S.; Haque, A.; Akter, S.; Hossain, M.B.; Albeshr, M.F.; Arai, T. Seasonal pattern of taxonomic diversity a functional groups of macro-benthos from a sub-tropical mangrove estuary. *J. Mar. Sci. Eng.* **2023**, *11*, 1453. [CrossRef]
51. Sharmin, S.; Rahman, S.; Naser, M.; Hoque, S. Macro benthic fauna in relation to limnological variables in a migratory b visiting lake at Jahangirnagar University, Bangladesh. *J. Biodivers. Conserv. Bioresour. Manag.* **2018**, *4*, 99–106.
52. Pawar, P.R.; Kulkarni, B.G. Diversity indices of selected macrobenthos in Karanja Creek (District-Raigad), Maharashtra, w coast of India. *J. Indian Fish. Assoc.* **2007**, *34*, 1–9.
53. Hussin, W.M.R.W.; Ab Lah, R. Community structure and taxonomic diversity of macrobenthic communities in Merchang lago Malaysia. *Aquac. Aquar. Conserv. Legis.* **2020**, *13*, 3593–3604.
54. Shivarudrappa, S.; Briggs, K. Macrobenthos community succession in the northern Gulf of Mexico hypoxic regions: Testing Pearson-Rosenberg model. *J. Mar. Res.* **2017**, *75*, 18–46.
55. Pacheco, A.S.; Laudien, J.; Thiel, M.; Oliva, M.; Arntz, W. Succession and seasonal variation in the development of subti macrobenthic soft-bottom communities off northern Chile. *J. Sea Res.* **2010**, *64*, 180–189.
56. Matin, A.; Hossain, M.B.; Iqbal, M.; Billah, M.M.; Asif, A.A.; Billah, M.M. Diversity and abundance of Macrobenthos i subtropical estuary, Bangladesh. *Species* **2018**, *19*, 140–150.
57. Kosari, S.; Nadushan, R.M.; Faremi, R.; Khanghah, K.E.; Mashinchian, A. Macrobenthos as bioindicator of ecological status in Yekshabe creek-estuary, Persian Gulf. *Iran. J. Fish. Sci.* **2011**, *20*. [CrossRef]
58. Gerami, M.H.; Patimar, R.; Negarestan, H.; Jafarian, H.; Mortazavi, M.S. Temporal variability in macroinvertebrates divers patterns and their relation with environmental factors. *Biodiversitas J. Biol. Divers.* **2016**, *17*. [CrossRef]
59. Dutertre, M.; Hamon, D.; Chevalier, C.; Ehrhold, A. The use of the relationships between environmental factors and bent macrofaunal distribution in the establishment of a baseline for coastal management. *ICES J. Mar. Sci.* **2013**, *70*, 294–308. [CrossR
60. Shi, Y.; He, Y.; Shin, P.K.; Guo, Y.; Zhang, G.; Wen, Y.; Zhang, G.; Xu, W.; Sun, J. Responses of biological traits of macrobent fauna to a eutrophication gradient in a semi-enclosed bay, China. *Mar. Environ. Res.* **2023**, *189*, 106072.
61. Das Ghosh, B.; Debnath, S.; Sarkar, U.K.; Das, B.K.; Puthiyottil, M.; Johnson, C.; Karnatak, G. Ecological assessment o subtropical floodplain wetland of the Ganga basin in the context of changing climate using GIS tools. *J. Water Clim. Change* **20** *14*, 4220–4235.

Joseph, S.J.P.; Gunasekaran, P.D.; Subramani, N. Influence of environmental parameters on the community structure of phytoplankton from river confluence of Cuddalore, Tamil Nadu, India. *Environ. Adv.* **2022**, *7*, 100170.
Molina, A.; Duque, G.; Cogua, P. Influences of environmental conditions in the fish assemblage structure of a tropical estuary. *Mar. Biodivers.* **2020**, *50*, 5.

diversity

MDPI

Article

Assessment of Financing for Biodiversity Conservation in Mexico: Links Between Biodiversity and Climate Change Adaptation Funds

Miriam Sosa [1] and Antonina Ivanova [2,*]

1 Department of Economics, Universidad Autónoma Metropolitana (UAM-Iztapalapa), Mexico City 09340, Mexico; msosac87@hotmail.com
2 Department of Economics, Universidad Autónoma de Baja California Sur, La Paz 23080, Mexico
* Correspondence: aivanova@uabcs.mx; Tel.: +52-6121400442

Abstract: At COP16 in Cali, Colombia, significant progress was made in biodiversity conservation efforts. In this regard, financing has been considered a key issue for achieving the objectives. The overview of Mexico's experience with biodiversity finance in this study presents the experience of an emerging economy, which must finance pressing development priorities and biodiversity and climate action at the same time. Therefore, it is very important to find synergies in the available finance and look for new innovative options. The large overlap between the climate and biodiversity agendas and the international commitments derived from these also presents an opportunity to accelerate biodiversity funding. The methodology applied is the Systematic Literature Review (SLR). The study presents the national strategy on biodiversity in Mexico (ENBioMex), the financial needs of the country, and the existing biodiversity financing, stressing the Global Environmental Facility (GEF), the Biodiversity Finance Initiative (BIOFIN), and the Adaptation Fund in Mexico. The discussion section centers on analyzing the existing results and outlining some proposals to enhance the existing instruments, looking for innovation and synergies. In the authors' opinion, the financing of Ecosystem-based Adaptation (EbA) is the main instrument that can link biodiversity conservation and adaptation to climate change impacts, at the same time providing a sustainable way of life and guaranteeing the well-being of communities, but it is not adequately used. Finally, we present some concluding remarks and future research topics.

Keywords: biodiversity conservation; financing; climate change adaptation; GEF; BIOFIN; Adaptation Fund; Mexico

1. Introduction

We are witnessing an unprecedented crisis of biodiversity loss on a global scale. Ecosystems that have thrived for millennia are now teetering on the brink as forests are cleared, wetlands drained, and irretrievable coral reefs bleached. This rapid disappearance of habitats is not just a loss of beauty or nature but has devastating consequences for humanity [1].

There is a growing concern that the deterioration of the natural environment is accelerating, with biodiversity declining at an unprecedented rate. WWF estimates that at least 10,000 species become extinct every year, a number that underscores the fragile interdependence of life on Earth [2]. For the billions of people who depend on these ecosystems for food, clean water, medicine and climate regulation, this crisis threatens their very survival [3]. Biodiversity erosion destabilizes economies, exacerbates poverty, and deepens inequality, creating a cascade of challenges that humanity must urgently address.

The global economy is fundamentally reliant on natural ecosystems. According to the World Economic Forum (WEF), over half of the global Gross Domestic Product (GDP) is moderately or highly dependent on nature and its services, making it vulnerable to ecosystem degradation [4]. In its Global Risks Report 2024, the WEF identifies biodiversity loss and ecosystem collapse as the third most severe perceived global risk over the next decade [5].

According to the 6th Assessment Report of the Intergovernmental Panel on Climate Change (IPCC), threats to species and ecosystems in oceans, coastal regions and on land, particularly in biodiversity hotspots, present a global risk that will increase with every additional tenth of a degree of warming. The transformation of terrestrial and ocean/coastal ecosystems and loss of biodiversity, exacerbated by pollution, habitat fragmentation and land use changes, will threaten livelihoods and food security [6].

At the Conference of the Parties 16 of the Convention on Biological Diversity (COP16) in Cali, Colombia, significant progress was made in biodiversity conservation efforts. A total of 119 countries, representing 61% of the Parties, submitted national biodiversity targets, supported by 44 countries that provided National Biodiversity Strategies and Action Plans (NBSAPs). A new Program of Work was adopted to enhance the role of Indigenous Peoples and local communities in biodiversity conservation, recognizing their rights and traditional knowledge [7].

Additionally, a process was established to identify and update Ecologically Significant Marine Areas (EBSAs) using advanced science. A groundbreaking financial mechanism, the "Cali Fund," was created to share benefits from Digital Sequence Information (DSI), with at least 50% of funds allocated to Indigenous Peoples and local communities. These measures reflect growing global commitment to addressing biodiversity loss [8].

Parties consider a new Strategy for Resource Mobilization (The Strategy for Resource Mobilization is an essential component of global efforts to halt biodiversity loss, fostering alignment between financial systems and conservation goals while addressing the significant funding gaps in biodiversity protection) to help secure USD 200 billion annually by 2030 from all sources to support biodiversity initiatives worldwide, in line with Target 19 of the Kunming-Montreal Global Biodiversity Framework (KMGBF), aimed at halting or reversing biodiversity loss. Target 18 of the KMGBF also addresses the reduction of harmful incentives by at least USD 500 billion per year by 2030.

In this regard, financing has been considered a key issue for achieving the objectives; thus, an additional USD 163 million has been committed to the Global Biodiversity Framework Fund (GBFF), raising its total funding to approximately USD 396 million. Supported by governments, private entities, and philanthropic organizations, the GBFF focuses on financing impactful projects in developing regions, prioritizing nations with vulnerable ecosystems, including small island states and transitioning economies [8]. The fund accepts contributions from governments, the private sector, and philanthropies, and finances high-impact projects in developing regions, with emphasis on supporting countries with fragile ecosystems, such as small island states and economies in transition. To date, 11 donor countries as well as the Government of Quebec have pledged nearly US USD 400 million to the GBF Fund, with USD 163 million pledged during COP 16.

On the other hand, the 29th session of the Conference of the Parties (COP29) to the United Nations Framework Convention on Climate Change (UNFCCC) held in Baku, Azerbaijan, from 11 to 22 November 2024 and its main theme is enhancing climate finance ambition, enabling action and results, including finance.

The Adaptation Fund works alongside biodiversity funds to support projects that help vulnerable communities in developing countries respond to climate change. These initiatives are tailored to each country's specific needs, priorities, and perspectives, often leveraging nature-based solutions such as reforestation, mangrove restoration, and the recovery of habitats like coral reefs. An important component is ecosystem-based adaptation, which presents the opportunity to link climate change and biodiversity finance.

Since 2010, the Adaptation Fund has allocated over USD 1 billion to resilience and adaptation efforts, implementing 150 localized projects in highly vulnerable communities worldwide and benefiting over 38 million people. The fund also introduced Direct Access and Enhanced Direct Access mechanisms (Direct Access is a financing mechanism introduced by the Adaptation Fund that allows countries to directly access funding for climate adaptation projects without relying on intermediaries, such as international organizations. Countries can access these funds through accredited national implementing entities (NIEs), which are responsible for managing and implementing projects in line with national priorities. Enhanced Direct Access (EDA) takes this approach further by empowering national entities to not only manage funding but also make decisions on how to allocate resources at the local level. This decentralization strengthens country ownership and allows for more tailored, community-specific interventions, as well as the inclusion of local stakeholders in the decision-making process), enabling countries to directly access funding and manage projects through accredited national entities [9].

Based on the above, it is very important to tackle the insufficiency of financing in both areas, creating synergies between biodiversity conservation and climate change adaptation finance, including innovation options, and the aim of this article is to present a contribution on how to advance in this task, assessing the case of Mexico.

The article presents and analyzes the biodiversity protection strategies in Mexico, the financing instruments, and how synergy could be created between biodiversity protection and climate change adaptation. In the result section, Section 3.1 presents the national strategy on biodiversity in Mexico (ENBioMex) and the financial needs of the country, Section 3.2 is focused on the existing biodiversity financing, and Section 3.3 shows the role of GEF, BIOFIN and the Adaptation Fund in Mexico. The discussion section centers on analyzing the existing results and outlining some proposals to enhance the existing instruments, looking for innovation and synergies. Finally, we present some concluding remarks and future research topics.

2. Materials and Methods

The methodology is the Systematic Literature Review (SLR) adopted from [10] based on six stages: formulation of research questions, search criteria and identification, screening process, title and abstract screening, quality assessment and data extraction. The aim of SLR in this study was to find clarity of scholarly communication, validity where the literature is defensible against bias, and audibility of the literature to obtain accurate results, and to develop a strategic recommendations approach.

The SLR methodology provides guidelines to ensure comprehensive, unbiased, and reproducible results. The review focused on identifying and analyzing studies related to biodiversity conservation and financing in Mexico. Keywords used included "Biodiversity Conservation", "Climate Change Adaptation", and "Ecosystem-based Adaptation", combined with geographical and financial terms like "Mexico", "Financing", and "National Strategy". Searches were conducted in Scopus, Web of Science, and Google Scholar. Gray literature (GL) was included from government repositories such as INEGI, CONABIO and SEMARNAT.

The inclusion criteria comprised peer-reviewed articles, reports, and studies in English or Spanish, directly addressing biodiversity and climate finance, published between 2016 and 2024 and with contextual relevance to our study (Table 1).

Table 1. Inclusion and exclusion criteria.

Criterion	Eligibility	Exclusion
Literature type	Journal (original articles)	Journals (reviews), chapters in book, conference proceedings,
Grey literature	International and Mexican documents and databases	N/A
Timeline	2016–2024	<2016
Language	English Spanish	N/A
Index	Scopus Web of Science Google Scholar	N/A
Countries	Mexico	N/A
Contextual relevance	Climate change adaptation financing Biodiversity financing	N/A

The exclusion of irrelevant papers was performed through a multistep criteria-based rigorous screening process which was purposefully designed to ensure that the selected publications were reasonably representative of the prevailing situation. As shown in Figure 1, the initial search yielded 67 documents that were broadly relevant to the study. Thereafter, these documents were screened to exclude 26 publications by only including those that were published between 2016 and 2024 (Stage 1) and did not meet the inclusion criteria of contextual relevance to this study (Stage 2). The cut-off dates were preferred because documents produced before 2016 were reasoned to be dated and unrepresentative of current perspectives. The screening yielded 31 publications, which provided this study's final compilation of relevant publications.

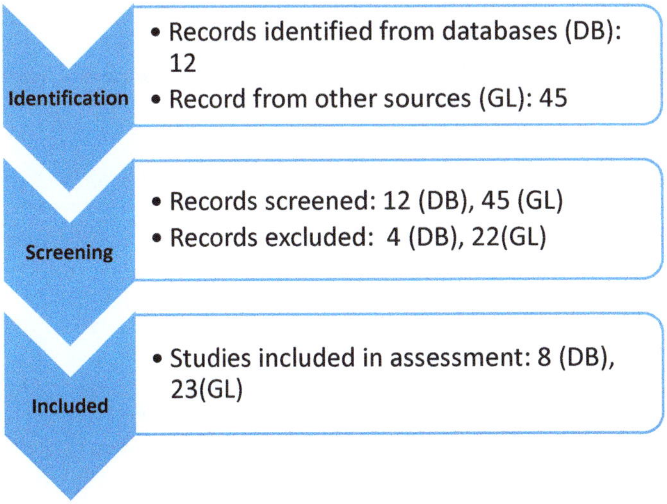

Figure 1. Screening process.

To ensure the relevance and reliability of our sources, we selected recently published documents with strong data and empirical evidence that effectively support and justify the significance of our research focus. Additionally, we examined the latest biodiversity reports in Mexico from SEMARNAT, CONANP, PNUD, INEGI, Biodiversidad Mexicana, BIOFIN, and the Adaptation Fund. These sources provided a clear understanding of the current state of biodiversity financing in Mexico, including financial needs and existing funding gaps.

Furthermore, we analyzed datasets from the Global Environment Facility (GEF) and the Adaptation Fund in Mexico to assess their funding mechanisms, operational structures, primary beneficiaries (by sector, country, or region), and key projects financed over the past decade. Lastly, we conducted an in-depth examination of the most recent project proposals submitted to biodiversity and adaptation funds, evaluating their status, approval trends, and, where applicable, the reasons behind project rejections.

For the statistical analysis and graphical representation, the study involved downloading and systematically organizing relevant databases on biodiversity funds in Mexico. This process included classifying the data to identify key mechanisms and instruments, categorizing the types of solutions implemented, and pinpointing the institutions responsible for managing these funds. The analysis also focused on highlighting the focal areas of investment, identifying the primary beneficiary countries, and examining recently supported projects in Mexico. The structured approach ensured a comprehensive understanding of biodiversity financing dynamics and trends, supporting informed recommendations for future strategies.

For the analysis of adaptation funds in Mexico, a comprehensive database of adaptation projects was obtained and meticulously organized. The data were classified to provide insights into the number of approved projects, their duration, geographical distribution, and sectoral focus. This systematized approach enabled the generation of detailed graphs illustrating the allocation of funds across regions and sectors, offering a clear visualization of financial priorities and trends.

Additionally, an in-depth examination was conducted on each Mexican project to assess their current status. For projects that were not approved, the analysis identified specific reasons for rejection, highlighting critical gaps and challenges in meeting funding criteria. This thorough evaluation provides valuable insights into the effectiveness and alignment of adaptation funding with national and regional priorities, underscoring opportunities for improving project design and resource allocation.

3. Results

3.1. National Strategy on Biodiversity (ENBioMex)

Since Mexico joined the Convention on Biological Diversity (CBD) in 1992, the National Biodiversity Commission (CONABIO) has been in charge of providing scientific and technical oversight for projects and negotiations under the Convention. The Biodiversity Cooperation Directorate (DCB) coordinates efforts to implement Mexico's National Biodiversity Strategy and supports subnational studies and strategies [11].

In 2016, following a four-year review, Mexico published the National Strategy on Biodiversity and Action Plan 2030 (ENBioMex), a comprehensive framework outlining priorities for understanding, conserving, restoring, and sustainably managing biodiversity and its ecosystem services across short-, medium-, and long-term horizons [11].

In 2022, the Kunming-Montreal Global Biodiversity Framework (GBF) was adopted, requiring all Party nations to revise their biodiversity strategies to align with its objectives. This further emphasizes Mexico's ongoing commitment to biodiversity conservation and sustainable use.

The Subnational Strategy on Biodiversity is coordinated by CONABIO, which collaborates with state governments and various sectors of society through the State Biodiversity Strategies (SBSs) initiative. This initiative aims to strengthen local human and institutional capacities for planning and managing biological resources at the state level while supporting Mexico's commitments under the Convention on Biological Diversity (CBD).

As part of the SBSs initiative, diverse local and regional actions are implemented in partnership with state governments to promote the conservation, protection, and sustainable use of biodiversity. These efforts also contribute to fulfilling additional national and international obligations. Key collaborations include work with the Association of State Environmental Authorities and the Central-Western Mexico Biocultural Corridor, fostering integrated approaches to biodiversity management [11].

The ENBioMex has a mission, a vision for 2030, and 14 guiding principles. The Action Plan is made up of the following strategic axes: Conservation and restoration, Sustainable use and management, and Attention to pressure factors that seek to increase actions that have a positive impact on biodiversity and reduce the direct causes of biodiversity loss. Finally, but most importantly is the Integration and Governance pillar, which seeks to reinforce the implementation of actions by strengthening coordination between actors and sectors, harmonizing the legal framework, and promoting integration and cooperation (Figure 2) [9].

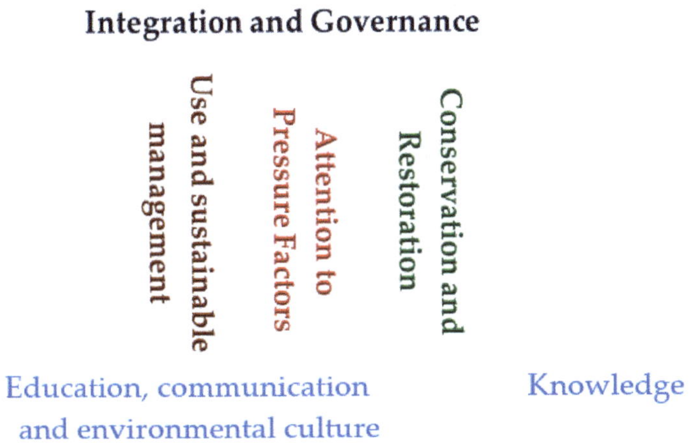

Figure 2. Strategic axes of Mexico's National Biodiversity Strategy and Action Plan 2016–2030. Source: produced by the authors with CONABIO Information.

The ENBioMex strategy integrates nature conservation, restoration, and sustainable use with essential pillars, including knowledge, governance, and education. However, its implementation has been inconsistent, reflecting shifts in administrative priorities. Institutions such as CONABIO, CONANP, and CONAFOR operate with a certain degree of autonomy but are subject to SEMARNAT oversight [11].

At the sub-national level, the principal policy instruments concerning biodiversity are the Strategies for the Conservation and Sustainable Use of Biodiversity (ECUSBE). These strategies are developed in a participatory manner at the state level in accordance with the prevailing national strategy. As of 2023, 18 such strategies have been adopted and published [11].

3.2. Importance of Financing for Biodiversity Protection in Mexico

Mexico, while comprising only 1% of the Earth's surface, hosts over 10% of the world's known plant and animal species, highlighting its critical role in global biodiversity conservation [12]. Safeguarding this ecological wealth is a key priority both nationally and internationally. Also, Mexico is one of the 17 world's megadiverse nations, hosting 10–12% of all known species globally [13].

Mexico ranks third in mammal diversity with more than 564 species, 30% of which are endemic, and second in reptile diversity, with 864 species, 45% of them unique to the country. Mexico also possesses a remarkable biocultural heritage, serving as the center of origin for globally utilized species and preserving its rich natural legacy through the stewardship of Indigenous peoples and local communities. There are more than 3000 medicinal plants in the country, and more than 5000 species of flora (23%) have a traditional use [14].

In Mexico, land use is primarily concentrated on primary activities, including agriculture, extensive livestock farming, forestry, and fisheries. These activities are managed at the community level and by ejidos, which are legal entities that represent collective land ownership. To ensure the conservation of biodiversity in these territories, it is essential to consider the social governance aspects involved [15].

Indigenous Peoples, who often inhabit and protect Mexico's most biodiverse areas, are vital environmental stewards. They have led efforts to defend ecosystems from harmful projects but face ongoing marginalization, threats, and violence. Between 2017 and 2021, 131 environmental defenders were killed in Mexico, 54 in 2021 alone—nearly half of them Indigenous—alongside frequent forced disappearances [16].

Mexico faces significant challenges that drive biodiversity loss, including overfishing, habitat loss from urbanization and agriculture, pollution, and river alteration [17]. Nearly 2600 species in Mexico are classified as endangered, threatened, or under special protection. This encompasses 97 mammal species, 232 amphibian species, and 306 fish species [18].

To aggravate the problem, Total Environmental Protection Spending in favor of biodiversity has been decreasing since 2013, allocating only 14.1 billion Mexican pesos in 2022, marking the lowest investment in a decade (0.7% of GDP) [15] (see Figure 3).

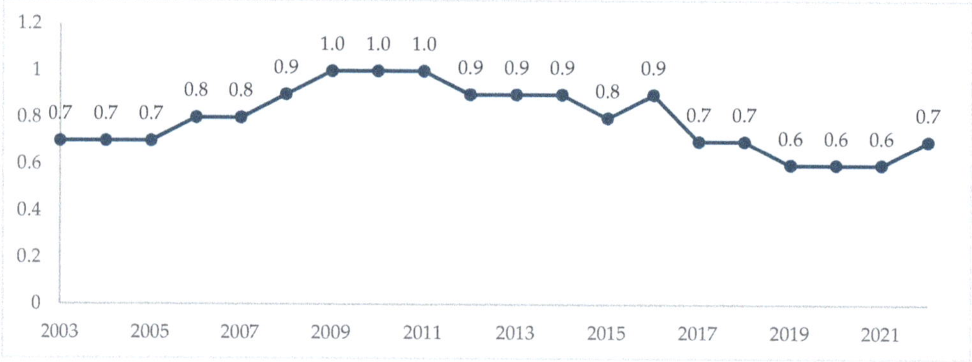

Figure 3. Public Expenditure on Environmental Protection as % of GDP, 2003–2022. Source: [19].

Figure 4 shows how public spending on environmental protection is distributed by sector. Air, Environment and Water appears to be the sector most benefited (24%), followed by Water (20%), Waste (18%) and Wastewater (10%). Spending on biodiversity is only 3%, and on environmental protection 10.3%. We can appreciate that biodiversity is in last place in this spending.

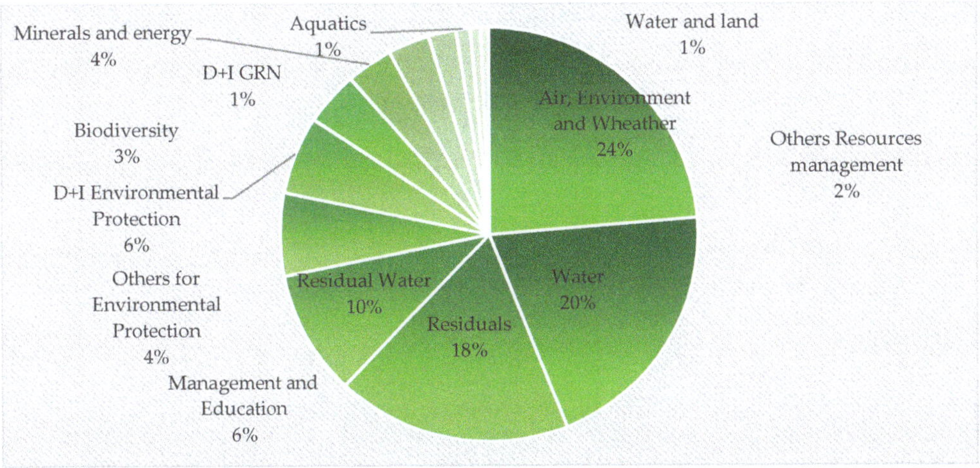

Figure 4. Total Public Sector Environmental Protection Expenditure by Functional Classification, 2022. Source. [20].

Despite the notable expansion of Mexico's national protected areas, the environmental sector's budget, including funding for pivotal agencies like CONABIO and CONANP, has experienced a decline over the past decade (Figure 5).

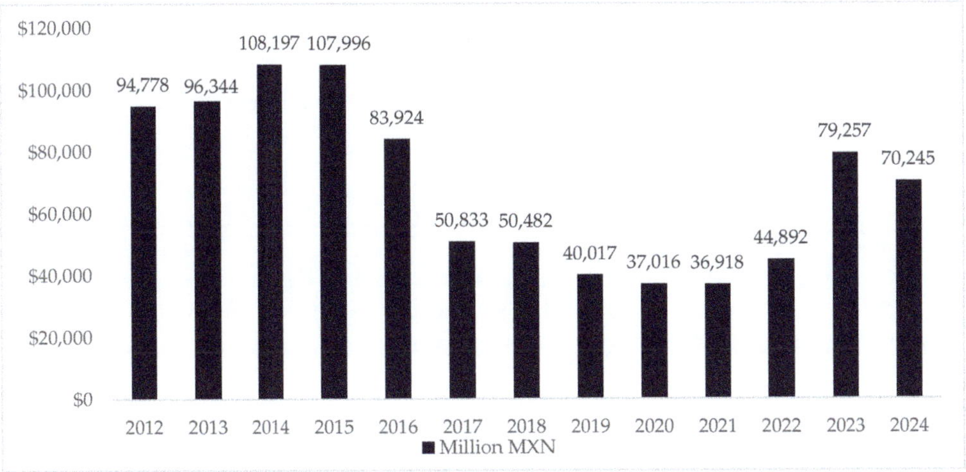

Figure 5. Approved budget for Branch 16: Environment and Natural Resources for the period 2012–2024, in million MXN. Source: [11].

Mexico has significantly expanded its national protected areas, now 232 sites and covering nearly 95 million hectares—over a third of the country's land area [13]. However, the 2023 budget for CONANP was 7% lower than the previous year, amounting to less than USD 1 per hectare of protected land. In general, the total budget assigned to Branch 16—Environment and Natural Resources—for 2025, is just over MXN 44,370 million, while in 2024, the amount assigned was MXN 70,245 million.

According to [21] the percentage of Protected Natural Areas with a management program in place that maintain or achieve an effectiveness index at an outstanding or high level according to the "i-effectiveness" was reported as 49.6% in 2020 and 50% in 2022 and 2024.

3.2.1. Financial Needs Related to ENBioMex Strategic Axes

According to UNDP, the estimated annual financial need for biodiversity in Mexico during 2017–2020 was approximately USD 461.9 million annually, reflecting a 46.7% increase compared to biodiversity spending in 2015. This financial assessment encompassed several key areas [21]:

- Protected Natural Areas (PNA): A funding shortfall of USD 60 million per year, as estimated by CONANP.
- Payment for Environmental Services (PES): A funding requirement of USD 202.1 million annually, identified by CONAFOR to meet demand.
- National Biodiversity Strategy and Action Plan (NBSAP–ENBioMex): An estimated annual need of USD 191.4 million for effective implementation.

ENBIOMEX's Finance Needs:

- Strategic priority 1 Knowledge: USD 55.4 million (12%)
- Strategic priority 2 Conservation and restoration: USD 350.8 million (76%) [Includes the financial gap of PNA calculated by CONANP and the estimate of finance needs to cover the demand for CONAFOR's PES]
- Strategic priority 3 Sustainable use and management: USD 25.8 million (5.6%)
- Strategic priority 4 Attention to negative drivers and threats: USD 19.9 million (4.3%)
- Strategic priority 5 Education, communication and environmental culture: USD 3.6 million (0.8%)
- Strategic priority 6 Integration and governance: USD 6.2 million (1.3%)

The assessment also revealed that 76% of the required resources pertain to recurrent expenditures, while the remaining 24% are allocated for investment spending [22].

3.2.2. Financial Needs Beyond ENBioMex

There are other needs that must be considered, i.e., some actions indirectly affecting biodiversity should be considered, for instance the Environmental Degradation and Resource Depletion Total Costs. As shown in Table 2, in 2022, the National Institute of Geography and Statistics (INEGI) reported that environmental degradation and resource depletion imposed costs equivalent to 4.11% of national GDP, of which degradation represented the equivalent of 3.63%, while depletion was equivalent to 0.48% [15].

Table 2. Composition of Depletion and Environmental Degradation Costs in 2022.

Concept	Depletion and Environmental Degradation Costs (Millions of Current MXN)	%GDP
Hydrocarbons	72,674.8	0.3
Forest resources	36,214.9	0.1
Groundwater	32,881.9	0.1
Depletion costs	114,771.6	0.5
Soil degradation	154,194.9	0.5
Municipal solid waste	113,632.4	0.4
Untreated wastewater	62,298.8	0.2
GHG emissions	738,950.1	2.5
Degradation costs	1,069,076.2	3.6
Total depletion and environmental degradation costs	1,210,847.8	4.1

Source: [18].

3.3. *Financial Funds Focused on Biodiversity in Mexico*

The main economic instruments and mechanisms for environmental management in Mexico identified as of 2018 can be classified into market-based, fiscal, regulatory, risk-related and environmental funds [22].

Market-Based Instruments include diverse solutions such as carbon markets, eco-labels, sustainability standards for products and services, forest bonds, environmental funds, green banking, permits and licenses, access and hunting fees, sustainable species trade, ecotourism, environmental compensation, carbon taxes, payment for environmental services (water and forests), incentives for conservation and sustainable businesses, and ecological fiscal transfers. These are managed by institutions like CONANP, SEMARNAT, CONAFOR, SHCP, PROFEPA, FMCN, GRAF, Profauna, and Aguas de Saltillo [22].

Risk Instruments involve environmental and climate insurance schemes, such as those for coral reef restoration and livestock protection in agriculture. Grants include Official Development Assistance (ODA), donations, and private funding mobilized through mechanisms like international biodiversity financing and philanthropy [23].

Regulatory Instruments focus on intersectoral approaches, fishing quotas, and addressing harmful subsidies in agriculture. Key players include SEMARNAT, SAGARPA, CONAPESCA, EDF, and COBI [20].

Environmental Funds are overseen by CONABIO, SEMARNAT, CONAFOR, FMCN, subnational governments, and FONCET. These funds support biodiversity conservation, forest management, protected areas, climate change mitigation, and specific species protection, such as efforts for the Monarch Butterfly [23].

3.3.1. Global Environment Facility (GEF) in Mexico

GEF is a multilateral fund focused on the fight against biodiversity loss, climate change and pollution, and supporting land and ocean health. Financing allows developing countries to face challenges and achieve objectives related to environmental goals. The partnership incorporates 186 member governments, civil society, Indigenous Peoples, women and youth.

Operating for over 30 years, GEF has allocated more than USD 25 billion in funding and mobilized an additional USD 145 billion for country-driven priority projects. As Figure 5 shows the fund group includes the Global Environment Facility Trust Fund, Global Biodiversity Framework Fund (GBFF), Least Developed Countries Fund (LDCF), Special Climate Change Fund (SCCF), Nagoya Protocol Implementation Fund (NPIF), and the Capacity-building Initiative for Transparency Trust Fund (CBIT) [23]. These investments have been pivotal in addressing biodiversity loss, climate change, land degradation, and other pressing environmental challenges.

According to the GEF, USD 25.5 billion has been allocated to 6155 projects worldwide. The main agencies involved in funding these initiatives are the United Nations Development Program (UNDP), followed by the World Bank and the United Nations Environment Program (UNEP).

GEF's funding is managed through a group of specialized funds, each tailored to specific environmental goals. The Global Environment Facility Trust Fund serves as the primary financial mechanism, accounting for 91% of the total funding. The Least Developed Countries Fund (LDCF), which supports climate adaptation initiatives in the most vulnerable nations, contributes an additional 6%. The remaining funds, each representing approximately 1%, include:

- The Global Biodiversity Framework Fund (GBFF), which aligns with biodiversity targets under the Kunming-Montreal Framework;

- The Special Climate Change Fund (SCCF), addressing urgent climate adaptation and mitigation needs;
- The Nagoya Protocol Implementation Fund (NPIF), promoting the fair and equitable sharing of benefits derived from genetic resources;
- The Capacity-building Initiative for Transparency Trust Fund (CBIT), enhancing countries' ability to meet transparency requirements under the Paris Agreement (Figure 6).

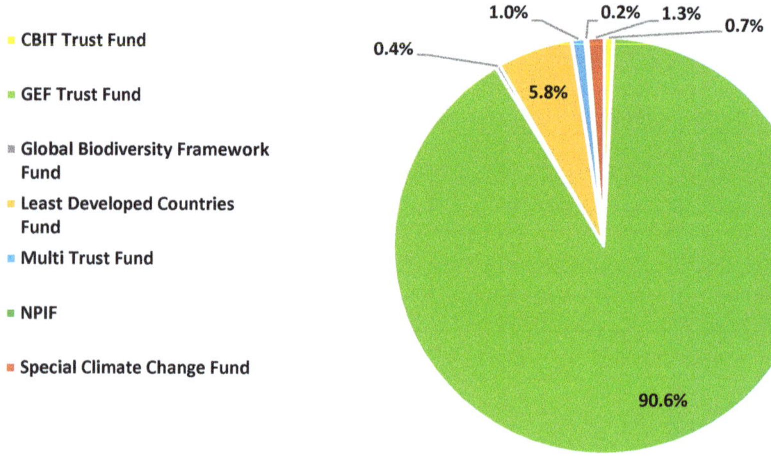

Figure 6. Main Focal Areas for GEF Project. Source [22].

This diversified structure enables GEF to target a wide range of environmental priorities while maximizing the impact of its financial resources. Its comprehensive approach underscores its vital role in bridging financial gaps and supporting global efforts to achieve sustainable development and environmental resilience.

Grant amounts are between USD USD 7000 and USD 306 million and the main funding source is GEF Trust Fund with 91% of grants. Geographically, most of the grants are designated to Global projects, but the most benefited countries are China, Brazil, India, Mexico, Indonesia, Russia, Colombia, Philippines and South Africa (Figure 7).

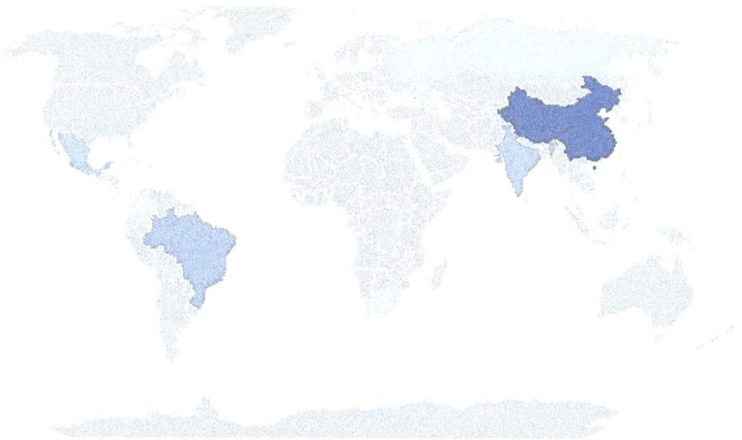

Figure 7. Countries benefiting most from GEF grants (USD millions). The darker color indicates greater access to funds. Source: Produced by the authors with data of [13,23].

Table 3 presents all the projects related to Biodiversity in Mexico approved by GEF Fund in the last ten years (2014–2024). The biggest project by GEF grant is "Sustainable Productive Landscapes", followed by "Mex30 × 30: Conserving Mexican biodiversity through communities and their protected areas" and "ORIGEN: Restoring Watersheds for Ecosystems and Communities". The agency which manages more projects is the UNDP, and the main funding source is GEF Trust Fund. The amount of the grants has increased in 2023 and 2024.

Over the years, GEF projects in Mexico have demonstrated significant advances in funding and implementation. The GEF Trust Fund emerges as the primary funding source, supporting the majority of initiatives. Recent years, particularly 2023 and 2024, have seen an increase in project size, with substantial grants allocated to projects such as "Mex30 × 30" (USD 16.6 million) and "ORIGEN" (USD 14.3 million). This reflects a growing emphasis on large-scale biodiversity conservation and restoration efforts.

The projects cover a wide range of focus areas, including ecosystem restoration, biodiversity conservation, sustainable agriculture, and tourism. For instance, "Promoting sustainability in the agave-mezcal value chain" highlights sustainable agricultural practices in biocultural landscapes, while "From bait to plate" focuses on marine biodiversity and food security. These initiatives underline the strategic diversification of biodiversity efforts in Mexico, addressing both terrestrial and marine ecosystems.

Key implementing agencies play a pivotal role in delivering these projects. The United Nations Development Programme (UNDP) and the Food and Agriculture Organization (FAO) lead multiple projects, demonstrating their centrality in biodiversity initiatives. Additionally, organizations like Conservation International manage impactful projects such as "Mex30 × 30" and "ORIGEN," underscoring their influence in driving conservation efforts.

Cofinancing stands out as a major strength of GEF projects in Mexico. Many initiatives secure cofinancing amounts significantly higher than the GEF grants. For example, "COBIOCOM" leveraged USD 252.8 million in cofinancing compared to its USD 8.9 million GEF grant, while "ORIGEN" obtained USD 145.1 million in cofinancing alongside its USD 14.3 million grant. This demonstrates the ability of GEF projects to attract additional financial resources, amplifying their impact.

Specific regions and ecosystems are prioritized in these projects, such as the Biocultural Corridor of Central West Mexico and priority landscapes in Oaxaca and Chiapas. Marine and coastal ecosystems also receive attention, with projects like "Mainstreaming Biodiversity Conservation in Mexico Tourism Sector" addressing biodiversity in tourism-rich coastal areas. However, certain critical areas, such as invasive species management and genetic diversity conservation, receive relatively smaller grants, indicating potential gaps in funding allocation [23].

The involvement of local communities is notable in initiatives like the Small Grants Program, which focuses on grassroots-level interventions. Despite their relatively smaller size, these projects emphasize community-driven biodiversity efforts, suggesting opportunities to scale such initiatives for broader impact.

Table 3. GEF Projects in Mexico related to Biodiversity (2014–2024).

Project	Agencies	GEF Grant	Cofinancing	Approval FY	Funding Source (Indexed Field)
Mex30×30: Conserving Mexican biodiversity through communities and their protected areas	Conservation International	16,672,477	115,549,532	2024	Global Biodiversity Framework Fund
Ecosystem restoration and sustainable livelihoods in the Biocultural Corridor of the Central West of Mexico (COBIOCOM)	Food and Agriculture Organization	8,932,420	252,808,974	2024	GEF Trust Fund
ORIGEN: Restoring Watersheds for Ecosystems and Communities	Conservation International	14,378,898	145,146,243	2024	GEF Trust Fund
Promoting sustainability in the agave-mezcal value chain through restoration and integrated management of biocultural landscapes in Oaxaca	United Nations Environment Programme	4,507,534	31,874,019	2023	GEF Trust Fund
Agriculture and Biodiversity in Mexico (AgribioMex): Mainstreaming biodiversity in the productive activities of rural landscapes	International Fund for Agricultural Development	8,974,312	75,763,312	2023	GEF Trust Fund
From bait to plate: strengthening sustainable fisheries to safeguard marine biodiversity and food security	Food and Agriculture Organization	9,005,609	41,680,250	2023	GEF Trust Fund
Global Biodiversity Framework Early Action Support (LAC-4)	United Nations Development Programme	1,917,811		2022	GEF Trust Fund
Green and Inclusive Recovery in Mexico (GreenMex): Making high-value ecosystems and rural livelihoods more resilient and sustainable in a post COVID-19 scenario	Food and Agriculture Organization	10,103,670	50,363,532	2022	GEF Trust Fund
Connecting Watershed Health with Sustainable Livestock and Agroforestry Production	The World Bank	13,761,468	99,013,829	2021	GEF Trust Fund
Seventh Operational Phase of the GEF Small Grants Programme in Mexico	United Nations Development Programme	4,481,210	12,233,787	2021	GEF Trust Fund
Mainstreaming Biodiversity Conservation Criteria in Mexico Tourism Sector with Emphasis on Biodiversity-rich Coastal Ecosystems	United Nations Development Programme	7,238,613	52,712,089	2020	GEF Trust Fund
Sustainable Productive Landscapes	The World Bank	21,862,385	54,295,600	2018	GEF Trust Fund
Conservation and Sustainable Use of Biological Diversity in Priority Landscapes of Oaxaca and Chiapas	Conservation International	7,219,450	47,456,966	2018	GEF Trust Fund
Securing the Future of Global Agriculture in the Face of Climate Change by Conserving the Genetic Diversity of the Traditional Agro-ecosystems of Mexico	Food and Agriculture Organization	5,329,452	36,185,188	2018	GEF Trust Fund
Sixth Operational Phase of the GEF Small Grants Programme in Mexico	United Nations Development Programme	4,429,223	6,333,389	2017	GEF Trust Fund

Project	Agencies	GEF Grant	Cofinancing	Approval FY	Funding Source (Indexed Field)
Strengthening of National Capacities for the Implementation of the Nagoya Protocol on Access to Genetic Resources and the Fair and Equitable Sharing of Benefits Arising from their Utilization to the Convention on Biological Diversity	United Nations Development Programme	2,283,105	8,938,579	2016	GEF Trust Fund
Strengthening Management of the PA System to Better Conserve Endangered Species and their Habitats	United Nations Development Programme	5,525,114	31,850,000	2014	GEF Trust Fund
Integrating the Management of Protection and Production Areas for Biodiversity Conservation in the Sierra Tarahumara of Chihuahua	United Nations Environment Programme	4,900,000	40,036,159	2014	GEF Trust Fund
Enhancing National Capacities to Manage Invasive Alien Species (IAS) by Implementing the National Strategy on IAS	United Nations Development Programme	5,354,545	26,050,760	2014	GEF Trust Fund

Source: produced by the authors with data of [21].

3.3.2. The Biodiversity Finance Initiative (BIOFIN) in Mexico

BIOFIN was initiated in 2014 at the CBD COP 11, by UNDP and the European Commission, in response to the urgent global need to divert more finance from all possible sources towards global and national biodiversity goals. Now present in 40 countries, BIOFIN is working with governments, civil society, vulnerable communities, and the private sector to catalyze investments in nature.

The UNDP's BIOFIN FNA initiative has taken a crucial first step by identifying financial needs and outlining a pathway for ENBioMex implementation until 2025. It has also fostered collaboration with subnational governments, the private sector, and civil society, facilitating dialogue, generating synergies, and aligning potential financing mechanisms to target biodiversity needs effectively.

Currently, BIOFIN MX is developing concrete resource mobilization proposals, focusing on identifying funding sources, mechanisms for resource allocation, instruments, expected financial outcomes, and necessary institutional arrangements.

So far BIOFIN MX has identified four types of financing solutions:

- Increasing resources from different sources (public, private and social),
- Efficient spending to improve results, aligning programs and budgets across sectors,
- Reducing future costs of negative impacts.
- Also, there are financial solution proposals like: integration of biodiversity (mainstreaming), Climate and Biodiversity Finance, Financial mechanisms of conservation, Sustainable business and impact investment, Greening the financial system, Support solutions (International finance vision and Economic and finance analysis) [24].

As of November 2024, BIOFIN has calculated the biodiversity expenditure, identified the biodiversity needs to be covered, calculated the financial gap for conservation, and developed a plan for financing solutions.

Currently, BIOFIN Mexico has a portfolio of more than 25 financing mechanisms grouped into five main financial solutions (FS): 1. Greening financial flows in cross-sectoral policies. 2. Strengthening financial mechanisms for climate change and biodiversity. 3. Strengthening biodiversity at the subnational level. 4. Bioeconomy. 5. Greening of development and commercial banks [25].

According to UNDP, in terms of climate finance and biodiversity, some mechanisms have been proposed and integrated in Mexico, particularly in some states [24].

1. The Sustainable Fund. The main purpose is to develop and implement a financial mechanism to channel resources toward sustainable development projects in the country, focusing on environmental and biodiversity goals, as a response to the elimination of public trusts.

In Jalisco, funding strategies have been implemented to address biodiversity, climate change, and other environmental degradation challenges. The plan integrates the following activities: (1) Assess the state's existing financing mechanisms, (2) Develop a financial solutions plan to meet biodiversity needs and align with state and national commitments, (3) Evaluate and align Jalisco's four public trusts (environmental, forestry, entrepreneurship, and agricultural), including legal, operational, and regulatory aspects, to optimize resource use and enhance their effectiveness [24].

2. Creation of the Green Investments Office (OIV) and the reinforcement of the Public Environmental Fund (FAP) in Mexico City. In this regard, BIOFIN has supported SEDEMA in strengthening the Public Environmental Fund to promote the allocation of resources through innovative financial mechanisms for projects in sustainable development, biodiversity conservation and sustainable use, and climate change adaptation and mitigation.

This FAP will serve as the focal point and single window for Mexico City to engage with national and international providers of financing and technical assistance in these areas. Additionally, institutional and financial capacities will be enhanced to transform the fund into an effective tool for selecting, evaluating, and financing environmental, climate change, and biodiversity projects [25].

3. Forest Carbon Credits, for which BIOFIN has proposed several initiatives. These include assessing the feasibility of complementary financing mechanisms to ensure the sustainability of forest projects, emphasizing the benefits of such initiatives to influence public policy and private sector strategies, and documenting the development process to provide replicable recommendations for scaling up forest carbon projects in Mexico.

Promising results have been achieved in Mexico City. The Public Environmental Fund (FAP) has been strengthened, leveraging financial resources from international sources such as the Inter-American Development Bank (IDB), the Spain-Mexico Mixed Fund, and the UK Embassy. Additionally, a fiduciary change resulted in savings of USD 3 million, which were redirected to conservation activities, achieving budget efficiency of USD 28 million. The reception and execution processes for international funding have also been systematized to enhance operational efficiency.

In addition, SEDEMA and Mexico City government officials have been trained and the legal and operational structure of the Public Fund for the Environment has been improved. Key areas included financial structuring, risk management, credit and financial instruments, international resource mobilization, audit and internal control, project appraisal methodologies, and development of regulatory and operational frameworks for funding. The process of strengthening the fund was systematized to enable its replication in other Mexican states [25].

Efforts in Jalisco included revising and updating fund operating rules and contracts while creating a linkage strategy to avoid duplication and enhance efficiency. Of 49 identified financial mechanisms, six were prioritized for implementation through a workshop with strategic stakeholders. A state financing solutions plan was developed, aligned with the State Biodiversity Strategy. Capacity-building initiatives were conducted for SEMADET, alongside establishing frameworks for investment selection and impact measurement for state environmental funds, ensuring alignment with the Sustainable Development Goals (SDGs).

3.3.3. The Adaptation Fund in Mexico

The Adaptation Fund provides financial support for projects that assist vulnerable communities in developing countries in adapting to climate change, with a particular focus on local needs and priorities. Since 2010, the Fund has allocated over USD 1.2 billion to 176 localized projects, which have benefited more than 45 million people. Furthermore, the Fund introduced Direct and Enhanced Direct Access, which enables countries to develop and manage projects through accredited national entities [26].

The Adaptation Fund is financed through government and private donations, as well as a 2% levy on Certified Emission Reductions (CERs) from Clean Development Mechanism projects under the Kyoto Protocol.

According to the Database of Adaptation Fund to November 2024, 233 projects were approved. The project duration is between 0.5 and 6 years. The regional share of the projects is 36% in Africa, 28% in Asia-Pacific, 21% in Latin America and the Caribbean and 4% in Eastern Europe (Figure 8). In terms of the duration of the projects, most of them are between 3 and 5 years (64%), but also a quarter are for less than 1 year [25].

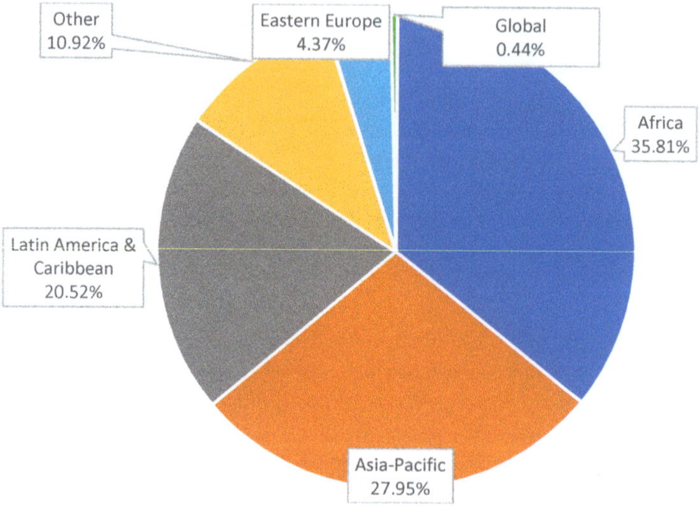

Figure 8. Adaptation Fund Projects by Region. Source: [26].

As observed in Figure 9, the majority of the projects are focused on Multi-sector (28), followed by Agriculture (24), Food security (23), Disaster risk reduction and early warning system (20), Water management (18), Rural Development (16). It is important to highlight that only 1 project of the Adaptation Fund in 2024 was oriented to ecosystem-based adaptation.

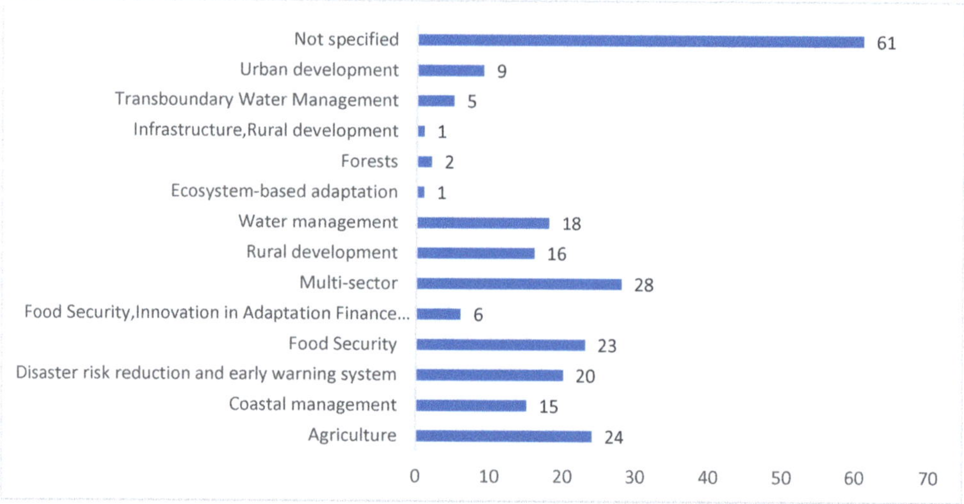

Figure 9. Adaptation Fund Projects by Sector, 2024. Source: Produced by the authors with data from [26].

Most of the projects are in the implementation phase (114), 73 projects have been completed and 38 have been approved. Grant amounts range from USD 10,000 to 14,000,000. Almost half of the projects are below USD 2.8 million and an important portion are between USD 8.4 and 11.2 million (Figure 10).

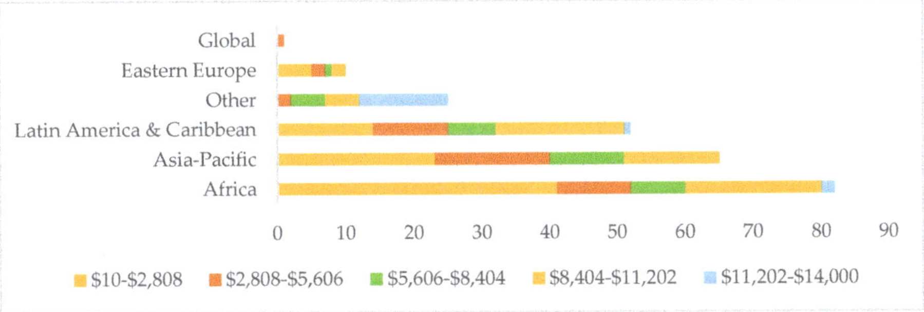

Figure 10. Adaptation Fund by Grant Amount and Region (thousands USD). Source: Produced by the authors with data from [27].

According to adaptationfund.org Mexico has received one support for USD 25,000 related to "Technical assistance grant for the environmental and social policy and gender policy (ta-esgp)" requested by Mexican Institute of Water Technology (IMTA) Mexico. The project was developed from 5 January to 31 October 2021 [28].

The support activities were related to the development of procedures/manuals/guidelines for screening projects for: (i) environmental, social and gender risks; (ii) conducting environmental, social and gender risk assessments of projects and formulating gender-sensitive risk management plans; (iii) public disclosure and gender-sensitive consultation; (iv) developing transparent, accessible, fair and effective mechanisms for receiving and addressing complaints of environmental or social harm and complaints related to gender inequality and other adverse gender impacts caused by projects/programs during implementation; and (v) training selected staff to carry out relevant tasks related to the implementation of the Fund's environmental and social policy and gender policy [27].

Other adaptation fund projects submitted but not accepted are: (1) "Restoration of Lake Texcoco through resilient actions" and (2) "Ha Ta Tukari, "Water for Life". In both cases, projects did not clarify climate vulnerabilities in the target areas, define the target beneficiary communities, sustainability, and cost-effectiveness of concrete activities, and compliance with the Environmental and Social Policy and Gender Policy of the Fund [28].

4. Discussion

It is well established that biodiversity loss and climate change are correlated and mutually reinforcing. A thriving nature keeps carbon stored where it naturally belongs and not in our planet's atmosphere. Biodiversity enhances adaptation capacity and resilience, including in disaster-risk reduction. Climate change, on the other hand, is one of the major drivers of biodiversity loss.

Available adaptation options can reduce risks to ecosystems and the services they provide, but they cannot prevent all changes and should not be regarded as a substitute for reductions in greenhouse gas emissions. Ambitious and swift global mitigation offers more adaptation options and pathways to sustain ecosystems and their services [6].

The overview of Mexico's experience with biodiversity finance in this study presents the experience of an emerging economy, which must finance pressing development priorities and biodiversity and climate action at the same time. However, the 2025 Federal Expenditure Budget of Mexico reflects significant cuts in the environmental sector, which complicates the outlook. The total budget allocated to Branch 16 (Environment and Natural Resources) was reduced by almost 40%, from MXN 70,245 million in 2024 to just MXN 44,370 million in 2025 [29]. Therefore, it is very important to find synergies in the available finance and look for new innovative options.

Mexico is an extraordinarily biodiverse country with a long history of sustainably using natural resources. The country's most important policy instrument on biodiversity protection, the National Biodiversity Strategy of Mexico, explicitly connects nature conservation, restoration and sustainable use with key supporting elements, such as knowledge, good governance and mainstreaming, and education.

The National Strategy on Biodiversity and Action Plan 2030 (ENBioMex) should be outlined as a positive achievement, as well as the definition of financing needs for biodiversity protection. The strategy is developed at both national and subnational (state) levels to establish the bases to promote, guide, coordinate and harmonize government and society efforts for conservation, sustainable use and the fair and equitable distribution of the benefits derived from the use of biological diversity components and their integration into sectoral priorities of the country [30].

CONABIO collaborates with state governments and various sectors of society through the State Biodiversity Strategies (EEB) initiative. This initiative aims to strengthen local human and institutional capacities for planning and managing biological resources at the state level while supporting Mexico's commitments under the Convention on Biological Diversity. Important achievements include the establishment of 232 Natural Protected Areas and Marine Protected Areas. Another positive development is the growing inclusion of Indigenous Communities, women, and youth as stewards of biodiversity conservation.

However, in the last decade, public funding has suffered a significant decrease, not only for biodiversity but in the environmental field in general. The incorporation of Mexico into global initiatives such as BIOFIN marks important steps to improve funding, but the effective implementation of the suggestions and recommendations of such initiatives remains a challenge. The total budget assigned to Branch 16—Environment and natural resources—for 2025 is just over MXN 44,370 million, while in 2024, the amount assigned was MXN 70,245 million [29]. Additionally, the ratio between the amount invested in environmental protection and the amount for environmental depletion and degradation is 6:1. Therefore, it is necessary to increase financing oriented to environmental depletion and degradation.

A positive development is that GEF funding for Mexico has increased between 2014 and 2024. Additionally, Mexico is one of the most benefited countries along with China, Brazil, and India.

The large overlap between the climate and biodiversity agendas and the international commitments derived from these also presents an opportunity to accelerate biodiversity funding. Biodiversity-related commitments derived from climate instruments, such as Mexico's nationally determined contribution (NDC) under the Paris Agreement, can help leverage resources. In the Adaptation Fund projects by region, Latin America and the Caribbean are in third place after Africa and Asia and the Pacific.

However, our results show that only 1 project of 229 financed by the Adaptation Fund in 2024 was oriented to ecosystem-based adaptation (EbA). It is reported that in the case of Mexico, some projects were not accepted because they did not clarify climate vulnerabilities in the target areas, define the beneficiary communities, sustainability, and cost-effectiveness of concrete activities, and compliance with the Environmental and Social Policy and Gender Policy of the Fund.

In our opinion, the financing of EbA is the main instrument that can link biodiversity conservation and adaptation to climate change impacts while providing a sustainable way of life and guaranteeing the well-being of communities, but it is not adequately used. We assume this is because community-based conservation (CbA) has been primarily championed by development practitioners, and EbA by environment/conservation practitioners. Additionally, EbA is a subset of Nature based Solutions (NbS), which is an umbrella term for conservation, restoration and sustainable use of nature to address societal challenges. Thus, EbA harnesses biodiversity and ecosystem services to increase resilience and reduce the vulnerability of human communities and natural systems to climate change.

That is why it is necessary to implement some actions that would foster EbA, and in general biodiversity financing, addressing the following issues: Gender imbalance in access to climate and EbA information; Difference in risk perceptions (what is at risk and why); Including Indigenous knowledge and traditional governance systems and participatory processes; Providing funding for project upkeep and management; Defining limits and thresholds under which EbA might not deliver expected benefits; Addressing the mismatch of governance (borders, jurisdictions) vs. problem scales (climate hazards and ecosystems).

It is necessary to strengthen Mexico's biodiversity financing through improvements in its investment targeting, effective coordination among institutions, empowerment of local communities, and by increasing support for innovative production schemes in coordination with other actors such as the Ministry of Agriculture, Livestock, Rural Development, Fisheries and Food. In this context, it is important to communicate the financing targets and models to all actors working in their implementation to ensure broad understanding; create incentives for interagency coordination; build capacity for all actors to help them meet their new functions; strengthen of the role that state governments play in the harmonization of public policies that affect their regions; and strengthen the role that local communities play to identify and develop intervention strategies that really address local needs and specificities.

In summary, Mexico's experiences indicate the importance of political will to promote interagency coordination that leads to comprehensive intervention strategies in contexts with different historical, social, environmental and economic values.

5. Conclusions

Maintaining planetary health is essential for human and societal health and a precondition for climate resilient development. Effective ecosystem conservation of Earth's land, freshwater and ocean areas, including all remaining areas with a high degree of naturalness and ecosystem integrity, will help protect biodiversity, build ecosystem resilience and ensure essential ecosystem services.

In these concluding remarks we summarize some main issues based on our results and discussion of the Mexican case to formulate recommendations for the biodiversity conservation and climate adaptation financing agendas, which are of utmost importance to overcome the funding deficit in countries of the Global South.

The host government's theme for COP 16 on Biodiversity was 'Making Peace with Nature'. Encouraging was the decision on climate change and biodiversity loss, as well as the attention to the need to align related international efforts. These crises are inextricably linked, and we cannot tackle one without the other.

One important outcome was that COP 16 of CBD committed to enhancing policy coherence, including a potential joint work program of the three Rio conventions, namely the Convention on Biological Diversity (CBD), the United Nations Framework Convention on Climate Change (UNFCCC), and the United Nations Convention to Combat Desertification (UNCCD). The details of this collaboration should be further specified both in international fora and in academic research.

The subjects for future research are to explore how to more effectively bring these agendas together, to create synergies tackling biodiversity loss, planetary health, and climate change adaptation issues.

An obligatory task would be to generate precise methodologies for Monitoring and Evaluating outcomes and benefits of EbA projects and other financing oriented to attend biodiversity and climate adaptation priorities.

Because of the decrease in governmental funding, it is imperative to explore financial innovation in market-based instruments, risk instruments, and outcome-based and impact projects. Another challenge is to develop new varieties of blended finance, combining governmental, private and international funding. In this context, it is very useful to share information on cases of success in innovative financing.

It is important to make information on the requisites to apply for Adaptation Fund Financing more transparent and accessible and provide more orientation on Ecosystem-based Adaptation that is directly related to biodiversity conservation.

In Table 4, we highlight the main contributions that are very important to the countries of the Global South to bridge the funding gap for conserving biodiversity, adapting to the impacts of climate change, and creating constructive synergies. Table 4 integrates key points from the discussion and conclusion sections, offering a consolidated view for researchers, policymakers, and implementation stakeholders.

Table 4. Main contributions to synergies between biodiversity and climate change adaptation financing.

Category	Main Contributions
Future Research	Explore methodologies for monitoring and evaluating Ecosystem-based Adaptation (EbA) outcomes and benefits. Investigate synergies between biodiversity conservation and climate adaptation finance. Analyze successful case studies of innovative biodiversity financing mechanisms. Develop comprehensive guidelines for applying to adaptation funds, with a focus on EbA. Assess the impact of blending governmental, private, and international funding for biodiversi projects. Examine gender imbalances in access to biodiversity and climate finance.
Innovative Strategies	Mainstream biodiversity financing in cross-sectoral policies. Develop financial mechanisms for climate adaptation, such as green investments and bioeconom promotion. Strengthen local governance and community participation in biodiversity conservation. Create incentives for interagency coordination and capacity building. Promote risk-sharing instruments like environmental insurance for conservation projects.
Implementation Challenges	Address governance mismatches (e.g., between administrative borders and ecosystem scales). Ensure consistent financial flows for project upkeep and management. Overcome barriers to interagency coordination and policy harmonization. Increase transparency and accessibility of adaptation fund requirements. Resolve challenges in scaling up EbA projects, including community involvement and long-ter sustainability.

Source: Produced by the authors.

To conclude, it is imperative to stress the importance of communication, capacity building, and promotion of coordination between government agencies in order to foster biodiversity conservation and to converge environmental and development agendas. It is necessary to prioritize these needs in Mexico and in the countries of the Global South.

Author Contributions: Conceptualization, M.S. and A.I.; methodology, A.I.; software, M.S.; validation, M.S. and A.I.; formal analysis, A.I. and M.S.; investigation, M.S.; resources, A.I.; data curation, M.S.; writing—original draft preparation, M.S. and A.I.; writing—review and editing, A.I.; visualization, M.S. All authors have read and agreed to the published version of the manuscript.

Funding: This research received no external funding.

Institutional Review Board Statement: Not applicable.

Data Availability Statement: The original contributions presented in this study are included in the article. Further inquiries can be directed to the corresponding author.

Conflicts of Interest: The authors declare no conflicts of interest.

Abbreviations

BIOFIN	The Biodiversity Finance Initiative
CBD	Convention on Biological Diversity
CbA	Community-based Adaptation
CBIT	Capacity-building Initiative for Transparency Trust Fund
CERs	Certified Emission Reductions (CERs)
CONABIO	National Commission on Biodiversity of Mexico
CONAFOR	National Commission on Forestry in Mexico
CONANP	National Commission on Natural Protected Areas in Mexico
COP	Conference of the Parties
DSI	Digital Sequence Information
EbA	Ecosystem-based Adaptation
EBSAs	Ecologically Significant Marine Areas
ENBioMex	National Strategy on Biodiversity of Mexico
GBFF	Global Biodiversity Framework Fund
GBFF	Global Biodiversity Framework Fund
GDP	Gross Domestic Product
GEF	Global Environmental Facility
GL	Grey Literature
IMTA	Mexican Institute of Water Technology
IPCC	Intergovernmental Panel on Climate Change
KMGBF	Kunming-Montreal Global Biodiversity Framework
LDCF	Least Developed Countries Fund
NbS	Nature-based Solutions
NBSAPs	National Biodiversity Strategies and Action Plans
NPIF	Nagoya Protocol Implementation Fund
ODA	Official Development Assistance
SAGARPA	Ministry of Agriculture, Livestock and Fisheries of Mexico
SBSs	State Biodiversity Strategies
SCCF	Special Climate Change Fund
SEMARNAT	Ministry of Environment and Natural Resources of Mexico
SLR	Systematic Literature Review
UNCCD	United Nations Convention to Combat Desertification
UNDP	United Nations Development Program
UNEP	United Nations Environment Program
UNFCCC	United Nations Framework Convention on Climate Change
WEF	World Economic Forum

ferences

De León, L.F.; Silva, B.; Avilés-Rodríguez, K.J.; Buitrago-Rosas, D. Harnessing the omics revolution to address the global biodiversity crisis. *Curr. Opin. Biotechnol.* **2023**, *80*, 102901. [CrossRef] [PubMed]
WWF. Well... This Is the Million Dollar Question. And One That's Very Hard to Answer. 2024. Available online: https://wwf.panda.org/discover/our_focus/biodiversity/biodiversity/ (accessed on 24 October 2024).
Halder, M.; Jha, S. The Current Status of Population Extinction and Biodiversity Crisis of Medicinal Plants. In *Medicinal Plants: Biodiversity, Biotechnology and Conservation*; Jha, S., Halder, M., Eds.; Sustainable Development and Biodiversity; Springer: Singapore, 2023; Volume 33. [CrossRef]
WEF. Global Competitiveness Report Special Edition 2020: How Countries are Performing on the Road to Recovery. 2020. Available online: https://www.weforum.org/publications/the-global-competitiveness-report-2020/ (accessed on 6 November 2024).

5. WEF. Global Risks Report 2024. 2024. Available online: https://www.weforum.org/publications/global-risks-report-20 (accessed on 20 August 2024).
6. IPCC. Fact Sheet—Biodiversity. Climate Change Impacts and Risks. SIXTH ASSESSMENT REPORT, Working Group II—Impa Adaptation and Vulnerability. 2023. Available online: https://www.ipcc.ch/report/ar6/wg2/downloads/outreach/IPCC_A _WGII_FactSheet_Biodiversity.pdf (accessed on 10 August 2024).
7. COP-16. COP-16 Calí, Colombia, Paz con la Naturaleza. 2024. Available online: https://www.cop16colombia.com/es/ (access on 9 September 2024).
8. CBD. Biodiversity COP 16: Important Agreements Reached Towards Making Peace with Nature. 2024. Available onli https://www.cbd.int/article/agreement-reached-cop-16 (accessed on 11 August 2024).
9. NDC Partnership. Adaptation Fund. 2024. Available online: https://ndcpartnership.org/knowledge-portal/climate-fun explorer/adaptation-fund#:~:text=The%20Adaptation%20Fund%20finances%20projects,country%20needs,%20views%20 and%20priorities (accessed on 12 September 2024).
10. Xiao, Y.; Watson, M. Guidance on Conducting a Systematic Literature Review. *J. Plan. Educ. Res.* **2019**, *39*, 93–112. [CrossRef]
11. CONABIO (Comisión Nacional para el Conocimiento y Uso de la Biodiversidad). Para Alcanzar Sus Objetivos y Cumplir S Funciones, la CONABIO Financia Proyectos de Temas Sobre el Conocimiento y Uso de la Biodiversidad. 2023. Available onli https://www.gob.mx/conabio/acciones-y-programas/proyectos-56730 (accessed on 12 February 2024).
12. Guzmán, S.; Velasco, A.; Barbosa, O. Biodiversity Finance in Mexico. A Fair Share of Biodiversity Finance? Perspectives fr Developing Countries. ODI Country Study. 2024. Available online: https://media.odi.org/documents/Biodiversity_finance_ Mexico_country_study.pdf (accessed on 30 October 2024).
13. GEF. Global Biodiversity Framework Fund. 2024. Available online: https://www.thegef.org/what-we-do/topics/glob biodiversity-framework-fund (accessed on 10 November 2024).
14. SEMARNAT. México, Segundo Lugar del Mundo en Bioculturalidad. 2018. Available online: https://www.gob.mx/semarn articulos/mexico-segundo-lugar-del-mundo-en-bioculturalidad?idiom=es (accessed on 10 November 2024).
15. PNUD Mexico. Biodiversity Expenditure Review. September, 2024. PNUD. Available online: www.biodiversityfinance.o mexico (accessed on 10 November 2024).
16. Acevedo-Ortiz, M.A.; Lugo-Espinosa, G.; Ortiz-Hernández, Y.D.; Pérez-Pacheco, R.; Ortiz-Hernández, F.E.; Martínez-Tom S.H.; Tavera-Cortés, M.E. Nature-Based Solutions for Conservation and Food Sovereignty in Indigenous Communities of Oaxa *Sustainability* **2024**, *16*, 8151. [CrossRef]
17. Lozano, R.E.; Sundaram, A. The Deadly Costs for Mexico's Indigenous Communities Fighting Climate Change. Opini 2023. Available online: https://www.latimes.com/opinion/story/2023-02-26/us-mexico-border-indigenous-climate-chan environment (accessed on 20 February 2024).
18. SEMARNAT. Conoce las Categorías de Riesgo de la NOM 059 SEMARNAT-2010 Para Especies de Flora y Fauna. 2021. Av able online: https://www.gob.mx/semarnat/articulos/conoce-las-categorias-de-riesgo-de-la-nom-059-semarnat-2010-pa especies-de-flora-y-fauna (accessed on 6 March 2024).
19. IUCN. The IUCN Red List od Endangered Species. 2024. Available online: https://www.iucnredlist.org/ (accessed on February 2024).
20. INEGI CEEM. Cuentas Económicas y Ecológicas de México 2022. Comunicado de Prensa Número 755/23 1 de Diciembre de 2(Página 1/7. Available online: https://www.inegi.org.mx/contenidos/saladeprensa/boletines/2023/CEEM/CEEM2022.p (accessed on 20 February 2024).
21. CONANP. Programa Nacional de Áreas Naturales Protegidas 2020—2024. Comisión Nacional de Áreas Naturales Pro gidas. 2024. Available online: https://www.conanp.gob.mx/datos_abiertos/DES/PNANP2020-2024.pdf (accessed on February 2024).
22. UNDP. BIOFIN Executive Summary Phase I: Results and Biodiversity Finance Solutions. The Biodiversity Finance Initiati United Nations Development Programme. 2018. Available online: https://www.biofin.org/knowledge-product/executi summary-phase-i-results-and-biodiversity-finance-solutions (accessed on 24 October 2024).
23. GEF. GEF Projects Database. 2024. Available online: https://www.thegef.org/projects-operations/database (accessed on February 2024).
24. UNDP. Plan de Soluciones de Financiamiento BIOFIN 2020. 2020. Available online: https://www.biofin.org/sites/default/fil content/knowledge_products/PSF-BIOFIN.pdf (accessed on 25 October 2024).
25. UNDP México. Análisis de Gasto Público Federal a Favor de la Biodiversidad 2022. Proyecto 00108628 Iniciativa Finanzas Biodiversidad BIOFIN México Fase II. 2024. Available online: https://www.biofin.org/sites/default/files/content/knowledg products/050924_BER_Disen%CC%83o.pdf (accessed on 8 November 2024).
26. Adaptation Fund. Helping Developing Countries Build Resilience and Adapt to Climate Change. 2024. Available onli https://www.adaptation-fund.org/ (accessed on 7 November 2024).

Adaptation Fund Board. Project and Programme Review Committee Thirtieth Meeting Bonn, Germany, 11–12 October 2022. Available online: https://www.adaptation-fund.org/wp-content/uploads/2022/10/AFB.PPRC_.30.21-Proposal-for-Mexico-2.pdf (accessed on 20 March 2024).
Adaptation Fund Board. Technical Assistance Grant for the Environmental and Social Policy and Gender Policy (ta-esgp)—Mexican Institute of Water Technology (IMTA) Mexico. Project and Programme Review Committee. 2020. Available online: https://www.adaptation-fund.org/wp-content/uploads/2020/11/AFB.PPRC_.26-27.3-TAESGP-Grant-proposal-for-Mexico_final.pdf (accessed on 15 November 2024).
Greenpeace México. El Proyecto de Presupuesto para 2025 Recorta Todavía Más los Recursos para Proteger el Medio Ambiente. 2024. Available online: https://www.greenpeace.org/mexico/noticia/54846/el-proyecto-de-presupuesto-para-2025-recorta-todavia-mas-los-recursos-para-proteger-el-medio-ambiente (accessed on 6 December 2024).
Biodiversidad Mexicana. La Estrategia Nacional sobre Biodiversidad de México (ENBioMex). 2016. Available online: https://www.biodiversidad.gob.mx/pais/enbiomex (accessed on 12 February 2024).

diversity

MDPI

Article

Phytosociological Research and Meaningful Learning: Constructivist Approaches for Teaching Vegetation

Ana Cano-Ortiz [1], Carmelo Maria Musarella [2] and Eusebio Cano [3,*]

[1] Department of Didactics of Experimental, Social and Mathematical Sciences, University of Complutense (UCM), 28040 Madrid, Spain; acano07@ucm.es
[2] Department of AGRARIA, "Mediterranea" University of Reggio Calabria, Loc. Feo di Vito snc, I-89122 Reggio Calabria, Italy; carmelo.musarella@unirc.it
[3] Department of Animal and Plant Biology and Ecology Section of Botany, University of Jaen, Campus Las Lagunillas s/n, 23071 Jaén, Spain
* Correspondence: ecano@ujaen.es

Abstract: Phytosociology reached its peak development during the 20th century, particularly after 1992 with the implementation of the EU Habitats Directive, which incorporated plant associations into habitat classifications. The objective of this study is to assess the diversity of Mediterranean grasslands using the phytosociological methodology, focusing specifically on the PBTs and BsHl associations. Although both types of grasslands are of interest for livestock farming, we emphasize those belonging to *Poetea bulbosae* (PbTs), which are included in the priority habitat 6220 and are of high value for sheep grazing. The results indicate that Vca-Vco = 0.390 for PBTs and 0.525 for BsHl, the latter due to the absence of companion species diversity in BsHl. A strong linear correlation was observed for both associations, with $R^2 > 0.8$ for PbTs and $R^2 > 0.9$ for BdHl. Given the importance of grassland diversity research, its teaching should be predominantly practical and conducted outside the classroom. To achieve this, an inquiry-based constructivist methodology is employed, which is particularly suitable for learning the phytosociological method, allowing students to take ownership of their learning process. The aim of this teaching approach is to expose students to field research methodologies, enhancing their scientific training. The results are highly positive, with a 100% success rate in students' comprehension of the two grassland types after the teaching-learning process.

Keywords: association; grasslands; diversity; learning; holistic; botany; ecology

1. Introduction

1.1. Phytosociological Background

Various methodologies have been developed to study vegetation and its associations. Humboldt's physiognomic-ecological perspective, combined with subsequent contributions from other researchers, has significantly influenced the epistemological foundations of modern Vegetation Science and Phytosociology. Since Grisebach introduced the concept of "formation" in 1838, it has primarily been maintained within a physiognomic context. During the 1910 International Botanical Congress in Brussels, plant formation was defined as the "expression of specific living conditions, organized by associations differing in floristic composition but coinciding in seasonal conditions and biological forms".

Phytosociology, an ecological science derived from Geobotany, focuses on the study of phytocoenoses—plant communities—by examining their floristic composition, environmental relationships, biogeographical distribution, and temporal dynamics. By synthesizing these data through inductive and statistical methods based on phytosociological inventories, a universal hierarchical typology is established [1]. This typology facilitates the identification of plant associations, which represent the fundamental units of the system. Each association must demonstrate characteristic and differential species that are statistically faithful to specific environments, assuming a stabilized succession within which the association exists. Furthermore, every association is embedded within a hierarchical system encompassing alliance, order, and class [2].

Acquiring this knowledge is essential for developing specialized scientific competencies [3,4].

Throughout the 20th century, European phytosociological knowledge expanded significantly due to intensive research efforts. Phytosociological research has been published in hundreds of articles, as noted by Ivanova [5]. This author highlights 1168 articles based on Braun-Blanquet's methodology, and among the 187 references on vegetation studies, we emphasize the contributions of prestigious researchers such as Biondi, Blasi, Braun-Blanquet, Guinochet, Rivas-Martínez, Mucina, Tüxen, and Van der Maarel, demonstrating the continued relevance of phytosociology today.

Based on floristic composition, distribution, and ecological factors, two major groups of plant communities can be distinguished:

1. Non-endangered habitats: Communities with broad distribution, lacking endemic or rare species, and not under significant threat.
2. Fragile habitats: Communities with limited distribution, endemic species, or unique synecology, which are highly susceptible to degradation or disappearance, necessitating conservation measures.

This phytosociological methodology is applicable to any plant community. In this study, the focus is on grasslands, which are of particular interest for livestock farming and as CO_2 sinks. Grassland research should be recognized and supported globally due to its substantial cultural, educational, and economic value. Understanding grassland ecology, dynamics, and diversity is crucial for sustainable local development and highlights the importance of these ecosystems as valuable societal resources.

As with plant associations, the criteria for diagnosing higher-rank syntaxa—such as alliances, orders, and classes—are based on the principle of fidelity. Consequently, characteristic and differential species can also be identified at these higher taxonomic levels.

The alliance represents the rank immediately superior to the association and encompasses similar plant associations defined by characteristic species of the alliance. These species are typically absent or occur with low abundance in other alliances, reflecting distinct ecological niches. Within an alliance, species may be classified as characteristic, transgressive, differential, or introgressive. Additionally, associations with diverse ecological traits can coexist within the same alliance.

Floristically similar alliances are grouped into orders, which, in turn, are aggregated into vegetation classes. The characteristic species of a class possess broad ecological amplitude, making them less stenotopic than species associated with lower-rank syntaxa.

Sinphytosociology represents an entirely new discipline. Although its conceptual foundations have been recognized since the early 20th century, contemporary phytosociologists have systematically developed the principles of community integration as a defining feature of the landscape. Key contributors to this field include Braun-Blanquet, Tüxen, Bolós, Gèhu, and Rivas-Martínez, who pioneered the philosophical underpinnings of this innovative methodology for landscape analysis [6].

Since its inception, numerous authors have further refined and expanded this methodology [7–11]. The concepts of sigmetum, sinassociation, and vegetation series are intrinsically linked to territorial climate conditions, rendering vegetation series inherently climatophilous in nature [12,13]. Each vegetation series is characterized by a unique dynamic-catena process [14], with the climax stage representing the territorial climax, determined by the interplay of thermoclines and ombroclimatic conditions. For example, territories may support rainforests when water availability depends on rainfall. Conversely, when water availability is driven by soil conditions, the series is classified as edaphophilous. These are further subdivided into edaphohygrophilous series (e.g., riparian ecosystems) or edaphoxerophilous series (e.g., communities on rocky outcrops) [15].

In this study, we examine the exceptional diversity of grasslands in Mediterranean climates, which constitute a vital resource for humanity. These ecosystems support diverse livestock systems, with specific animal species depending on particular grass types. This interdependence underscores the ecological and economic value of grasslands as a key ecosystem service [16].

Our analysis focuses on the ecology, dynamics, floristic composition, and diversity of various grassland types, situating them within broader grassland classifications. Particular attention is given to their role as EU-designated habitats, including subnitrophilous and nitrophilous grasslands dominated by *Trifolium, Medicago, Taeniatherum, Hordeum,* and *Poa bulbosa* communities. These grasslands are especially significant for sheep farming, contributing substantially to grazing value and sustainable livestock systems [10,17].

Subnitrophilous-nitrophilous grasslands are widespread in dehesas and agricultural landscapes. However, in agricultural settings, these grasslands are often heavily modified due to herbicide application. To address this, several authors advocate for enhanced rural education within agricultural environments, emphasizing the importance of understanding natural resources and adopting a more practical, outdoor-focused approach to education [18–23].

Subnitrophilous grasslands and manure-enriched meadows, where approximately 50% of the botanical families belong to Fabaceae and Poaceae, hold significant grazing value. These grasslands are particularly suited for cattle, deer, and horse grazing due to their structural characteristics and high biomass/biovolume, which align well with the oral systems of these animals. Moreover, they act as effective CO_2 sinks, contributing to climate change mitigation efforts [24].

A noteworthy case involves grassland communities dominated by *Poa bulbosa, Trifolium subterraneum, Biserrula pelecinus,* and *Astragalus sesameus,* classified under the EU's Habitat 6220. While these species are valuable for grazing, their limited biomass/biovolume makes them primarily suitable for sheep. These grasslands, along with other communities, form the structural foundation of Spanish dehesas and Portuguese montados, classified as Habitat 6310. This habitat, protected under EU regulations, represents a sociocultural and economic system shaped by historical land-use practices [25].

The main objective of this research is to highlight the diversity of Mediterranean grasslands, establish a relationship between the type of grassland and the type of livestock, and propose the phytosociological method for their study.

1.2. Vegetation Education

This study aims to enhance vegetation education through the application of the phytosociological method. Originally developed in Central Europe and now widely applied across European countries, this method plays a pivotal role in diagnosing and assessing habitats. Such knowledge is crucial for environmental managers and educators to foster behaviors that support sustainability and promote environmental education [26–28].

Vegetation studies have reached their greatest advancement in Europe, primarily through the efforts of the Central European school of phytosociology, culminating in initiatives such as the EU's Habitats Directive 92/43 [29]. This highlights the importance of integrating such knowledge into educational curricula at all levels. García Barros et al., Cañas, and Cañas et al. [3,4,30] have emphasized the role of environmental activities in promoting scientific competencies among students. Similarly, Miño et al. [31] advocate for the inclusion of biological texts in education to enhance the comprehension of ecological principles. They further stress the importance of training future educators in these critical subjects [32,33], particularly in addressing urgent global challenges such as climate change [34–36].

The effectiveness of environmental education is influenced not only by the context in which it is delivered but also by the level of environmental literacy that students attain. As Da Silva et al. [37] argue, understanding habitats requires an interdisciplinary approach.

The development of phytosociology necessitates integrating knowledge from diverse disciplines, including soil science, climatology, physical-chemical factors, floristics, and even artistic elements. This interdisciplinary framework equips students with a comprehensive education, embodying a holistic approach that prepares them for meaningful integration into cultural, educational, and socioeconomic contexts [38].

Holistic education emphasizes empowering students to take ownership of their lives [39,40]. It seeks to achieve the integral formation of individuals across multiple dimensions: biological, ecological, psychological, and sociological. Recent advancements in this paradigm incorporate cognitivism, aligning it with contemporary constructivist principles [41]. This approach promotes the development of students' full potential and supports comprehensive education and human development [42].

Rivas et al. [43] highlight the necessity of educating university students on sustainability, climate change, and ecosystem services through a holistic lens. A significant gap exists in students' understanding of ecosystem services [44], which translates into broader societal challenges, including food insecurity [45]. Everjoy et al. [46] link reduced rainfall in semi-arid regions, as a consequence of climate change, to increased poverty, advocating for educational campaigns to raise awareness among farmers about these impacts [47]. Vilches and Gil-Pérez [48] similarly argue that sustainability education must be a critical objective in addressing the ongoing environmental crisis.

Within this framework, discovery-based inquiry methodologies enable students to tailor their learning experiences, fostering active participation and personal investment in their educational journey [49]. This paradigm not only prepares students to understand complex environmental systems but also equips them to contribute meaningfully to sustainability efforts and societal progress. The aim of this teaching approach is to promote the transmission of the research methodology on vegetation to students, using an essentially practical inquiry methodology. For this purpose, we apply constructivist theory, as the teaching of the phytosociological method is ideal for this teaching model.

2. Materials and Methods

2.1. Analysis of the Phytosociological Research Method

The phytosociological methodology was employed to study vegetation associations, with a particular focus on the alpha and beta diversity of the *Trifolio subterranei–Periballion* communities (*Poo bulbosae–Trifolietum subterranei*, PbTs) and *Hordeion leporini* communities (*Bromo scoparii–Hordeetum leporini*, BsHl) [50–53], as well as their conservation status. The Shannon diversity index was applied to evaluate the diversity of characteristic species (Ca), companion species (Co), and scenarios where the association comprised only two characteristic species (Ca2). Indices calculated include Shannon_Ca, Shannon_Co, Shannon_Ca2, and Shannon_Ca-Co (the difference between characteristic and companion diversity). This last index provides insights into the dynamics and conservation status of the associations.

Given the ecological significance of these vegetation associations under the EU Habitats Directive 92/43, the dynamics of the associations were further analyzed through comparative assessments of the abundance and frequency of characteristic and companion species.

The phytosociological method involves two key phases:

1. Analytical Phase: This phase requires the meticulous selection of sampling plots to ensure homogeneity and representativeness.
2. Synthetic Phase: Data collected in the analytical phase are integrated to develop a comprehensive understanding of the plant community.

Sampling involves the delineation of homogeneous plots where the total coverage of the phytocoenosis is recorded. Coverage, expressed as a percentage, represents the proportion of soil covered by vegetation. Plot sizes vary according to the type of phytocoenosis and biome:

- Forests: Mediterranean zones require plot sizes of 300–500 m^2, while tropical and subtropical zones necessitate larger areas of approximately 1000–2000 m^2.
- Shrublands: Plot sizes range from 50–200 m^2.
- Grasslands: Optimal plot sizes are between 0.5–2 m^2.

The minimum area method is recommended to maximize floristic diversity within the smallest possible surface. For example, in a plot measuring 100 m^2, if 45 m^2 are covered by vegetation, the coverage degree is 45% [7,54,55].

2.2. Teaching the Phytosociological Method to Students

The teaching of these techniques is designed for university students, particularly those in Botany and Geobotany courses, as well as for future educators and environmental managers working within the framework of the EU Habitats Directive.

To evaluate the effectiveness of this educational intervention, a 17-question survey was administered to students of Botany (131 students) and Geobotany (12 students) before and after the instructional process. The survey included questions designed to assess the students' understanding of:

- The concept of vegetation associations.
- Methods for calculating diversity indices.
- The ecological and conservation implications of phytosociological studies.
- Practical applications of the EU Habitats Directive.

This approach not only facilitates the development of technical and analytical skills but also emphasizes the integration of ecological theory with practical conservation strategies.

Questions posed to students before and after the teaching process included the following:

1. To what extent do you believe there is a relationship between diversity and conservation?
2. Do you think it is important to understand an ecosystem's diversity to assess its conservation status?
3. Do you consider that, in a plant community, a higher number of species indicates better conservation?
4. Do you know how to calculate the minimum area?
5. Do you know how to perform a phytosociological inventory?
6. Do you know how to calculate Alpha, Beta, and Gamma diversity?
7. What sciences are involved in the concept of biodiversity?
8. Define the concept of biodiversity.
9. Do you know the differences between characteristic and companion species?
10. Do you know what sustainable development means?
11. Define sustainable development.

12. The set of terrestrial ecosystems that, by sharing similar climatic conditions, host the same types of living beings is called?
13. What is the action plan to conserve biodiversity?
14. What is biodiversity?
15. How is biodiversity measured?
16. What are endemic species?
17. What is a biodiversity hotspot?

This study results from the application of field-based methodologies to university students, assessing their knowledge through the aforementioned questions. The focus is on fostering a deeper understanding of vegetation and stimulating observational skills by guiding students on how to conduct systematic observations and interpret their findings effectively.

Student motivation is heavily influenced by the instructor's ability to navigate natural settings, interpret ecological observations, and accurately convey the complexities of natural ecosystems. Observations must faithfully represent the ecological reality, ensuring that students grasp the key principles underlying biodiversity and conservation.

Practical, outdoor education has been widely advocated by various authors [21,56] for its ability to enhance hands-on learning and critical thinking. Fieldwork group sizes should not exceed 20 students to minimize distractions and maximize individual engagement with the proposed activities. Following field observations, group discussions are conducted to collectively interpret findings, fostering a collaborative learning environment.

As part of the learning process, students are tasked with creating diagrams or illustrations of the ecosystems where the studied vegetation associations occur. This exercise reinforces their ability to synthesize observations into visual representations, a critical skill in ecological studies.

The teaching methodology prioritizes student-centered learning, with the instructor serving as a motivator and guide. For students entering the program with limited foundational knowledge, often influenced by educational and social disparities, specific didactic strategies are employed. These include:

- Formulating new hypotheses.
- Conducting additional observations.
- Interpreting observed phenomena through "feedback" and "feedforward" mechanisms.

This approach aims to address knowledge gaps and preempt errors in reasoning, promoting a deeper understanding of ecological principles. Moreover, it underscores the importance of providing a robust foundational scientific education that extends beyond traditional classroom settings [57,58].

This methodology aligns with emerging trends in constructivist education, which emphasize active learning and the construction of knowledge. By integrating observational activities with group discussions and personalized feedback, students are equipped to critically engage with the complexities of biodiversity and sustainability. This holistic approach fosters not only academic competence but also the development of transferable skills essential for addressing real-world environmental challenges.

Practical sessions must be planned with the same meticulous care as theoretical classes, ensuring that students have opportunities to observe, analyze, and experiment independently. These sessions often represent students' first direct engagement with real-world laboratory and fieldwork, experiences that frequently ignite an interest in scientific research.

As highlighted previously, the ideal structure involves small groups of 15–20 students, with practical sessions running parallel to theoretical instruction. This approach fosters a close professor-student relationship, emphasizing critical observation and discussion. By applying theoretical principles to practical contexts, whether in the field or the laboratory, students develop a more profound understanding of scientific concepts and sharpen their reasoning abilities. This dynamic reinforces the retention and application of knowledge while also improving the quality and depth of information through hands-on activities. Practical sessions thus serve as a vital component of scientific education, allowing students to experience firsthand the phenomena under study and acquire essential skills for their academic and professional development [18].

To ensure the effectiveness of practical sessions, students must understand the objectives and theoretical foundations of each activity beforehand. The instructor begins by delivering a brief introductory presentation that fosters dialogue and contextualizes the session's goals. Students then undertake independent and responsible work, conducting observations and recording their findings in practice notebooks under the instructor's supervision.

Fieldwork and laboratory activities proceed according to a carefully designed protocol that balances structure with opportunities for student exploration. Sessions conclude with a group discussion to critically analyze results, reinforcing the connection between observation, data interpretation, and broader scientific principles. This structured yet flexible methodology helps students internalize key concepts while encouraging active participation and critical thinking.

By integrating theoretical instruction with practical experiences, this pedagogical approach not only enhances learning outcomes but also cultivates students' confidence and competence in scientific inquiry.

3. Results

3.1. *Phytosociological Analysis*

The application of the phytosociological method allows for the diagnosis of a specific plant community, a process carried out through syntaxonomy. Adhering to the international standards established in the Phytosociological Nomenclature Code [1] is essential, as it provides a hierarchical system of syntaxonomic ranks. The association functions as the foundational unit of this system.

Syntaxonomy should be understood as the systematics of plant communities, involving a hierarchical classification system that is recognized for a given phytocoenosis. The core unit of syntaxonomy is the plant association. Within this classification, various syntaxonomic units of vegetation are defined based on the presence or absence of specific botanical species. These units include subassociation, association, alliance, order, and class.

The association, as the fundamental unit of phytosociology, is analogous to the species in taxonomic classification. It represents the primary building block of phytosociology, derived from vegetation inventories conducted under ecologically homogeneous conditions. Associations must exhibit a distinct and original floristic composition that is statistically faithful to their biogeographical distribution and habitat. Subassociations and variants are subordinate to the association.

A specific phytocoenosis can be classified as an association only if it contains a minimum number of characteristic and companion species. The characteristic species, along with those consistently present, form the core framework of the community. The greater the proportion of constant species relative to the total species composition, the more homogeneous the association becomes. A higher proportion of characteristic species indicates a more clearly defined association, both from a floristic and ecological standpoint.

An association may become fragmented due to factors such as limited space, unfavorable habitat, human or animal interventions, or other disturbances. A characteristic species is one that consistently appears in a community, distinguishing it from others from a floristic, ecological, and chorological perspective. To define an association, a plant must have a minimum distribution range. If the plant's range is excessively large, it may belong to a higher-ranking syntaxon. Conversely, if its range is smaller than the required threshold for the association, the plant may be classified within a subassociation.

Thus, within any phytocoenosis, taxa may be characteristic of subassociations, associations, alliances, orders, and classes. This classification is closely tied to a plant's ecological specialization: the more specialized a plant is, the smaller its distribution range, which typically correlates with lower-ranking syntaxa. In contrast, less specialized plants have broader distribution ranges and may characterize higher-ranking syntaxa [2,55].

Furthermore, there are species that do not define the community but are referred to as companion species. These species, typically from neighboring associations, may overlap or intrude into the studied association due to their relatively low ecological specialization. These plants are considered introgressive. Similarly, species from the studied association may encroach upon neighboring communities, making them transgressive species.

Differential species are those that exist within a specific community (or association) and contribute to its differentiation from other communities. If subsequent studies reveal species that differ from the original diagnostic description, these species are labeled as differential species (Figure 1).

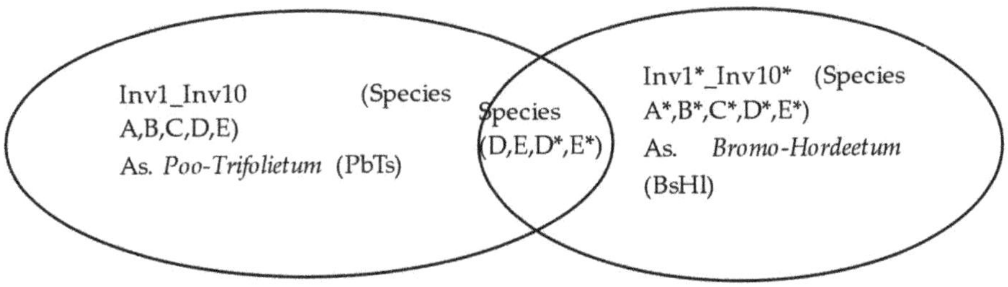

Figure 1. Differential and common species between the *Poa bulbosae–Trifolietum subterranei* and *Bromus scoparii–Hordeetum leporini* associations. * Differential species.

Considering the two neighboring associations, *Poa bulbosae–Trifolietum subterranei* (PbTs) and *Bromus scoparii–Hordeetum leporini* (BsHl), their characteristic species differ due to the distinct ecological niches they occupy. However, common species (D, E, D*, E*) appear in both associations, as they are capable of thriving under a broader range of ecological conditions.

For PbTs, the characteristic species are *Poa bulbosa* and *Trifolium subterraneum*, while for BsHl, they are *Bromus scoparius* and *Hordeum leporinum.* The transition zone between these two associations is referred to as an ecotone, which contains species from both associations. In this context, the BsHl introgressive generalist species (*Coleostephus myconis, Sherardia arvensis, Leontodon taraxacoides, Ornithopus compressus, Plantago lagopus, Echium plantagineum, Medicago arabica, Medicago polymorpha*) appear within the PbTs association.

The ecotone is an intermediate zone where gradual changes in ecological factors occur, promoting an increase in plant diversity, as species from both associations coexist. In this intermediate zone, plants exhibiting both ecological and chorological (spatial) characteristics, as well as dynamic features, are considered differential species. These species are particularly useful for diagnosing subassociations (Figure 2).

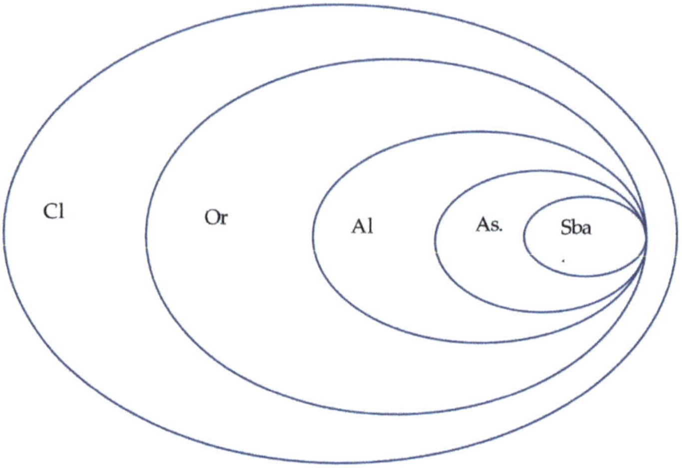

Figure 2. Hierarchical arrangement of syntaxonomic ranks: Cl = Class, Or = Order, Al = Alliance, As = Association, Sba = Subassociation.

3.2. *Diversity Measures*

The diversity study of *Poa bulbosae–Trifolietum subterranei* reveals the highest diversity values in inventory 3, with a Shannon_T of 2.671, which represents the total diversity of the association. However, when analyzing diversity based on the number of characteristic and companion species, changes are observed in inventories 3, 4, and 8 (Table 1). A notable case arises when the association is considered solely based on the two guiding species. In these instances, Shannon_Ca2 exceeds Shannon_Ca in all cases.

Table 1. Diversity indices for *Poo bulbosae–Trifolietum subterranei*. Shannon_T = Total diversity. Shannon_Ca = Diversity of characteristic species. Shannon_Ca2 = Diversity of the directive characteristic species of the association. Shannon_Co = Diversity of companion species. Shannon_Ca-Shannon_Co (Shannon_Ca-Co) = Difference between the diversity values of characteristic species and companion species.

	Inv1	**Inv2**	**Inv3**	**Inv4**	**Inv5**	**Inv6**	**Inv7**	**Inv8**	**Inv9**	**Inv1**
Shannon_T	2.131	2.339	2.671	1.972	2.603	2.005	1.762	1.908	1.765	1.74
Shannon_Ca	1.566	1.756	1.911	1.011	2.051	1.518	1.082	1.082	1.089	1.09
Shannon_Ca2	1.696	1.895	2.258	1.834	2.059	1.542	1.587	1.754	1.579	1.56
Shannon_Co	1.369	1.594	2.059	1.609	1.772	1.082	1.061	1.332	1.055	1.07
Shannon_Ca-Co	0.197	0.162	−0.148	−0.598	0.279	0.436	0.021	−0.25	0.034	0.02

For the *Bromus scoparii–Hordeetum leporini* grassland, the highest Shannon_T value is observed in inventory 2, where Shannon_Ca2 is less than Shannon_Ca. In this case, inventories Inv3–Inv9 show a Shannon_Co of 0 (Table 2).

The analysis of the difference between the mean values of characteristic and companion species, denoted as Vca-Vco, yields a value of 0.390 for the *Poa bulbosae–Trifolietum subterranei* grassland and 0.525 for the *Bromus scoparii–Hordeetum leporini* grassland.

The linear regression analysis for *Poa bulbosae–Trifolietum subterranei* (Figure 3) reveals a strong relationship, with an R^2 value greater than 0.8. For the *Bromus scoparii–Hordeetum leporini* grasslands, the R^2 value is even higher at greater than 0.9 in the Sh_T–Sh_Ca case. However, for the Sh_T–Sh_Co case, the R^2 value is low (Figure 4, due to the absence of companion especie in inventories Inv3–Inv9.

Table 2. Diversity index for *Bromo scoparii–Hordeetum leporini*. Shannon_T = Total diversity. Shannon_Ca = Diversity of characteristic species. Shannon_Ca2 = Diversity of the directive characteristic species of the association. Shannon_Co = Diversity of companion species. Shannon_Ca-Shannon_Co (Shannon_Ca-Co) = Difference between the diversity values of characteristic species and companion species.

	Inv1	Inv2	Inv3	Inv4	Inv5	Inv6	Inv7	Inv8	Inv9	Inv10
Shannon_T	2.566	2.674	2.463	2.216	2.321	2.288	2.305	2.085	1.818	2.624
Shannon_Ca	2.408	2.463	2.38	2.117	2.223	2.192	2.213	1.967	1.662	2.39
Shannon_Ca2	1.305	1.461	0.9596	0.5297	0.9949	0.9923	0.984	0.9596	0.5297	1.481
Shannon_Co	0.673	1.04	0	0	0	0	0	0	0	1.079
hannon_Ca-Co	1.735	1.423	2.38	2.117	2.223	2.192	2.213	1.967	1.662	1.311

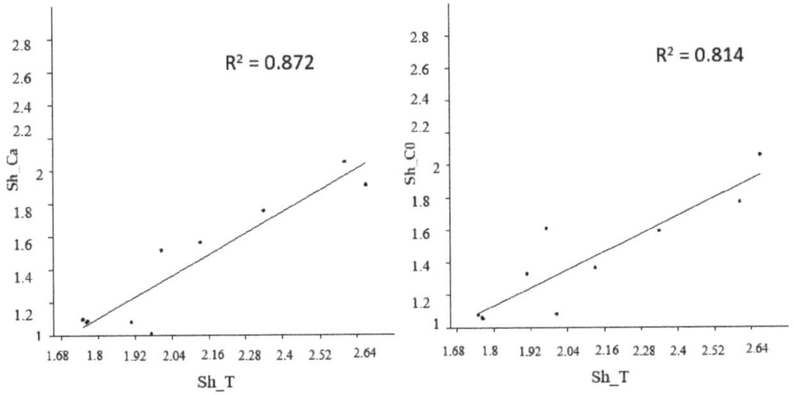

Figure 3. Linear regression analysis between Shannon values of total species with characteristic and companion species (*Poo bulbosae–Trifolietum subterranei*).

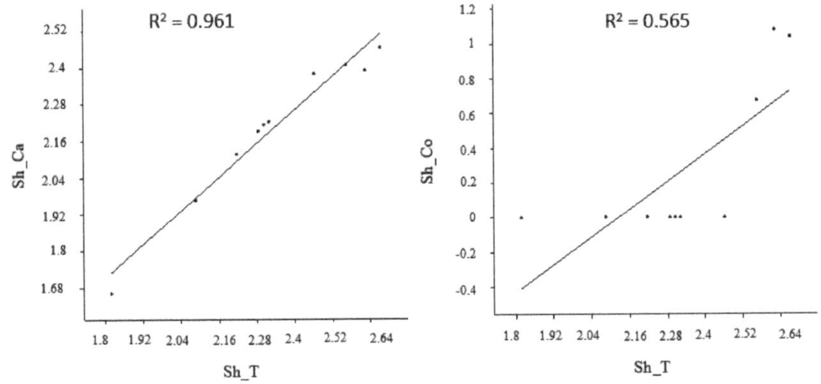

Figure 4. Linear regression analysis between Shannon values of total species with characteristic and companion species (*Bromo scoparii–Hordeetum leporini*). Sh_T = Total species diversity. Sh_Ca = Diversity of characteristic species. Sh_Co = Companion species diversity.

3.3. Learning Outcomes

The assessment of students' knowledge regarding vegetation, prior to engaging in theoretical and practical instruction, revealed significant deficiencies across both the Botany and Geobotany groups. Among the Botany students, there were no affirmative responses to questions 4, 5, and 6, while only minimal positive responses were recorded for questions

13 and 14. In contrast, Geobotany students provided affirmative answers to all questions; however, their responses to questions 4 and 5 were limited, and no responses were recorded for questions 6 and 13 (Table 3 and Figure 5).

Table 3. Responses obtained from students before and after learning. I-Gb % = Percentage of Geobotany students' responses before learning. I-Bg % = Percentage of Botany students' responses before learning. F-Gb % = Percentage of Geobotany students' responses after learning. F-Bg % = Percentage of Botany students' responses after learning.

	I-Gb %	I-Bg %	F-Gb %	F-Bg %
1. Very high	100	100	100	100
2. Very important	100	100	100	100
3. No	58	48	100	100
4. Yes	25	0	100	76
5. Yes	8	0	100	76
6. Yes	0	0	100	38
7. Correct	100	99	100	100
8. Correct defined	92	66	100	98
9. Yes	83	25	100	100
10. yes	100	100	100	100
11. Correct defined	100	93	100	100
12. Correct	100	55	100	100
13. Correct	0	1	67	60
14. Correct	100	72	100	100
15. Correct	100	2	100	100
16. Correct	100	97	100	100
17. Correct	100	67	100	100

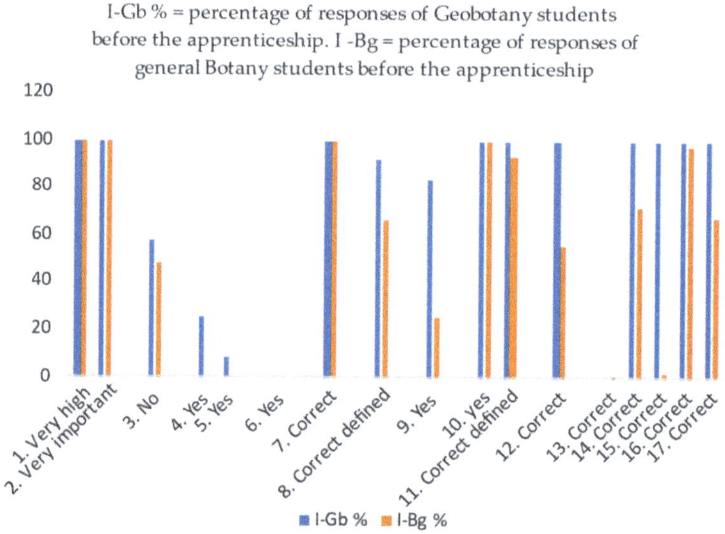

Figure 5. Graphical representation of students' responses to the 17 questions posed before instruction.

After the instruction focused on two specific vegetation associations, used as prototypes for learning the phytosociological method, 100% of the Geobotany students demonstrated proficiency in the methodology, except for question 13, where the knowledge level reached only 75%.

For the Botany students, 13 questions were answered correctly, while the remaining four (questions 4, 5, 6, and 13) showed response percentages ranging from 40% to 80% (Figure 6).

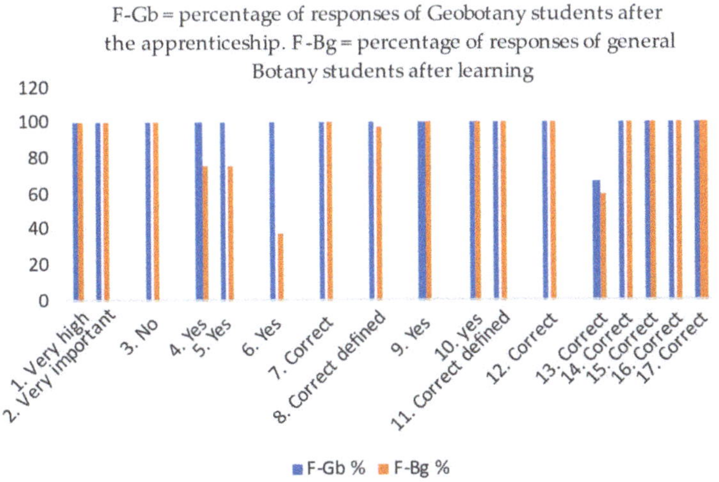

Figure 6. Graphical representation of students' responses to the 17 questions posed after instruction.

4. Discussion

4.1. Diversity Analysis

Regarding the study of diversity [52], we analyzed the Shannon index for both characteristic and companion species. The relationship between the abundance of characteristic and companion species showed negative values in inventories 3, 4, and 8 for the *Poo bulbosae–Trifolietum subterranei* (PbTs) association. However, linear regression analysis revealed consistently high R^2 values greater than 0.8, demonstrating the influence of companion species on the total species count. These inventories are undergoing transformation into neighboring communities of *Bromo scoparii–Hordeetum leporini* (BsHl) due to the introgression of plants from that association [17], likely a consequence of poorly managed grazing. As a result, these pastures, classified under habitat type 6220, are now considered endangered [29,53].

Positive values observed in the *Bromo scoparii–Hordeetum leporini* (BsHl) association are further supported by linear regression analysis, which shows an $R^2 > 0.9$ for Shannon diversity between total species (Sh_T) and characteristic species (Sh_Ca). However, for Shannon diversity between total species (Sh_T) and companion species (Sh_Co), the R^2 value is lower ($R^2 = 0.565$, Figure 6). This is attributed to the low number of companion species, with Sh_Co values in inventories Inv3–Inv8 being equal to 0.

4.2. Analysis of the Teaching-Learning Process

The results from the pre-teaching assessment show significant deficiencies in the Botany group, with somewhat lesser shortcomings in the Geobotany group. This is understandable, as these students come from secondary education (Bachillerato) backgrounds, where exposure to botanical or geobotanical content is limited or nonexistent.

The Botany students' responses to questions related to diversity, phytosociology, and ecosystems were predominantly negative or absent. In contrast, the Geobotany students demonstrated a greater number of correct responses, as they had prior exposure to Botany content through earlier courses. However, they still showed gaps in their understanding, particularly with questions 3, 4, 5, and 6.

By diagnosing the students' knowledge levels, we were able to address deficiencies in their understanding of Botany and Geobotany through inquiry-based methodologies, all while being guided and supervised by the instructor. The students studied two grassland communities from phytosociological, ecological, and floristic perspectives.

We applied a blended learning approach (Flipped Classroom), project-based learning (PBL), STEN, STEAM, and collaborative learning [21,56–59]. The results indicate a 100% success rate in most questions, except for questions 4, 5, 6, and 13, where the success rate ranged from 40% to 80%. Therefore, practical, out-of-class teaching methods proved to be highly effective. Students demonstrated improved observation and interpretation skills regarding plant associations, grasped the concept of ecological niches, and learned to distinguish between characteristic and companion species [2].

Based on these results, the instructor applies the "feedforward" method to address potential errors in the applied methodology or to confirm the findings [59]. In this case, the results are validated, as the BsHl association shows a very low total count of companion species, suggesting a healthy conservation status, as previously demonstrated by Cano-Ortiz et al. [53].

These findings underscore the importance of new educational spaces where students acquire knowledge that they can later share with society [60,61]. Through these practical studies of plant associations, their diversity, conservation status, and integration into specific ecosystems, students engage in an integrated learning process that connects the fields of science, technology, engineering, art, and mathematics (STEAM) (Figure 7).

Figure 7. Image of a holm oak dehesa (*Pyro rotundifoliae–Quercetum rotundifoliae* habitat 6310) where the BsHl grasslands and the PbTs grasslands of habitat 6220 are developed.

5. Conclusions

5.1. Conclusion on the Research

The study of the two Mediterranean grassland associations reveals the differences in floristic diversity between the two grassland types. It highlights the total species diversity compared to the diversity of characteristic species within the association and the companion species. Of particular significance is the relationship between characteristic species and companion species. Specifically, the greater the diversity of characteristic species, the lower the number of companion species, which suggests greater stability within the association. These findings are relevant to livestock farming, as the Poo bulbosae-Trifolietum subterranei (PbTs) pastures are particularly important in dehesas and Iberian montados for sheep farming, while the Bromo scoparii-Hordeetum leporini nitrophilous pastures support cattle and horses.

5.2. Conclusion on the Teaching and Learning Process

The teaching model based on constructivist theory has proven to be highly effective, as it equips students with the skills and motivation necessary to manage and educate about economically and socially important grasslands. Through this model, students gain an understanding of the hierarchical system of syntaxonomic units. The combination of field inventories and subsequent laboratory analysis enables students to develop a comprehensive understanding of the floristic diversity of the association and its conservation status.

The multidisciplinary, integrated learning approach has created an educational environment where students take ownership of their learning. Starting from a baseline of limited knowledge, significant progress has been made in students' understanding of both Botany and Geobotany. In conclusion, it can be stated that the learning process has successfully enhanced students' grasp of geobotanical concepts. Students are now able to differentiate between specific grassland associations and understand their floristic diversity.

Author Contributions: Conceptualization, A.C.-O. and E.C.; methodology, A.C.-O.; software, C.M.M.; validation, A.C.-O., C.M.M. and E.C.; formal analysis, A.C.-O.; investigation, A.C.-O.; resources, E.C.; data curation, C.M.M.; writing—original draft preparation, A.C.-O.; writing—review and editing, A.C.-O.; visualization, C.M.M.; supervision, E.C.; project administration, E.C.; funding acquisition, E.C. All authors have read and agreed to the published version of the manuscript.

Funding: This research received no external funding.

Institutional Review Board Statement: Ethical approval is not required since the survey was anonymous and with prior consent of the students, and privacy law is preserved.

Data Availability Statement: Informed consent for participation was obtained from all subjects involved in the study. And the approval of the Ethics Committee of the Complutense University is available with Ref. No. 115_CE20241212_17_SOC.

Conflicts of Interest: The authors declare no conflicts of interest.

ferences

Theurillat, J.P.; Willner, W.; Fernández-González, F.; Bültmann, H.; Čarni, A.; Gigante, D.; Mucina, L.; Weber, H. International Code of Phytosociological Nomenclature. *Appl. Veg. Sci.* **2020**, *24*, e12491. [CrossRef]

Cano Ortiz, A.; Piñar Fuentes, J.C.; Ighbareyeh, J.M.H.; Quinto Canas, R.; Cano, E. Aspectos Didácticos en la Enseñanza de Conceptos Geobotánicos. *IJHSSE* **2021**, *8*, 271–276. [CrossRef]

Cañas, A.; Martín-Díaz, M.; Nieda, J. *Competencia en el Conocimiento y la Interacción en el Mundo Físico. La Competencia Científica*; Alianza Editorial: Madrid, Spain, 2007.

García Barros, S.; Martínez Losada, C.; Rivadulla López, J. Actividades de textos excolares. Su contribución al desarrollo de la competencia científica. *Enseñanza Cienc.* **2021**, *39*, 219–238. [CrossRef]

Ivanova, N. Global Overview of the Application of the Braun-Blanquet Approach in Research. *Forests* **2024**, *15*, 937. [CrossRef]

Rivas Martínez, S. *Memoria del Mapa de Series de Vegetación de España 1:400.000. Ministerio de Agricultura, Pesca y Alimentación*; I.C.O.N.A.: Madrid, Spain, 1987; p. 268.

Rivas-Martínez, S.; Díaz, T.E.; Fernández-González, F.; Izco, J.; Loidi, J.; Lousa, M.; Penas, A. Vascular PlantCommunities of Spain and Portugal. *Itinera Geobot.* **2002**, *15*, 5–922.

Rivas-Martínez, S. Mapa de series, geoseries y geopermaseries de vegetación de España. Parte I. *Itinera Geobot.* **2007**, *17*, 5–436.

Rivas-Martínez, S. Mapa de series, geoseries y geopermaseries de vegetación de España. Parte II. *Itinera Geobot.* **2011**, *18*, 425–800.

Pinto Gomes, C.J.; Paiva Ferreira, R. *Flora e Vegetaçao Barrocal Algarvio, Tavira-Portimao. Publ. Comissao de Coordinaçao e Desenvolvimento Regioal do Alvarve*; Comissão e Coordenação e Desenvolvimento Regional do Algarve: Faro, Portugal, 2005; pp. 1–353.

Lazare, J.-J. Phytosociologíe dynamico-catenal et gestión de la biodiversité. *Acta Bot. Gall.* **2009**, *156*, 49–61.

Raposo, M.; Mendes, P.; Cano Ortiz, A.; Pinto Gomes, C.J. Séries de Vegetação prioritárias para a conservação no centro e sul de Portugal continental. *Botanique* **2016**, *1*, 113–148.

13. Quinto Canas, R.; Vila-Viçosa, C.; Paiva Ferreira, R.; Cano Ortiz, A.; Pinto Gomes, C.J. The Algarve (Portugal) climatophil vegetation series: A base documentto theplanning, management and nature conservation. *Acta Bot. Gall.* **2012**, *159*, 289–2 [CrossRef]
14. Cano, E.; Musarella, C.M.; Cano-Ortiz, A.; del Río, S.; Rodríguez Meireles, C.; Raposo, M.; Pinto Gomes, C.J. Contribution the iberian thermomediterranean oak woods (Spain, Portugal): The importance of their teaching for the training of experts environmental management. *Plant Biosyst.* **2024**, *58*, 1285–1305. [CrossRef]
15. Cano, E.; Musarella, C.M.; Cano-Ortiz, A.; Piñar Fuentes, J.C.; Rodríguez Torres, A.; del Río, S.; Pinto Gomes, C Quinto Canas, R.; Spampinato, G. Geobotanical Study of the Microforests of Juniperus oxycedrus subsp. badia in the C tral and Southern Iberian Peninsula. *Sustainability* **2019**, *11*, 1111. [CrossRef]
16. Chang, H.; Xiong, K.; Zhu, D.; Zhang, Z.; Zhang, W. Ecosystem Services Value Realization and Ecological Industry Design Scenic Areas of Karst in South China. *Forests* **2024**, *15*, 363. [CrossRef]
17. Rivas Martínez, S. La vegetación de *Hordeion leporini* es España. *Doc. Fitosoc.* **1978**, *9*, 377–392.
18. Tilbury, D. Environmental education for sustainability: Defining the new focus of environmental education in the 1990s. *Envi Educ. Res.* **1995**, *1*, 195–212. [CrossRef]
19. Palacios, J.; Amud, N.; Mendoza, D. *Implementación de Huertas Escolares como Estrategia de Enseñanza-Aprendizaje de la Biolo de Grado Sexto en la Institución Educativa Agrícola de Urabá del Municipio de Chigorodó y de Grado Séptimo de la Institución Edu tiva Rural Zapata, de Necoclí, Departamento de Antioquia.* (Tesis magistral Universidad Pontificia Bolivariana, Medellín). 2C Available online: https://repository.upb.edu.co/bitstream/handle/20.500.11912/2950/T.G.%20JULIO%20%C3%89DINSO 20PALACIOS%20Y%20OTROS.pdf (accessed on 20 November 2024).
20. Monroe, M.C.; Krasny, M.E. Across the Spectrum: Resources for Environmental Education. *N. Am. Assoc. Environ. Ed* 2015. Available online: https://www.academia.edu/27500674/Foundation_of_Environmental_Education (accessed on November 2024).
21. Caraballo Vidal, I.; Pezelj, L.; Ramos-Álvarez, J.J.; Guillen-Gamez, F.D. Level of Satisfaction with the Application of Collaborative Model of the Flipped Classroom in the Sport of Sailing. *Educ. Sci.* **2024**, *14*, 150. [CrossRef]
22. Köpeczi-Bócz, T. The Impact of a Combination of Flipped Classroom and Project-Based Learning on the Learning Motivation University Students. *Educ. Sci.* **2024**, *14*, 240. [CrossRef]
23. Rodríguez Angarita, T.E.; Mora Penagos, W. El Sentido Educativo de los Espacios no Convencionales de Educación (ENCE) T Jardines Botánicos: Un Estudio de Caso. Enseñanza de las Ciencias, nº Extraordinario 2017, pp. 3411–3416. Available onli https://raco.cat/index.php/Ensenanza/article/view/337075 (accessed on 20 November 2024).
24. Jia, L.; Wang, M.; Yang, S.; Zhang, F.; Wang, Y.; Li, P.; Ma, W.; Sui, S.; Liu, T.; Wang, M. Analysis of Agricultural Carbon Emissi and Carbon Sinks in the Yellow River Basin Based on LMDI and Tapio Decoupling Models. *Sustainability* **2024**, *16*, 468. [CrossR
25. Cano-Ortiz, A.; Piñar Fuentes, J.C.; Leiva Gea, F.; Igbareyech, J.M.H.; Quinto Canas, R.J.; Rodrigues Meireles, C.I.; Raposo, Pinto Gomes, C.J.; Spampinato, G.; Del Río González, S.; et al. *How to Reduce the Supply of Nutrients to the Soil, Incre Water Reserves, and Mitigate Climate Change in Agroforestry for Carbon and Ecosystem Management*; Jhariya, M.K., Meena, R Banerjee, A., Kumar, S., Raj, A., Eds.; Elsevier: Amsterdam, The Netherlands, 2023; pp. 223–232. Available online: ht //shop.elsevier.com/books/agroforestry-for-carbon-and-ecosystem-management/jhariya/978-0-323-95393-1 (accessed on November 2024).
26. Álvarez, P.; Vega, P. Actitudes ambientales y conductas sostenibles. Implicaciones para la educación ambiental. *Rev. Psicodidáct* **2009**, *4*, 245–260.
27. Cantú, P. Educación ambiental y la escuela como espacio educativo para la promoción de la sustentabilidad. *Rev. Ed* **2014**, *18*, 39–52.
28. Sandoval, M. Comportamiento sustentable y educación ambiental: Una visión desde las prácticas culturales. *Rev. Latinoam. Psi* **2012**, *44*, 181–196.
29. EEC. Council directive 92/43/CEE of 21 May 1992 on the conservation of natural habitats and of wild fauna and flora. *Off.* **1992**, *206*, 7–50.
30. Cañal, P. ¿Cómo evaluar la competencia científica? *Investig. Esc.* **2012**, *78*, 5–17.
31. Miño, M.H.; Toia, S.N.; Pérez, G.M.; Gutierrez, T.N.; González Galli, L.M.; Meinardi, E.M. Comparación del conocimier metacognitivo sobre la lectura de textos de Biología entre estudiantes de la Ciudad de Buenos Aires. *Rev. Electrónica Enseñar Cienc.* **2021**, *1*, 114–134.
32. Tovar-Gálvez, J.C. Design of Intercultural Teaching Practices for Science Education based on evidence. *Enseñanza Cie* **2021**, *39*, 99–115. [CrossRef]
33. Fernández, I.M.; Pires, D.M.; Villamañán, R.M. Educación científica con enfoque ciencia- tecnología-sociedad-ambiente: Constr ción de un instrumento de análisis de las directrices curriculares. *Form. Univ.* **2014**, *7*, 23–32. [CrossRef]
34. Bello Benavides, L.O.; Cruz Sánchez, G.E.; Meira Cartera, P.A.; González Gaudiano, E. El cambio climático en el bachillera Aportes pedagógicos para su abordaje. *Enseñanza Cienc.* **2021**, *39*, 137–156. [CrossRef]

Amblar Frances, P.; Casado Calle, M.J.; Pastor Saavedra, A.; Ramos Calzado, P.; Rodríguez Camino, E. *Guide to Regionalized Climate Change Scenarios over Spain from IPCC-AR5 Results*; AEMET: Madrid, Spain, 2017. [CrossRef]
IPCC. Climate Change Synthesis Report The Intergovernmental Panel on Climate Change. 2023. Available online: https://www.ipcc.ch/report/ar6/syr/downloads/report/IPCC_AR6_SYR_FullVolume.pdf (accessed on 20 November 2024).
Da Silva Siltori, P.F.; Lourenzani, W.L.; Satolo, E.G.; Ferreira Caldana, A.C.; Salati Marcondes de Moraes, G.H.; Batista Martins, V.W.; Simon Rampasso, I. Training Future Managers to Address the Challenges of Sustainable Development: An Innovative, Interdisciplinary, and Multiregional Experience on Corporate Sustainability Education. *World* **2024**, *5*, 155–172. [CrossRef]
Guerrero Fernández, A.; Rodríguez Marín, F.; López Lozano, L.; Solís Ramírez, E. Alfabetización ambiental en la formación inicial docente: Diseño y validación de un cuestionario. *Enseñanza Cienc.* **2022**, *40*, 25–46. [CrossRef]
Forbes, S.H. *Holistic Education: An Analysis of Its Ideas and Nature*; Brandon (Vermont, EE. UU.): Foundation for Educational Renewal; Resource Center for Redesigning: San Diego, CA, USA, 2003.
Hare, J. *La Educación Holística: Una Interpretación para los Profesores de los Programas del IB*; Organización del Bachillerato Internacional: Geneva, Switzerland, 2010; pp. 1–8. Available online: https://es.scribd.com/document/177983712/La-educacion-holistica-John-Hare (accessed on 8 November 2024).
María Domíguez, S. Del Holismo al Constructivismo. Los Grandes Maestros. Revista de Postgrado FACE-UC. **2014**, *6*, 39–51. Available online: http://www.arje.bc.uc.edu.ve/arj15/art03.pdf (accessed on 20 November 2024).
Añez de Bravo, M.A. Modelo de aprendizaje holístico del ser: Una propuesta pedagógica en orientación. *Rev. Estilos Aprendiz.* **2009**, *3*, 177–195. Available online: https://revistaestilosdeaprendizaje.com/article/view/884/1572 (accessed on 20 November 2024). [CrossRef]
De Rivas, R.; Vilches, A.; Mayoral., O. Secondary School Students' Perceptions and Concerns on Sustainability and Climate Change. *Climate* **2024**, *12*, 17. [CrossRef]
Chen, Y.; Hong, C.; Yang, Y.; Li, J.; Wang, Y.; Zheng, T.; Zhang, Y.; Shao, F. Mining Social Media Data to Capture Urban Park Visitors' Perception of Cultural Ecosystem Services and Landscape Factors. *Forests* **2024**, *15*, 213. [CrossRef]
Torres, D.; Capote, J. Agroquímicos un problema ambiental global: Uso del análisis químico como herramienta para el monitoreo ambiental. *Ecosistemas* **2004**, *13*, 2–6.
Magwegwe, E.; Zivengwa, T.; Zenda, M. Adaptation and Coping Strategies of Women to Reduce Food Insecurity in an Era of Climate Change: A Case of Chireya District. Zimbabwe. *Climate* **2024**, *12*, 126. [CrossRef]
Salinas, I.; Guerrero, G.; Satlov., M.; Hidalgo., P. Climate Change in Chile's School Science Curriculum. *Sustainability* **2022**, *14*, 15212. [CrossRef]
Vilches, A.; Gil-Pérez, D. The transition to Sustainability as an urgent objective for overcoming the current systemic crisis. *Eureka J. Sci. Educ. Dissem.* **2016**, *13*, 395–407. Available online: http://hdl.handle.net/10498/18296 (accessed on 20 November 2024).
García Cué, J.L.; Santizo Rincón, J.A.; Alonso García, C.M. Instrumentos de medición de estilos de aprendizaje. *Rev. Estilos Aprendiz.* **2009**, *3*, 3–21. Available online: https://revistaestilosdeaprendizaje.com/article/view/886/1574 (accessed on 20 November 2024). [CrossRef]
Cano Ortiz, A.; Pinto Gomes, C.J.; Esteban, F.J.; Rodríguez Torres, A.; Gonñi, J.; De la Haza, I.; Cano, E. Biodiversity of *Hordeion leporini* in Portugal: A phytosociological and edaphic analysis. *Acta Bot.Gall.* **2009**, *156*, 33–48. [CrossRef]
Cano Ortiz, A.; Biondi, E.; Pinto Gomes, C.J.; Del Río González, S.; Cano, E. Soil and phytosociological caracterisation of grasslands in the western mediterranean. *Am. J. Plant Sci.* **2014**, *5*, 3213–3240. [CrossRef]
Cano Ortiz, A. Teaching about biodiversity from phytosociology: Evaluation and conservation. *Plant Sociol.* **2023**, *60*, 25–37. [CrossRef]
Cano Ortiz, A.; Piñar Fuentes, J.C.; Musarella, C.M.; Cano, E. Educational and socio-economic strategies for the sustainability of iberian mediterranean grasslands. *Tansylvanian Rev.* **2024**, *32*, 16588. Available online: https://www.transylvanianreviewjournal.com/index.php/TR/issue/view/65 (accessed on 20 November 2024).
Braun-Blanquet, J. *Fitosociología*; Ed. Blume: Madrid, Spain, 1979; pp. 1–820.
Géhu, J.-M.; Rivas-Martínez, S. Notions fondamentales de Phytosociologie. In *Syntaxonomie [Berichte der Internationalen Symposien der Internationalen Vereinigung für Vegetationskunde 1980]*; Dierschke, H., Ed.; Springer: Dordrecht, The Netherlands, 1981; Volume LI, pp. 5–33.
Montaner-Villalba, S.; Santiago, R.; Bergmann, J. Aprender al revés. Flipped Learning 3.0 y metodologías activas en el aula. Barcelona: Faidós Educación pp. 240. *Rev. Interuniv. Investig. Tecnol. Educ. (RIITE)* **2018**, *7*, 98–99. [CrossRef]
Cantó Doménech, J.; Hurtado Soler, A.; Vilches Peña, A. Educación científica más allá del aula. *Alambique Didáctica Cienc. Exp.* **2013**, *74*, 76–83.
Nguyen, H.T.T. Implementing Feedforward-based Collaborative Assessment at Higher Education. *Athens J. Educ.* **2024**, *11*, 335–352. [CrossRef]

59. Satrústegui Moreno, A.; Mateo González, E. Mejora del pensamiento crítico en alumnos de ESO a través del aprendizaje bas
en problemas en un entorno STEAM. *Rev. Estilos Aprendiz.* **2023**, *16*, 19–32. [CrossRef]
60. de los Santos, J.S.N. Espacios educativos para promover la sustentabilidad y la seguridad escolar. Un estudio entre Méxic
Canadá. *Rev. Estilos Aprendiz.* **2023**, *16*, 78–88. [CrossRef]
61. García Ochoa, C. Percepción de la imagen de marca y nuevos espacios de aprendizaje. Un estudio de caso del nuevo Centro
Formación ESIC Madrid. *Rev. Estilos Aprendiz.* **2023**, *16*, 62–77. Available online: https://revistaestilosdeaprendizaje.com/arti
view/5816 (accessed on 20 November 2024). [CrossRef]

diversity

cle

nderstanding Wildlife Biodiversity Awareness: Rural ildren's Knowledge, Attitudes and Perceptions of onservation in the Selva Maya

athan Pérez-Flores [1] and Jorge Omar López-Martínez [2,3,*]

[1] Departamento de Observación y Estudio de la Tierra, la Atmósfera y el Océano, El Colegio de la Frontera Sur, Av. Centenario km 5.5, Chetumal 77014, Quintana Roo, Mexico; johnspf77@gmail.com
[2] Secretaría de Ciencias, Humanidades, Tecnología e Innovación (SECIHTI), Benito Juárez, Ciudad de Mexico 03940, Mexico
[3] Centro de Investigación en Ciencias de Información Geoespacial (CentroGeo), Mérida 97302, Yucatan, Mexico
* Correspondence: lmjorgeomar@gmail.com

Abstract: We live in an era where future generations are experiencing a serious loss of knowledge, sensitivity, and interest in biodiversity. This situation is thought to be unique to urbanised countries with limited green spaces. However, with the increasing urbanisation of rural areas, it is necessary to investigate what is happening in places of high biodiversity. In this study, we investigated the knowledge, attitudes and perceptions of children and adolescents in the rural municipality of José María Morelos in southeastern Mexico. We analysed 1119 drawings by 117 children in their first three years of school and interviewed 280 students in their last three years of primary, secondary and high school. Younger children drew more native species, with the majority being mammals. Of the students surveyed, secondary school students had the highest level of species knowledge and considered deer and jaguar to be the most important species. They considered hunting to be the main cause of animal depletion and had a positive attitude towards nature conservation. Based on our findings, we recommend that educational strategies be implemented at all school levels to promote knowledge of the less emblematic species that inhabit the Selva Maya.

Keywords: biophilia; ecological knowledge; education for conservation; nature connection; Selva Maya

1. Introduction

In recent decades, conservationists have expressed growing concern about the lack of interest, sensitivity, and knowledge among younger generations regarding local flora and fauna [1]. Urbanisation and the technological boom have been postulated as the main drivers of the decline in local knowledge [2,3]. The perception of urban life as synonymous with progress has reduced children's and adolescents' opportunities and interest in interacting with nature. Limited access to natural spaces and constrained time for outdoor exploration [2,4,5] hinder the development of emotional connections with the natural world [6]. Additionally, the increasing prevalence of electronic media, such as television, video games, and smartphones, has contributed to children and adolescents spending more time indoors and less time outdoors, further weakening their bond with nature [3,7].

Most studies examining the loss of knowledge and sensitivity towards wildlife among children and youth have been conducted in developed Western countries, focusing primarily on the urban population [5,8–11]. In developing nations, where many people live in

 https://doi.org/10.3390/d17030152

rural areas under different socio-ecological conditions (e.g., ethnic diversity, high poverty rates, low educational levels, and high biodiversity), such studies are scarce [12–14].

Many authors suggest that children and adolescents living in rural areas have more knowledge, sensitivity, and connection to nature than their counterparts living in urban areas [15,16]. It is hypothesised that spending more time in nature generates stronger ties to it [17,18]. However, this hypothesis cannot be universally applied due to the great cultural and ethnic diversity of rural regions worldwide. Surprisingly, some studies have begun to observe that children in certain rural areas are losing their connection to nature, probably due to increasing urbanisation and exposure to digital devices [13].

The Selva Maya is the second largest forest after the Amazon, located between Belize, Guatemala and southern Mexico [19]. Its human settlements are located around protected natural areas, allowing people to coexist with diverse wildlife species [20]. This intimate contact with nature would imply that the children of these communities have knowledge and sympathy for wild species, which would translate into a greater willingness to conserve them [21]. However, this information is currently unknown, as studies on the perception and knowledge of wildlife in the Selva Maya have focused on a single species (e.g., the jaguar) and its different conservation problems, and have mainly been directed towards adults [22]. Based on the above, we consider it fundamental to assess the knowledge, attitudes and perceptions about wildlife held by children and young people living in rural communities in Latin America, especially in regions considered biodiversity hotspots.

Several investigations have used species identification to evaluate children's and adolescents' knowledge, sensitivity, and connection with wildlife [23]. Understanding species-specific knowledge (e.g., naming species, recognising habitats, and understanding interactions) and emotional attitudes (e.g., liking or disliking specific animals) is crucial. These factors strongly predict attitudes toward wildlife and conservation support [24–26]. Some studies have found a positive correlation between increased knowledge and sympathy for species with increased willingness to conserve them [27,28]. Therefore, information on the knowledge, attitudes, and perceptions of Selva Maya children and adolescents towards native fauna is important, as they may be future decision-makers in conserving and using their natural resources [29].

Our study occurred in the rural municipality of José María Morelos (JMM) in south-eastern Mexico. Its population is overwhelmingly poor (~80%), with low levels of formal education, and is composed mainly (41%) of children and adolescents between 0 and 19 years old [30]. As JMM surrounds the Bala'an Ka'ax Flora and Fauna Protection Area (BKFFPA), several projects for species conservation (e.g., jaguar and Baird's tapir) and community ranger training (e.g., wildlife monitoring, firefighting, ecosystem restoration) have been developed in recent years. However, these activities have mainly been targeted towards adults and some adolescents (16–19 years), rarely involving children [31]. Children and adolescents in JMM receive information about local or global wildlife from what they are taught at school, what they see on social networks, and the knowledge transmitted at home (SFH pers. comm.). Therefore, the study of children's knowledge, perceptions and attitudes towards local fauna will allow the design of specific educational strategies that integrate their interests and concerns about their environment, encouraging a more conscious and active relationship in favour of conservation from an early age.

The objectives of this study were as follows: (1) to evaluate knowledge and preferences for local fauna among lower primary school children (1st to 3rd grades) using drawings; (2) to identify the species recognized by children and adolescents (4th to 12th grades); and (3) to determine their awareness of conservation actions led by local stakeholders, assess their perceptions of species decline (causes), and evaluate their pro-conservation attitudes through semi-structured surveys.

2. Materials and Methods

2.1. Study Site

The Selva Maya is Mesoamerica's largest contiguous tropical forest, covering 100,000 km and extending from southern Mexico to northwestern Belize and northern Guatemala. Within the Selva Maya, in the centre of the Yucatan Peninsula, is the Bala'an Ka'ax Flora and Fauna Protected Area (BKFFPA), covering 128,390.16 hectares with an area of influence of 230,552.6 ha (19°19′ N–89°03′ W) (Figure 1). The BKFFPA forms part of the Sian Ka'an–Calakmul Biological Corridor encompassing four regions across three states in the Yucatan Peninsula: Hopelchén in Campeche; Bacalar and JMM in Quintana Roo; and Tekax in Yucatán. Within the BKFFPA are 37 ejidos, and conservation efforts are led by the National Commission of Natural Protected Areas (CONANP) in coordination with community rangers made up of residents of the ejidos within the area of influence of the reserve. Our study focused on communities belonging to JMM (19°14′–88°24′, 20°19′–89°19′). José María Morelos has a population of 39,165 inhabitants (49.3% females and 50.7% males), and 31% are between 0 and 14 years old. José María Morelos has a great ethnic diversity, with the Maya, Ch'ol and Mam being the most represented indigenous groups. Fifty-seven per cent of the population is moderately poor, and 25.1% is extremely poor, with low levels of formal education (29.5% with 1 to 6 years of schooling, 37.8% with 6 to 9 years and 20.5% with 10 to 12 years) [30]. The primary economic activities in the communities surrounding the BKFFPA include agriculture through the traditional slash-and-burn system (milpa), timber harvesting, beekeeping and, to a lesser extent, cattle ranching [32]. Subsistence hunting remains an essential cultural and livelihood practice in the region. Over the past decade, a Mennonite community has settled near the BKFFPA, resulting in approximately 1000 ha of forest being deforested.

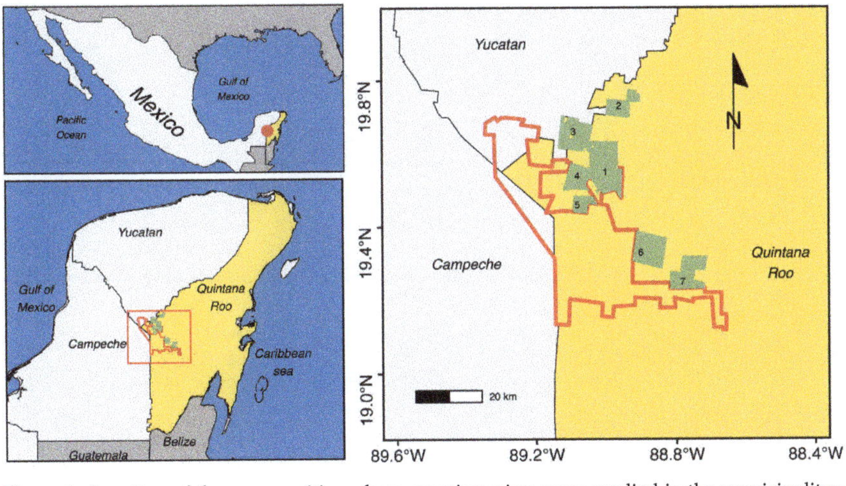

Figure 1. Location of the communities where questionnaires were applied in the municipality of José María Morelos, Quintana Roo, Mexico. The red polygon corresponds to the BKFFPA area, and green polygons with numbers correspond to the ejidos of the schools: (1) Othón P. Blanco, (2) La Candelaria, (3) Puerto Arturo, (4) Sabana San Francisco, (5) San Felipe, (6) San Isidro Poniente, (7) Venustiano Carranza.

2.2. Data Collection

Data were collected from schools in seven communities adjacent to the BKFFPA: San Isidro Poniente (primary), Venustiano Carranza (primary), Sabana San Francisco (primary), Puerto Arturo (secondary and high school), Candelaria (high school), San Felipe Primero (primary) and Othón P. Blanco (primary and secondary). This included five primary schools, two secondary schools, and two high schools (Figure 1). To obtain the data, we had the approval of the principals and teachers of the schools in which the environmental education workshops were given. Primary education is almost universal in these communities, each having at least one primary school. However, secondary and high schools are not as widespread, requiring students at these levels to travel several kilometres to neighbouring communities. This contributed to achieving spatial representativeness across the seven communities [33]. These communities were selected because they had previously been part of a wildlife monitoring project using camera traps, and some residents are part of the community rangers.

For this cross-sectional study, students were categorised based on their years of schooling: lower primary (1 to 3 years), upper primary (4 to 6 years), secondary (7 to 9 years), and high school (10 to 12 years). The age range of the students was as follows: lower primary school 6 to 8 years, upper primary school 9 to 12 years, secondary 13 to 15 years and high school 16 to 18 years (Table 1). Lower primary students were asked to draw their favourite animals to explore their knowledge of the local fauna. Semi-structured interviews were conducted for upper primary, secondary, and high school students. These interviews included closed-ended, open-ended, and multiple-choice questions to gather comprehensive information. The topics addressed included demographics, knowledge about wildlife species and their importance, perceptions of animals' representative of their community, views on species that had disappeared and the perceived reasons for their decline, suggested actions to protect animals, and opinions on the main actors responsible for conservation efforts within their community. The activities were carried out during the second semester of 2019, as part of environmental education workshops to teach children the fundamentals of wildlife conservation in the Selva Maya.

Table 1. Socio-demographic characteristics of the 397 participants surveyed for their knowledge and perceptions about wildlife and its conservation in seven communities of José María Morelos, Quintana Roo, Mexico.

Education Level	Years of Schooling	Age Range (Years)	Female	Male
Lower primary school	1–3	6 to 8	68	49
Upper primary school	4–6	9 to 12	37	40
Secondary school	7–9	13 to 15	38	40
High School	10–12	16 to 18	64	61
Total			207	190

2.3. Data Analysis

The drawings of the lower primary school students were digitised, identified and classified into four main classes: reptiles and amphibians, birds, aquatic animals and mammals. They were also categorised according to their geographical distribution: local or native, non-local (those animals without distribution in Yucatan Peninsula), domestic and exotic (those animals not distributed in the Americas). Each class and category's absolute and relative frequencies were calculated and presented as percentages. On the other hand, the semi-structured interviews were organised in a spreadsheet to identify patterns of responses based on the thematic analysis. Trends in the perception of local fauna, perceived reasons for the decline of specific species, and conservation proposals were analysed using simple descriptive statistics. To assess whether there were statistically significant differences in the frequency of animals mentioned between the genders (boys and girls) in lower primary school and to determine [34] whether there were differences in the responses obtained from the semi-structured surveys, Fisher's exact test and a

Monte Carlo simulation with 10,000 iterations were used to obtain an approximate p-value in a contingency table relating the gender of the participants to the animals mentioned. Together, these statistical tools allowed us to identify significant differences and relevant patterns in participants' knowledge, perception and mention of fauna, highlighting the importance of factors such as gender, education level and community in the analysis. All analyses were performed in R version 4.4.2 [35].

3. Results

3.1. Drawings

A total of 1119 drawings belonging to 97 animal species were made by 117 lower primary school children. Of the total number of species reported, girls reported 89 species in 655 drawings, while boys reported 79 species in 464 drawings (Fisher test, $p = 0.0001$). Of the 97 species reported, 64% were identified as local, 22.5% as domestic, 10.5% as exotic, and 3% as non-local (Figure 2). Boys drew 70% of local species compared to 60% of the species drawn by girls, 10% more (Figure 2). The most represented group was mammals, with 61% of the drawings, followed by birds (22.8%), amphibians and reptiles (6.5%), aquatic animals (7.5%) and insects (2.2%) (Figure 3). Of the total number of species drawn by the girls, 64% belonged to the group of mammals and 19.5% to the group of birds. Boys, on the other hand, drew fewer mammals (57%) but more birds (27%) (Figure 3). The five most frequently drawn species were deer (49%), peccary (32%), jaguar (32%), rabbit (29%) and cattle (29%). The animal with the highest number of drawings was the deer, 41% of girls and 59% of boys. For girls, jaguars and cows were the other animals with the most drawings (34%). For boys, crocodile (41%), agouti (37%) and peccary (37%) were preferred.

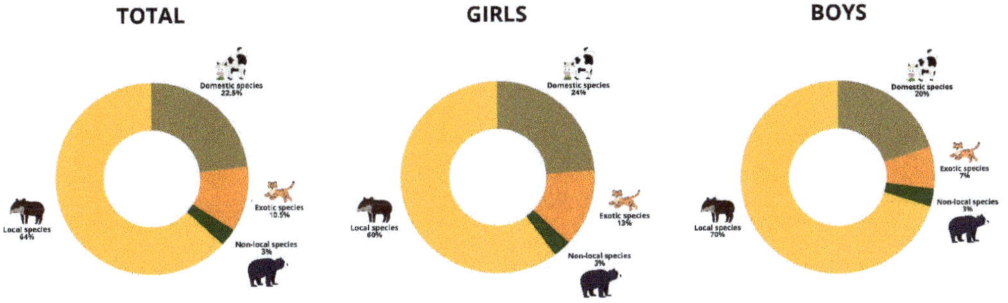

Figure 2. The percentage of each species category according to its geographical distribution (local or native, non-local, domestic and exotic) drawn by girls and boys in lower primary school. The total values represent the percentage of the most frequently drawn groups, regardless of gender.

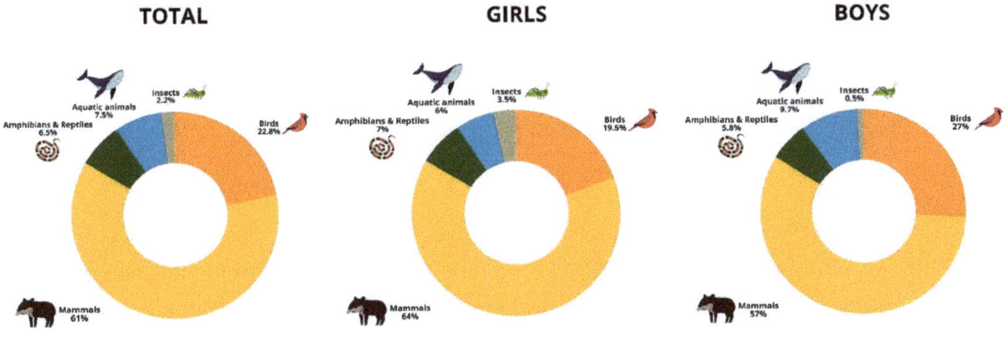

Figure 3. Percentage of animal groups drawn by girls and boys in lower primary school. The total values represent the percentage of the most frequently drawn groups, regardless of gender.

3.2. Knowledge of Wildlife and Conservation Actors

A total of 280 students were interviewed, of whom 77 were in upper primary school, 78 in secondary school and 125 in high school. The results are presented below based on the responses obtained. The students mentioned 47 species of animals; we took as a single species when the students wrote in general terms: bird, bat, lizard, bee, etc. Sixty-three per cent of the students interviewed stated they knew more than five native fauna species. Secondary school students showed a higher knowledge (86%) compared to primary school students (43%) ($p = 0.0059$) and high school adolescents (62%). However, the difference in knowledge of fauna between secondary and high school students was not statistically significant ($p = 0.6$). On the other hand, boys (54%) have a slightly higher knowledge of native fauna than girls (46%), although this is not significant. The difference is minimal between boys (52%) and girls (48%) with four to nine years of schooling but tends to increase slightly in the last three years of schooling (high school) (56% boys and 44% girls), although in no case was the difference statistically significant ($p = 0.56$).

When students were asked if they knew the authority responsible for environmental protection (in our case, we used the National Commission for Natural Protected Areas), 49% answered that they did. Children in upper primary school (57%) were the most familiar with it, followed by those in secondary school (56%) and finally those in high school (41%). Overall, the percentage of males (51%) who were aware of the authority was slightly higher (49%) than that of females (Table 1). In upper primary and secondary school, the frequency was higher for girls than for boys (upper primary 55%, secondary 52%). However, the frequency found in high school males is considerably higher than in females (59% men compared to 41% women). One of the most important figures in these communities is the community ranger. Fifty-eight per cent of students said they knew them. Upper primary and secondary school students showed the highest knowledge with 61% and 62%, respectively. High school students were much less aware (41%). Males are more aware of community rangers (56%) than females (44%) (Table 2).

Table 2. Children and adolescents' knowledge of the principal actors in the conservation of local fauna in their region.

Question		Overall	Years of Schooling		
		Overall $n = 280$	4 to 6 n (%)	7 to 9 n (%)	10 to 12 n (%)
Do you know the authorities in charge of protecting wildlife?					
	Yes	139 (49%)	44 (57%)	44 (56%)	51 (41%)
	Female yes	68 (49%)	24 (55%)	23 (52%)	21 (41%)
	Male yes	71 (51%)	20 (45%)	21 (48%)	30 (59%)
	No	141 (51%)	33 (43%)	34 (44%)	74 (59%)
Do you know what a community ranger is?					
	Yes	146 (58%)	47 (61%)	48 (62%)	51 (41%)
	Females yes	65 (44%)	19 (60%)	21 (44%)	25 (49%)
	Males yes	81 (56%)	28 (40%)	27 (56%)	26 (51%)
	No	134 (42%)	30 (39%)	30 (38%)	74 (59%)

3.3. Perception and Concerns About Wildlife in the Community

3.3.1. Do You Think Wild Animals Represent Your Community?

The vast majority of students (84%) thought that wild animals represented their community. Secondary school students had the highest frequency (91%), followed by high school (86%) and upper primary school (74%). There was no overall difference according to gender, but in upper primary, boys tended to have a higher frequency (56% compared to 44% of girls). In contrast, in secondary school, girls had the highest frequency (54% compared to 46% of boys) (Table 2).

3.3.2. Which Animal Do You Think Is the Most Important?

Jaguar (37%) and Deer (33%) were the species that students considered to be the most important. In the upper primary, the frequency of the two species was the same (34%). There was a big difference in secondary school, where deer was considered more important (43%) than the jaguar (25%). In high school, jaguar was considered the most important (46% vs. 27% deer) (Table 3).

Table 3. Species considered most important in the forest by children and adolescents in seven communities of José María Morelos, Quintana Roo, Mexico. Local species (excluding deer and jaguar).

Education Level	Deer	Jaguar	Local Species	Non-Local Species
per primary school (4 to 6 years of schooling)	34%	34%	18%	14%
condary school (7 to 9 years of schooling)	42%	24%	14%	20%
gh School (10 to 12 years of schooling)	28%	46%	13%	13%

3.3.3. Do You Feel That There Are Fewer Wild Animals than Before?

Students perceived fewer animals than before (84%), and the percentage increased as they progressed through school: 69% in upper primary, 83% in secondary school, and 93% in high school. There was a gender difference in upper primary, with 57% of boys compared to 43% of girls (Table 4).

Table 4. Upper primary, secondary and high school students' perceptions of the importance and decline of wild animals in their communities.

uestion		Years of Schooling			
		Overall n = 280	4 to 6 n (%)	7 to 9 n (%)	10 to 12 n (%)
you think wild animals represent your community?					
	Yes	238 (85%)	57 (74%)	71 (91%)	110 (88%)
	Female yes	119 (50%)	25 (44%)	38 (54%)	56 (51%)
	Male yes	119 (50%)	32 (56%)	33 (46%)	54 (49%)
	No	42 (15%)	20 (26%)	7 (9%)	15 (12%)
hich animal do you think is the most important?					
	Jaguar	104 (37%)	26 (34%)	19 (24%)	58 (46%)
	Deer	92 (33%)	26 (34%)	33 (42%)	35 (28%)
	Species of the Yucatan Peninsula	42 (15%)	14 (18%)	11 (14%)	16 (13%)
	Species not of the Yucatan Peninsula	42 (15%)	11 (14%)	15 (20%)	16 (13%)
you feel that there are fewer wild animals than before?					
	Yes	235 (84%)	53 (69%)	64 (82%)	118 (94%)
	Female yes	116 (50%)	23 (43%)	32 (50%)	61 (52%)
	Male yes	119 (50%)	30 (57%)	32 (50%)	57 (48%)
	No	45 (16%)	24 (31%)	14 (18%)	7 (6%)
you think wild animals in your community are endangered?					
	Yes	151 (54%)	50 (65%)	34 (44%)	67 (54%)
	Female yes	66 (50%)	20 (40%)	14 (41%)	32 (48%)
	Male yes	85 (50%)	30 (60%)	20 (59%)	35 (52%)
	No	129 (46%)	27 (35%)	44 (56%)	58 (46%)

Table 4. *Cont.*

Question		Years of Schooling		
	Overall n = 280	4 to 6 n (%)	7 to 9 n (%)	10 to 12 n (%)
What do you think is the main reason why they are disappearing?				
Hunting	220 (79%)	55 (72%)	66 (85%)	99 (79%)
Fires	46 (16%)	21 (27%)	7 (9%)	18 (15%)
Deforestation	8 (3%)	1 (1%)	3 (4%)	4 (3%)
Agriculture and livestock farming	6 (2%)	0 (0%)	2 (2%)	4 (3%)

3.3.4. Do You Think Wild Animals in Your Community Are Endangered?

Just slightly more than half of students (54%) believed that animals in their community are threatened or endangered. The highest frequency was found among upper primary school students (65%), followed by high school students (54%) and then secondary school students (44%). There was a notable difference between the sexes at all levels, with males having the highest frequencies compared to females (60% upper primary, 59% secondary school, 52% high school) (Table 4).

3.3.5. What Do You Think Is the Main Reason Why Animals Are Disappearing?

Students thought that hunting was the main reason for the disappearance of animals (79%). Hunting was selected as the main cause at all school levels: upper primary by 72%, secondary (85%), and high school (79%) (Table 4).

3.4. *Attitudes*

3.4.1. Do You Think You Can Do Something to Protect Wild Animals?

Almost all students (87%) felt they could do something to protect wildlife. Generally, there was no significant gender difference (48% females compared to 52% males). Only primary school boys had a higher frequency (55%) than girls (45%) (Table 5).

Table 5. Upper primary, secondary and high school students' attitudes towards protecting wild animals in their community.

Question		Years of Schooling		
	Overall n = 280	4 to 6 n (%)	7 to 9 n (%)	10 to 12 n (%)
Do you think you can do something to protect wild animals?				
Yes	243 (87%)	56 (73%)	74 (95%)	113 (90%)
Female yes	118 (48%)	25 (45%)	36 (49%)	57 (51%)
Male yes	125 (52%)	31 (55%)	38 (51%)	56 (49%)
No	37 (13%)	21 (27%)	4 (5%)	12 (10%)
What actions can you take to protect wild animals?				
Take care of the forest	111 (40%)	38 (49%)	32 (41%)	41 (33%)
Do not throw trash in the forest	30 (11%)	7 (9%)	9 (12%)	14 (11%)
Teaching others to take care of animals	139 (49%)	32 (42%)	37 (47%)	70 (56%)

3.4.2. What Actions Can You Take to Protect Wild Animals?

The main actions that students mentioned to protect wild animals were teaching others to care for them (49%) and caring for the forest (40%). For upper primary children, caring for the forest was the most common action (49%), while for secondary school (47%) and high school (56%), it was teaching others to care for the animals (Table 5).

4. Discussion

The results of this study show that students' knowledge of the local fauna of the Selva Maya varies according to school level. In middle school, adolescents have the greatest knowledge about the species, with a downward trend during high school. Despite this, the results suggest that students maintain a strong connection with nature. This is evidenced by their ability to identify and value the animals that inhabit their communities, particularly emblematic species such as the deer and jaguar, which they frequently cited as the most important and symbolic of their cultural and natural environment. The connection to nature is further reinforced by the emotional concern expressed by students at the declining number of animals in their communities due to hunting. Research on environmental identity highlights that emotional engagement, such as concern for wildlife, is a significant indicator of connection to nature [36]. In addition, students' positive attitudes toward protecting local wildlife and their habitats reflect a personal and social sense of responsibility toward the environment, which is consistent with theoretical frameworks in environmental education that emphasise that awareness and concern are precursors to pro-environmental behaviour [37]. Their identification with iconic species such as jaguar and deer suggests cultural resonance and recognition of the ecological importance of these animals, which are key elements of connection with nature, according to studies on the relationship with the environment [38]. Given that children and adolescents are still in a cognitive and emotional development phase, these findings strongly support the need to design and implement a permanent environmental education program adapted to different school levels. Such a program would seek to broaden knowledge about less popular species of native fauna and deepen the existing connection with nature, fostering greater awareness and long-term conservation behaviour.

4.1. Drawings

Several ecological studies have used children's drawings as a technique to obtain detailed information about children's knowledge, attitudes and feelings about nature [39–44]. It has been shown that children as young as five can provide reliable information on ecological issues, provided they are asked clearly and appropriately [45]. We found drawings to be a good approach to exploring JMM lower primary children's knowledge of local species. Children drew more local species, which is crucial, as knowledge and empathy are determining factors in establishing a connection with nature [26,46–48].

An encouraging result is the low percentage of exotic species (zebra, elephant, lion, tiger, giraffe) that were drawn by the population of children participating in this study. This finding contrasts with what has been observed in other studies conducted in developed countries and urbanised environments, where children have a greater knowledge of charismatic species (e.g., giant pandas, polar bears, great apes, elephants and rhinoceroses, etc.) than of local fauna [48]. We also observed that boys' knowledge of native species was higher than girls', with a difference of 10%. We consider that this situation may be influenced by the fact that boys tend to have more contact with nature, as they accompany their parents from a very young age in various productive activities in the forest (milpa, cattle raising, beekeeping), so that they can observe animals during these tours. At the same time, girls stay at home to help their mothers (JSPF pers. obs.).

We found that boys have a higher affinity for birds and mammals, which has been extensively documented in other studies [49–52]. The mammal group had the highest number of drawings for both sexes due to their high charisma and the fact that they are the basis of many children's toys and books [53]. Furthermore, it should be considered that mammals are one of the most important groups in Selva Maya communities, as in some communities, they are the main source of animal protein and are even kept as pets (e.g., coatis, deer, kinkajous, monkeys) [54]. In this sense, young children may have seen several mammal species in their homes, which could have influenced their knowledge.

One of our findings was that girls drew more mammals than boys, which previous studies have attributed to girls preferring cuddlier and cuter animals, while boys prefer animals that provoke fear and disgust [55,56]. On the other hand, boys drew more bird species than girls; this finding differs from recent work that found no difference in knowledge of bird species between girls and boys [57]. We believe that the knowledge of birds that the children of JMM have is the result of their daily life, as when they go out to play, they can observe a great diversity of species because birds are attracted to the fruit trees that people are used to plant in these communities (JPF pers. obs.).

Interestingly, the two species most frequently drawn by the lower primary school children were deer and peccary or "jabalí". These species are culturally important in Maya and mestizo communities in southeastern Mexico [54], and are among the species with the highest hunting pressure in the communities in the Yucatan Peninsula [22,58]. In the Selva Maya, subsistence hunting is a traditional activity in which knowledge is passed from generation to generation, and children participate from an early age (8 to 12 years old) [59]. Hunting could, therefore, be a factor favouring children's knowledge of native fauna, as children are more exposed to local biodiversity during this activity [60,61].

The jaguar is the most emblematic species of the Selva Maya and has been used in many conservation campaigns by different NGOs; however, it was the third most drawn species for children. This result is similar to what Mascote et al. [62] found; when they asked elementary school children (aged 6 to 12) what species live in the jungle, they ranked the jaguar in fourth place. This would largely result from the few times that children have direct contact with a jaguar, as it is a species that is difficult to see in the wild. This coincides with Wyndham [63] who mentions that children acquire knowledge first about nearby and easily observable elements of their environment.

4.1.1. Knowledge About Species

Students from JMM only know about 7% of the animal taxa found in the Mexican part of the Selva Maya (almost 700 species) [31]. Several authors have mentioned that knowledge of animal species increases with age and education [51,64,65]. Our results confirm that the highest peak of knowledge occurs in secondary school; however, we observed a significant decrease (almost 20%) when students move from secondary to high school. This decline in species knowledge after age 14 coincides with that observed by Randler [51] in German adolescents. This decline may be related to puberty when young people have different concerns and motivations. During the workshops in high schools, we observed that some young people already had computers and smartphones. This could lead to losing "contemplation" and "being" with nature due to spending more time on social networks or digital games [6]. Another factor that could influence adolescents' knowledge of their environment is the quality of the information provided at this school level. This is very general because few teachers (three) cover a wide range of subjects. In addition, the teaching system relies on information transmitted by television and satellite networks [33]. In this sense, it is necessary to increase efforts to find ways for young people to maintain their connection with nature and interest in local wildlife.

We found a general pattern in the knowledge shown by males at all levels, who had greater knowledge of native species than females, similar to what has been observed in previous studies [1,60,64]. In the case of this study, these differences are due to the socio-cultural environment in which the children live. As we have already mentioned, in the rural communities of the Selva Maya, girls have certain limitations in exploring the environment, as they generally help their mothers with household chores (cooking, washing, and caring for backyard animals).

4.1.2. Knowledge About Conservation Actions

Children are one of the population sectors that show the greatest interest in species conservation [66]. As a result, in recent years, several governmental institutions (universities, research institutes, federal and state authorities, etc.) and NGOs have implemented various conservation education projects in communities near natural protected areas (JPF pers. obs.). In this sense, the present study evaluated how these institutions' work has reached children in the region. We found that only half of the students knew about CONANP and its activities, and just over half knew who the community guardians were. With this in mind, we consider it relevant to carry out dissemination activities on the work that CONANP and local agents are doing for the benefit of conservation. In this way, it will be easier for the environmental agenda to permeate the minds of students and promote the development of pro-conservation attitudes.

Knowledge of these topics decreases as the level of education increases, so it is advisable to increase efforts to ensure that this knowledge does not become diluted. We found gender differences in knowledge about CONANP. Girls in upper primary and secondary school had higher frequencies than boys. This could be because topics related to government institutions are discussed at home when most family members are together (breakfast, lunch or dinner), so girls may hear more about these authorities at these times. Males in high school had much more knowledge about CONANP than females. We believe that males know more about CONANP because they are involved in productive activities related to government subsidies from a very early age, so they increase their knowledge of these institutions by talking about them during the working day or by socialising with other males (family, friends, neighbours).

About the figure of the community ranger, males also had more knowledge than females. This is because all the community rangers were men, and so boys were able to observe the guardians' work more closely and even participate in some of them by accompanying a family member or neighbour. We confirmed this when we went out into the field with the community rangers, where some adolescents aged between 17 and 19 were already very interested and participating in these activities.

4.2. Perceptions

4.2.1. Do You Think Wild Animals Represent Your Community?

Most students responded that they felt represented by local animals positively and encouragingly. During the workshops, the students mentioned some characteristics that they found amazing about the animals, such as the strength of the tapir, the speed of the deer, or the intelligence of the jaguar. These perceptions could result from past and present interactions with these species [67], which could help develop positive attitudes towards their conservation [68].

Our results suggest that adolescents in secondary and high school had a greater positive affect than children in upper primary school. This may be because secondary and high school students are in the adolescent stage when they begin to express their concern for the welfare of animals [69]. However, some studies have found that concern for environmental issues declines during adolescence [70]. Responses vary according to the socioecological context in which adolescents find themselves. Regarding gender, females in secondary and high school had the highest percentages, consistent with other studies that mention that females are more empathetic and have more positive attitudes towards animals than males [55].

4.2.2. Which Animal Do You Think Is the Most Important?

In the Selva Maya, as in many regions, conservation messages are based on a few iconic and emblematic species (e.g., pandas, big cats, eagles, whales, elephants, rhinos, etc.) [64,71]. Finding the jaguar in the top five most important animals for students in the Selva Maya was unsurprising. It was surprising to find the deer in the first place (although some students do not distinguish between white-tailed deer and brocket deer) as an important species since, as mentioned above, it is traditionally used in the Yucatan Peninsula [59]. These results are similar to those observed by Schuttler et al. [61], who found that children who hunted or had family members who practice hunting chose deer as their favourite species.

A worrying finding is that students in JMM perceive the same importance between local fauna and fauna not from the Yucatan Peninsula (many of these exotic species). Generating excessive information about attractive or emblematic animals (exotic or local) may condition students' knowledge and perceptions of other local species [48]. For example, recent conservation projects in the Selva Maya that exalt the jaguar and downplay the importance of other species ("calling them prey") diminish their ecological importance and make them appear complementary rather than primary actors. This situation could likely lead to a loss of connection with the less charismatic local species and, therefore, a loss of will to conserve them [61].

4.2.3. Do You Think There Are Fewer Wild Animals than Before?

Given the accelerated loss of biodiversity that we are currently experiencing, it is necessary to ask children and adolescents how they perceive the loss of wild animals in their community. The vast majority responded that they perceived fewer animals than before. We believe this response could be conditioned by the species they know best, which are generally the most popular and the largest (jaguar, puma, deer, peccary, tapir), and have lower population densities [72–74]. Future research could, therefore, investigate which species are thought to be declining and why.

The increase in children's awareness was gradual as the level of schooling increased. We assume this is because as children age, they begin to know more about the biodiversity problems and understand what actions they need to take to address them [75]. Regarding gender, we only noticed a large difference between boys and girls in upper primary, where the frequency of boys was higher. This finding should be taken with caution, as we believe it is sometimes difficult for children to accurately perceive the presence and abundance of flora and fauna species at this age [76].

4.2.4. Do You Think Wild Animals in Your Community Are Endangered?

Knowing whether children and adolescents are concerned about species extinction is important, as it provides a bridge for them to care about the natural world [77]. In our case, the situation is striking in that many students think animals are not in danger of extinction. This response could be due to two reasons: firstly, the low level of knowledge of basic aspects of species biology and ecology, which we were able to verify during the workshops. Secondly, because it is difficult for local people to perceive that an animal is threatened (even if it is an emblematic species [78]), they sometimes tend to have a negative attitude towards these species because of the damage they cause them (damage to crops, predation on livestock, etc.). Although the level of concern was low, we observed an increase in concern as the level of schooling increased, so it was noticeable that high school students have received more information and acquired more knowledge. This is consistent with findings from other studies, which have observed a positive correlation between increased knowledge and greater concern for species conservation [27,28,79].

When comparing between sexes, males showed greater concern than females at all school levels. This differs from other studies documenting that females are more concerned about conservation than males [76,80]. Thus, it is necessary that in Selva Maya communities, we begin to provide girls with the same opportunities as boys so that we can highlight the pro-conservation qualities of women, which consist of a better connection to nature and a greater concern for environmental issues than men [81].

4.2.5. What Do You Think Is the Main Reason Why Animals Are Disappearing?

Currently, the wildlife of the Selva Maya is threatened by various climatic factors, such as droughts, floods, hurricanes, and natural fires [58,82–84], as well as by anthropogenic factors, such as deforestation, livestock and agriculture, hunting, overexploitation, introduction of invasive species, and pollution [85–88]. This situation has attracted the attention of scholars who have approached this problem from different perspectives, studying human–wildlife conflict, climate change impacts and disease transmission. However, to our knowledge, no study has focused on how local communities perceive the decline of wildlife, especially children and adolescents.

According to the perceptions of the children who participated in the workshops, hunting is the main reason for the disappearance of animals in their community, similar to what Gavrilakis et al. [89] found among children in Greece. This is important, as unsustainable hunting is one of the main factors threatening wildlife globally [90]. The fact that children perceive hunting as the main cause is an interesting finding, as we mentioned earlier, hunting is a traditional activity in Selva Maya communities and is practised from a very young age [59]. Given this situation, we can assume that these future generations will be less attracted to this type of activity and that in the future, hunting skills may be reduced and, consequently, hunting pressure on endangered species may decrease [91]. Similarly, fires are one of the greatest threats to the biodiversity of the Selva Maya, so children and adolescents may be familiar with these events. As a result, government institutions have encouraged the creation of community fire brigades. In the case of the BKFPPA, the brigades are made up of people from the communities surrounding the reserve. Thus, the children receive information about fighting forest fires through collective oral communication. It should be remembered that forest fires have acute and long-term effects on the composition of the landscape [92] and alter the dynamics of wildlife populations by increasing mortality [93]. It is therefore necessary to continue transmitting the culture of fire-fighting from generation to generation among the inhabitants of the Selva Maya, in order to have a better attitude towards nature conservation.

4.3. *Attitudes*

4.3.1. Do You Feel You Can Do Something to Protect Wild Animals?

Research has shown that the more time children spend in natural environments, the more likely they are to develop pro-environmental or conservationist attitudes throughout their lives [18,94,95]. However, most of this research has been developed with adults and urban populations, where natural spaces are increasingly shrinking [64,96]. This study provided a unique opportunity to explore the attitudes of children and adolescents in rural populations surrounded by nature. Our results showed that most children in the Selva Maya have positive attitudes. For example, they favour conserving and protecting wild animals in their community. These results are consistent with those of Duron-Ramos et al. [96], who found that children from rural communities were more empathetic and protective of nature than children living in urban environments. However, we found no difference in pro-conservation attitudes between the sexes at different school levels. As for the comparison between school levels, secondary and high school adolescents had a higher frequency than upper primary school children, which could be because young children have less interest, knowledge and information about animals and their natural environment [64].

4.3.2. What Actions Can You Take to Protect Wild Animals?

Although students' knowledge of local fauna is not very deep, they recognise that to protect animals, they need to teach others to protect them and the forest (habitat). This affective attitude towards nature is crucial in developing people's conservation behaviour and attitudes [67,97]. These attitudes may be more influential than species knowledge in influencing people's involvement in biodiversity conservation [67,98]. Therefore, some researchers have proposed cultivating positive feelings (pleasure, empathy, love or pity) towards animals to promote wildlife conservation [99].

5. Conclusions

There is a global belief that the loss of knowledge, interest and sensitivity towards nature occurs only in children and adolescents living in urban environments where proximity to nature is disappearing. However, our results show this is already happening in rural communities in one of Latin America's most important biodiversity hotspots, the Selva Maya. We therefore recommend developing specific educational strategies for each school level to promote knowledge and empathy for local species and their conservation (actors and institutions). Finally, governmental institutions and NGOs that support the conservation of biodiversity in the Selva Maya must avoid implementing conservation education projects that focus on a single species or on emblematic species, since we are unintentionally preventing the awakening of children's interest in non-charismatic species (invertebrates, reptiles and amphibians, birds).

Author Contributions: Conceptualization, J.P.-F. and J.O.L.-M.; methodology, J.P.-F. and J.O.L.-M.; formal analysis, J.P.-F. and J.O.L.-M.; data curation, J.P.-F. and J.O.L.-M.; writing—original draft preparation, J.P.-F.; writing—review and editing, J.O.L.-M.; funding acquisition, J.O.L.-M. All authors have read and agreed to the published version of the manuscript.

Funding: This research was funded by Comisión Nacional de Áreas Naturales Protegidas (CONANP) through the following programs: Program for the Protection and Restoration of Ecosystems and Species at Risk 2019: Program for the Protection and Restoration of Ecosystems and Species at Risk 2019: Implementation of conservation schemes for large mammals and restoration of water bodies in Bala'an K'aax.

Data Availability Statement: The raw data supporting the conclusions of this article will be made available by the authors on request.

Acknowledgments: The authors thank Sandra Flores Hernández, Abelardo Chan Ek and Wilberth Ek Moreno (CONANP-BKFFPA), and the community rangers thank you for their valuable assistance. We thank the school staff for allowing us to give the workshops and participate in this important project. Of course, we thank all the girls, boys, and young people who kindly participated. Also, we thank Jesus Dominguez for his help in cleaning up the drawings. Also, to the National Council of Humanities, Sciences and Technologies (CONAHCYT) for the support granted through the project: evaluation, monitoring and maintenance of ecosystem processes and services in Mexican socio-ecological systems (No. 7035).

Conflicts of Interest: The authors declare no conflicts of interest.

References

1. Huxham, M.; Welsh, A.; Berry, A.; Templeton, S. Factors Influencing Primary School Children's Knowledge of Wildlife. *J. B Educ.* **2006**, *41*, 9–12. [CrossRef]

Soga, M.; Yamanoi, T.; Tsuchiya, K.; Koyanagi, T.F.; Kanai, T. What Are the Drivers of and Barriers to Children's Direct Experiences of Nature? *Landsc. Urban Plan.* **2018**, *180*, 114–120. [CrossRef]
Charan, G.S.; Kalia, R.; Khurana, M.S.; Narang, G.S. From Screens to Sunshine: Rescuing Children's Outdoor Playtime in the Digital Era. *J. Indian Assoc. Child Adolesc. Ment. Health* **2024**, *20*, 11–17. [CrossRef]
Turner, W.R.; Nakamura, T.; Dinetti, M. Global Urbanization and the Separation of Humans from Nature. *Bioscience* **2004**, *54*, 585. [CrossRef]
Skar, M.; Gundersen, V.; O'Brien, L. How to Engage Children with Nature: Why Not Just Let Them Play? *Child. Geogr.* **2016**, *14*, 527–540. [CrossRef]
Chawla, L. Childhood Nature Connection and Constructive Hope: A Review of Research on Connecting with Nature and Coping with Environmental Loss. *People Nat.* **2020**, *2*, 619–642. [CrossRef]
Mjaavatn, P.E. Changes in Children's Preferences for Outdoor Activities: A Longitudinal Study. *Leisure* **2016**, *40*, 225–244. [CrossRef]
Collado, S.; Corraliza, J.A. Children's Restorative Experiences and Self-Reported Environmental Behaviors. *Environ. Behav.* **2015**, *47*, 38–56. [CrossRef]
Krettenauer, T.; Bauer, K.; Sengsavang, S. Fairness, Prosociality, Hypocrisy, and Happiness: Children's and Adolescents' Motives for Showing Unselfish Behaviour and Positive Emotions. *Br. J. Dev. Psychol.* **2019**, *37*, 505–518. [CrossRef]
Keith, R.J.; Given, L.M.; Martin, J.M.; Hochuli, D.F. Urban Children's Connections to Nature and Environmental Behaviors Differ with Age and Gender. *PLoS ONE* **2021**, *16*, e0255421. [CrossRef]
Price, E.; Maguire, S.; Firth, C.; Lumber, R.; Richardson, M.; Young, R. Factors Associated with Nature Connectedness in School-Aged Children. *Curr. Res. Ecol. Soc. Psychol.* **2022**, *3*, 100037. [CrossRef]
Oh, R.Y.R.; Fielding, K.S.; Nghiem, T.P.L.; Chang, C.-C.; Shanahan, D.F.; Gaston, K.J.; Carrasco, R.L.; Fuller, R.A. Factors Influencing Nature Interactions Vary between Cities and Types of Nature Interactions. *People Nat.* **2021**, *3*, 405–417. [CrossRef]
Mustapa, N.D.; Maliki, N.Z.; Aziz, N.F.; Hamzah, A. The Differences of Nature Experiences between Urban and Rural Children. *Malays. J. Soc. Space* **2018**, *14*, 225–237. [CrossRef]
Ji, R.; Li, S.; Shao, Y. A Study on the Characteristics of Children's Natural Activities in the Neighborhood and Their Influencing Factors: Evidence from Hangzhou, China. *Int. J. Environ. Res. Public Health* **2022**, *19*, 16087. [CrossRef]
Hand, K.L.; Freeman, C.; Seddon, P.J.; Recio, M.R.; Stein, A.; van Heezik, Y. The Importance of Urban Gardens in Supporting Children's Biophilia. *Proc. Natl. Acad. Sci. USA* **2017**, *114*, 274–279. [CrossRef]
Summers, J.K.; Smith, L.M.; Case, J.L.; Linthurst, R.A. A Review of the Elements of Human Well-Being with an Emphasis on the Contribution of Ecosystem Services. *Ambio* **2012**, *41*, 327–340. [CrossRef]
Gifford, R.; Nilsson, A. Personal and Social Factors That Influence Pro-Environmental Concern and Behaviour: A Review. *Int. J. Psychol.* **2014**, *49*, 141–157. [CrossRef]
Hinds, J.; Sparks, P. Engaging with the Natural Environment: The Role of Affective Connection and Identity. *J. Environ. Psychol.* **2008**, *28*, 109–120. [CrossRef]
Pérez-Flores, J.; Borges-Ramírez, M.M.; Vargas-Contreras, J.A.; Osten, J.R. Inter-Annual Variation in the Microplastics Abundance in Feces of the Baird's Tapir (*Tapirus bairdii*) from the Selva Maya, México. *Sci. Total Environ.* **2024**, *941*, 173659. [CrossRef]
Rodriguez Solorzano, C.; Fleischman, F. Institutional Legacies Explain the Comparative Efficacy of Protected Areas: Evidence from the Calakmul and Maya Biosphere Reserves of Mexico and Guatemala. *Glob. Environ. Change* **2018**, *50*, 278–288. [CrossRef]
Xu, J.; Jiang, A. Effects of Nature Contact on Children's Willingness to Conserve Animals under Rapid Urbanization. *Glob. Ecol. Conserv.* **2022**, *38*, e02278. [CrossRef]
Piña-Covarrubias, E.; Chávez, C.; Doncaster, C.P. Knowledge of Wildlife, Hunting, and Human-Felid Interactions in Maya Forest Communities of the Northern Yucatán Peninsula, Mexico. *Hum. Ecol. Interdiscip. J.* **2022**, *50*, 1035–1045. [CrossRef]
Van Weelie, D.; Wals, A. Making Biodiversity Meaningful through Environmental Education. *Int. J. Sci. Educ.* **2002**, *24*, 1143–1156. [CrossRef]
Jacobs, M.H.; Vaske, J.J.; Roemer, J.M. Toward a Mental Systems Approach to Human Relationships with Wildlife: The Role of Emotional Dispositions. *Hum. Dimens. Wildl.* **2012**, *17*, 4–15. [CrossRef]
Lundberg, P.; Vainio, A.; MacMillan, D.C.; Smith, R.J.; Veríssimo, D.; Arponen, A. The Effect of Knowledge, Species Aesthetic Appeal, Familiarity and Conservation Need on Willingness to Donate. *Anim. Conserv.* **2019**, *22*, 432–443. [CrossRef]
Wilson, C.; Tisdell, C. Knowledge of Birds and Willingness to Support Their Conservation: An Australian Case Study. *Bird Conserv. Int.* **2005**, *15*, 225–235. [CrossRef]
Liordos, V.; Kontsiotis, V.J.; Kokoris, S.; Pimenidou, M. The Two Faces of Janus, or the Dual Mode of Public Attitudes towards Snakes. *Sci. Total Environ.* **2018**, *621*, 670–678. [CrossRef]
Liordos, V.; Foutsa, E.; Kontsiotis, V.J. Differences in Encounters, Likeability and Desirability of Wildlife Species among Residents of a Greek City. *Sci. Total Environ.* **2020**, *739*, 139892. [CrossRef]

29. Dominguez Contreras, E.; Krasny, M.E. Young Children Contribute to Nature Stewardship. *Front. Psychol.* **2022**, *13*, 945
[CrossRef]
30. INEGI. *Panorama Sociodemográfico de Quintana Roo: Censo de Población y Vivienda 2020*; Instituto Nacional de Estadistica y Geogra
Aguascalientes, Mexico, 2020.
31. FD-MSDRM; CONAP; CONANP; SERMANAT. *Estrategia Integral Selva Maya 2030*; Somos Selva Maya: Ciudad de Guatem
Guatemala, 2021.
32. Pérez Cortez, S.; Matus Pérez, E.S. El Tapir Tapirus Bairdii En La Región Sureste Del Área de Protección de Flora y Fauna Bala
Ka'ax, Quintana Roo, México. *Therya* **2010**, *1*, 137–144. [CrossRef]
33. Posadas, M.G. Educação a distância para comunidades rurais: Conquistas e desafios no ensino superior. *Cad. Pesqui.* **2023**, *5*
34. Mehta, C.R.; Patel, N.R. A Network Algorithm for Performing Fisher's Exact Test in r × c Contingency Tables. *J. Am. Stat. As*
1983, *78*, 427. [CrossRef]
35. R Core Team. *R: A Language and Environment for Statistical Computing, Version 4.4.2*; R team: Vienna, Austria, 2013.
36. Clayton, S.; Myers, G. *Conservation Psychology: Understanding and Promoting Human Care for Nature*, 2nd ed.; Wiley-Blackw
Hoboken, NJ, USA, 2015; ISBN 9781118874608.
37. Hungerford, H.R.; Volk, T.L. Changing Learner Behavior through Environmental Education. *J. Environ. Educ.* **1990**, *21*, 8–
[CrossRef]
38. Nisbet, E.K.; Zelenski, J.M.; Murphy, S.A. The Nature Relatedness Scale. *Environ. Behav.* **2009**, *41*, 715–740. [CrossRef]
39. Lundqvist, C.; Rugland, E.; Clench-Aas, J.; Bartonova, A.; Hofoss, D. Children Are Reliable Reporters of Common Sympto
Results from a Self-Reported Symptom Diary for Primary School Children: Self-Reported Symptom Diary. *Acta Paediatr.* **2010**,
1054–1059. [CrossRef]
40. Aaron, R.; Witt, P. Urban Students' Definitions and Perceptions of Nature. *Child. Youth Environ.* **2011**, *21*, 145–167. [CrossRef]
41. Drissner, J.R.; Haase, H.-M.; Wittig, S.; Hille, K. Short-Term Environmental Education: Long-Term Effectiveness? *J. Biol. Ed*
2014, *48*, 9–15. [CrossRef]
42. Massey, W.V.; Thalken, J.; Szarabajko, A.; Neilson, L.; Geldhof, J. Recess Quality and Social and Behavioral Health in Element
School Students. *J. Sch. Health* **2021**, *91*, 730–740. [CrossRef]
43. Morón-Monge, H.; Hamed, S.; Morón Monge, M. del C. How Do Children Perceive the Biodiversity of Their Nearby Environm
An Analysis of Drawings. *Sustainability* **2021**, *13*, 3036. [CrossRef]
44. Moula, Z.; Palmer, K.; Walshe, N. A Systematic Review of Arts-Based Interventions Delivered to Children and Young People
Nature or Outdoor Spaces: Impact on Nature Connectedness, Health and Wellbeing. *Front. Psychol.* **2022**, *13*, 858781. [CrossR
45. Snaddon, J.L.; Turner, E.C.; Foster, W.A. Children's Perceptions of Rainforest Biodiversity: Which Animals Have the Lion's Sh
of Environmental Awareness? *PLoS ONE* **2008**, *3*, e2579. [CrossRef] [PubMed]
46. Wells, N.M.; Lekies, K.S. Nature and the Life Course: Pathways from Childhood Nature Experiences to Adult Environmentali
Child. Youth Environ. **2006**, *16*, 1–24. [CrossRef]
47. Stokes, M.; Newton, N.; Kaur, A. Stalking, and Social and Romantic Functioning among Adolescents and Adults with Auti
Spectrum Disorder. *J. Autism Dev. Disord.* **2007**, *37*, 1969–1986. [CrossRef] [PubMed]
48. Ballouard, J.-M.; Brischoux, F.; Bonnet, X. Children Prioritize Virtual Exotic Biodiversity over Local Biodiversity. *PLoS ONE* **20**
6, e23152. [CrossRef]
49. Lindemann-matthies, P. The Influence of an Educational Program on Children's Perception of Biodiversity. *J. Environ. Educ.* **20**
33, 22–31. [CrossRef]
50. Littledyke, M. Science Education for Environmental Awareness: Approaches to Integrating Cognitive and Affective Doma
Environ. Educ. Res. **2008**, *14*, 1–17. [CrossRef]
51. Randler, C. Morningness–Eveningness, Sleep–Wake Variables and Big Five Personality Factors. *Pers. Individ. Dif.* **2008**,
191–196. [CrossRef]
52. Shapiro, H.G.; Peterson, M.N.; Stevenson, K.T.; Frew, K.N.; Langerhans, R.B. Wildlife Species Preferences Differ among Child
in Continental and Island Locations. *Environ. Conserv.* **2017**, *44*, 389–396. [CrossRef]
53. Patrick, P.; Byrne, J.; Tunnicliffe, S.D.; Asunta, T.; Carvalho, G.S.; Havu-Nuutinen, S.; Sigurjónsdóttir, H.; Óskarsdóttir, G.; Traca
R.B. Students (Ages 6, 10, and 15 Years) in Six Countries Knowledge of Animals. *Nord. Stud. Sci. Educ.* **2013**, *9*, 18–32. [CrossR
54. García Del Valle, Y.; Naranjo, E.J.; Caballero, J.; Martorell, C.; Ruan-Soto, F.; Enríquez, P.L. Cultural Significance of Wild Mamm
in Mayan and Mestizo Communities of the Lacandon Rainforest, Chiapas, Mexico. *J. Ethnobiol. Ethnomed.* **2015**, *11*, 36. [CrossR
55. Prokop, P.; Zvaríková, M.; Zvarík, M.; Ježová, Z.; Fedor, P. Charismatic Species Should Be Large: The Role of Admiration a
Fear. *People Nat.* **2024**, *6*, 945–957. [CrossRef]
56. Schlegel, J.; Rupf, R. Attitudes towards Potential Animal Flagship Species in Nature Conservation: A Survey among Students
Different Educational Institutions. *J. Nat. Conserv.* **2010**, *18*, 278–290. [CrossRef]
57. Härtel, T.; Vanhöfen, J.; Rosenberger, A.; Heil, F.; Ginter, M.; Randler, C. Assessing Determinants and Trends in Bird Spec
Identification Skills among Students: A Comparative Analysis with Early 2000s Data. *Int. J. Sci. Educ.* **2024**, 1–25. [CrossRef]

Santos-Fita, D.; Naranjo, E.J.; Rangel-Salazar, J.L. Wildlife Uses and Hunting Patterns in Rural Communities of the Yucatan Peninsula, Mexico. *J. Ethnobiol. Ethnomed.* **2012**, *8*, 38. [CrossRef]
Ramírez-Barajas, P.J.; Calmé, S. Subsistence Hunting and Conservation. In *Biodiversity and Conservation of the Yucatán Peninsula*; Springer International Publishing: Cham, Switzerland, 2015; pp. 333–351. ISBN 9783319065281.
Peterson, M.N.; Chesonis, T.; Stevenson, K.T.; Bondell, H.D. Evaluating Relationships between Hunting and Biodiversity Knowledge among Children: Childhood Hunting and Biodiversity Knowledge. *Wildl. Soc. Bull.* **2017**, *41*, 530–536. [CrossRef]
Schuttler, S.G.; Stevenson, K.; Kays, R.; Dunn, R.R. Children's Attitudes towards Animals Are Similar across Suburban, Exurban, and Rural Areas. *PeerJ* **2019**, *7*, e7328. [CrossRef]
Mascote, C.; Castillo, A.; Peña-Mondragón, J.L. Perceptions and Knowledge of the Jaguar among Children in Communities Neighboring the Montes Azules Biosphere Reserve in Chiapas, Mexico. *Trop. Conserv. Sci.* **2016**, *9*, 194008291667940. [CrossRef]
Wyndham, F.S. The Transmission of Traditional Plant Knowledge in Community Contexts. In *Ethnobiology and Biocultural Diversity*; Stepp, J.R., Wyndham, F.S., Zarger, R.K., Eds.; University of Georgia: Athens, GA, USA, 2002.
Kellert, S.R. Attitudes toward Animals: Age-Related Development among Children. In *Advances in Animal Welfare Science 1984*; Springer: Dordrecht, The Netherlands, 1985; pp. 43–60. ISBN 9789401087131.
Bjerke, T.; Østdahl, T. Animal-Related Attitudes and Activities in an Urban Population. *Anthrozoos* **2004**, *17*, 109–129. [CrossRef]
Galli, F.; De Campos, C.B.; Sarriera, J.C. Comportamiento proambiental en la infancia: Un análisis de niños del sur de Brasil. *Rev. Latinoam. Psicol.* **2014**, *45*, 459. [CrossRef]
Zhang, W.; Goodale, E.; Chen, J. How Contact with Nature Affects Children's Biophilia, Biophobia and Conservation Attitude in China. *Biol. Conserv.* **2014**, *177*, 109–116. [CrossRef]
Turpie, J.K. The Existence Value of Biodiversity in South Africa: How Interest, Experience, Knowledge, Income and Perceived Level of Threat Influence Local Willingness to Pay. *Ecol. Econ.* **2003**, *46*, 199–216. [CrossRef]
Yanez, R.E.; Fees, B.S.; Torquati, J. Preschool Children's Biophilia and Attitudes toward Nature: The Effect of Personal Experiences. *Int. J. Early Child. Environ. Educ.* **2017**, *5*, 57–67.
Olsson, D.; Gericke, N. The Adolescent Dip in Students' Sustainability Consciousness—Implications for Education for Sustainable Development. *J. Environ. Educ.* **2016**, *47*, 35–51. [CrossRef]
Clucas, B.; McHugh, K.; Caro, T. Flagship Species on Covers of US Conservation and Nature Magazines. *Biodivers. Conserv.* **2008**, *17*, 1517–1528. [CrossRef]
Arita, H.; González, G.C. LOS MAMÍFEROS DE MÉXICO: DISTRIBUCIÓN Y ESTADO DE CONSERVACIÓN. *Rev. Mex. Mastozool.* **1997**, *2*, 33. [CrossRef]
Medellin, R.A.; Equihua, C.; Chetkiewicz, C.-L.; Crawshaw, P.G., Jr.; Rabinowitz, A.; Redford, K.H. *El Jaguar en el Nuevo Milenio*; Ediciones Cientficas Universitarias: Fondo de Cultura Economica, Mexico, 2002; ISBN 9789681666170.
Naranjo, E.J.; Amador-Alcalá, S.A.; Falconi-Briones, F.A.; Reyna-Hurtado, R.A. Distribución, Abundancia y Amenazas a Las Poblaciones de Tapir (*Tapirus bairdii*) y Pecarí de Labios Blancos (*Tayassu pecari*) En México. *Therya* **2015**, *6*, 227–249. [CrossRef]
Jalil, N.J.; Sharif, Z.M. Factors Affecting the Awareness of Biodiversity Conservation among Students in Malaysia Private University. *Int. J. Eng. Technol.* **2018**, *7*, 791–795. [CrossRef]
Montgomery, L.N.; Gange, A.C.; Watling, D.; Harvey, D.J. Children's Perception of Biodiversity in Their School Grounds and Its Influence on Their Wellbeing and Resilience. *J. Adventure Educ. Outdoor Learn.* **2024**, *24*, 187–201. [CrossRef]
Myers, O.E.; Saunder, C.D. Animals as Links toward Developing Caring Relationships with the Natural World. In *Children and Nature: Psychological, Sociocultural, and Evolutionary Investigations*; MIT Press: Cambridge, MA, USA, 2002.
Wojciechowski, F.J.; Kaszycka, K.A.; Otadoy, J.B. Utilizing Local Community Knowledge of the Philippine Tarsier in Assessing the Bilar Population Endangerment Risk, and Implications for Conservation. *J. Nat. Conserv.* **2021**, *62*, 126028. [CrossRef]
Liordos, V.; Kontsiotis, V.J.; Georgari, M.; Baltzi, K.; Baltzi, I. Public Acceptance of Management Methods under Different Human-Wildlife Conflict Scenarios. *Sci. Total Environ.* **2017**, *579*, 685–693. [CrossRef]
Czech, B.; Devers, P.K.; Krausman, P.R. The Relationship of Gender to Species Conservation Attitudes. *Wildl. Soc. Bull.* **2001**, *29*, 187–194.
Peng, X.-Y.; Fu, Y.-H.; Zou, X.-Y. Gender Equality and Green Development: A Qualitative Survey. *Innov. Green Dev.* **2024**, *3*, 100089. [CrossRef]
Edwards, D.P.; Socolar, J.B.; Mills, S.C.; Burivalova, Z.; Koh, L.P.; Wilcove, D.S. Conservation of Tropical Forests in the Anthropocene. *Curr. Biol.* **2019**, *29*, R1008–R1020. [CrossRef] [PubMed]
Popradit, A.; Srisatit, T.; Kiratiprayoon, S.; Yoshimura, J.; Ishida, A.; Shiyomi, M.; Murayama, T.; Chantaranothai, P.; Outtaranakorn, S.; Phromma, I. Anthropogenic Effects on a Tropical Forest According to the Distance from Human Settlements. *Sci. Rep.* **2015**, *5*, 14689. [CrossRef] [PubMed]
Young, H.S.; McCauley, D.J.; Galetti, M.; Dirzo, R. Patterns, Causes, and Consequences of Anthropocene Defaunation. *Annu. Rev. Ecol. Evol. Syst.* **2016**, *47*, 333–358. [CrossRef]

85. Pérez-Flores, J.; Hernández-Pérez, E.; Weissenberger, H.; Carballo-Soto, J.; López-Castro, E.M.; Ávila-Flores, F.J. The Effects of Decline in Tourism during the COVID-19 Lockdown on the Wild Mammal Community in Forest Surrounding a Theme Park the Mexican Caribbean. *J. Nat. Conserv.* **2024**, *81*, 126693. [CrossRef]
86. Pérez Flores, J.; Weissenberger, H.; López-Cen, A.; Calmé, S. Environmental Factors Influencing the Occurrence of Unheal Tapirs in the Southern Yucatan Peninsula. *Ecohealth* **2020**, *17*, 359–369. [CrossRef]
87. Borges-Ramírez, M.; Escalona-Segura, G.; Huerta-Lwanga, E.; Iñigo-Elias, E.; Rendón-von Osten, J. Black Vulture (*Coragyps atrat* Feathers as Bioindicators of Exposure to Metals and Metalloids Contamination in Urban, Semi-Urban, and Rural Areas fr Campeche State, Mexico. *Environ. Res.* **2024**, *247*, 118241. [CrossRef]
88. Sánchez-Soto, M.F.; Gaona, O.; Vigueras-Galván, A.L.; Suzán, G.; Falcón, L.I.; Vázquez-Domínguez, E. Prevalence and Transn sion of the Most Relevant Zoonotic and Vector-Borne Pathogens in the Yucatan Peninsula: A Review. *PLoS Negl. Trop. Dis.* **20** *18*, e0012286. [CrossRef]
89. Gavrilakis, C.; Stamouli, E.; Liarakou, G. Primary School Students' Awareness of and Attitudes toward Local Threatened Anim *Hum. Dimens. Wildl.* **2023**, *29*, 159–175. [CrossRef]
90. Gross, M. Hunting Wildlife to Extinction. *Curr. Biol.* **2019**, *29*, R551–R554. [CrossRef]
91. Kik, A.; Duda, P.; Bajzekova, J.; Baro, N.; Opasa, R.; Sosanika, G.; Jorge, L.R.; West, P.; Sam, K.; Zrzavy, J.; et al. Hunting Sk and Ethnobiological Knowledge among the Young, Educated Papua New Guineans: Implications for Conservation. *Glob. E Conserv.* **2023**, *43*, e02435. [CrossRef]
92. Dennison, P.E.; Brewer, S.C.; Arnold, J.D.; Moritz, M.A. Large Wildfire Trends in the Western United States, 1984–2011. *Geop Res. Lett.* **2014**, *41*, 2928–2933. [CrossRef]
93. Albery, G.F.; Turilli, I.; Joseph, M.B.; Foley, J.; Frere, C.H.; Bansal, S. From Flames to Inflammation: How Wildfires Affect Patte of Wildlife Disease. *Fire Ecol.* **2021**, *17*, 23. [CrossRef]
94. Thompson, C.W.; Aspinall, P.; Montarzino, A. The Childhood Factor: Adult Visits to Green Places and the Significance Childhood Experience. *Environ. Behav.* **2008**, *40*, 111–143. [CrossRef]
95. Cheng, J.C.-H.; Monroe, M.C. Connection to Nature: Children's Affective Attitude toward Nature. *Environ. Behav.* **2012**, *44*, 31– [CrossRef]
96. Duron-Ramos, M.F.; Collado, S.; García-Vázquez, F.I.; Bello-Echeverria, M. The Role of Urban/Rural Environments on Mexic Children's Connection to Nature and pro-Environmental Behavior. *Front. Psychol.* **2020**, *11*, 514. [CrossRef]
97. Mulder, M.B.; Schacht, R.; Caro, T.; Schacht, J.; Caro, B. Knowledge and Attitudes of Children of the Rupununi: Implications Conservation in Guyana. *Biol. Conserv.* **2009**, *142*, 879–887. [CrossRef]
98. Martín-López, B.; Montes, C.; Benayas, J. The Non-Economic Motives behind the Willingness to Pay for Biodiversity Conservati *Biol. Conserv.* **2007**, *139*, 67–82. [CrossRef]
99. Castillo-Huitrón, N.M.; Naranjo, E.J.; Santos-Fita, D.; Estrada-Lugo, E. The Importance of Human Emotions for Wild Conservation. *Front. Psychol.* **2020**, *11*, 1277. [CrossRef]

diversity

MDPI

y

ow Ancient Trees and Botanical Indicators Evidence Both ıange and Continuity

D. Rotherham

Advanced Wellbeing Research Centre, Sheffield Hallam University, Sheffield S1 1WB, UK; ianrotherham36@gmail.com

Abstract: Knowledge of ancient trees and tree-forms helps inform understanding of landscape continuity and change. Information analysis of tree-form and growth rates may be combined with precise aging through dendro-chronology and carbon dating. Until recently, much of the information has been neglected, and indeed, there was an absence of accepted methodologies. Assessments of both coppice trees and pollards, for example, now suggest that trees achieve great age (for example, 500 years+ to 1500 years or more). These trees reflect both natural processes and human management. Examined within a robust conceptual and analytical framework, these trees generate insight into landscape evolution over centuries. Tree analysis combined with archival sources and site archaeological information, including mapping of 'botanical indicator species', soils, and other sediments, generate reliable timelines of human–environment interactions. Drawing together diverse approaches and insights into landscape evolution helps the formulation of new concepts of historical ecology and environmental history. Importantly, such emerging paradigms trigger new ways to demonstrate how understanding past landscape evolution both informs knowledge of contemporary ecologies and may guide future site planning. However, there is a rider to these observations since case studies in Great Britain highlight the vulnerability of such historic landscapes and show how they are being rapidly erased from the countryside.

Keywords: ancient trees; ancient woods; botanical indicator species; landscape evolution; landscape change; continuity; land-use planning

1. Introduction

Research on coppice trees and pollards indicates their potential to achieve considerable age, in some cases from 500 years+ to in excess of 1500 years [1–6]. Such aged specimens result from a mix of natural processes and human, often traditional, management. If these trees are considered through a suitably robust conceptual and analytical framework, they may provide information on landscape evolution over many centuries. Case studies drawn from locations in Great Britain and subject to long-term action research and assessment have been presented in order to demonstrate the vulnerability of these historic landscapes. The studies indicate that these unique resources are being erased from the countryside, and the rate of attrition has increased dramatically in recent years [7–12].

Ancient woodlands can be evidenced through research on trees and on treescapes (such as ancient woods, wood-pastures, and wooded commons) and on botanical indicator species associated with them. Such indicator species distributions can be mapped and analysed in terms of their rates of recolonisation into new or disturbed sites together with their ability to persist or not, in the event of significant change. Indicators help inform understanding of both continuity and antiquity, especially when combined with other site-related evidence [13].

Particularly in Great Britain, but also in North America, there has been considerable interest in the concepts of 'ancient' woods and 'old-growth' forest, e.g., [14–23]. Attempts are then made to relate these to contemporary management, to conservation, and to local and national planning issues though the Ancient Woodland Inventory [23]. The baseline for such approaches emerged from concepts of treed landscapes that emerged in Britain in the 1970s led by discussions between Cambridge University scholar Oliver Rackham and British government agency woodland advisor George Peterken (Peterken pers. comm. and [24]). This early dialogue was a first attempt to rationalise 'woodlands' and 'woods' in the historic landscape context. Additionally, site-based studies evidenced these landscapes with discernible long timelines were 'ancient' (defined as either by Rackham as pre-1600 AD or by Peterken as pre-1700 AD) but not 'primeval' wildwood, e.g., [25,26]. These were eco-cultural sites of anciently treed countryside but with long histories of human utilisation and impacts. Such observations challenged the wider, popular misconceptions of ancient woods being *de facto* 'wildwood' [27,28].

However, even as wider awareness of the importance of ancient woodlands was recognised, the misunderstanding of their origins has remained. Many approaches to woodland management are fundamentally flawed by a lack of understanding of their historic context [29–31].

With these issues in mind, research in Britain and also across Europe addressed the issues in order to develop a robust interrogation of forest and woodland histories and dynamics to provide information for managers and policy-makers [32]. This paper presents an overview of the assertions and a model for historical and ecological insight into more effectively framed woodland antiquity and ecological continuity for future management. The policy context of the discussion is the importance placed on the designation of 'ancient woodland' in, for example, British land-use planning, e.g., [33,34]. This takes the discussion of woodland histories and origins beyond an academic exercise to a matter of statutory recognition, protocols, and protection. Finally, though, it is asserted that woodland heritage and archaeology are widely neglected and contemporary management, even by conservation organisations, is rapidly eroding the resource.

1.1. *What Is Ancient Woodland?*

A fundamental concept relating to nature conservation and land-use planning, especially in Great Britain, but elsewhere in Europe too, is that of 'ancient woodland'. However, this designation presents many challenges when applied in practice, and definitions vary between countries and are affected also, by biogeographical influences. Furthermore, an understanding of the landscape history context is essential in evaluating ancient status.

'Ancient Woodland' in England is defined as an area that has been wooded continuously since at least 1600 AD. Some definitions give 1700 AD as the preferred date, a choice explained by Peterken [35], and more recently [24], and by Goldberg et al. [36]. The definitions are essential but also problematic. Across Great Britain, for example, there are different cut-offs in dates between England and Wales (1600 AD), Scotland (1750 AD), and Northern Ireland (1830 AD) [37]. This raises issues of both science and history since a woodland might be deemed ancient in Gretna Green (Southwest Scotland) but not so in nearby Carlisle (Northwest England) under a system based on the production of the first reliable maps, the subjective understanding of countryside history, and an assumption for England and Wales that there was little if any tree planning for woodland creation before the date of 1600 AD.

The 2023 report on the Ancient Woodland Inventory in Scotland expresses very clearly what it is, and how it should be applied with caution. The below is modified from the entry in the report by Hall [38]. The Ancient Woodland Inventory (AWI) is a provisional guide to the location of ancient woodland. It contains three main categories of woodland, all of which are likely to be of value for their biodiversity and cultural value by virtue of their antiquity:

i. Ancient woodland (1a and 2a)

Interpreted as semi-natural woodland from maps of 1750 (1a) or 1860 (2a) and continuously wooded to the present day. If planted with non-native species during the twentieth century they are referred to as plantations on ancient woodland sites (PAWS).

ii. Long-established woodlands of plantation origin (LEPO) (1b and 2b)

Interpreted as plantations from maps of 1750 (1b) or 1860 (2b) and continuously wooded since. Many of these sites have developed semi-natural characteristics, especially the oldest ones, which may be as rich as ancient woodland.

iii. Other woodlands on 'Roy' woodland sites (3)

Shown as unwooded on the first-edition maps but as woodland on the Roy maps. Such sites have, at most, had only a short break in continuity of woodland cover and may still retain features of ancient woodland.

iv. A note of caution

The AWI was derived from the Roy maps (*c.*1750) and the Ordnance Survey first edition (c. 1860). It is not definitive and should be used with care. When evaluating woods, it is important to conduct the following:

(a) Examine the site on the ground, looking for archaeological, biological, and other indicators of antiquity and of its current biodiversity value. (b) Examine old maps. The Ordnance Survey first edition and Roy maps are available online. Woods not shown in the AWI, but present on the historic maps, are likely to be ancient and should be treated as such unless evidence is available to the contrary. (c) Seek specialist advice if in doubt.

The idea of 'continuously wooded' should not imply continuous physical cover of trees and shrubs as open land (both temporary and permanent) is an important component of a woodland landscape and a core element of a wood-pasture or wood-meadow. Furthermore, in most and maybe all ancient woodlands, trees and shrubs have periodically been cut down as part of anthropogenic management. In England (Natural England, 2008), guidance states that an area is still considered ancient if it has been maintained by replanting or regrowth from, for instance, coppicing. In the latter case, if a working wood has been cut-over many times in its history, it will have very old (and frequently unrecognised) coppice stools but often few old standard trees [16–18]. Agency guidance in Britain advocates that woodland experiencing long periods over the last four centuries when the land was in effective open (grassland, heath, moor, or arable) is defined as 'recent' woodland. Natural England asserts that such a wooded site may retain high value for nature conservation, but it is not ancient. Reinterpretation from an ecological historical perspective of site-based studies and core criteria provides more nuanced interpretations of the landscape history [39]. Evidence of former human utilisation is also significant (Figures 1 and 2).

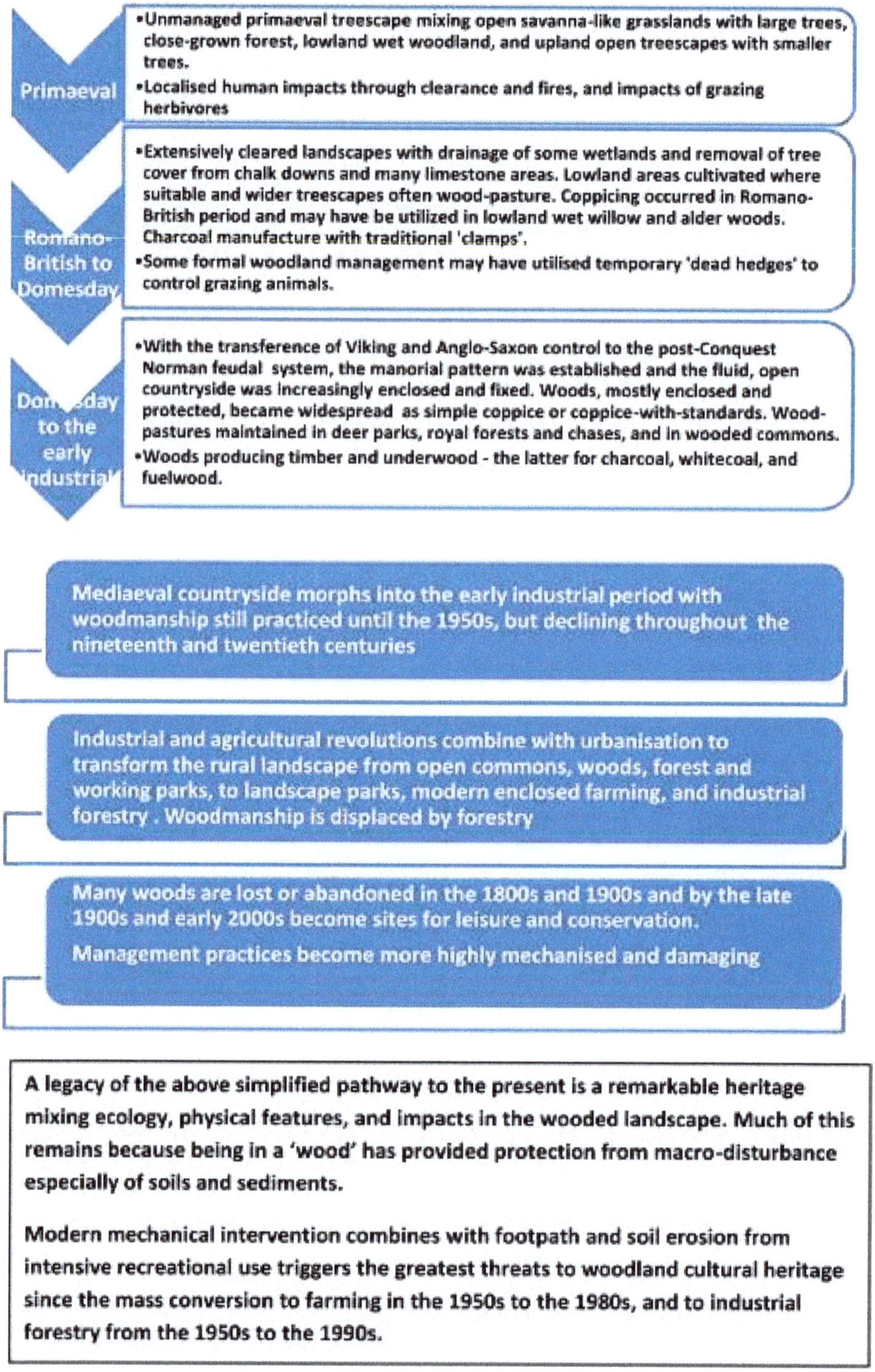

Figure 1. Overview timeline of typical English woodland history.

Figure 2. Wood-collier's monument in Ecclesall Woods, Sheffield.

1.2. The Value of Ancient Woodlands

In nature conservation terms, ancient woodlands are of very high ecological and landscape importance, though their remarkable heritage values are frequently overlooked [40–42]. The Natural England 2008 standing advice noted how relatively undisturbed woodland sites often have heritage features of historical, archaeological, and landscape importance [43]. In terms of biodiversity, ancient woods have many associated species which are uncommon and often threatened. The 2008 guidance stresses the following as a basis for site conservation:

(1) Woodlands are exceptionally rich in biodiversity including rare species and habitats.
(2) They provide renewable resources such as hardwood timber and other woodland products [but note later concerns].
(3) Woods may include survivals from original natural 'forests', but this requires caution in its interpretation. Countryside considered either 'forest' (in Britain is a medieval hunting area and not necessarily treed or indeed, 'natural' [44]).
(4) Remaining or relict sites provide reservoirs of biodiversity so wildlife species may recolonise into new woodlands.
(5) Woodlands form integral parts of the historic landscapes.
(6) In the often-protected boundaries of ancient woodlands, the countryside may hold historic features little altered by modern cultivation, or other macro-disturbances such as urbanisation.
(7) Woods help provide a sense of place and imagination for a specific location.
(8) They have importance as features in the landscape.

It is widely accepted that woodlands of various types, together with their trees, provide some of the most important wildlife habitats in Britain and Europe. For nature conservation, the site condition has great influence, so that mature, mixed woodland having ancient and veteran trees, dead branches and timber both standing and fallen, together with open rides, clearings, and glades scattered throughout, is especially valuable. Topographic variation also influences diverse ground vegetation with distinctive field layers of shrubs and no individual component dominant. Such habitat combinations within a woodland macro-habitat generates the widest variety of conditions for woodland wildlife.

1.3. Ancient Woodland Indicators

The ability to reliably identify ancient woodland sites is increasingly important. Land is under pressure from urban development and from agricultural intensification. In Britain, these pressures and the financial imperatives for developers and farmers to encroach onto woodland sites lead to divergent attitudes to definition and the underpinning evidence-base. Government agencies, local authorities, community action groups, and other conservation bodies try to identify and protect woodlands. This action is in part by their designation as ancient sites which in Britain involves being added to the government's 'Ancient Woodland Inventory', which affords a degree of protection through specific planning guidance. The counter approach is by developers who pay ecological consultants in order to undermine such designation and thereby open up land for financially lucrative usage such as for housing. This conflict of interests presents problems since amongst other issues, neither the Ancient Woodland Inventory nor the concept of 'ancient woodland' were designed to be tested in the quasi-legal arena of, say, a public inquiry. Their origins were as an intellectual inquiry into the nature of woodlands. Recent guidance and a national review of ancient woodlands has begun to address some of the issues, e.g., [23,24,45]. However, the human footprint in ancient woodland is still generally overlooked (Figures 3 and 4). A major problem with the original designations was that they ignored all sites of less than an area of two hectares and thus excluded many smaller sites and those along topographic or historic linear features like streams, rivers, and, for example, medieval parish boundaries. Updates to the inventory seek to include these within the designations, but the need for multi-disciplinary research remains challenging.

Figure 3. Historic usage—wood-collier's wigwam hut, Parkwood Springs, Sheffield, late 1800s.

Figure 4. Historic usage—English Lake District charcoal burners in the early 1900s.

Identification of ancient woodland in Britain involves map-based reviews along with field evidence such as botanical indicator species. The latter are usually vascular plants more commonly found in ancient woods than in recent sites. The supporting evidence for the interpretation of indicators is from regional site species lists as discussed by [46]. Triggered by the pioneering work of Peterken [24] and Rackham [14], to identify ancient woods from map-based and field evidence the Nature Conservancy Council (a predecessor of the UK government English agency, Natural England) began the process of producing provisional county lists of ancient woodland sites [47]. Following the gathering of data and interrogation of species lists, the presence of suites of indicator species is taken as evidence for ancient status. Interrogation with local and regional context is essential and indicators vary in status biogeographically. Indeed, no single plant species is perfect as an ancient woodland indicator; the degree of species association with ancient woodland varies across landscapes with geology, topography, climate, and land-use historical factors. These considerations led to researchers assessing more critically such use of indicator lists, e.g., [36] and conducting this in relation to county-based site lists of the Ancient Woodland Inventories with assumed continuous tree cover since 1600 AD or earlier. These inventories included ancient, semi-natural woods along with plantations on ancient woodland sites (PAWS) [36]. The county-level reports are available online with digital maps showing ancient woodland boundaries.

This review-based paper considers the identification and designation of ancient woodland and asks whether current approaches are fit for purpose in the broadest sense.

2. Materials and Methods

The paper considers the following:

(a) Methodological issues including approaches to survey, information gathering, and analysis [the studies have involved field survey, archival research, citizen science, GIS manipulation, and LiDAR analysis].

(b) Review of the selected literature [on indicators, colonisation rates, issues of tree forms and aging].

(c) Emerging concepts [shadow woods (suggested to be remnant, unenclosed, medieval wooded commons [39]) and indicators.

(d) Synthesis of ongoing research.

The long-term research involved stakeholder workshops, interviews, and questionnaires, site-based field research, in-depth case studies, and extensive literature review [38,44,48,49].

The approaches to site identification and designation discussed above are then examined in relation to the case study and action research examples.

3. Results

3.1. *Evidence for Ancient Woodlands Using Botanical Indicators*

The variety of plants and trees in woodland is strongly influenced by soil type and underlying geology, wetness, and land-use history [13,16]. However, the patterns of sunlight reaching the woodland floor are also important along with influencing factors such as aspect, dominant tree species in the canopy, natural treefall, and woodland management [15]. So, for example, heavily shaded sites such as conifer monocultures and beech-dominated woodland are not conducive to diverse ground floras. In turn, without diverse flowering plants, invertebrate variety is much reduced and consequently, populations of small mammals and of birds will be restricted too. Trees such as beech especially, but also sycamore, come into leaf early in spring and cast a very dense shade [50,51]. Consequently, the ground layer under such a canopy may be totally eradicated and the soil exposed to massive erosion [52–56].

Mixed composition and the age of woodland trees and shrubs are generally accepted as necessary for high nature conservation value in most sites along with structural diversity with clearings and glades formed by treefall or loss of major limbs from large, mature trees, e.g., [17]. Such openings in the woodland canopy with dynamic successional processes are very significant in maintaining diversity. Traditional woodland management (coppicing and timber extraction), implemented systematically over long periods, result in variety and cycles of light conditions [54,55]. Compartments which have been recently coppiced are open to the penetration of sunlight to the woodland floor, whereas later in the cycle with regrown springwood, conditions are much darker. Management history therefore impacts the presence and apparency of indicator plants in woodland and affects invertebrates and other biodiversity both indirectly through vegetation and directly via the microclimate (Figures 5 and 6). Along with seasonality, these factors must be considered if assessing woodland survey data.

Figure 5. Example of a typical ancient woodland botanical indicator—bluebell (*Hyacinthoides non-scripta* (L.) Chouard ex Rothm.)—in the Moss Valley Woodlands, Sheffield.

Figure 6. Bluebell and stitchwort (*Stellaria holostea* L.) in ancient woodland ground vegetation the Moss Valley Woodlands, Sheffield.

3.2. Traditional Management Impacts

Traditional coppice management involves cutting and removing 'wood' from the woodland. The intensity of such historic activity should not be underestimated since it included 'tidying up' coppice stools and having a 'clean' housekeeping approach in the working wood. Dead wood such as fallen limbs from trees and 'ramel' left after extraction of timber or underwood were cleared away and used. This clearance of dead wood adversely affected associated saproxylic invertebrates and fungi through the removal of habitats. However, in the medieval wood for instance, the habitat was extensive and there was still abundant dead wood in the tree and shrub canopy, though this may have been reduced in the early industrial coppice woods such as in the study areas around Sheffield, in England. Furthermore, the stage of dead wood decay also influences invertebrate associations, with some species in recent dead wood and others in well-rotted material. Whether wood is standing or fallen to the forest floor is also significant. Other factors include ground flora flowering plants, along with flowering trees and shrubs, being important nectar and pollen sources for the flighted adult stages of deadwood insects. This means that mature hawthorn and wild fruit trees like crab apple and wild pear, for example, in the woodland edge, along glades, or rides, are important food sources for adult deadwood beetles, craneflies, and hoverflies whose larvae feed on the ancient trees.

Specific activities had varying types and scales of impact on soils and woodland plants. So, for example, the production of traditional and industrial products like charcoal and 'white-coal' (kiln-dried wood used in metal smelting) are activities that involved stripping vegetation and soil from the woodland floor. These were used to cover charcoal clamps and white-coal kilns and controlled oxygen penetration to the burn. Such usage occurred in some cases over many centuries with densities of charcoal hearths recorded in the Sheffield region up to 350 in around a hundred hectares of managed woodland which also held around 150 white-coal kilns. These intensities of usage resulted in managed woods from the early industrial period today having very limited flowering plant diversity and soils depauperate in upper horizons. Many of these woodland soils were replaced almost entirely by shallow layers of pure charcoal dust seated on a clay C-horizon [12,27].

3.3. *Woodland Types*

The various influences on plant species in ancient woodlands interact rather intimately. With this in mind, in understanding indicator plants and their interpretation, it is essential to recognise the diverse woodland types and their origins. These are broadly, as follows:

(a) Medieval and industrial coppice.
(b) Park and wood-pastures including ancient forest, wooded common, and 'shadow woods' [39] (Figure 7).
(c) Linear remnants and fragments [23,57].

Figure 7. Wood-pasture veteran oak (*Quercus robur* L.) at Chatsworth Park, Derbyshire, England.

Generally speaking, ancient coppice woods were, from the medieval period, enclosed and protected from intensive stock grazing and tended to be botanically species-rich but often lacking in deadwood habitat and associated biodiversity.

Wood-pastures have often had histories of open grazing with the consequent reduction in ground flora diversity, but more abundant deadwood habitat associated with ancient pollard and maiden trees (features generally lacking in ancient coppice woods) [56,57]. Wood-pasture sites tend to be biodiversity hotspots for rare lichens, fungi, and invertebrates [58,59].

Woodland historical status may be assessed by means of botanical indicators [60–62], whereas woodland vegetation is often subject to more rigorous but less historically informed approaches [63]. The use of botanical indicators relates particularly to coppice woods because of their botanical richness [30,33,34,64–72] whereas wood-pastures are better evaluated with invertebrates, lichens, and fungi. Ongoing research over a forty-year period with long-term action research involving expert stakeholders, together with detailed regional audits has allowed a review of evidence [29,46,49]. The two-year project on the *Woodland Heritage Manual* [40] evolved into a series of expert stakeholder workshops examining and reviewing woodland indicators, woodland inventories, and associated landscape issues. The research produced major reviews of the literature, of evidence-bases, practitioner approaches [46,48,49], and a toolkit [29]. The study generated a review of the use and interpretation of botanical indicator species and lists, and the promotion of holistic evidence-based evaluations.

3.4. Ancient Woodland Inventories

In England, Ancient Woodland Inventories (lists of known ancient woods as determined by the criteria described earlier) are held by Natural England, each covering a specific county. This provides a first step in identifying whether woodland is already recognised and listed as ancient. The inventories were originally based on work conducted in the 1980s and 1990s by the Nature Conservancy Council (now in England called Natural England) producing inventories for England, Scotland, and Wales. The documents are now held and administered by the relevant statutory conservation agency in each British devolved country, and also in their county or regional offices. Ancient Woodland Inventories are essentially provisional with the maps and sources interrogated varying regionally. In England and Wales, the main cartographic sources used were 1:25,000 maps from the 1920s and 1930s, the first edition one-inch Ordnance Survey maps (generally mid-1800s), and aerial photographs, along with any available survey reports. Recent reviews have been aided by use of LiDAR imagery and GIS computer mapping. The original surveys and reviews generally did not consider wood-pastures and parkland. Evaluation of individual woods may not always be clear from the regional inventory summaries [33,38,40].

Most inventories were based on research during the late 1970s and early 1980s, with ground-truthing where possible. The indicators used were generally intuitive lists produced by known local experts and target sites were identified from first edition Ordnance Survey maps (dating around 1830 to 1840). Earlier documentation, such as estate plans and surveys, was used if available. These might support the initial assessment and then likely sites had ecological field surveys during the subsequent period of ten to fifteen years. Depending on surveyors and their experience, the results varied in quality. The landmark 2008 *Woodland Heritage Manual* [40] was the first time sites had wider evaluations of factors like soils, woodland archaeology and heritage, and historic context. The development of the Ancient Woodland Inventories was a first step and proved useful over the following period of thirty to forty years. However, it was not intended as a statutory tool and proved to be not 'fit for purpose' in that context. As the methodology was neither comprehensive nor robust, it was challenged by developers and their consultants. To complicate things further, the initial schedules were generally 'conservative' and omitted sites such as smaller woods. Identification of the limitations of the original process was a key driver of the methodological reviews to develop evidence-based assessments [31].

Glaves et al. [46,48,49] concluded that historic sources and archaeological evidence providing broader context were often overlooked. Consultation with users noted that even approaches like evaluation of map-based, place names, for example, were not widely used. Other issues identified were the lack of awareness by surveyors of the likely age of multi-stemmed coppices and clonal trees. These are often present in ancient coppice woods, but surveyors were frequently looking for large, old-looking, standard trees, as are found in wood-pastures but much less commonly in woods. Since inventories largely record former medieval coppice woods, which because of their management history often lack veteran standard trees, this is a serious omission. There is a further complication of there being a lack of accepted, robust approaches to the aging of old coppice stools. Additionally, many conservation organisations remain reluctant to recognise such trees as ancient or veteran, and yet for evidencing ancient woods for formal scheduling and protection from development, presence of verifiable, ancient, coppice stools is confirmation of woodland going back centuries. Moreover, such trees evidence both site antiquity and past management. Trees which are naturally self-coppicing or clonal (e.g., *Ilex aquifolium* L., *Alnus glutinosa* (L.) Gaertn., *Tilia cordata* Mill.) can provide additional timeline information, in some cases for well over a thousand years.

3.5. Robust Methodologies

Interest in the effective evaluation of woodland histories necessitated robust, evidence-based methods to assess and define ancient woodland, and an integrated approach was developed [29]. This involves logical steps through a decision-making tree for the evidence-based process. Evolved from *The Woodland Heritage Manual* [40], the approach has been tested successfully in planning inquiries and other conflict situations. Carefully considered steps and decisions address and evaluate information such as site documentation, maps, place names, ecology, pedology, archaeology, and history. Developed from established ancient woodland definitions, this uses a broader information base than formerly available to provide a logical framework for woodland assessment with evidence combined to confirm woodland continuity and status. Evidence varies in the level of certainty and robustness but combined it may support, confirm, or indicate woodland continuity or not [72–76]. Using mixed sources combined lends more confidence to any interpretation. Many texts on woodlands and forests are researched and written from either an ecological or an historical perspective and relatively few cross the divider to employ mixed sources. This has a further consequence in that quite frequently the insights and knowledge are then not picked up by researchers in other disciplines [77–82]. Joined-up approaches to forest and woodland management begin to join otherwise disparate research and perceptions into a more unified whole [83–86]. The emerging discipline of 'historical ecology' also helps to cross the divide [87–89]. Confidence in designating a site as ancient woodland depends on two factors:

(1) Evidence of woodland continuity or of gaps in woodland cover.
(2) Reliability evidence for continuity or gap.

Most woods will only have some evidence types available, and confirmation of ancient status may rest on subjective judgment based on objectively gathered information.

3.6. Intelligent Interrogation

The evaluation concludes with 'intelligent interrogation' of lists, sources such as maps and archives, indicators, field survey data, and any other relevant information. If required, a formulaic outcome is achieved by assigning subjective numerical weightings to particular factors in the analysis. It may be helpful to generate a numeric index or gradation based on scrutiny and assessment of evidence such as lists of botanical indicators or observations on ancient coppice stools. Evidence is assessed and evaluated in terms of site history, map information of land-use, archaeological and heritage information relating to human activity, with documentation, archives and other historical sources, and both vegetation communities and botanical indicators.

Overall, there is a wide diversity of information and sources to assess. Site evaluation may include soils and other sediments and the evidence they hold which may include paleo materials such as pollen, pollution profiles, and sediment downwash giving information on tree species and on other flowering plants and land-use history. Working and worked trees can be very informative, and site location, biogeography, and landscape context are significant influences. All these factors are taken into account in the final assessment. However, in many of the earlier site evaluations, the trees themselves were often not assessed in detail, and this was in part because of a lack of recognition but compounded by the absence of accepted methodologies.

3.7. *The Importance of Ancient and Veteran Trees in Woodland*

The presence of ancient and veteran trees in woodland adds a highly valuable habitat component for many uncommon and rare species, particularly invertebrates. However, and frequently overlooked, is that the ancient trees can, if effectively recognised and assessed, provide key confirmation of 'ancient' status by proving a woodland has existed on the site beyond the date accepted for ancient woodland. In this context, the definition becomes important and the terms 'ancient' and 'veteran' are accepted as below [40].

Ancient: This term relates to a tree that is very old, in the declining stages of life for that species, and usually with relatively large girth compared with other trees of the same species. This does depend on how it has grown and factors like altitude, aspects, soil, and geographic location which affects climate. Ancient trees are often visually distinctive, but recognition depends on the experience of the surveyor. One category of ancient or veteran trees is that of worked trees (i.e., those once managed to produce particular crops and products), which can be hugely informative but also sometimes difficult to recognise or age. These may include coppices, clones, stubs, shreds and most easily recognised, pollards. Ancient specimens of smaller tree species are also problematic and often missed in surveys.

Veteran: Varying from species to species, a veteran tree is one usually in a mature phase of its lifecycle. This stage will generally be associated with the development of important wildlife and habitat features such as hollowing, holes, wounds, and large dead branches and decay fungi. The specimens will generally be old trees, but perhaps due to environmental stress premature aging characteristics may develop in younger, middle-aged trees. This varies from species to species and with levels of environmental stress experienced. Stress might be factors such as drought, waterlogging, low levels of available nutrients, extremes of soil pH, and intensive grazing, etc.

Ancient and veteran trees are important in terms of their presence or absence, but also, if they are present, in their growth form. This may be a vital piece of information in understanding past woodland management. Furthermore, what is defined as woodland and then as ancient is also very significant. This is often more complex than is generally assumed. As noted earlier, most sites recognised by means of botanical indicators and historical sources are medieval coppice woods, and whilst current reviews across England are more inclusive, many sites atypical, historical, timelines are omitted from ancient woodland designations. The consequences for conservation and also for site management are serious.

Two main important considerations emerge from the current research. Firstly, there is a need for widely accepted, robust indicators which need to be considered within their geographical, historical, and environmental contexts. Secondly, such indicators must be integrated into a broader framework for integration and interrogation [29].

4. Discussion

Emerging Concepts About Ancient Woodlands

With the exceptions of obviously pollarded trees, 'worked' trees, 'modified' trees, and naturally contorted trees (such as by animal grazing and/or extreme weather) have often been overlooked in woodland assessments. Except for the pioneering work of Donald Pigott [1], historically coppiced and self-coppicing trees such as ancient limes, alders, and willows (*Salix* spp.) have generally been neglected. There are ongoing studies on some species like sweet chestnut (*Castanea sativa* Mill.) and coppiced oak, e.g., [87]. There is also a tendency to overlook smaller species and to neglect peripheral habits such as '*Shadow Woods*' [39]. This is especially the case in upland areas, but ancient coppiced trees are widely ignored in the lowlands too. Whilst understanding and aging these specimens may be challenging, they provide intimate connections to human exploitation of the countryside and sometimes to extreme weather events. In these respects, such trees have important stories to tell. Many are older than 'standard' trees, more generally recognised as significant, and therefore provide important evidence of woodland being ancient and moreover of its management history.

As noted earlier, there are distinct categories of woodland with coppice woods being treed sites enclosed from the wider countryside, named and managed over subsequent centuries. Trees were protected from large grazing herbivores at least during the early years of the coppice cycle or after major tree-felling to allow regeneration of the wood. The other broad woodland type is wood-pasture of varying sorts and treed landscapes mixing tree cover with open ground grazed by large herbivores, both wild and domesticated. The most widely recognised forms of wood-pastures derive largely from medieval land-use and are parks, chases, forests, and wooded commons. However, there are complications with changing land-use and human impacts over long periods. These have led to the emerging concept in recent times of 'shadow woods' [39] (Figure 8).

Figure 8. Bluebell as an indicator in ancient shadow wood, Peak District, England.

The concept of 'shadow woods' as 'lost Domesday landscapes' [39] is one of the most exciting outcomes from ongoing research. It is argued that these are remnants of once extensive, Domesday wood-pastures which survived unenclosed into the medieval period as wooded commons. Furthermore, it is suggested that the enclosed woods managed for centuries as coppice woods, originating from landscapes like these, and having survived episodes of medieval, Tudor, and parliamentary enclosures, some areas of former wood-pasture common remain. Research into this phenomenon has included observations nationwide and detailed site studies in the England Peak District and Pennines. Interestingly, some Peak District moors, for example, are shown on nineteenth-century maps as 'wood-pastures', and paleo evidence indicates greater tree cover at that time [88]. Relict sites were identified through detailed field surveys of botanical indicators, smaller veteran trees, soils, and other evidence. Not being managed 'woods', however, these sites often have limited, associated documentary evidence.

Other complicating factors in woodland assessment include changes between wood-pasture and enclosed coppice, and sometimes back again over time. Other sites are effectively lost as woods because of land-use changes. Enclosed woods often suffered damage, loss or destruction over the centuries following their establishment, and so, 'lost woods' are sites that were enclosed and named as woods, but where the tree cover was subsequently removed. This might be by conversion to farmland, to urban development, or by infrastructure such as roads. In some cases, the sites were simply opened up and reverted to a grazed landscape. Some ancient woods, still exist but for various reasons have been 'lost' as woods from memory and even from maps. An example is a Peak District wood now shrouded by secondary birch growth, but with over 1000 veteran coppice trees overlooked by contemporary ecological surveyors mapping ancient trees and woods [39,89].

When woods are lost, destruction may be total with tree cover and vegetation completely removed. However, in many cases the 'ghost' of a lost wood can still be seen in the landscape where physical anthropogenic features like woodbanks, walls, charcoal hearths, lanes, etc., survive. However, in many cases, veteran trees and ancient woodland botanical indicators mark the existence former woodland now etched in the modern landscape. Even in intensively farmed or highly urbanised areas, indicators and other evidence survive alongside field names, lane names, and place names connected to one-time woodland.

For the range of woodland types in the landscape, a big challenge is being able to recognise and understand them. Firstly, we must see them in the landscape and searching for lost woodlands, hidden ecologies, and forgotten histories requires patience and persistence. Yet once the surveyor is able to see the evidence and read the signs, then new insights into the countryside's past and present emerge. They are there all along but often they remain unseen.

Key questions considered in this paper relate to the development of unenclosed wood-pastures, the origins of enclosed 'ancient coppice woods', and the nature of the countryside within which such enclosure occurred. Management of enclosed woods by formal coppice was known to the Romans and they undoubtedly applied this to British woodlands especially in the southeast of England where the Weald was a significant charcoal manufacturing area. Charcoal is essential for, amongst other things, metal smelting and working, and so a sustainable supply of cut poles for coaling was important [90]. Furthermore, examination of the archaeological evidence of fenland prehistoric trackways, for example, shows large amounts of 'brushwood' being used which suggests ample supplies of small-bore underwood. However, simply cutting back of willow and hazel might be the source rather than formally coppiced, enclosed woods.

It seems that most English coppice woods originated after Domesday (1086 AD) though some might have originated in the Saxon countryside [39,91]. This further challenges the persistent myth that today's ancient woods are direct descendants of the so-called 'wildwood', which they are not. Indeed, one of the reasons that ancient woods are so unique and valuable in conservation terms is that they hold evidence of human occupation and land-use sometimes going back millennia, and these bear evidence of non-woodland phases. Probably the closest to wildwood or primeval countryside lies in some of the most ancient of the medieval forests or deer parks, and perhaps in some of the shadow wood fragments of wooded commons. These were extensive tracts of wood-pasture with scattered trees and areas of scrubland, with probably large patches of closed-canopy forest, and of course, the whole landscape was much wetter than today. Due to the latter observation, there would be a far greater extent of wet woodlands dominated by willow and alder from the lowland fens to the upland moors, which were wood-pastures with scattered smaller trees and willow in wet fen areas. It was from this extensive landscape of large wood-pastures that coppice woods were enclosed and protected. A driver for this practice was the fixed manorial system imposed by the Norman feudal overlords and rising human populations. Since each manor or parish was largely self-reliant in terms of resources such as underwood and timber, as wood-pasture was cleared and cultivated for arable production to feed local people, the woodland was reduced to near critical levels. Protection from incursions by local peasantry and their grazing animals allowed a system of renewable timber and underwood supply called 'coppice-with-standards', and the 'woods' enclosed by a bank and ditch, a stone wall, a hedge or a dead hedge was then named in the landscape. The evidence today is in names like 'The Manor Wood', 'The Lady's Wood', or 'The Lord's Wood', for example [92,93].

Once enclosed, the 'woods' were managed over centuries as simple coppice or coppice-with-standards, and those surviving today are the 'ancient woodlands'. In this context, the botanical indicators of ancient woodlands are plants that provide robust evidence of antiquity and continuity back to these medieval landscapes. The Domesday survey (1086 AD) shows that these woods were enclosed from extensive wood-pasture in the pre-Domesday Saxon countryside. The areas which remained unenclosed were mostly a mix of royal forest and wooded commons [39,89,90].

From extensive literature review, stakeholder workshops, and from a questionnaire survey of experts and practitioners across England [46,48,49], the application of botanical indicators of ancient woodland was shown to be universal in the country. In particular, the indicators help identify potential locations of ancient woodland and confirm the ancient woodland status of former medieval and industrial coppices.

Issues remain in terms of interpretating what such status implies and how this relates to the ideas of the nature of the primeval landscapes proposed by Frans Vera [91]. Clearly, a value of anciently wooded landscapes relates to our emotional attachments to 'ancient' woods, as fragments of a perceived primeval 'wildwood'. Nevertheless, as already noted, such direct lineage is largely misconceived. However, it may be that the woodland-associated indicators (including plants and invertebrates) might provide connections to former landscapes and ecologies. Field research in landscapes that have not suffered major disruption such as deep ploughing, chemical fertilisation, or urbanisation suggest the presence of presently unrecognised and unrecorded 'ghosts' or 'shadow woods'. These are marked out by ground flora, veteran trees, and typical woodland soils. Often unnamed (unlike typical ancient woods), these are probably ancient, wooded commons and forestlands, and they remained unenclosed during major phases of land enclosure in early medieval or early industrial times [2,8,94,95].

A key process in transforming the pre-Domesday wood-pastures was the piecemeal enclosure of land in the early medieval period, something recognised in the 1235 Act of Commons or Statute of Merton. This legislation passed by King Henry III probably reflected what was already happening rather than itself being a driver of change. It was this Act, however, that empowered the Lord of the Manor to enclose and protect the various constituents of the manorial lands. So, the still fluid and generally unimproved English countryside of the Saxon era was replaced by enclosed 'woods', common fields, heaths and commons, parks, fens, bogs, forest and other features. At this time, these were now fixed in place, named, and accounted for [39].

Therefore, in England at this time, woods were recorded, enclosed, named, and protected, but other treed landscapes remaining in the wider countryside were subject to the vagaries of agricultural and then urban encroachment. This situation gave rise to the two major resources of ancient, wooded landscapes which remain in Britain today and as described earlier. These are the relatively widespread coppices of various types descended from sites enclosed and named post-Domesday, and the medieval parks and forests. However, there is growing evidence of shadow woods which are unrecognised and unenclosed and still survive in commons, heaths, and other unimproved countryside, and as fragments elsewhere in places like roadsides.

A question often raised is that of what should be carried out to conserve veteran or ancient 'worked' trees, now 'retired'. These specimens include ancient pollards (resulting entirely form historic human management) and coppices of varying ages, in particular. Pollards are problematic because if left uncut, they eventually become top-heavy and tend to break up. Furthermore, if trees are abandoned from traditional management and then re-cut, there is a high failure rate and many trees die. Research through the auspices of the Ancient Tree Forum, the Tree Council, and partners has improved practice and increased success rates [94,96]. However, with abandoned coppices, the issues are somewhat different and the use of big machines on steeply sloping ancient woods, for example to 're-coppice' sites [95], is very damaging to site archaeology and heritage. Like pollards, if abandonment has been for some decades, it is likely that re-coppicing may result in death. Additionally, if the coppicing last took place say, two hundred years ago, the regrown trees are both vulnerable to shock in re-cut but have also become an important part of the historic landscape and the tangible heritage. Some trees such as alder and small-leaved lime will naturally self-coppice and regrow if a stem collapses, and the suggestion is to allow nature to take its course. More recently, abandoned coppice can be reworked, but this should be by hand and not by vehicle-mounted machines, and older trees should be left. Two global authorities, the late Professor Donald Pigott and the late Professor Oliver Rackham, both commented on this issue to me when they implored me to try to dissuade agencies and conservation bodies from intervening with what in their minds were historically significant trees.

5. Conclusions

In terms of their significance, the ancient woods of Britain are hugely important for nature conservation and heritage reasons. However, perhaps the special importance of an ancient wood is the feeling of walking in the footsteps of the ghosts of people that lived and worked our woods over thousands of years. These 'ghosts' have left their mark on the wooded landscape, on the soils, and even on the vegetation itself, and these landscapes are strongly eco-cultural rather than being some sort of wildwood. The human element in ancient woods may be evidenced back over thousands of years with field boundaries, settlement sites and enclosures, carved stones, and burial sites from prehistory. However, alongside extensive medieval and early industrial heritage (such as charcoal pits and platforms), in recent centuries modern impact has included felling and replanting with exotic trees (both broadleaves and confers) and contemporary management for amenity and recreational usage. The human cultural influence interacts with nature to produce what we now define as 'ancient woods' and which result from unique and unrepeatable timelines. As stewards of the environment, we now have a responsibility to protect ancient, semi-natural woodlands for future generations. Furthermore, the discovery that shadows and ghosts of woodlands persist today beyond the 'wood' boundary is exciting and challenging.

In Great Britain, recognition and designation of ancient woodland status gives a modicum of protection under national planning guidance. However, as discussed, the process of designation still leaves some important questions unanswered. Either oversight or even misrepresentation at planning inquiries has been shown to lead to serious, long-term damage to woodlands because they were not recognised at 'ancient' by the landowner, the local authority, the government agency (Natural England), or by the consultants for the developer [6–8].

Such oversights and the other complicating issues are problematic. However, it is clear that the 'precautionary principle' should apply, and if there is evidence to support the status of a site even if not currently on the Ancient Woodland Inventory, then the onus is upon the developer to prove it is not ancient. Current practice often means that local community action groups and the Woodland Trust (Britain's leading woodland conservation charity) or the Wildlife Trust have to prove the site is ancient at every inquiry or planning application, even when a site is listed on the inventory. The present situation raises serious issues of whether a local planning authority charged with a responsibility under PPG 9 (Planning Policy Guidance 9) [23,97], and later guidance [98], is actually able to discharge its duty, and this is even more onerous when the local authority itself is a landowner with a significant financial interest in a consent being granted. Furthermore, in many cases, the local authority planning officer simply does not have the skill or experience to decide on these technical issues. Increasingly they also lack access to adequate in-house professional support.

Finally, the process of identification and designation largely overlooks historic landscape, heritage, and woodland archaeology issues. Yet, as argued elsewhere, it is this eco-cultural element of the ancient wood that strongly influences the ecological features associated with antiquity [16,27,38]. However, following from the last paragraph regarding local authority and agency competence in deciding these issues, in most cases observed there was little if any input to case inquiries or assessments by archaeological or heritage specialists [7,8]. This situation leaves the identification and protection of ancient woodlands in England, and probably elsewhere too, as at best a work in progress.

Detailed case studies and long-term action research evidence serious issues with a flawed planning system [7,8,92]. It is argued that the historic timelines of ancient woodlands govern the contemporary ecology, and that the archaeological features associated with human usage of these eco-cultural landscapes amount to unique and irreplaceable heritage. Furthermore, this remarkable resource remains largely unrecognised and extremely vulnerable to the impacts of contemporary management (Figures 9 and 10).

This account has focused largely on examples from Great Britain but of course, similar processes and phenomena are occurring around the world but particularly across Europe. Europe-wide examples are presented in the two volumes *Ancient Woodlands and Trees: A Guide for Landscape Planners and Forest Managers*, published in 2018 [32] and the follow-up book [99], which include national case studies, for Austria and Turkey for instance. Having worked with colleagues and attended conferences across Europe, it is clear that in many cases the necessary research and archiving and inventory of woodland and forest heritage (both tangible and intangible) are undertaken more effectively. For example, the work of Elisabeth Johann in Austria has been exemplary in impacting awareness, guidance, and practical conservation [100,101]. More widely across Europe, such approaches are actively promoted by IUFRO (The International Union or Forest Research Organisations) and ESEH (The European Society for Environmental History) [80 (Agnoletti)]. These integrated approaches to treescapes, their history, and future management are generally lacking in Great Britain.

Figure 9. Forestry extraction through archaeological site at Whitwell Wood, Derbyshire.

Figure 10. Rough Standhills PAWS clear-felled with unrecorded archaeology erased, Sheffield.

Funding: The author received no funding towards this research.

Acknowledgments: Numerous colleagues and community volunteers have worked on projects for nearly forty years, and I thank the individuals who triggered and inspired my initial interest.

Conflicts of Interest: The author declares no conflicts of interest.

References

1. Pigott, C.D. Estimation of the age of lime-trees (*Tilia* spp.) in parklands from stem diameter and ring counts. *Arboric. J.* **1989**, 289–302. [CrossRef]
2. Rotherham, I.D. Historical Woodland Ecology: Through the Lens of Ancient Trees and Botanical Indicators. In *Woodlands: Ecol Management and Threats*; Nova Science Publishers: New York, NY, USA, 2024; pp. 209–233, Chapter 10.
3. Pigott, D. *Lime-Trees and Basswoods: A Biological Monograph of the Genus Tilia*; Cambridge University Press: Cambridge, UK, 2(
4. Pigott, C.D. Biological Flora of the British Isles No. 174: *Tilia cordata* Miller. *J. Ecol.* **1991**, *79*, 1147–1207. [CrossRef]
5. Pigott, C.D. Factors controlling the distribution of *Tilia cordata* at the northern limits of its geographical range. 4. Estimated a of the trees. *New Phytol.* **1989**, *112*, 117–121. [CrossRef]
6. Rotherham, I.D. (Ed.) Evidencing change and continuity in woodland landscapes. In *Woodlands: Ecology, Management and Thre* Nova Science Publishers: New York, NY, USA, 2024; pp. 153–172, Chapter 7.
7. Rotherham, I.D. Comment: The new 'Locust Years' of ancient-woodland destruction. *Br. Wildl.* **2024**, *35*, 244–250.
8. Rotherham, I.D. How modern management erases 'Ancient Woods'. *ECOS* **2024**, *45*. Available online: https://www.ecos.org. ecos-45-3-how-modern-management-erases-ancient-woods/ (accessed on 1 January 2020).
9. Bannister, N.R. *Woodland Archaeology in Surrey: Its Recognition and Management*; Surrey County Planning Department: Kingst upon-Thames, UK, 1996.
10. Forestry Commission; Natural England. Ancient Woodland and Veteran Trees: Assessment Guide to Potential Impacts in Relat to Planning Decisions. 2015. Available online: https://www.forestry.gov.uk/pdf/150330AWAssessmentGuide2.pdf/$FILE/1 30AWAssessmentGuide2pdf (accessed on 20 January 2025).
11. Rackham, O. *The Last Forest. The Fascinating Account of Britain's Most Ancient Forest*; J.M. Dent: London, UK, 1989.
12. Rotherham, I.D. (Ed.) A case-study approach to reconstructing evidence for relict ancient woodlands from ecological indicat and archival sources. In *Woodlands: Ecology, Management and Threats*; Nova Science Publishers: New York, NY, USA, 20 pp. 301–329, Chapter 15.
13. Rackham, O. *Trees and Woodland in the British Landscape*; J.M. Dent & Sons Ltd: London, UK, 1976.
14. Rackham, O. *Ancient Woodland: Its History, Vegetation and Uses in England*; Edward Arnold: London, UK, 1980.
15. Rackham, O. *The History of the Countryside*; J.M. Dent: London, UK, 1986.
16. Rackham, O. *Woodlands*; Collins New Naturalist: London, UK, 2006.
17. Peterken, G.F. *Woodland Conservation and Management*; Chapman and Hall: London, UK, 1981.
18. Peterken, G.F. *Natural Woodland: Ecology and Conservation in Northern Temperate Regions*; Cambridge University Press: Cambrid UK, 1996.
19. Maloof, J. *Nature's Temples. A Natural History of Old-Growth Forests*; Princeton University Press: Princeton, NJ, USA; Oxford, U 2023.
20. Davis, M.B. (Ed.) *Eastern Old-Growth Forests. Prospects for Rediscovery and Recovery*; Island Press: Washington, DC, USA, 1996.
21. Bolgiano, C. *The Appalachian Forest*; Stackpole Books: Mechanicsburg, PA, USA, 1998.
22. Cronon, W. *Changes in the Land. Indians, Colonists, and the Ecology of New England*; Hill and Wang: New York, NY, USA, 1983.
23. Sansum, P.; Bannister, N.R. A Handbook for updating the Ancient Woodland Inventory for England. Natural England Comm sioned Reports NECR248. In *Ancient Woodland Inventory Handbook for England*; Natural England: Peterborough, UK, 2018.
24. Peterken, G.F. Ancient Woodland in Concept and Practice. In *Ancient Woodlands and Trees: A Guide for Landscape Planners a Forest Managers*; Çolak, A.H., Kirca, S., Rotherham, I.D., Eds.; Turkish Academy of Sciences, Ankara and IUFRO: Vienna, Aust 2018; pp. 15–24.
25. Pigott, C.D. The History and Ecology of Ancient Woodlands. *Landsc. Archaeol. Ecol.* **1993**, *1*, 1–11.
26. Day, P. Origins of Medieval Woodland. *Landsc. Archaeol. Ecol.* **1993**, *1*, 12–25.
27. Rotherham, I.D.; Ardron, P.A. The Archaeology of Woodland Landscapes: Issues for Managers based on the Case-study Sheffield, England and four thousand years of human impact. *Arboric. J.* **2006**, *29*, 229–243. [CrossRef]
28. Rotherham, I.D. The implications of perceptions and cultural knowledge loss for the management of wooded landscapes: A U case-study. *For. Ecol. Manag.* **2007**, *249*, 100–115. [CrossRef]
29. Rotherham, I.D. A Landscape History Approach to the Assessment of Ancient Woodlands. In *Woodlands: Ecology, Managem and Conservation*; Wallace, E.B., Ed.; Nova Science Publishers Inc.: New York, NY, USA, 2011; pp. 161–184.

Barnes, G.; Williamson, T. *Rethinking Ancient Woodland. The Archaeology and History of Woods in Norfolk*; Studies in Regional and Local History; University of Hertfordshire Press: Hatfield, UK, 2015.

Williamson, T. *An Environmental History of Wildlife in England 1650–1950*; Bloomsbury: London, UK, 2013.

Çolak, A.H.; Kirca, S.; Rotherham, I.D. (Eds.) *Ancient Woodlands and Trees: A Guide for Landscape Planners and Forest Managers*; Turkish Academy of Sciences, Ankara and IUFRO: Vienna, Austria, 2018.

Natural England. *State of the Natural Environment Report*; Natural England: Sheffield, UK, 2008.

Kirby, K.; Goldberg, E. *Ancient Woodland: Guidance Material for Local Authorities*; English Nature: Peterborough, UK, 2006.

Peterken, G.F. Habitat conservation priorities in British and European woodlands. *Biol. Conserv.* **1977**, *11*, 223–236. [CrossRef]

Goldberg, E.; Kirby, K.; Elisabeth Hall, J.E.; Latham, J. The ancient woodland concept as a practical conservation tool in Great Britain. *J. Nat. Conserv.* **2007**, *15*, 109–119. [CrossRef]

Mitchell, R.J.; Bailey, S.; Beaton, J.K.; Bellamy, P.E.; Brooker, R.W.; Broome, A.; Chetcuti, J.; Eaton, S.; Ellis, C.J.; Faren, J.; et al. *The Potential Ecological Impact of Ash Dieback in the UK*; JNCC Report No. 483; JNCC: Peterborough, UK, 2014.

Hall, J.E. *A Guide to Understanding the Scottish Ancient Woodland Inventory (AWI)*; NatureScot: Inverness, Scotland, 2023. [CrossRef]

Rotherham, I.D. *Shadow Woods: A Search for Lost Landscapes*; Wildtrack Publishing: Sheffield, UK, 2017.

Rotherham, I.D.; Jones, M.; Smith, L.; Handley, C. (Eds.) *The Woodland Heritage Manual: A Guide to Investigating Wooded Landscapes*; Wildtrack Publishing: Sheffield, UK, 2008.

Rotherham, I.D. The cultural heritage of woods and forests. In *Ancient Woods, Trees and Forests: Ecology, History and Management*; Çolak, A.H., Kirca, S., Rotherham, I.D., Eds.; Pelagic Publishing: Exeter, UK, 2023; pp. 15–30.

Rotherham, I.D. Worked trees and ecological indicators in wooded landscapes. In *Ancient Woods, Trees and Forests: Ecology, History and Management*; Çolak, A.H., Kirca, S., Rotherham, I.D., Eds.; Pelagic Publishing: Exeter, UK, 2023; pp. 108–123.

Natural England. *Standing Advice Ancient Woodland*; Natural England: Peterborough, UK, 2008.

Rotherham, I.D. The Historical Ecology of Medieval Deer Parks and the Implications for Conservation. In *The Medieval Deer Park: New Perspectives*; Liddiard, R., Ed.; Windgather Press: Macclesfield, UK, 2007; pp. 79–96.

Davies, R. *A revision of the Ancient Woodland Inventory for Surrey*; Surrey Biodiversity Information Centre: Woking, UK, 2011.

Glaves, P.; Rotherham, I.D.; Wright, B.; Handley, C.; Birbeck, J. *A Report to the Woodland Trust. A Survey of the Coverage, Use and Application of Ancient Woodland Indicator Lists in the UK*; Hallam Environmental Consultants Ltd., Biodiversity & Landscape History Research Institute, and Geography, Tourism & Environment Change Research Unit, Sheffield Hallam University: Sheffield, UK, 2009.

Spencer, J.W.; Kirby, K.J. An inventory of ancient woodland for England and Wales. *Biol. Conserv.* **1992**, *62*, 77–93. [CrossRef]

Glaves, P.; Rotherham, I.D.; Wright, B.; Handley, C.; Birbeck, J. *A Report to the Woodland Trust. Field Surveys for Ancient Woodlands: Issues and Approaches*; Hallam Environmental Consultants Ltd., Biodiversity & Landscape History Research Institute, and Geography, Tourism & Environment Change Research Unit, Sheffield Hallam University: Sheffield, UK, 2009.

Glaves, P.; Rotherham, I.D.; Wright, B.; Handley, C.; Birbeck, J. *A Report to the Woodland Trust. The Identification of Ancient Woodland: Demonstrating Antiquity and Continuity*; Hallam Environmental Consultants Ltd., Biodiversity & Landscape History Research Institute, and Geography, Tourism & Environment Change Research Unit, Sheffield Hallam University: Sheffield, UK, 2009.

Kirby, K. *Woodland Flowers. Colourful Past, Uncertain Future*; Bloomsbury Wildlife: London, UK, 2020.

Peterken, G. *Trees and Woodlands*; Bloomsbury Wildlife: London, UK, 2023.

Kittredge, J. *Forest Influences. The Effects of Woody Vegetation on Climate, Water, and Soil*; Dover Publications Inc.: New York, NY, USA, 1948.

Rotherham, I.D.; Jones, M. Seeing the Woodman in the Trees—Some preliminary thoughts on Derbyshire's ancient coppice woods. *Peak Dist. J. Nat. Hist. Archaeol.* **2000**, *2*, 7–18.

Rotherham, I.D.; Jones, M. The Impact of Economic, Social and Political Factors on the Ecology of Small English Woodlands: A Case Study of the Ancient Woods in South Yorkshire, England. In *Forest History: International Studies in Socio-Economic and Forest Ecosystem Change*; Agnoletti, M., Anderson, S., Eds.; CAB International: Wallingford, UK; Oxford, UK, 2000; pp. 397–410.

Packham, J.R.; Harding, D.J.L.; Hilton, G.M.; Stuttard, R.A. *Functional Ecology of Woodlands and Forests*; Chapman & Hall: London, UK, 1992.

Alexander, K.N.A. Historic parks and pasture-woodlands: The National Trust resource and its conservation. *In The National Trust and Nature Conservation: 100 Years On; Bullock, D.J., Harvey, H.J., Eds.; Biol. J. Linnaean Soc.* **1995**, *56*, 155–175. [CrossRef]

Kirby, K.J.; Thomas, R.C.; Key, R.S.; McLean, I.F.G.; Hodgetts, N. Pasture-woodland and its conservation in Britain. *Biol. J. Linn. Soc.* **1995**, *56*, 135–153. [CrossRef]

Harding, P.T.; Wall, T. (Eds.) *Moccas: An English Deer Park*; English Nature: Peterborough, UK, 2000.

Harding, P.T.; Rose, F. *Pasture-Woodlands in Lowland Britain—A Review of Their Importance for Wildlife Conservation*; Institute of Terrestrial Ecology, Monks Wood Experimental Station: Huntingdon, UK, 1986.

Peterken, G.F. Identifying ancient woodland using vascular plant indicators. *Br. Wildl.* **2000**, *11*, 153–158.

61. Rose, F. Indicators of Ancient Woodland; the Use of Vascular Plants in Evaluating Ancient Woods for Nature Conservation *Wildl.* **1999**, *10*, 241–251.
62. Spencer, J. Indications of antiquity: Some observations of the nature of plants associated with ancient woodland. *Br. Wildl.* **1** *2*, 90–102.
63. Rodwell, J.S. (Ed.) *British Plant Communities: Vol I: Woodlands and Scrub (Vol. 1)*; Cambridge University Press: Cambridge, 1991.
64. Peterken, G.F. A method for assessing woodland flora for conservation using indicator species. *Biol. Conserv.* **1974**, *6*, 239– [CrossRef]
65. Hill, M.O.; Mountford, J.O.; Roy, D.B.; Bunce, R.G.H. *Ellenberg's Indicator Values for British Plants*; ECOFACT 2a Technical An Centre for Ecology and Hydrology: Wallingford, UK, 1999.
66. Jones, M. *Sheffield's Woodland Heritage*, 4th ed.; Wildtrack Publishing: Sheffield, UK, 2009.
67. Jones, M. *The Making of the South Yorkshire Landscape*; Wharncliffe Books: Barnsley, UK, 2000.
68. Hermy, M.; Honnay, O.; Firbank, L.; Grashof-Bokdam, C.; Lawesson, J.E. An ecological comparison between ancient and ot forest plant species of Europe, and the implications for forest conservation. *Biol. Conserv.* **1999**, *91*, 9–22. [CrossRef]
69. Verheyen, K.; Guntenspergen, G.R.; Biesbrouck, B.; Hermy, M. An integrated analysis of the effects of past land use on forest h colonization at the landscape scale. *J. Ecol.* **2003**, *91*, 731–742. [CrossRef]
70. Kirby, K.J. *A Woodland Survey Handbook*; Nature Conservancy Council: Peterborough, UK, 1988.
71. Kirby, K.J. *Forestry Operations and Broad-Leaved Woodland Conservation*; Nature Conservancy Council: Peterborough, UK, 198
72. Kirby, K.J. *Ancient Woodland: Guidance Material for Local Authorities*; English Nature: Peterborough, UK, 2002.
73. Kirby, K.J. Table of Ancient Woodland Indicator Plants. In *The Wild Flower Key*; Rose, F., Ed.; Penguin Books: London, UK, 2
74. Kirby, K.J. Woodland Indicators—Some Experiences from Natural England. In Proceedings of the a Woodland Indicator Worksh Sheffield, UK, 14 May 2008; Biodiversity & Landscape History Research Institute: Sheffield, UK, 2008.
75. Watkins, C. *Trees, Woods and Forests. A Social and Cultural History*; Reaktion Books: London, UK, 2014.
76. Muir, R. *Ancient Trees, Living Landscapes*; Tempus: Stroud, UK, 2005.
77. Fowler, J. *Landscapes and Lives. The Scottish Forest Through the Ages*; Canongate: Edinburgh, UK, 2002.
78. Hayman, R. *Trees, Woodlands and Western Civilization*; Hambledon & London: London, UK, 2003.
79. Perlin, J. *A Forest Journey*; Harvard University Press: Cambridge, MA, USA, 1989.
80. Linnard, W. *Welsh Woods and Forests: History and Utilzation*; National Museum of Wales: Cardiff, UK, 1982.
81. Agnoletti, M.; Anderson, S.; Johann, E.; Kulvik, M.; Saratsi, E.; Kushlin, A.; Mayer, P.; Montiel, C.; Parrotta, J.; Rotherham, The Introduction of Historical and Cultural Values in the Sustainable Management of European Forests. *Glob. Environ.* **200** 172–193. [CrossRef]
82. Dargavel, J.; Johann, E. *Science and Hope: A Forest History*; White Horse Press: Cambridge, UK, 2013.
83. Wallace, E.B. (Ed.) *Woodlands: Ecology, Management and Conservation*; Nova Science Publishers Inc.: New York, NY, USA, 201
84. Woitsch, J. (Ed.) *European Forests—Our Cultural Heritage*; Nová tiskárna Pelhřimov & Institute of Ethnology of the Czech Acade of Sciences Pelhřimov: Prague, Czech Republic, 2021.
85. Crumley, C.L.; Lennartsson, T.; Westin, A. *Issues and Concepts in Historical Ecology. The Past and Future of Landscapes and Regi* Cambridge University Press: Cambridge, UK, 2018.
86. Decocq, G. (Ed.) Historical Ecology. In *Learning from the Past to Understand the Present and Forecast the Future of Ecosystems*; IS Wiley: New York, NY, USA, 2022.
87. Vrška, T.; Janík, D.; Pálková, M.; Adam, D.; Trochta, J. Below- and above-ground biomass, structure and patterns in anci lowland coppices. *iForest* **2016**, *10*, 23–31. [CrossRef]
88. Rotherham, I.D. Out of the shadows: Searching for lost Domesday landscapes. In *Histories of People and Landscape: Essay History and Landscape in Memory of David Hey*; Hoyle, R.W., Ed.; University of Hertfordshire Press: Hertfordshire, UK, 2 pp. 87–108.
89. Rotherham, I.D. *Ancient Woodland: History, Industry and Crafts*; Shire Publications: Oxford, UK, 2013.
90. Rotherham, I.D. Forest & Wood as Historic Archives of People, Place & Past. In *European Forests—Our Cultural Heritage*; N tiskárna Pelhřimov & Institute of Ethnology of the Czech Academy of Sciences; Woitsch, J., Ed.; Pelhřimov: Prague, Cz Republic, 2021; pp. 11–28.
91. Vera, F.W.M. *Grazing Ecology and Forest History*; CABI Publishing: Oxon, UK, 2000.
92. Rotherham, I.D. Challenges for the restoration of cultural values in UK woodlands. *For. Ecol. Manag.* **2022**, *503*, 119756. [CrossR
93. Rotherham, I.D. Ancient trees and botanical indicators as evidence for change and continuity in landscape evolution. In *Histor Ecology. Learning from the Past to Understand the Present and Forecast the Future of Ecosystems*; Decocq, G., Ed.; ISTE, Wiley: N York, NY, USA, 2022.
94. Lonsdale, D. *Ancient and Other Veteran Trees: Further Guidance on Management*; Tree Council: London, UK, 2013.

Anon. *Socio-Economic Factors Influencing Coppice Management in Europe*; EuroCoppice Working Group 5, COST Action FP1301 EuroCoppice; Albert Ludwig University: Freiburg, Germany, 2017.
Read, H. *Veteran Trees: A Guide to Good Management*; English Nature: Peterborough, UK, 2000.
ODPM. *Planning Policy Statement 9: Biodiversity and Geological Conservation*; Office of the Deputy Prime Minister: London, UK, 2005.
McKernan, P.; Goldberg, E. *A Review of the Revision of the Ancient Woodland Inventory in the South East*; Natural England Research Report NERR042: Peterborough, UK, 2011.
Çolak, A.H.; Kirca, S.; Rotherham, I.D. (Eds.) *Ancient Woods, Trees and Forests: Ecology, History and Management*; Pelagic Publishing: London, UK, 2023.
Johann, E. The diversity of ancient woodlands in Austria; historical developments and contemporary social importance. In *Ancient Woods, Trees and Forests: Ecology, History and Management*; Çolak, A.H., Kirca, S., Rotherham, I.D., Eds.; Pelagic Publishing: London, UK, 2023; pp. 194–234.
Johann, E. Old-Growth Forests in the Eastern Alps: Management and Protection. In *Countryside History*; Rotherham, I.D., Moody, J.A., Eds.; The Life and Legacy of Oliver Rackham; Pelagic Publishing: London, UK, 2024; pp. 124–137.

diversity

MDPI

Article

Beauty Bias? Exploring the Influence of Attractiveness on Conservation Intentions for Plants and Their Pollinators

Pavol Prokop [1,2,*], Simona Todáková [3] and Jana Fančovičová [3]

[1] Department of Environmental Ecology and Landscape Management, Faculty of Natural Sciences, Comenius University, Ilkovičova 6, 842 15 Bratislava, Slovakia
[2] Institute of Zoology, Slovak Academy of Sciences, Dúbravská Cesta 9, 845 06 Bratislava, Slovakia
[3] Faculty of Education, Trnava University, Priemyselná 4, 918 43 Trnava, Slovakia; todakova99@gmail.com (S.T.); fankaj@gmail.com (J.F.)
* Correspondence: pavol.prokop@savba.sk; Tel.: +421-60296236

Abstract: Plants are an essential component of ecosystems; however, their significance is frequently underestimated, resulting in less effective conservation efforts. One approach to address plant blindness (PB) or plant awareness disparity (PAD) is to establish connections between plants and animals, as animals inherently attract more human attention. In this study, we employed an online questionnaire to manipulate visual materials featuring plant flowers alone, plant flowers accompanied by pollinators, and pollinators in isolation. We assessed participants' willingness to protect (WTP) and the perceived attractiveness of both plants and their pollinators. Our findings revealed that pollinators presented alone received higher WTP scores than plants. Moreover, the visual association between pollinators and plants enhanced WTP plants. Conversely, plants were consistently perceived as more attractive than pollinators, irrespective of whether the flowers were displayed alone or alongside pollinators. The perceived attractiveness of both plants and pollinators was significantly correlated with WTP. Notably, colourful pollinators such as *Selasphorus rufus*, *Vestiaria coccinea*, and *Danaus plexippus* positively influenced WTP plants, while the remaining five species (predominantly invertebrates) did not exhibit a similar effect. We propose that establishing a connection between visually appealing pollinators and plants can help mitigate PB/PAD. These pollinators should be utilised as umbrella species to enhance human attention and interest in pollination processes and plant biology.

Keywords: plant awareness disparity; plant blindness; attitudes toward plants; willingness to protect plants

1. Introduction

Citizens play a crucial role in environmental conservation, natural resource management, and environmental protection [1]. Public participation in citizen science projects has proven effective in local monitoring efforts, particularly in tracking invasive species and informing control strategies [2,3]. Environmental education is fundamental in empowering citizens, with evidence linking awareness campaigns to behavioural shifts toward sustainable practices [4,5]. For instance, urban green initiatives, such as citizen-led tree-planting programs, contribute to improved ecosystem services and climate resilience in cities [6,7]. Moreover, collaborative governance models that include citizens in decision-making processes often lead to more equitable and effective environmental policies [8–10]. Thus, fostering citizen engagement in nature protection is not only a practical necessity but also a moral imperative for ensuring the sustainability of natural systems.

Living organisms, however, do not receive equal support from citizens when it comes to conservation. Citizens tend to prioritise animal conservation over plant conservation, as animals often evoke stronger emotional connections and are perceived as more charismatic [11]. This bias can lead to a "plant blindness" (PB) effect [12] or plant awareness disparity (PAD) [13]. Plant blindness, defined by Wandersee and Schussler [12] as "the inability to see or notice the plants in one's own environment" (p. 82), can result in significant issues, including a failure to recognise the ecological and societal importance of plants [14]. To address the ableist connotations of this term, Parsley [13] proposed plant awareness disparity, which emphasises that equating disability with negative traits is problematic. Parsley et al. [15] identified four dimensions of plant awareness disparity: attention, attitude, knowledge, and relative interest. Building on this, Pany et al. [16] introduced the concept of plant awareness (PA), which shifts the focus from a deficit-oriented view to a more positive perspective aimed at enhancing plant awareness. They suggested that both plant blindness and plant awareness disparity can be seen as manifestations of a lack of plant awareness, and they similarly categorised dimensions of plant awareness into attention, knowledge, attitudes, and the recognition of plants as living organisms.

People often greatly undervalue the ecological importance of plants [17,18], despite their foundational role in ecosystems. Consequently, conservation efforts frequently focus disproportionately on fauna, potentially neglecting critical threats to plant biodiversity and ecosystem services [19]. This bias extends to scientific research, where studies in major conservation journals prioritise animals over plants [20] and even botanists bias their research toward blue-coloured plants, irrespective of their conservation status [21]. To effectively address the urgent threats facing plant biodiversity, it is essential to raise public interest and awareness about plants, ensuring that conservation efforts include all living organisms.

The aesthetic appeal of living organisms plays a significant role in conservation efforts. Research indicates that both plants [22,23] and animals [24,25] perceived as beautiful elicit greater public interest and a higher willingness to support conservation initiatives. For example, the high aesthetic preferences for animals in zoos correlate strongly with their presence in conservation programs and visitors' willingness to provide financial support for their conservation, suggesting that beauty enhances the perceived value of species and influences conservation priorities [26,27]. Conversely, negative emotions such as fear and disgust can adversely affect conservation efforts for less aesthetically appealing species [28]. Research on the associations between aesthetic appeal and willingness to protect species is far more comprehensive for animals than for plants, calling for further investigation into the role of aesthetics in plant conservation efforts.

Attention bias toward animals stems from innate human preferences for moving stimuli, which animals provide [29–31], as well as a lack of direct interaction with plants in modern lifestyles [19,32–34]. Humans are better at detecting animals than plants when shown images in quick succession [35–37]. Research with university students showed better recall of animal names than plant names, even when the names were equally familiar and nameable [38]. Student teachers in Sanders et al.'s [18] study frequently overlooked plants depicted in visual images, while animals received significantly more attention. Even when plants were mentioned, they were often cited simply as sources of food for animals, supporting the idea that plants are generally valued less than animals, which are perceived as superior in ecological importance. However, certain plant characteristics, such as flowers with vivid colours and/or pleasing scents, significantly enhance human attention and positive attitudes toward plants [23,39,40]. Stressed plants also seem to attract human attention; people were able to recognise water-stressed plants with wilted leaves presented with non-stressed plants with a similar accuracy to abused animals presented with nonabused animals [41]. Fostering human interest and curiosity in plants appears to be an effective strategy for reducing PAD/PA/PB [42], especially considering that the objects that attract the most attention within the visual field are those associated with positive attitudes [43].

Researchers have made several important attempts, which would be helpful in mitigating at least some of the four components of PAD [42–44]. For instance, integrating authentic plant-based research to undergraduate student's curriculum strongly influenced attitudes and knowledge domains of PAD [45]. Engagement in hands-on activities, such as gardening [46,47], and teacher-guided physical contact and dialogues about exceptional or medicinal plants [48] fostered a more intimate connection with plants. The Pet Plant Project, where students grew and monitored an unknown plant from seed, also resulted in the enhanced appreciation of plants in most students [49]. Integrating plant awareness into educational frameworks is essential for fostering a deeper connection with nature, ultimately contributing to the achievement of sustainability goals [17,50].

Brkovic et al. [42] emphasised that fostering a stronger attachment between people and plants should not rely predominantly on assessing knowledge about plants; it should also involve examining behavioural intentions, which serve as proxies for actual behaviours. The willingness to protect living organisms is a widely recognised behavioural intention that reflects an individual's commitment to environmental stewardship and biodiversity conservation [51].

Previous research has shown that both people's memory [37] and their interest in plants [52] increased when plants were presented alongside animals. However, concerning the latter, the visual presence of most animals that serve as pollinators largely did not enhance plant attractiveness or the willingness to protect (WTP) plants [52]. This finding may be attributed to the fact that most pollinators are insects, which are generally less popular among people than vertebrates [53–55]. However, the perceived attractiveness of pollinators has not yet been considered. This issue is crucial, given that perceived attractiveness supports WTP living organisms [24,25,27,28,51,56,57]. In this study, we investigated how the perceived attractiveness of pollinators influences the willingness to protect plants and pollinators by experimentally manipulating the presence of pollinators near flowers using visual materials.

2. Materials and Methods

2.1. Participants and Procedure

The research, conducted online between November 2023 and January 2024 using an online questionnaire, involved participants recruited through the university's website in Trnava, Slovakia. A total of 251 volunteer respondents participated, with a majority being women (N = 206, 82%). The ages of the respondents ranged from 11 to 54 years, with a mean age of 25 years (SE = 0.51). Median age was 23. The percentage representation of age cohorts was as follows: 28% (age category 11–20, 76% female), 52% (age category 21–30, 84% female), 14% (age category 31–40, 89% female), and 6% (age category 41–54, 80% female). Our goal was to gather data from participants across diverse age groups. To achieve this, the questionnaire was distributed via the university's webpage and through social networks, inviting all potential participants regardless of age, sex, or education level to complete it. To reach younger participants, the questionnaire was also sent to six science teachers at various primary and secondary schools, who were asked to share it with their students. Importantly, the respondents were not informed about the research hypotheses.

2.2. Description of Research Tool

The research tool utilised was an online questionnaire created using the Google Forms application. This questionnaire was divided into two main sections. The first section served as an introduction, familiarizing participants with our intention to collect data on human preferences for living organisms. Participants began by providing informed consent and answering basic demographic questions, such as sex and age. The second section focused on gathering data regarding the attractiveness of and willingness to protect various organisms: plants presented alone, pollinators presented alone, and plants in the presence of pollinators. This section included 24 photos sourced from Google, with sixteen of these images previously used in different research [52]. The photos were displayed in random order to minimise bias. The photos were randomised during the development of the questionnaire. However, Google Forms does not provide a built-in feature for randomizing question order during administration. Therefore, the randomised order was fixed at the time of questionnaire creation and remained consistent for all participants who accessed the survey. This means that while each participant viewed the images in a predetermined random order, there was no further randomisation applied during their individual sessions. Below each image, participants were asked two questions related to attractiveness and willingness to protect (WTP) the depicted plants and pollinators. The syntax of each question was crafted to be clear and concise, minimizing ambiguity. Therefore, we utilised straightforward language and avoided jargon to ensure comprehension across diverse demographic groups. Prior to administering the questionnaire, we asked a sample of six science teachers to assess the clarity and relevance of the questions as well as their ability to capture the intended constructs of attractiveness and WTP. Feedback from teachers was analysed to refine the questions further. To assess self-perceived attractiveness, participants responded to the following questions: "How do you like the plant in the picture?" and "How do you like the animal in the picture?" WTP was evaluated through the following questions: "Do you think this plant should be protected by law?" and "Do you think this animal should be protected by law?" Respondents rated their answers using a five-point Likert scale: for attractiveness ratings, options ranged from "I don't like it at all" to "I like it extremely"; for WTP ratings, options ranged from "definitely not" to "definitely yes".

2.3. Selection of Images for the Questionnaire

In assessing the attractiveness and willingness to protect plants presented alone and in the presence of pollinators, we selected eight plants, eight pollinators, and eight images depicting pollinators alongside plants. The pollinators included both vertebrates and invertebrates, with a complete list of species provided in Table 1. We predominantly used non-native species for two reasons: first, to avoid any confounding effects related to familiarity with the species; second, because Slovakia lacks vertebrate pollinators, such as bats or birds. Our study aimed to use an unrestricted list of potential pollinators, not limited to insects alone. When choosing the photos, we ensured that the pollinator was positioned similarly and at the correct angle in images where it appeared with a specific plant. However, we found that the backgrounds in some photos were distracting and did not match those of the images featuring plants with pollinators. To address this issue, we used free software from Photoroom (https://www.photoroom.com) accessed on 2 October 2023 to remove the distracting backgrounds. In the modified images, we retained only the branches on which the animals were standing. These edited images were then incorporated into the questionnaire.

Table 1. List of plants and pollinators used in the questionnaire. Native species are marked with "n".

	Species Name	Tree/Shrub/Herbaceous	Described by
Plants alone	Hybrid Fuchsia (*Fuchsia hybrida*)	Shrub	Ch. Plumier, 1696–1697
	Gray's Lobelia (*Lobelia grayana*)	Shrub	E. Wimm., 1948
	Durian nyekak (*Durio kutejensis*)	Tree	G. de Orta, 1563
	Candlestick Banksia (*Banksia attenuata*)	Tree	R. Brown, 1810
	Common yarrow (*Achillea millefolium*) n	Herbaceous	C. Linné, 1753
	Wall hawkweed (*Hieracium murorum*) n	Herbaceous	J. Loudon, 1829
	Purple coneflower (*Echinacea purpurea*)	Herbaceous	C. Moench, 1794
	Wood cranesbill (*Geranium sylvaticum*) n	Herbaceous	W. Withering, 1796
		Invertebrate/Vertebrate	
Pollinators alone	Rufous hummingbird (*Selasphorus rufus*)	Vertebrate	J. F. Gmenil, 1788
	Scarlet honeycreeper (*Vestiaria coccinea*)	Vertebrate	G. Forster, 1780
	Cave nectar bat (*Eonycteris spelaea*)	Vertebrate	G. E. Dobson, 1871
	Honey possum (*Tarsipes rostratus*)	Vertebrate	P. Gervais & J. Verreaux, 1842
	Bee-eating beetle (*Trichodes apiarius*) n	Invertebrate	C. Linné, 1758
	Summer butterbur blacket (*Cheilosia canicularis*) n	Invertebrate	G. W. F. Panzer, 1801
	Monarch butterfly (*Danaus plexippus*)	Invertebrate	C. Linné, 1758
	Honey bee (*Apis mellifera*) n	Invertebrate	C. Linné, 1758
Plants with Pollinators	Hybrid Fuchsia/Rufous hummingbird	-	-
	Gray's Lobelia/Scarlet honeycreeper	-	-
	Durian nyekak/Cave nectar bat	-	-
	Candlestick Banksia/Honey possum	-	-
	Common yarrow/Bee-eating beetle	-	-
	Wall hawkweed/Summer butterbur blacket	-	-
	Purple coneflower/Monarch butterfly	-	-
	Wood cranesbill/Honey bee	-	-

2.4. Statistical Analyses

Two Generalised Linear Mixed Models (GLMMs) were employed. In the first GLMM examining the influence of Species display arrangement (SDA) (plant alone, pollinator alone, plant presented together with pollinator), sex and participant age on WTP plants and pollinators, the WTP was defined as an ordinal dependent variable. Categorical predictors were the effect of SDA and sex. Scores of attractiveness (plant alone, plant with a pollinator, and pollinator) and age were defined as continuous predictors. We defined attractiveness and age as continuous predictors due to their significant role in influencing conservation attitudes. Research indicates that the aesthetic appeal of living organisms elicits greater public interest and greater willingness to support conservation initiatives [23,25]. Additionally, Fančovičová et al. [23] reported a negative correlation between the age of the participants and their willingness to protect the plants, suggesting that age may influence WTP. ID of participant ($N = 251$) was treated as the random effect. Simple correlations (Pearson) between plant attractiveness presented alone and WTP plants presented alone and between pollinator attractiveness presented alone and WTP pollinators presented alone were examined with summed attractiveness and WTP data per each participant. These data achieved normality after Box–Cox transformation. The optimal lambda values obtained for our variables are as follows: Attractiveness of pollinators: $\lambda = 0.85$; Attractiveness of plants: $\lambda = 1.79$; WTP pollinators: $\lambda = 1.34$; WTP plants: $\lambda = 0.76$. The application of these transformations significantly improved the normality of the data distributions, as assessed by Shapiro–Wilk test. A subsample of attractiveness and willingness to protect (WTP), treated as ordinal dependent variables, was analysed using an additional GLMM. In this model, herbaceous plants were compared to trees (pooled with shrubs) as a categorical predictor, with participant ID included as a random effect. In the second GLMM, we examined the influence of SDA (plant alone, pollinator alone, plant presented together with pollinator), sex, and participant age on perceived attractiveness of plants and pollinators (ordinal dependent variable). WTP was not included in this model because attractiveness is assumed to influence WTP but not vice versa. ID of participant (N = 251) was treated as the random effect. Differences between SDA were performed with Bonferroni post-hoc test. Differences in WTP and perceived attractiveness between species were performed with series of GLMMs (WTP plant, plant attractiveness, WTP pollinators, pollinator attractiveness), where WTP or attractiveness were defined as ordinal dependent variables, species (plant or pollinator) was defined as categorical predictor, and ID of participant was treated as the random effect. Participant age and sex were not included in these analyses, because previously described analyses showed that their influence on WTP/attractiveness is small. Data from pictures showing both pollinators and plants together were not included in these analyses. All statistical tests were performed with the jamovi project (2024) [58].

3. Results

3.1. Factors Influencing WTP Plants and Pollinators

WTP plants and pollinators was significantly influenced by the SDA (Table 2, Figure 1). Participants demonstrated a greater willingness to protect pollinators than plants, regardless of whether the plants were presented alone or alongside a pollinator. Plants presented alone received the lowest WTP scores. The attractiveness of both plants and pollinators, defined as a continuous predictor, was significantly correlated with WTP for both groups (estimate = 1.08). Notably, a significant interaction between variables indicated that female participants exhibited a higher WTP pollinators than male participants. However, the effects of age (estimate = −0.01) and sex were not significant (Table 2).

Table 2. GLMM on willingness to protect (WTP) plants and pollinators.

	X^2	df	*p*
Attractiveness	1553.895	1.00	<0.001
Age	0.864	1.00	0.353
SDA	121.458	2.00	<0.001
Sex	1.612	1.00	0.204
SDA × Sex	9.378	2.00	0.009

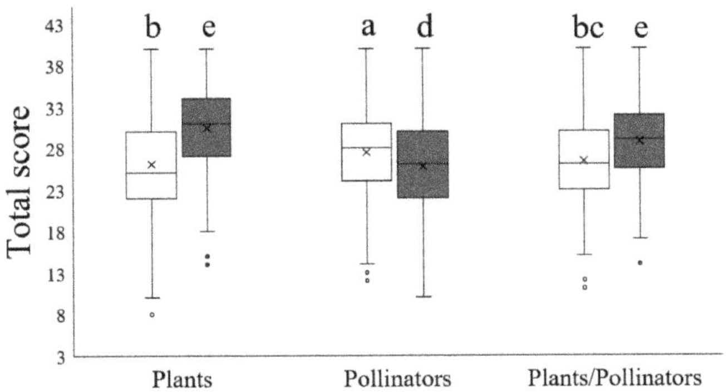

Figure 1. Differences in WTP (open boxes) and perceived attractiveness (grey boxes) with respect to SDA. Box plots represent medians (line), means (asterisk), 1st and 3rd quartile, and minimum and maximum

values. Dots are outliers. The letters denote significant differences (WTP: a vs. b, $p < 0.001$, a vs. bc, $p < 0.001$, b vs. bc, $p = 0.04$; attractiveness: d vs. e, $p < 0.001$, e vs. e, $p = 0.93$).

Simple correlations showed that there was a significant positive correlation between the attractiveness of plants and the WTP plants (r = 0.58, $p < 0.001$) as well as between the attractiveness of pollinators and the WTP pollinators (r = 0.65, $p < 0.001$).

3.2. Factors Influencing Perceived Attractiveness of Plants and Pollinators

The effect of SDA significantly influenced the perceived attractiveness of plants and pollinators (Table 3). Pollinators were perceived as least attractive compared with plants. Plants were similarly attractive regardless of whether presented alone or with pollinators (Figure 1).

Table 3. GLMM on perceived attractiveness of plants and pollinators.

	χ^2	df	p
SDA	50.706	2.00	<0.001
Sex	0.049	1.00	0.826
Age	0.283	1.00	0.595

3.3. Differences in WTP and Attractiveness Between Plants

Differences in WTP plants and their attractiveness were significant (GLMM, $\chi^2 = 203$ and 484, df = 7, both $p < 0.001$, respectively). All species received lower WTP scores than perceived attractiveness scores (Figure 2). Fuchsia received the highest WTP/attractiveness scores, while banksia was perceived as the least attractive.

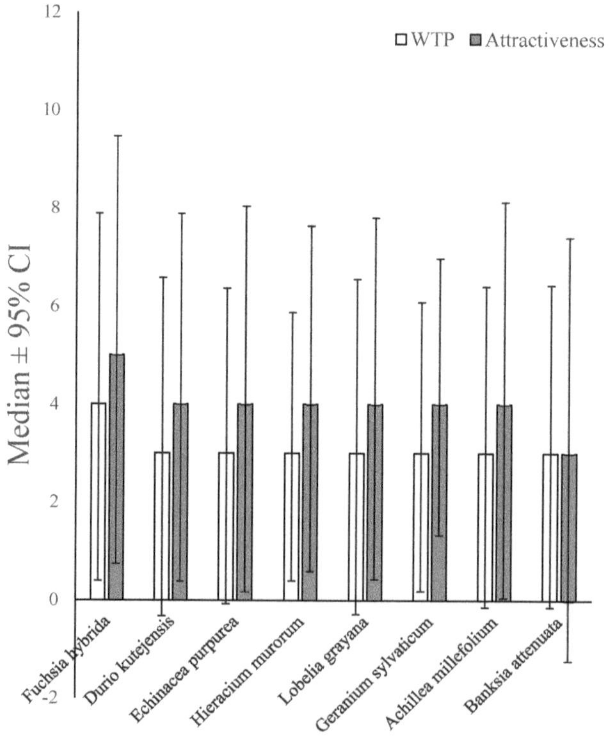

Figure 2. Descriptive statistics of plants' WTP and attractiveness scores.

3.4. Differences in WTP and Attractiveness Between Pollinators

Differences in WTP pollinators and their attractiveness were significant (GLMM, $\chi^2 = 580$ and 553, df = 7, both $p < 0.001$, respectively). In contrast to plants, WTP scores were never smaller than the perceived attractiveness scores. *C. canicularis*, *T. rostratus*, and *T. apiarius* received lower WTP/attractiveness scores than other species (Figure 3). Additional GLMMs with the invertebrate/vertebrate pollinator defined as a categorical predictor instead of the pollinator species showed that plants presented with vertebrate pollinators received significantly higher WTP and attractiveness scores than plants with invertebrate pollinators ($\chi^2 = 20$ and 15.7, df = 1, both $p < 0.001$, respectively).

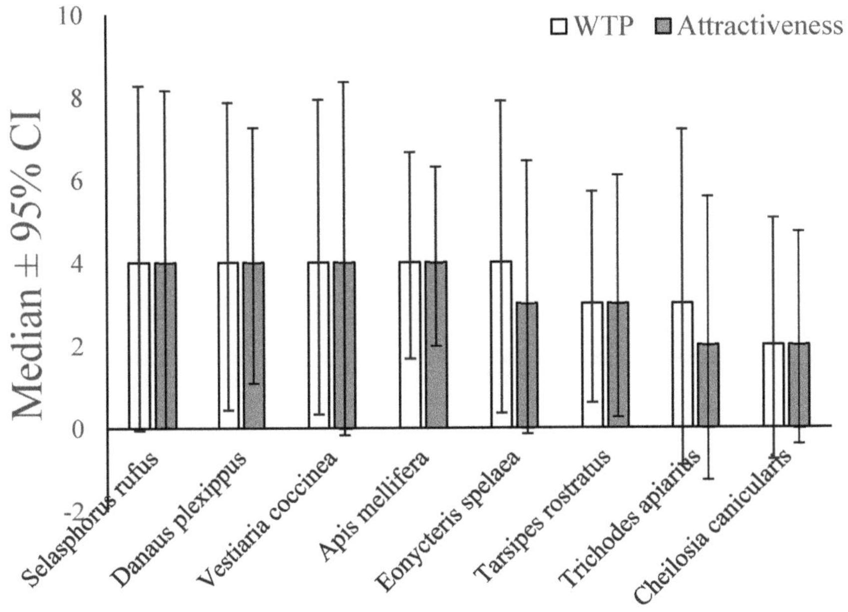

Figure 3. Descriptive statistics of pollinators' WTP and attractiveness scores.

3.5. Differences in WTP and Attractiveness Between Plants Presented with Pollinators

Differences in WTP and the attractiveness of plants presented with pollinators were significant (GLMM, $\chi^2 = 257$ and 336, df = 7, both $p < 0.001$, respectively). Plants associated with two of four vertebrates (fuchsia with *S. rufus*, durian with *E. spelaea*) received higher WTP scores than plants with remaining vertebrates (*L. grayana* with *Vestiaria coccinea* and *B. attenuata* with *T. rostratus*) and plants associated with invertebrates (Figure 4). An additional GLMM analysis was conducted using invertebrate versus vertebrate pollinators as a categorical predictor instead of individual pollinator species. The results showed that plants associated with vertebrate pollinators received significantly higher WTP scores ($\chi^2 = 90$, df = 1, $p < 0.001$). However, the attractiveness scores were not affected by the animal class ($\chi^2 = 0.253$, df = 1, $p = 0.61$).

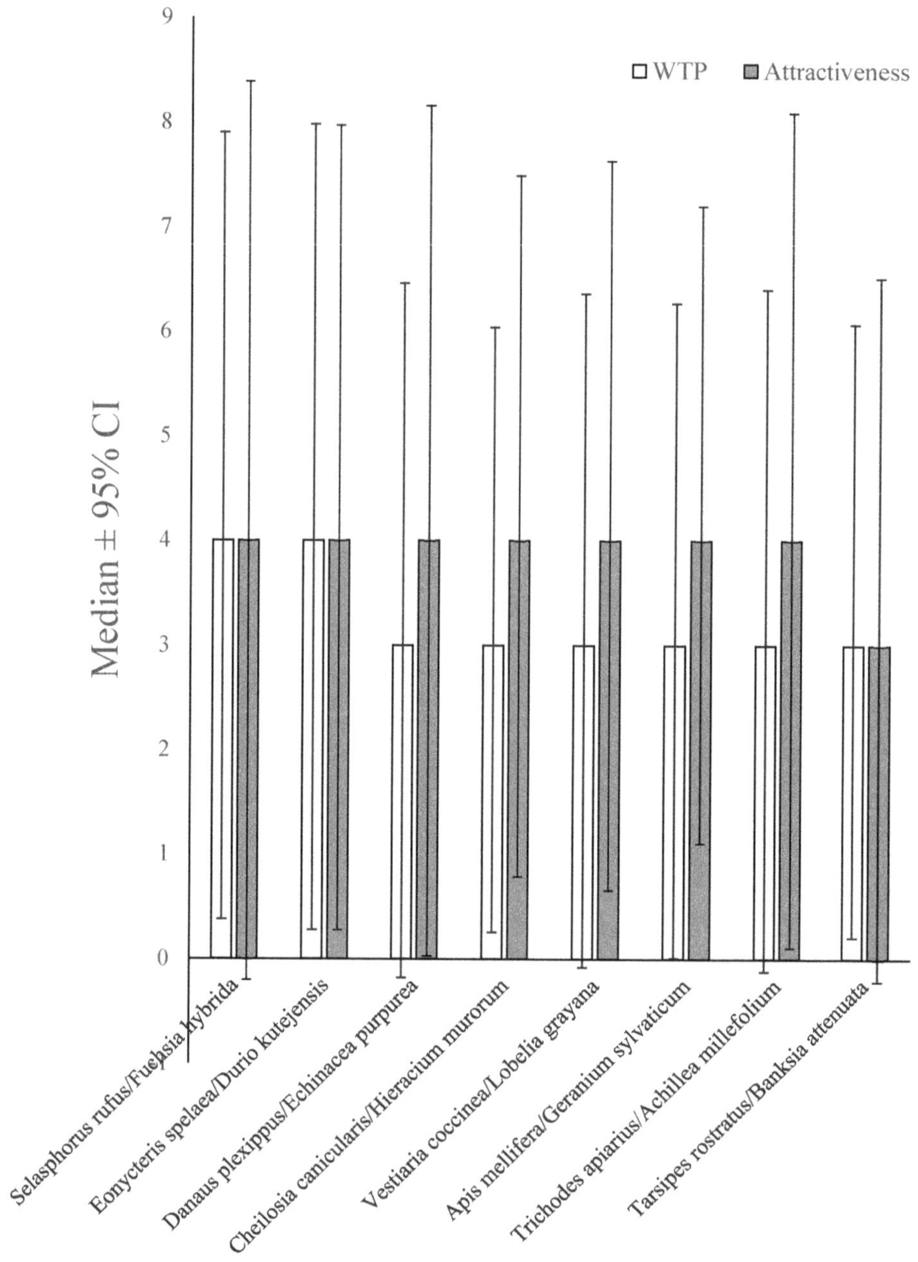

Figure 4. Descriptive statistics of WTP and attractiveness scores of plants presented with pollinators.

3.6. Differences in WTP and Attractiveness Between Herbaceous Plants and Trees

Herbaceous plants were perceived as more attractive (median = 4, 95% CI [3.90, 4.01]) than trees (median = 4, 95% CI [3.58, 3.73]) (GLMM, $\chi^2 = 29.7$, df = 1, $p < 0.001$). In contrast, WTP herbaceous plants was significantly lower (median = 3, 95% CI [3.06, 3.20]) compared to the WTP trees (median = 3, 95% CI [3.32, 3.46]) (GLMM, $\chi^2 = 40.1$, df = 1, $p < 0.001$).

4. Discussion

We demonstrated that the WTP plants and animals differs, with pollinators significantly influencing the WTP plants, but not their attractiveness. Additionally, there were notable differences in WTP and perceived attractiveness between various plant and pollinator species.

Our primary research question was whether the presence of pollinators on flowers influences the willingness to protect plants, given that animals are generally perceived as more attractive than plants [29,30]. Additionally, we aimed to determine whether pollinators' presence affects plants' perceived attractiveness, as perceived attractiveness is known to support the WTP living organisms [24,25]. Our findings indicate that the presence of pollinators on flowers significantly contributes to WTP plants; however, this effect appears to be limited to a few specific species of pollinators. Pollinators received higher WTP scores than plants presented alone or when plants were shown alongside pollinators. Paradoxically, despite the attractiveness scores of plants being higher than those of pollinators, participants exhibited lower WTP plants than pollinators.

Contrary to our expectations, pollinators did not enhance the perceived attractiveness of plants. Research showed that the perceived attractiveness and WTP are correlated [24,25,52,56,57]. However, the correlations between attractiveness and WTP, when assessed separately for plants and pollinators, suggest that attractiveness is more closely associated with WTP in the case of pollinators than in plants. Pollinators attract greater attention from participants, supporting the superiority of animals over plants [19,29,30]. These findings reinforce that human visual attention has evolved to prioritise animate stimuli due to ancestral survival needs, such as predator avoidance and prey detection. In contrast, inanimate stimuli—referring to motionless objects such as plants—are not prioritised by humans [59].

Herbaceous plants were rated higher for attractiveness but received lower WTP scores compared to trees and shrubs. Some researchers suggest that trees possess characteristics that make them easier to identify, leading to greater public familiarity with them than with herbs [60]. While it is possible that better knowledge motivates people to prioritise the protection of trees over herbs, it is unclear whether participants could discern whether they were viewing a tree or herb based on images predominantly showing flowers. If they could, we propose that admiration—an unconscious response to large objects, including trees—may underlie the higher WTP scores for trees [61].

Although the overall influence of pollinators on plants was found to be significant, only a limited number of pollinator species should be considered. Comparisons of the median WTP scores for plants with and without pollinators revealed that three vertebrate pollinators (*S. rufus*, *V. coccinea*, *E. spelaea*) and one invertebrate pollinator (*D. plexippus*) positively contributed to the WTP plants. Notably, in independent research, *V. coccinea* and *E. spelaea* also positively influenced the WTP plants [52], suggesting that their impact on WTP is not incidental. Vertebrates, being phylogenetically closer to humans, are generally perceived as more appealing than invertebrates [62], which may explain why vertebrate pollinators enhanced WTP scores for plants more than their invertebrate counterparts. These findings highlight the need to focus on specific pollinator species when assessing their impact on plant conservation efforts. Conservation strategies can be tailored to enhance public engagement and support for plant protection initiatives by understanding which pollinators significantly influence WTP.

The mean attractiveness scores of plants were generally high; however, their WTP scores consistently remained lower. To enhance the WTP plants through visual cues of pollinator–plant interactions, it is essential to improve the perceived attractiveness of pollinators. Our data indicated that certain species were perceived as less attractive—

specifically, some insects (*C. canicularis, T. apiarius*) and small mammals (*T. rostratus*). It would be misguided to exclude these species from environmental education activities; instead, we should utilise more attractive species as umbrella species to enhance the appeal of other pollinators. It is evident that highly colourful species such as *S. rufus*, *V. coccinea*, and *D. plexippus* naturally attract human attention. However, if relatively dull species like *A. mellifera* are also found to be attractive to humans, this suggests that factors beyond mere colouration contribute to their appeal [63]. To effectively increase the understanding of pollination processes and plant signalling, which ultimately enhances pollination success, educational programmes must incorporate physical experiences and a deeper exploration of these concepts. Doing so can foster a greater appreciation for pollinators and plants, thereby improving conservation efforts.

Most research indicates a tendency for greater appreciation of plant sciences among females [64]. However, several studies have found no significant differences between genders in areas such as factual knowledge about plants [47,64,65], interest in plants [30,66], and attitudes toward plants [67]. Similarly, the present study did not reveal robust differences between males and females in the aesthetic appreciation of plants, except for a higher WTP pollinators among females. This finding is surprising, considering that females generally show a lower preference for less popular animals compared to males [30]. Research suggests that females are often more inclined to protect nature, demonstrating higher levels of environmental concern and pro-environmental behaviours. For instance, studies have shown that women exhibit stronger pro-environmental attitudes and intentions than men [68]. This tendency may stem from traditional socialisation processes that encourage caregiving and nurturing roles among females, fostering a deeper connection to nature and environmental stewardship [69].

The results of our research are primarily useful for the development of biology curricula, specifically in creating illustrations for textbooks and workbooks that depict the reproductive organs of plants. Lower secondary school students (ages 10–11) in Slovakia are expected to understand the biology and functions of plant reproductive organs. They should be able to identify basic flowering plant species found in various ecosystems, such as meadows and forests, and comprehend the role of pollinators in maintaining plant diversity. These topics are further explored in the sixth grade, where students learn about the dependency of entomophilous plants on their pollinators and the significance of pollinators in agricultural plant production. They also study the detailed structure of plant flowers, and the processes involved in pollination. In grades 8 and 9, students delve into anthropogenic impacts on biodiversity and biological balance, examining organisms that thrive in urban environments and the co-evolutionary processes between plants and their pollinators. Consequently, lower secondary school biology explicitly requires an understanding of animal–plant interactions, which necessitates visual materials that depict these fascinating biological processes. However, it remains unclear whether existing visual materials in biology textbooks adequately represent the relationship between pollinators and plants. Research indicates that photographs of plants are less numerous and diverse than those of animals [70]. Therefore, integrating images of both plants and their pollinators can help bridge the gap between the often-overlooked importance of plants and the frequently emphasised role of animals. To enhance educational programs at various levels—primary, secondary, and even tertiary education—we recommend incorporating visually engaging materials that illustrate these associations. For instance, primary school curricula could include activities that allow children to explore local ecosystems, observe pollinators in action, and understand their vital role in plant reproduction. This experiential learning approach fosters a deeper appreciation for biodiversity from an early age. At the secondary level, educators could implement project-based learning methodologies where students investigate local plant–pollinator interactions through field studies or citizen science projects. Such hands-on experiences not only enrich students' understanding but also encourage critical thinking about ecological relationships. Moreover, integrating findings from previous research on effective visual learning strategies can further support these recommendations. Studies have shown that visually appealing materials enhance student engagement and retention of information [71]. By aligning our educational resources with these insights, we can create a more comprehensive biology curriculum that emphasises the interconnectedness of life forms.

Given that our research was conducted in Slovakia, we recognise that applying our findings requires consideration of regional differences. For example, not all visually appealing pollinators are present or equally recognisable across all areas. Species such as the Rufous hummingbird and Scarlet honeycreeper are limited to specific geographic regions. Cultural differences may influence perceptions of beauty [72] and the importance of pollinators among different populations [73], potentially limiting the appeal of certain species in diverse regions. Less conspicuous species, like the Summer butterbur blacket, may be less familiar or understood compared to globally recognised pollinators such as the Monarch butterfly. Additionally, certain pollinators, like the Cave nectar bat, may face negative perceptions in some regions due to traditional myths or cultural beliefs [74]. Science educators should design activities that ensure charismatic species, such as birds or large butterflies, do not overshadow less visually appealing yet equally important pollinators, like beetles or flies. By striking this balance, educational programs can foster a more inclusive and accurate understanding of pollinator diversity and its ecological significance.

In our study, we opted to use certain exotic, unfamiliar species to mitigate familiarity bias among participants, which could influence their responses regarding the attractiveness and willingness to protect various organisms. We acknowledge that employing unfamiliar species may not elicit the same natural responses from participants as native species would. This discrepancy can affect the generalisability of our findings to real-world contexts, where participants primarily interact with native flora and fauna. Societal attitudes toward non-native species often exhibit bias, as reflected in differing preferences and willingness to protect (WTP), since emotional and cultural connections with native species may differ from those associated with non-native species [24,25]. However, species such as Hybrid Fuchsia and Purple Coneflower are frequently grown in gardens, which may help bridge the gap in eliciting natural responses despite their non-native status. Therefore, we encourage future research to explore participant responses to both native and non-native species to better understand how familiarity influences perceptions across different contexts.

Our study clearly shows that certain pollinators significantly enhance the WTP plants; others may not contribute positively to this perception. Visually appealing species, such as *S. rufus* and *D. plexippus*, or species with similar appearance, can be used as umbrella species in conservation messaging to attract public interest. By cultivating a deeper understanding of pollination processes and plant signalling through experiential learning [71], we can cultivate greater appreciation for both pollinators and plants among local communities. For example, community-driven citizen science projects that emphasise the beauty and ecological roles of these pollinators can enhance social cohesion and promote sustainable practices. Contextualisation of the role of plants and pollinators in ecological services in local ecosystems can enhance students' critical thinking about ecological relationships while promoting pro-environmental behaviours [75]. For example, by creating pollinator gardens with plants known to attract these species, we can empower communities to take active roles in conservation efforts. Although an individual's capacity to conserve pollinators may seem small, their actions in private spaces, such as gardens, allotments, and areas where they work or volunteer, can have a significant impact [76]. Furthermore, their perceptions influence the management of public spaces, which collectively cover vast areas with substantial potential to support pollinators. Furthermore, educating people about the benefits of reduced mowing, leaving certain areas unmown, and creating patches of bare ground for ground-nesting bees can further promote plant and pollinator diversity, ultimately benefiting both the ecosystem and local communities [77].

Limitation

The age range of our sample was notably broad, with a mean age of 25; however, it is important to recognise that the distribution is skewed towards younger participants. Specifically, the older age groups were underrepresented, with only 6% of the participants falling into the 41–54 age category. This limitation restricts our ability to generalise findings across all age cohorts and may result in a biased understanding of plant and animal knowledge among different demographic groups. Fančovičová et al. [23] reported a negative correlation between the WTP plants and the age of the participants, while this study did not show associations between age and the WTP plants. One possible reason for the failure to replicate previous results could be the skewed age distribution of the participants. Another possibility is that the plant species used in Fančovičová et al. [23] were very different from those in the present research. We propose that the latter explanation is more likely, since Prokop and Fančovičová [52], using identical species of plants, did not show correlations between WTP and age.

It is important to note that the sample in this study was biased, with a significant overrepresentation of female respondents (82% of total participants). Therefore, while the results suggest no strong differences related to gender, they should be interpreted with caution until further research includes a more balanced sample.

5. Conclusions

In conclusion, the willingness to protect plants, and ultimately, the issues of plant blindness and plant awareness disparity [12,13], can be partly improved by visually connecting plants with their pollinators. However, the attractiveness of pollinators, which appears to be crucial for enhancing the willingness to protect plants, varies significantly. Only certain species, particularly colourful vertebrates and butterflies, are effective candidates for this enhancement. Utilising attractive pollinators as umbrella species in botany education can increase student interest in botany and human awareness of insect pollinators. These insects are often overlooked yet play a vital role in the functioning of ecosystems on our planet. By initially focusing educational efforts on visually appealing pollinators, we can foster a greater appreciation for their ecological importance and encourage conservation efforts. This approach serves as a crucial first step, creating an opportunity to engage students and the public with less visually appealing yet equally vital species within ecosystems. This "umbrella effect" can help broaden awareness and support for the conservation of all organisms that contribute to ecological balance.

Author Contributions: Conceptualization, P.P. and J.F.; methodology, P.P., S.T. and J.F.; data collection, J.F.; software, S.T. and J.F.; formal analysis, P.P.; investigation, P.P., S.T. and J.F.; resources, J.F.; writing—original draft preparation, P.P.; writing—review and editing, P.P., S.T. and J.F.; visualization, P.P., S.T. and J.F.; supervision, P.P. and J.F.; project administration, J.F. All authors have read and agreed to the published version of the manuscript.

Funding: This research was funded by Cultural and Educational Grant Agency of the Ministry of Education, Science, Research and Sport of the Slovak Republic KEGA no. 008TTU-4/2023, Scientific Grant Agency of the Ministry of Education, Science, Research and Sport of the Slovak Republic VEGA no. 1/0211/25, and by the institutional grant of Trnava University no. B4/2024.

Institution Review Board Statement: The study was conducted in accordance with the Declaration of Helsinki and approved by the Ethics Committee of Trnava University (protocol code KEIV 07/2024, 13 December 2024).

Data Availability Statement: Data will be made available on request.

Acknowledgments: We are grateful to the Editor and two anonymous referees for their constructive comments.

Conflicts of Interest: The authors declare no conflicts of interest.

erences

McKinley, D.C.; Miller-Rushing, A.J.; Ballard, H.L.; Bonney, R.; Brown, H.; Cook-Patton, S.C.; Soukup, M.A. Citizen science can improve conservation science, natural resource management, and environmental protection. *Biol. Conserv.* **2017**, *208*, 15–28. [CrossRef]
Crall, A.W.; Newman, G.J.; Stohlgren, T.J.; Holfelder, K.A.; Graham, J.; Waller, D.M. Assessing citizen science data quality: An invasive species case study. *Conserv. Lett.* **2011**, *4*, 433–442. [CrossRef]
Larson, E.R.; Graham, B.M.; Achury, R.; Coon, J.J.; Daniels, M.K.; Gambrell, D.K.; Suarez, A.V. From eDNA to citizen science: Emerging tools for the early detection of invasive species. *Front. Ecol. Environ.* **2020**, *18*, 194–202. [CrossRef]
Kollmuss, A.; Agyeman, J. Mind the gap: Why do people act environmentally and what are the barriers to pro-environmental behavior? *Environ. Educ. Res.* **2002**, *8*, 239–260. [CrossRef]
Jorgenson, S.N.; Stephens, J.C.; White, B. Environmental education in transition: A critical review of recent research on climate change and energy education. *J. Environ. Educ.* **2019**, *50*, 160–171. [CrossRef]
Aronson, M.F.; Lepczyk, C.A.; Evans, K.L.; Goddard, M.A.; Lerman, S.B.; MacIvor, J.S.; Vargo, T. Biodiversity in the city: Key challenges for urban green space management. *Front. Ecol. Environ.* **2017**, *15*, 189–196. [CrossRef]
Oikonomaki, E.; Papadaki, I.; Kakderi, C. Promoting Green Transformations through Smart Engagement: An Assessment of 100 Citizen-Led Urban Greening Projects. *Land* **2024**, *13*, 556. [CrossRef]
Ansell, C.; Gash, A. Collaborative governance in theory and practice. *J. Public Adm. Res. Theory* **2008**, *18*, 543–571. [CrossRef]
Reed, M.S. Stakeholder participation for environmental management: A literature review. *Biol. Conserv.* **2008**, *141*, 2417–2431. [CrossRef]
O'Faircheallaigh, C. Public participation and environmental impact assessment: Purposes, implications, and lessons for public policy making. *Environ. Impact Assess. Rev.* **2010**, *30*, 19–27. [CrossRef]
Kellert, S.R. *The Value of Life: Biological Diversity and Human Society*; Island Press: Washington, DC, USA, 1996.
Wandersee, J.H.; Schussler, E.E. Preventing plant blindness. *Am. Biol. Teach.* **1999**, *61*, 82–86. [CrossRef]
Parsley, K.M. Plant awareness disparity: A case for renaming plant blindness. *Plants People Planet* **2020**, *2*, 598–601. [CrossRef]
Wandersee, J.H.; Schussler, E. Toward a theory of plant blindness. *Plant Sci. Bull.* **2001**, *47*, 2–9.
Parsley, K.M.; Daigle, B.J.; Sabel, J.L. Initial development and validation of the plant awareness disparity index. *CBE Life Sci. Educ.* **2022**, *21*, ar64. [CrossRef] [PubMed]
Pany, P.; Meier, F.D.; Dünser, B.; Yanagida, T.; Kiehn, M.; Möller, A. Measuring students' plant awareness: A prerequisite for effective botany education. *J. Biol. Educ.* **2022**, *58*, 1103–1116. [CrossRef]
Amprazis, A.; Papadopoulou, P. Plant awareness: At the dawn of a new era. *J. Biol. Educ.* **2024**, 1–11. [CrossRef]
Sanders, D.; Nyberg, E.; Brkovic, I. Putting plants in the picture. *Environ. Educ. Res.* **2024**, *31*, 1–10. [CrossRef]
Balding, M.; Williams, K.J. Plant blindness and the implications for plant conservation. *Conserv. Biol.* **2016**, *30*, 1192–1199. [CrossRef] [PubMed]
Prokop, P.; Masarovič, R.; Hajdúchová, S.; Ježová, Z.; Zvaríková, M.; Fedor, P. Prioritisation of charismatic animals in major conservation journals measured by the altmetric attention score. *Sustainability* **2022**, *14*, 17029. [CrossRef]
Adamo, M.; Chialva, M.; Calevo, J.; Bertoni, F.; Dixon, K.; Mammola, S. Plant scientists' research attention is skewed towards colourful, conspicuous and broadly distributed flowers. *Nat. Plants* **2021**, *7*, 574–578. [CrossRef]
Martín-López, B.; Montes, C.; Benayas, J. The non-economic motives behind the willingness to pay for biodiversity conservation. *Biol. Conserv.* **2007**, *139*, 67–82. [CrossRef]
Fančovičová, J.; Prokop, P.; Kubíčková, M. The effect of aposematic signals of plants on students' perception and willingness to protect them. *Sustainability* **2022**, *14*, 9121. [CrossRef]
Lipták, B.; Kouba, A.; Patoka, J.; Paunović, M.; Prokop, P. Biological invasions and invasive species in freshwaters: Perception of the general public. *Hum. Dimens. Wildl.* **2024**, *29*, 48–63. [CrossRef]
Lipták, B.; Kouba, A.; Zorić, K.; Salvaras, L.; Prokop, P.; Paunović, M. The Attractiveness of Freshwater Species Correlates Positively with Conservation Support. *Anthrozoös* **2023**, *36*, 971–984. [CrossRef]

26. Colléony, A.; Clayton, S.; Couvet, D.; Saint Jalme, M.; Prévot, A.C. Human preferences for species conservation: Animal charis trumps endangered status. *Biol. Conserv.* **2017**, *206*, 263–269. [CrossRef]
27. Landová, E.; Poláková, P.; Rádlová, S.; Janovcová, M.; Bobek, M.; Frynta, D. Beauty ranking of mammalian species kept in Prague Zoo: Does beauty of animals increase the respondents' willingness to protect them? *Sci. Nat.* **2018**, *105*, 69. [CrossR [PubMed]
28. Castillo-Huitrón, N.M.; Naranjo, E.J.; Santos-Fita, D.; Estrada-Lugo, E. The importance of human emotions for wildlife conse tion. *Front. Psychol.* **2020**, *11*, 1277. [CrossRef]
29. Kinchin, I.M. Investigating secondary-school girls' preferences for animals or plants: A simple 'head-to-head' comparison us two unfamiliar organisms. *J. Biol. Educ.* **1999**, *33*, 95–99.
30. Lindemann-Matthies, P. 'Loveable' mammals and 'lifeless' plants: How children's interest in common local organisms can enhanced through observation of nature. *Int. J. Sci. Educ.* **2005**, *27*, 655–677. [CrossRef]
31. Guerra, S.; Betti, S.; Sartori, L.; Zani, G.; Castiello, U. Plant awareness in the hand. *J. Environ. Psychol.* **2024**, *94*, 102246. [CrossR
32. Soga, M.; Tsuchiya, K.; Evans, M.J.; Ishibashi, S. The inequalities of the extinction of experience: The role of personal characteris and species traits in the distribution of people–plant interactions in Japan. *Ecol. Res.* **2019**, *34*, 350–359. [CrossRef]
33. Stroud, S.; Fennell, M.; Mitchley, J.; Lydon, S.; Peacock, J.; Bacon, K.L. The botanical education extinction and the fall of p awareness. *Ecol. Evol.* **2022**, *12*, e9019. [CrossRef]
34. Querol i Mercadé, J.; Fernandez-Llamazares, Á.; Garnatje, T.; Casadevall, A.; Garet, A.; Gallois, S. Beyond plant awaren disparity: Exploring intangible relationships with plants in the Catalan Pyrenees. *Plants People Planet* **2024**. [CrossRef]
35. Balas, B.; Momsen, J.L. Attention "blinks" differently for plants and animals. *CBE Life Sci. Educ.* **2014**, *13*, 437–443. [CrossR [PubMed]
36. Kanske, P.; Schönfelder, S.; Wessa, M. Emotional modulation of the attentional blink and the relation to interpersonal reacti *Front. Hum. Neurosci.* **2013**, *7*, 641. [CrossRef] [PubMed]
37. Zani, G.; Low, J. Botanical priming helps overcome plant blindness on a memory task. *J. Environ. Psychol.* **2022**, *81*, 101 [CrossRef]
38. Schussler, E.E.; Olzak, L.A. It's not easy being green: Student recall of plant and animal images. *J. Biol. Educ.* **2008**, *42*, 112– [CrossRef]
39. Haviland-Jones, J.; Rosario, H.H.; Wilson, P.; McGuire, T.R. An environmental approach to positive emotion: Flowers. *E Psychol.* **2005**, *3*, 147470490500300109. [CrossRef]
40. Hůla, M.; Flegr, J. What flowers do we like? The influence of shape and color on the rating of flower beauty. *PeerJ* **2016**, *4*, e2 [CrossRef] [PubMed]
41. Prokop, P.; Belzárová, K.; Tomanová Čergeťová, I. Compassion and the perceived rarity of plants can increase plant appreciat *People Nat.* 2025; *in press*.
42. Brkovic, I.; Sanders, D.; Nyberg, E. Investigating plant awareness: Methodologies, challenges and possibilities. *Plants Pe Planet* **2024**. [CrossRef]
43. Stagg, B.C.; Hetherington, L.; Dillon, J. Towards a model of plant awareness in education: A literature review and framew proposal. *Int. J. Sci. Educ.* **2024**, 1–21. [CrossRef]
44. Stagg, B.C.; Dillon, J. Plant awareness is linked to plant relevance: A review of educational and ethnobiological literat (1998–2020). *Plants People Planet* **2022**, *4*, 579–592. [CrossRef]
45. Ward, J.R.; Clarke, H.D.; Horton, J.L. Effects of a research-infused botanical curriculum on undergraduates' content knowle STEM competencies, and attitudes toward plant sciences. *CBE Life Sci. Educ.* **2014**, *13*, 387–396. [CrossRef] [PubMed]
46. Lohr, V.I.; Pearson-Mims, C.H. Children's active and passive interactions with plants influence their attitudes and actions tow trees and gardening as adults. *HortTechnology* **2005**, *15*, 472–476. [CrossRef]
47. Fančovičová, J.; Prokop, P. Plants have a chance: Outdoor educational programmes alter students' knowledge and attitu towards plants. *Environ. Educ. Res.* **2011**, *17*, 537–551. [CrossRef]
48. Strgar, J. Increasing the interest of students in plants. *J. Biol. Educ.* **2007**, *42*, 19–23. [CrossRef]
49. Krosnick, S.E.; Baker, J.C.; Moore, K.R. The Pet Plant Project: Treating plant blindness by making plants personal. *Am. Biol. Te* **2018**, *80*, 339–345. [CrossRef]
50. Amprazis, A.; Papadopoulou, P. Plant blindness: A faddish research interest or a substantive impediment to achieve sustaina development goals? *Environ. Educ. Res.* **2020**, *26*, 1065–1087. [CrossRef]
51. Schönfelder, M.L.; Bogner, F.X. How to sustainably increase students' willingness to protect pollinators. *Environ. Educ. Res.* **2** *24*, 461–473. [CrossRef]
52. Prokop, P.; Fančovičová, J. Enhancing attention and interest in plants to mitigate plant awareness disparity. *Plants* **2023**, *12*, 2 [CrossRef] [PubMed]
53. Kellert, S.R. Values and perceptions of invertebrates. *Conserv. Biol.* **1993**, *7*, 845–855. [CrossRef]

Polák, J.; Rádlová, S.; Janovcová, M.; Flegr, J.; Landová, E.; Frynta, D. Scary and nasty beasts: Self-reported fear and disgust of common phobic animals. *Br. J. Psychol.* **2020**, *111*, 297–321. [CrossRef]
Fukano, Y.; Soga, M. Why do so many modern people hate insects? The urbanization–disgust hypothesis. *Sci. Total Environ.* **2021**, *777*, 146229. [CrossRef]
Gunnthorsdottir, A. Physical attractiveness of an animal species as a decision factor for its preservation. *Anthrozoös* **2001**, *14*, 204–215. [CrossRef]
Harnish, R.J.; Nataraajan, R.; Tarka, P.; Slack, F.J. Attitudes toward protecting endangered species: The impact of perceived physical attractiveness of animals and political ideology. *Psychol. Mark.* **2023**, *40*, 73–88. [CrossRef]
The Jamovi Project. Jamovi (Version 2.5) [Computer Software]. 2024. Available online: https://www.jamovi.org (accessed on 20 November 2024).
New, J.; Cosmides, L.; Tooby, J. Category-specific attention for animals reflects ancestral priorities, not expertise. *Proc. Natl. Acad. Sci. USA* **2007**, *104*, 16598–16603. [CrossRef] [PubMed]
Lückmann, K.; Menzel, S. Herbs versus trees: Influences on teenagers' knowledge of plant species. *J. Biol. Educ.* **2014**, *48*, 80–90. [CrossRef]
Prokop, P.; Zvaríková, M.; Zvarík, M.; Ježová, Z.; Fedor, P. Charismatic species should be large: The role of admiration and fear. *People Nat.* **2024**, *6*, 945–957. [CrossRef]
Miralles, A.; Raymond, M.; Lecointre, G. Empathy and compassion toward other species decrease with evolutionary divergence time. *Sci. Rep.* **2019**, *9*, 19555. [CrossRef]
Hall, D.M.; Martins, D.J. Human dimensions of insect pollinator conservation. *Curr. Opin. Insect Sci.* **2020**, *38*, 107–114. [CrossRef] [PubMed]
Kubiatko, M.; Fančovičová, J.; Prokop, P. Factual knowledge of students about plants is associated with attitudes and interest in botany. *Int. J. Sci. Educ.* **2021**, *43*, 1426–1440. [CrossRef]
Wiegand, F.; Kubisch, A.; Heyne, T. Out-of-school learning in the botanical garden: Guided or self-determined learning at workstations? *Stud. Educ. Eval.* **2013**, *39*, 161–168. [CrossRef]
Palmberg, I.; Berg, I.; Jeronen, E.; Kärkkäinen, S.; Norrgård-Sillanpää, P.; Persson, C.; Vilkonis, R.; Yli-Panula, E. Nordic-Baltic student teachers' identification of and interest in plant and animal species: The importance of species identification and biodiversity for sustainable development. *J. Sci. Teach. Educ.* **2015**, *26*, 549–571. [CrossRef]
Fančovičová, J.; Prokop, P. Development and initial psychometric assessment of the plant attitude questionnaire. *J. Sci. Educ. Technol.* **2010**, *19*, 415–421. [CrossRef]
Li, Y.; Wang, B.; Saechang, O. Is female a more pro-environmental gender? Evidence from China. *Int. J. Environ. Res. Public Health* **2022**, *19*, 8002. [CrossRef]
Dhenge, S.A.; Ghadge, S.N.; Ahire, M.C.; Gorantiwar, S.D.; Shinde, M.G. Gender attitude towards environmental protection: A comparative survey during COVID-19 lockdown situation. *Environ. Dev. Sustain.* **2022**, *24*, 13841–13886. [CrossRef]
Link-Perez, M.A.; Dollo, V.H.; Weber, K.M.; Schussler, E.E. What's in a Name: Differential Labelling of Plant and Animal Photographs in Two Nationally Syndicated Elementary Science Textbook Series. *Int. J. Sci. Educ.* **2010**, *32*, 1227–1242. [CrossRef]
Ulbig, S.G. A picture is worth what? Using visual images to enhance classroom engagement. *Int. J. Instr. Media* **2010**, *37*, 185–201.
Broer, P.N.; Juran, S.; Liu, Y.J.; Weichman, K.; Tanna, N.; Walker, M.E.; Persing, J.A. The impact of geographic, ethnic, and demographic dynamics on the perception of beauty. *J. Craniofacial Surg.* **2014**, *25*, e157–e161. [CrossRef] [PubMed]
Hill, R.; Nates-Parra, G.; Quezada-Euán JJ, G.; Buchori, D.; LeBuhn, G.; Maués, M.M.; Pert, P.L.; Kwapong, P.K.; Saeed, S.; Breslow, S.J.; et al. Biocultural approaches to pollinator conservation. *Nat. Sustain.* **2019**, *2*, 214–222. [CrossRef]
Boso, À.; Álvarez, B.; Pérez, B.; Imio, J.C.; Altamirano, A.; Lisón, F. Understanding human attitudes towards bats and the role of information and aesthetics to boost a positive response as a conservation tool. *Anim. Conserv.* **2021**, *24*, 937–945. [CrossRef]
Baldock, K.C.R.; Goddard, M.A.; Hicks, D.M.; Kunin, W.E.; Mitschunas, N.; Morse, H.; Osgathorpe, L.M.; Potts, S.G.; Robertson, K.M.; Scott, A.V.; et al. A systems approach reveals urban pollinator hotspots and conservation opportunities. *Nat. Ecol. Evol.* **2019**, *3*, 363–373. [CrossRef] [PubMed]
Bueddefeld, J.; Ostrem, J.; Murphy, M.; Halpenny, E.; Orr, B. Petting bees or building bee boxes? Strategies for transformative learning. *Environ. Educ. Res.* **2022**, *28*, 560–580. [CrossRef]
Knapp, J.L.; Phillips, B.B.; Clements, J.; Shaw, R.F.; Osborne, J.L. Socio-psychological factors, beyond knowledge, predict people's engagement in pollinator conservation. *People Nat.* **2021**, 3, 204–220. [CrossRef]

diversity

MDPI

Article

Effects of Flow Velocity on the Dynamics of Juvenile Fish Habitats in River Meanders of the Irtysh River

Andrey A. Chemagin [1,*], Elena I. Popova [1] and Martin Schletterer [2,*]

[1] Tobolsk Complex Scientific Station of Ural Branch of the Russian Academy of Sciences (UrB RAS), Akademika Yuriya Osipova Street 15, 626150 Tobolsk, Russia
[2] Institute of Hydrobiology and Aquatic Ecosystem Management, University of Natural Resources and Life Sciences, Vienna (BOKU University), Gregor-Mendel-Straße 33, 1180 Vienna, Austria
* Correspondence: chemaginaa@yandex.ru (A.A.C.); martin.schletterer@boku.ac.at (M.S.)

Abstract: Understanding the spatial distribution of freshwater fish in heterogeneous aquatic environments is crucial for understanding riverine ecosystems and the rational use of aquatic biological resources. This study investigates the distribution patterns of juvenile fish in the lower reaches of the Irtysh River, including hydrodynamic conditions during different water level regimes. With hydroacoustic surveys, we assessed fish density and distribution in two wintering riverbed depressions during the spring flood and summer low water period. The main fish aggregations consisted of Cyprinidae and Percidae, with juveniles predominantly occupying areas with reduced flow velocities (0.15–0.75 m s^{-1}). Correlation analysis showed strong direct relationships between the area occupied by juvenile carp and perch and zones with specific flow velocities. The study highlights that hydrodynamic characteristics, particularly flow velocity, are key factors influencing the distribution and aggregation of juvenile fish. These findings underscore the importance of considering hydrodynamic factors and species-specific traits in understanding fish distribution patterns and in managing freshwater ecosystems effectively. This research contributes to the understanding of the multifunctional roles of riverbed depressions in supporting juvenile fish populations and emphasizes the importance of hydroacoustics to predict fish distributions in dynamic aquatic environments.

Keywords: fish; habitats; riverbed depression; monitoring; hydroacoustics; echosounder; AsCor; Irtysh River

1. Introduction

Studying patterns and determining factors influencing the spatial distribution of aquatic organisms in heterogeneous aquatic environments [1], including large rivers, is one of the main problems in ecology. Currently, the spatial distribution of freshwater fish is studied primarily on a large scale, while studies of local ecological factors and their dynamics are rare [1,2]. The distribution of organisms in the heterogeneous environment of a river is uneven [3] due to the selection of optimal strategies for using the habitat to increase individual survival [4–6].

The spatial distribution of freshwater fish is determined by various factors: physicochemical parameters [7–9], availability of resources and habitats [10–12], predator–prey interactions [13–15], intraspecific relationships, and individual habitat preferences [16].

Knowledge about spatial distribution patterns of fish is necessary for a full understanding of ecosystem functioning [1] and rational use of aquatic biological resources [17]. One of the optimal methods for observing the spatial distribution of fish in aquatic habitats are hydroacoustic surveys [2,18]. This method has proven effective for quantitative real-time observation of fish populations with high spatiotemporal resolution [19–21]. When used in conjunction with geographic information systems (GIS), it creates an ideal opportunity for mapping and modeling the spatial distribution of fish [1], including small pelagic ones [2]. At the same time, the variability of the river channel and its sinuosity [22] provide a wide variety of microhabitats to which fish evolutionarily adapt [23]. The swimming ability of fish is a key factor determining their survival in the natural environment and ability to make various movements: short and long-distance, ensuring dispersal, habitat selection, interaction with food objects and predators, as well as spawning migrations [24,25]. When the hydrological phases change, for example in the case of increased flow velocity characteristics, fish individuals have to change their swimming regime, exerting greater physical effort and intensifying metabolism for additional energy expenditure [26,27].

The aim of this study was to determine the distribution patterns of juvenile fish, considering hydrodynamics in the riverbed depression areas during different water level regimes (during flood and low water), specifically comparing high-velocity zones and juvenile fish concentration areas in a large river system.

2. Materials and Methods

2.1. Research Area

The Irtysh River is a left tributary of the Ob, originating on the southwestern slopes of the Mongolian Altai. Known as the Black Irtysh, it flows into Lake Zaysan, traversing China (525 km), Kazakhstan (1835 km), and Russia (2010 km), including Omsk Oblast (1132 km), Tyumen Oblast (639 km), and Khanty-Mansi Autonomous Okrug (239 km). The total length of the Irtysh from its source is 4370 km, making it the world's longest tributary [28]. The mean discharge of Irtysh River at Tobolsk is 2140 $m^3 s^{-1}$ and the mean annual flow amounts 39.1 $km^3 a^{-1}$; at Khanty-Mansiysk (about 20 km upstream the mouth into Ob river) about 2800 $m^3 s^{-1}$ and 88.3 $km^3 a^{-1}$, respectively.

The Irtysh River channel in its lower reaches has a highly meandering character, with multidirectional currents and circulations [29]. The sinuosity of the channel and features of the riverbed relief create heterogeneity in the environment's hydrodynamic characteristics [29]. At sharp bends (meanders), intense soil erosion processes can be observed, leading to the formation of riverbed depressions ("wintering riverbed depressions"). In these wintering holes, which serve as "temporarily limited biotopes", valuable fish species concentrate during the winter period, including sterlet (*Acipenser ruthenus* Linnaeus), Siberian sturgeon (*Acipenser baerii* Brandt) and nelma (*Stenodus leucichthys nelma* Pallas). However, it has been shown that fish aggregations in such areas are recorded year-round, allowing the biological role of riverbed depressions to be defined as multifunctional—serving as rearing areas for juveniles, feeding grounds, and wintering sites for fish [30].

The survey was conducted in two wintering riverbed depressions, located in the lower reaches of the Irtysh River (Tyumen Oblast, Western Siberia). Observations were made during the active phase of the spring–summer flood on 30 May 2022, and the summer–autumn low water period on 31 July 2022.

The Gornolsinkinskaya wintering riverbed depression, formed by circulatory currents on a sharply curved meander, is located at 58.731950° N, 68.698582° E (Figure 1). The study area covered 41 hectares, with depths reaching over 40 m (Figure 1c).

The Ukinskaya wintering riverbed depression was formed as a result of erosion and the channel processes of an omega-shaped meander of the Irtysh River, located at 58.857111° N, 68.742535° E (Figure 1). The area of the studied site was 75.9 hectares, with depths reaching more than 20 m (Figure 1c). This river section with fish aggregations was recently identified [31] and a justification for its inclusion in the official list of wintering riverbed depressions of the Irtysh River was made, with the aim of preserving aquatic biological resources by fish protection authorities and preventing illegal fishing, primarily of sturgeon and whitefish.

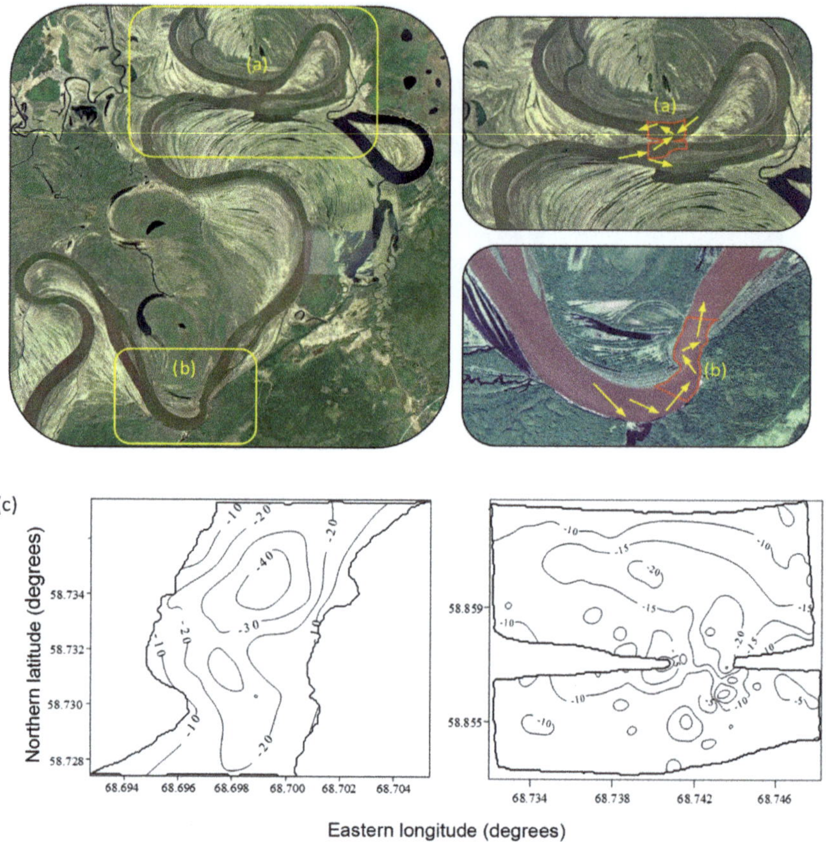

Figure 1. Location of the studied areas of wintering riverbed depressions in the Lower Irtysh: (**a**) Ukinskaya; (**b**) Gornoslinkinskaya. The red outline schematically shows the boundaries of the studied water areas; arrows schematically indicate the direction of the current. (**c**) Relief map of the bottom surface of the deep-water part of the Gornoslinkinskaya (left) and Ukinskaya (right) wintering riverbed depressions during the flood period.

2.2. Field Work and Analyses

The hydroacoustic survey for the assessment of fish density and distribution in the wintering riverbed depressions was carried out with the computerized hydroacoustic system "AsCor" with a vertical view (LLC «Promgidroakustika», Petrozavodsk, Russia) (Figure 2). The system's operation is based on the use of a serial Furuno LS 4100 echosounder (Furuno Inc., Ashihara-cho, Nishinomiya City, Hyogo, Japan), with operating frequencies of 50 and 200 kHz, with viewing angles of 46° and 14°, respectively, [32]. During the survey, recordings of fish distribution and tracks, along with navigation tracking data from an external satellite navigation receiver, were stored on the field tablet iX103 (Xplore Technologies, Austin, TX, USA). This system is used frequently in inland freshwater bodies [33].

Figure 2. Installation of the antenna-emitter of the "AsCor" hydroacoustic complex on a small craft. (**a**) Location of the antenna-emitter with vertical view in a fairing in the bow of the small craft for conducting echo survey. (**b**) Small craft with installed fairing containing the antenna-emitter for transportation. (**c**) Small craft with installed fairing containing the antenna-emitter for echo survey.

The hydroacoustic survey recording is processed frame by frame using a special software application «AsCor Expedition, 2014», with 1 frame consisting of 100 signal transmission-receptions. The processing program algorithm assists in determining fish body length using an equation used for mass fish species of the Lower Irtysh [19], as well as classifying registered fish into 4 groups of mass fish species of the Lower Irtysh based on the shape of the reflected echo signal: cyprinids (two-chambered swim bladder), percids (asymmetric swim bladder), pike and whitefish (symmetric swim bladder), unidentifiable (sturgeons and burbot, lacking a distinct swim bladder shape).

During the recording of the hydroacoustic survey file, the envelope curve of the echo signal with filtered noise is continuously analyzed. Based on the measured pulses, single and multiple targets are identified. This allows for the simultaneous use of echo counting, i.e., counting single targets, as well as echo integration, i.e., measuring the strength of the target for the entire aggregation and determining its density. This process accounts for the number, density (individuals/ha^{-1}), and size of fish (target strength) in situ.

To conduct the hydroacoustic survey in the studied river sections, movement was carried out according to established methods [34] along a grid of zigzag tacks (Figure 3). Before conducting the echo sounding survey, the antenna is calibrated using a standard copper sphere with D = 45 mm and TS = −39 dB [21].

To conduct the hydrometric survey in the studied areas of riverbed depressions, flow velocity measurements were taken using an ISP-M flow velocity meter (LLC «Gidrometeopribor» St. Petersburg, Russia) at points of predetermined cross-sections using the Google Earth Pro 6.3 application, with distances between points of 80–100 m. The flow velocity meter consists of a measuring part, a micro-propeller (120 mm diameter) with a horizontal axis of rotation, and a signal converter of the propeller, which translates the rotation frequency of the propeller into flow speed.

The marked points were transferred to a Garmin 62Stc navigator (Garmin Ltd., Schaffhausen, Switzerland), which was used to navigate to given points and control the location. To move from point to point, the ISP-M was put in transport position, and when the small craft was positioned at the specified points on course, flow velocity measurements were taken at a depth of 2 m. This depth corresponds to the zone where beams from the hydroacoustic complex with vertical echo sounding are deployed. The hydrometric propeller (ISP-M) on a metal rod was preliminarily installed in a rotary-extendable device located in the bow of the small craft before measurements (Figure 4). To avoid errors in flow velocity measurement, the absence of small craft drift was controlled using a GPS navigator and visually, relative to the shores. Flow velocity measurement at one point was performed for at least 3 min. The measured flow velocities stored in the device's memory were then exported as a TXT file to a computer via USB cable.

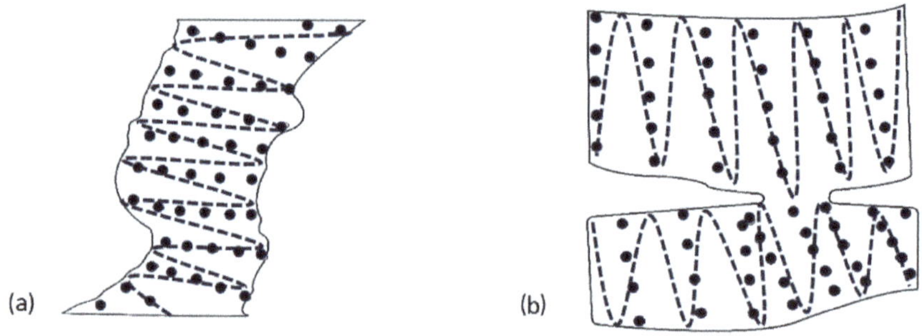

Figure 3. Survey schemes in the area of wintering riverbed depressions: (**a**) Gornoslinkinskaya; (**b**) Ukinskaya. Dotted line is a schematic representation of the zigzag tacks of a mobile sonar survey (the course of a small crafter); points are places of stationary hydrometric survey.

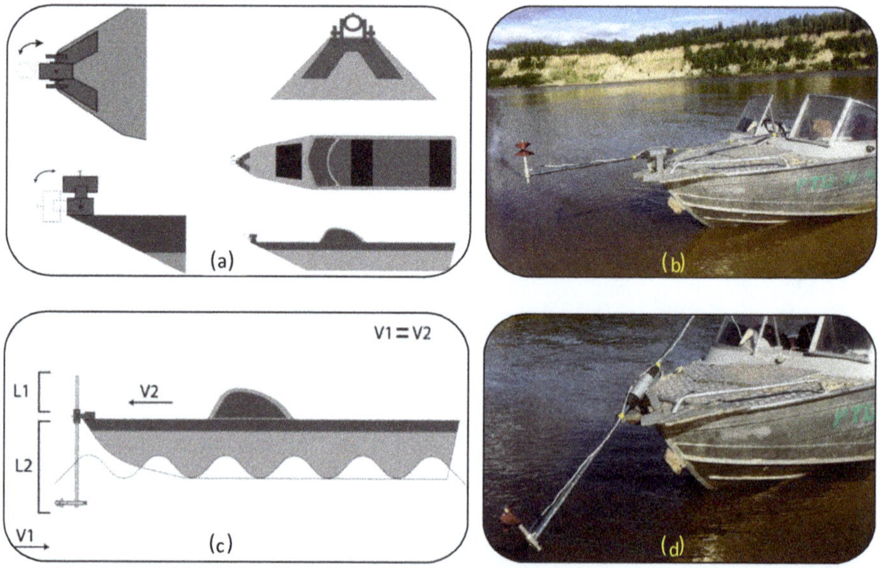

Figure 4. Diagram of installing a flow velocity meter on a small craft and the process of conducting a hydrometric survey. (**a**) Scheme of installing a metal rotary-extendable device in the bow of the small craft for transporting the flow velocity meter and conducting hydrometric measurements. (**b**) Small vessel with an installed flow velocity meter for transportation. (**c**) Scheme of conducting a hydrometric survey (L1—part of the metal rod above the small craft 's side; L2—part of the metal rod below the small craft 's side; V1—river flow velocity; V2—small craft speed). (**d**) Changing the position of the flow velocity meter for conducting a hydrometric survey.

Horizontal projections of the distribution of velocity zones and juvenile fish in the studied water areas were performed using Surfer 9.0 software (Golden Software, Golden,

CO, USA), with the geostatistical interpolation Kriging method [33,35], widely used in fishery and ecological research [36–38]. Preliminary calibration of the maps of the studied water areas was performed in the MapViewer 10.0 application, using imported satellite images of the area from Google Earth Pro 6.3. Analysis of the correlation (Pearson correlation coefficient) between fish distribution and zones of different velocity areas was performed using Statistica 10.0 software (StatSoft Inc., Tulsa, OK, USA). The normality of the data distribution was preliminarily determined (Kolmogorov–Smirnov and Shapiro–Wilk tests).

Data on flow speed zones and fish habitat areas were collected simultaneously for two riverbed depressions. For example, the values of habitat area use by Cyprinidae were compared (VS) with the values of zones for each class of flow speed.

We consider the distribution of fish over the entire water area of the riverbed—coastal zones and midstream. The midstream or thalweg are the fastest and deepest parts of the riverbed. The coastal zone of a river is the section along the coastline (close to the littoral), with flow velocities close to "0".

To control the species composition of the fish population, control fishing was conducted using a small-mesh (3–5 mm) net ("spider net", Supplementary S3).

Fishing was carried out to control the species composition of the fish population. Control fishing was conducted on 28 May 2022, 29 May 2022 (flood), 29 July 2022, and 30 July 2022 (low water). Fishing was carried out near the shores at a distance of up to 50 m, taking into account the possibility of anchoring a small craft. For one control fishing effort, the fishing gear was lowered into the water to a depth of up to 3 m, and the fish were allowed to settle for 30 min before the "spider-net" (dip net) was quickly lifted. A total of six catches were made on each side (left bank, right bank). During the flood as well as during low water periods, 12 catches were made in each riverbed depression. It is necessary to note that fishing was conducted in coastal zones with slowed current. Due to the impossibility of anchoring the small craft at greater depths in the main channel, control fishing of young fish was not carried out there. Apparently, the number of caught fish in the main channel would be significantly lower, and the overall fish density comes from hydroacoustic surveys.

3. Results

It was established that the area of the studied river sections was 75.9 and 41 hectares for the Ukinskaya and Gornoslinkinskaya riverbed depressions, respectively. During the spring–summer flood period, the water level in the Irtysh River, according to data from the nearest hydrological station "Uvat" (40 km downstream), was 437 cm (30 May 2022). When conducting research during the summer–autumn low water period, the water level was 203 cm (31 July 2022), i.e., the water level drop in the Irtysh River channel was 234 cm. During the hydroacoustic survey, it was found that during the flood period, the density of fish concentrations in the riverbed depressions was slightly lower compared to the period of low water in summer. During these periods, the fish density values in the Ukinskaya riverbed depression area were 4.83 and 5.61 thousand specimens/ha, and in the Gornoslinkinskaya riverbed depression area they were 3.98 and 4.37 thousand specimens/ha, respectively. Thus, there was an increase in density by 16% and 10%, respectively, (Table 1).

As a result of the taxonomic identification, it was established that the major part of the fish population in aggregations in riverbed depressions consists of representatives of the Cyprinidae and Percidae families, while fish from the Coregonidae, Esociidae, Acipenseridae, and Lotidae families were less represented. The proportions of cyprinids and perches in the Ukinskaya riverbed depression area during flood amounted to 51% and 33% of the total number of registered fish, while the proportions of whitefish and pike were 9%, and unrecognized fish (including sturgeon and burbot) were 7%. As the water level decreased, the proportions of registered taxonomic groups in the considered river section showed similar but slightly changed values compared to the previous period: 48%, 31%, 12%, and 9%, respectively, (Table 1). In the Gornoslinkinskaya riverbed depression area during the flood, the proportions of registered taxonomic groups were quite close to the values in the second studied river section, amounting to 51%, 34%, 8%, and 6%, respectively. During the low water period, the proportion of cyprinids and perches was 57% and 31%, while the proportions of whitefish and pike was 7.14%, and unrecognized groups were 4.59%. Thus, the main fish aggregations consisted of cyprinids and perches, with their total share in the taxonomic structure in the Ukinskaya channel pit area during flood and low water periods being 84% and 79%, respectively. In the Gornoslinkinskaya riverbed depression area, these values were 85% and 88%, respectively.

Table 1. Distribution of fish (fish density, thousand specimens/ha and share, %) in the area of riverbed depressions of the Lower Irtysh (Ukinskaya, Gornoslinkinskaya) during different hydrological phases.

Location	Ukinskaya	Gornoslinkinskaya	Ukinskaya	Gornoslinkinskaya
Hydrological Phase	Flood		Low Water	
Water level (at the nearest hydrological post "Uvat")	437		203	
Fish density, (thousand specimens/ha)	4.83	3.98	5.61	4.37
Cyprinids, (%)	50.91	51.33	48.01	57.33
Percids, (%)	33.08	33.85	30.81	30.94
Whitefish, pike (%)	9.18	8.33	12.34	7.14
Unrecognized fish, (sturgeons, burbots, %)	6.83	6.49	8.84	4.59
Area, (ha)	81	41	81	41
Percentage of utilized area, (cyprinids, %)	20.03	60	48.22	42.73
Utilized area, (cyprinids, ha)	15.20	24.60	36.60	17.52
Percentage of utilized area, (percids, %)	14.00	28.68	40.60	38.83
Utilized area, (percids, ha)	10.63	11.76	30.81	15.92

According to the data from the control fishing in the Ukinskaya and Gornoslinkinskaya riverbed depressions, the catch per unit effort (CPUE) during the flood period was 1–3 individuals, with a total of 12 catches in each riverbed depression. During the summer–autumn low water period, the CPUE was 2–5 individuals in the Ukinskaya riverbed depression and 2–6 individuals in the Gornoslinkinskaya riverbed depression. The average fish density in the specified riverbed depression areas during the flood period was 13.9 thousand individuals/ha, and during the low water period, it was 28.19 thousand individuals/ha in the Gornoslinkinskaya riverbed depression and 31.25 thousand individuals/ha in the Ukinskaya pit.

According to control fishing data, juvenile cyprinids were represented mainly by roach (*Rutilus rutilus* Linnaeus) and to a lesser extent by dace (*Leuciscus leuciscus* Linnaeus) and ide (*Leuciscus idus* Linnaeus), while juvenile perches were predominantly represented by perch (*Perca fluviatilis* Linnaeus) and slightly by pike-perch (*Stizostedion lucioperca* Linnaeus). Analysis of the size structure of the fish population in the studied areas revealed that the size group $\leq$10 cm was most abundant, with a share of 66% during flood and 74% during summer in the Ukinskaya riverbed depression. In the Gornoslinkinskaya riverbed depression, these values were 62% and 72% (Figure 5). The remaining portions of the fish population in the studied river sections during these periods are represented by groups of fish with body sizes >10 cm. Thus, the majority of fish in the area of riverbed depressions, both during the flood and during the low water period, are represented by juveniles of cyprinids and percids.

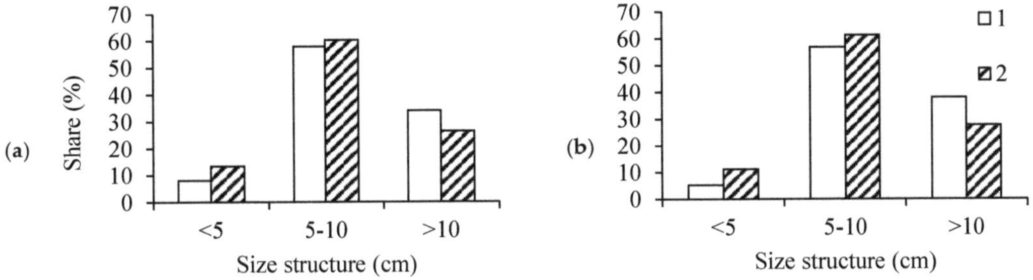

Figure 5. Size structure of fish in the areas of riverbed depressions: (**a**) Ukinskaya; (**b**) Gornoslinkinskaya. Graph bars 1 (in White)—Flood (30 May 2022) and 2 (Hatched)—low water period (31 July 2022).

As a result of the features of the channel relief on the river meander at the Gornoslinkinskaya riverbed depression, as well as due to the hydrophysical interaction of river flows during the ω-shaped evolution of the channel at the Ukinskaya riverbed depression, zones of hydrodynamic shadow areas are observed in the watercourse, with flow velocities close to "0". As a result of constructing horizontal projections of flow velocity distribution, it was established that a gradient of flow velocity is observed in the studied river sections, having peculiarities in the distribution of zones with different velocity areas. With a decrease in the water level in the river, a drop in flow velocities is also noted, resulting in an increase in the area of hydrodynamic shadow necessary for juvenile fish living in the river channel. In addition, there is a dynamic spatial distribution of points of high and low flow velocities in its horizontal projection in the studied river sections. Thus, in the waters of the Ukinskaya and Gornoslinkinskaya riverbed depressions during the flood, zones of velocities with characteristics reaching values of 1.2 and 1.3 m s^{-1}, respectively, were recorded (Table 2).

Table 2. Distribution of the area of aquatic zones with different flow velocities in the Ukinsky and Gornoslinkinsky riverbed depressions during different periods of the water level regime of the Irtysh River.

No.	Flow Velocity Class, (m s^{-1})	Area During Flood	Area During Low Water	Area Changes (+/−)		Area During Flood	Area During Low Water	Area Changes (+/−)	
		Ukinskaya				Gornoslinkinskaya			
		ha	ha	ha	%	ha	ha	ha	%
1.	1.3	0.960	–	−0.960	−100.00	–	–	–	–
2.	1.2	5.595	–	−5.595	−100.00	–	0.273	−0.273	−100.00
3.	1.1	2.477	–	−2.477	−100.00	–	2.470	−2.470	−100.00
4.	1	2.477	–	−2.477	−100.00	–	8.221	−8.221	−100.00
5.	0.9	4.765	–	−4.765	−100.00	–	8.028	−8.028	−100.00
6.	0.8	3.674	–	−3.674	−100.00	0.961	7.188	−6.227	−86.63
7.	0.75	4.137	0.355	−3.782	−91.42	4.596	3.655	0.942	25.76

Table 2. *Cont.*

No.	Flow Velocity Class, (m s^{-1})	Area During Flood	Area During Low Water	Area Changes (+/−)		Area During Flood	Area During Low Water	Area Changes (+/−)	
		Ukinskaya				Gornoslinkinskaya			
		ha	ha	ha	%	ha	ha	ha	%
8.	0.7	2.679	2.397	−0.282	−10.51	14.969	3.057	11.911	389.57
9.	0.65	1.766	3.171	1.405	79.56	11.386	3.543	7.842	221.32
10.	0.6	1.328	3.494	2.166	163.17	9.664	5.113	4.552	89.03
11.	0.55	1.162	3.193	2.031	174.85	4.194	2.693	1.501	55.75
12.	0.5	1.529	2.999	1.470	96.15	3.023	2.744	0.280	10.20
13.	0.45	1.114	3.633	2.519	226.11	4.710	3.695	1.015	27.46
14.	0.4	0.948	3.741	2.793	294.51	4.640	4.191	0.449	10.70
15.	0.35	0.818	3.590	2.773	339.00	5.024	4.222	0.803	19.01
16.	0.3	0.759	3.279	2.520	332.20	5.164	6.247	−1.082	−17.33
17.	0.25	0.699	2.214	1.515	216.65	3.154	1.610	1.545	95.96
18.	0.2	0.687	2.472	1.785	259.64	3.277	3.149	0.128	4.07
19.	0.15	0.842	3.644	2.803	333.02	0.516	3.027	−2.512	−82.97
20.	0.1	1.185	1.645	0.459	38.76	0.350	1.731	−1.382	−79.81
21.	0	1.399	1.172	−0.227	−16.23	0.271	1.043	−0.772	−74.02

During the low water period, the maximum flow velocities recorded in the studied water bodies reached values of 0.8 and 0.75 m s^{-1}. Consequently, at lower water levels, higher velocity sections (0.8–1.3 m s^{-1}) were either not recorded at all or their area significantly decreased, while the area of medium-speed sections (0.15–0.7 m s^{-1}) increased considerably. The area of zones with flow velocities in the range of 0–0.30 m s^{-1} in the Ukinskaya riverbed depression increased, whereas it decreased in the Gornoslinkinskaya riverbed depression (Table 2, Figure 6).

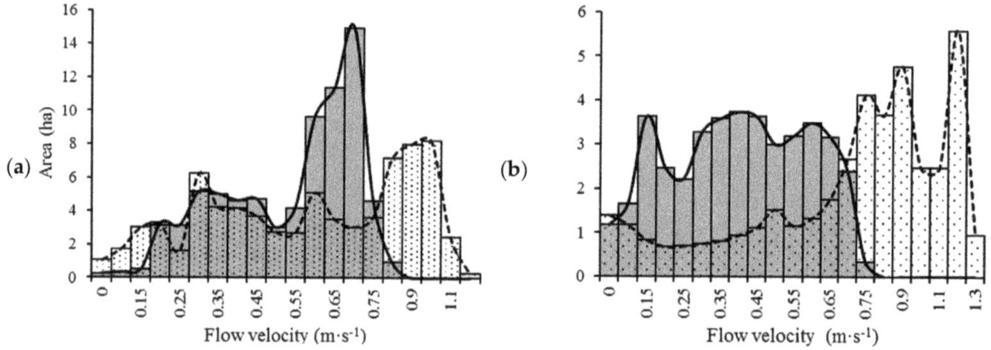

Figure 6. Dynamics of areas with different flow velocities in (**a**) Ukinskaya riverbed depression and (**b**) Gornoslinkinskaya riverbed depression during the flood period (white, dotted line) and low water period (gray, solid line).

As a result of the recorded dynamics of decreasing flow velocities and increasing areas with reduced velocity values, there is an observed increase in the area utilized by perch fish in the studied riverbed depressions. In the Ukinskaya riverbed depression, the area occupied by juvenile perch fish amounted to 14% and 1% of the total area of the riverbed depression, or 11 and 311 hectares, respectively, (Table 1). Similarly, an increase in the occupied area is noted for juvenile carp fish in this section, with the occupied area growing by more than two times: 20% and 48%, or 15 and 37 hectares during the flood and low water periods, respectively, (Table 1). The location of juvenile cyrpinid and percid fish aggregations in the Ukinskaya riverbed depression during the flood period was identical and was noted in the left bank part of the river, with flow velocity zones of 0–0.75 m s^{-1} ; additionally, a small number of fish were registered in the channel part of the river, also with reduced flow velocities of 0.40–0.75 m s^{-1} (Figure 7).

Table 3. Magnitude of correlation relationships between distributions of carp (CPR) and perch (PRC) abundance and flow velocity zones in riverbed depressions, taking into account the dynamics of values during the flood–low water period.

No.	Pair of Variables **	Fish (S_abs) × Current (S_rel)		Fish (S_rel) × Current (S_abs)		Fish (S_rel) × Current (S_rel)		Fish (S_abs) × Current (S_abs)	
	Velocity Class	CPR	PRC	CPR	PRC	CPR	PRC	CPR	PRC
1.	0	−0.495	−0.695	0.078	−0.411	0.422	−0.033	−0.743	−0.921
2.	0.10	−0.782	−0.733	−0.534	−0.588	0.007	−0.093	**−0.999** *	−0.882
3.	0.15	−0.722	−0.407	−0.656	−0.282	−0.286	0.083	−0.868	−0.523
4.	0.20	−0.228	0.267	−0.696	0.005	−0.463	0.334	0.086	0.488
5.	0.25	0.133	0.523	−0.140	0.589	0.045	0.762	0.525	0.871
6.	0.30	−0.285	0.195	−0.831	−0.289	−0.786	−0.041	−0.038	0.293
7.	0.35	−0.112	0.376	−0.591	0.136	−0.393	0.423	0.199	0.598
8.	0.40	−0.215	0.267	−0.640	0.111	−0.372	0.409	0.103	0.530
9.	0.45	−0.117	0.342	−0.506	0.258	−0.231	0.537	0.244	0.657
10.	0.50	−0.333	−0.014	−0.545	0.277	0.044	0.526	0.039	0.513
11.	0.55	−0.002	0.402	−0.286	0.477	−0.014	0.693	0.411	0.796
12.	0.60	0.563	0.889	−0.227	0.304	−0.139	0.565	0.639	0.858
13.	0.65	0.817	**0.994** *	0.056	0.497	0.202	0.717	0.822	**0.960** *
14.	0.70	**0.952** *	**0.975** *	0.190	0.513	0.356	0.639	0.903	**0.959** *
15.	0.75	0.417	−0.066	0.139	−0.267	0.512	−0.216	0.586	0.271
16.	0.80	−0.360	−0.685	−0.584	**−0.982** *	−0.214	−0.878	−0.460	−0.613
17.	0.90	−0.402	−0.750	−0.507	**−0.984** *	−0.123	−0.835	−0.540	−0.727
18.	1.00	−0.548	−0.719	−0.738	**−0.989** *	−0.537	**−0.989** *	−0.584	−0.632
19.	1.10	−0.240	−0.676	−0.186	−0.868	0.207	−0.608	−0.429	−0.755
20.	1.20	0.063	−0.412	0.652	−0.147	0.668	−0.126	0.050	−0.425
21.	1.30	0.078	−0.396	0.686	−0.101	0.686	−0.101	0.078	−0.396

*—Bold indicate significant correlation relationships (at $p < 0.05$); ** FISH (S_abs)—absolute index of the area of riverbed depressions occupied by carp and perch groups, ha; CURRENT (S_abs)—absolute index of the area of water zones with a certain current speed; FISH (S_rel)—relative index of the area of riverbed depressions occupied by carp and perch groups, %; CURRENT (S_rel)—absolute index of the area of water zones with a certain current speed.

During low water, the localization of aggregations changed; cyprinids were recorded in the channel and right bank parts of the watercourse at flow velocity ranges of 0.40–0.75 m s^{-1}. The left bank section of the riverbed depression's water area with zones of 0.30–0.55 m s^{-1} was additionally developed by juvenile perch fish. (Figure 7).

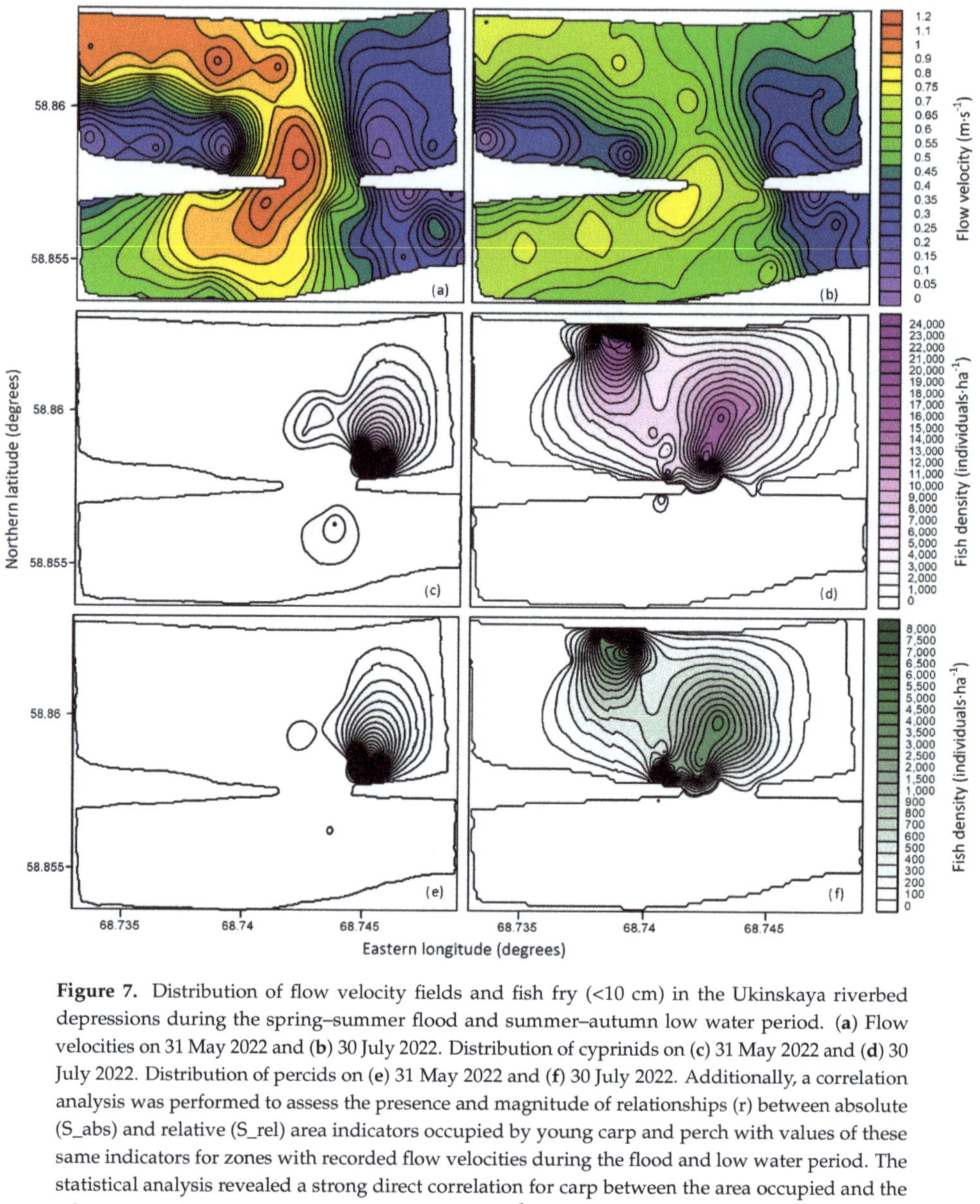

Figure 7. Distribution of flow velocity fields and fish fry (<10 cm) in the Ukinskaya riverbed depressions during the spring–summer flood and summer–autumn low water period. (**a**) Flow velocities on 31 May 2022 and (**b**) 30 July 2022. Distribution of cyprinids on (**c**) 31 May 2022 and (**d**) 30 July 2022. Distribution of percids on (**e**) 31 May 2022 and (**f**) 30 July 2022. Additionally, a correlation analysis was performed to assess the presence and magnitude of relationships (r) between absolute (S_abs) and relative (S_rel) area indicators occupied by young carp and perch with values of these same indicators for zones with recorded flow velocities during the flood and low water period. The statistical analysis revealed a strong direct correlation for carp between the area occupied and the relative area of zones with flow velocities of 0.70 m s^{-1} ($r = 0.952$, $p < 0.05$) (Table 3).

In the Gornoslinkinskaya riverbed depression, the effect of changes in habitat utilization is most vividly expressed not in the increase in usable area but in the shift in aggregation locations for fry. Thus, during the flood period, the highest concentrations of carp and perch fry were located near coastal areas of the left bank part of the riverbed depressions: carp in the zones of current velocities of 0.45–0.70 m s^{-1}, perch in the zones of lower velocities of 0.45–0.65 m s^{-1}.

During low water periods, aggregations of carp and perch were additionally found in the open channel part of the watercourse; both on the channel and on the right bank, individuals of both families were distributed mainly in areas with current velocities of

0.25–0.50 m s^{-1}. (Figure 8). It is noted that for young carp, there was a decrease in the area occupied from flood to low water: from 60% to 43%, or from 25 to 18 hectares. Conversely, for young perch, an opposite trend was observed—an expansion of the occupied area: from 29% to 39%, or from 12 to 16 hectares, respectively, (Table 1).

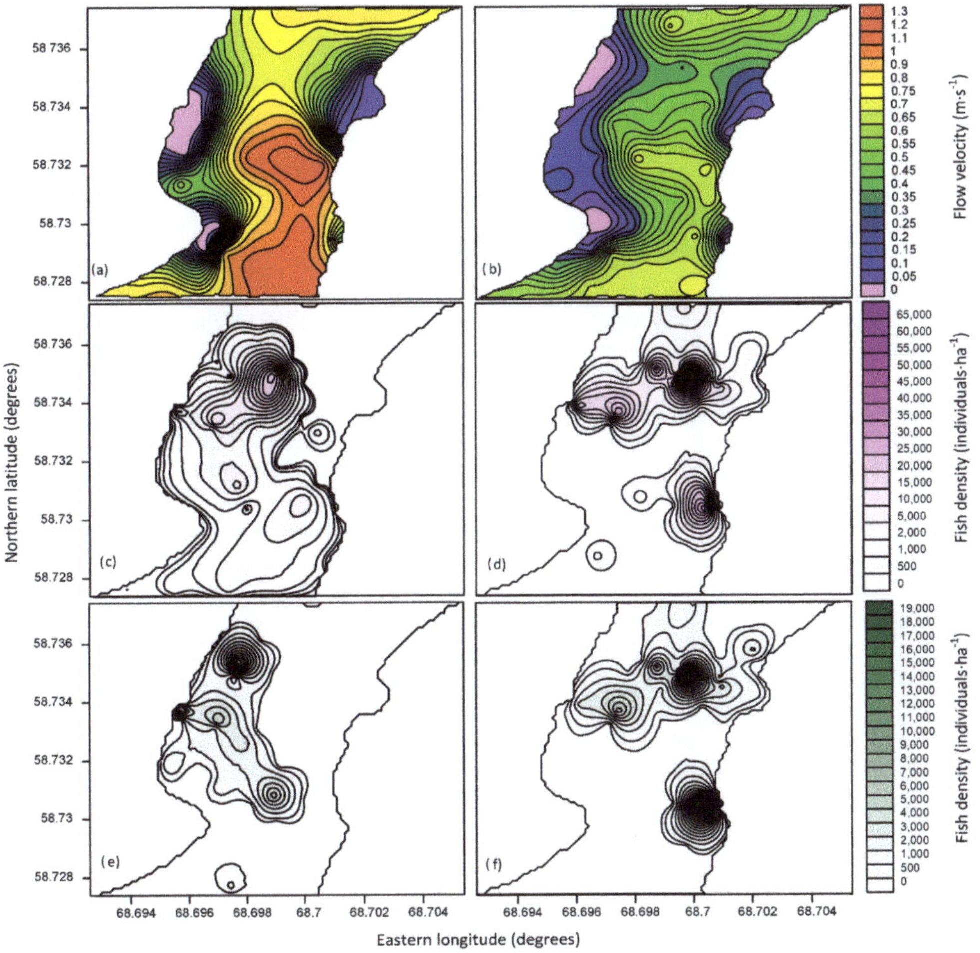

Figure 8. Distribution of flow velocity fields and fish fry (<10 cm) in the Gornoslinkinskaya riverbed depression during the spring–summer flood and summer–autumn low water period. (**a**) Flow velocities on 31 May 2022 and (**b**) 30 July 2022. Distribution of cyprinids on (**c**) 31 May 2022 and (**d**) 30 July 2022. Distribution of percids on (**e**) 31 May 2022 and (**f**) 30 July 2022.

For perch, a strong correlation was also found between the area occupied in the riverbed depression and the relative area of zones with flow velocities of 0.65 and 0.70 m s^{-1} (r = 0.994 and r = 0.975, $p < 0.05$) (Table 3). When analyzing the relationship between the relative area occupied and flow velocity areas, a strong inverse relationship was found only for perch with flow velocity areas of 0.80, 0.90, and 1.00 m s^{-1} (r = −0.982, r = −0.984, and r = −0.989, $p < 0.05$) (Table 3). The analyses of the relationship between the relative area occupied and relative areas of flow sections revealed a strong inverse relationship only for perch with a flow velocity area of 1.00 m s^{-1} (r = −0.989, $p < 0.05$) (Table 3). For indicators of occupied riverbed depression areas by carp and flow velocity areas at 0.10 m s^{-1}, a strong negative correlation was found ($r = -0.999$, $p < 0.05$), while for perch, a strong positive correlation was found, with flow velocity sections at 0.65 and 0.70 m s^{-1} ($r = 0.960$ and $r = 0.959$, $p < 0.05$) (Table 3). In all other cases, no significant correlation was detected.

The analysis established a strong correlation between the area occupied by juvenile fish in riverbed depressions and the relative area of fast-flowing sections, with values of 0.70 m s^{-1} for carp species and 0.65 to 0.70 m s^{-1} for perch species. A negative correlation was observed only for perch in areas with flow speeds of 0.80, 0.90, and 1.00 m s^{-1}. Thus, it has been reliably shown that the studied fish groups preferentially inhabit flow speeds of 0.65–0.70 m s^{-1} and tend to avoid faster sections with speeds of 0.80–1.00 m s^{-1}.

4. Discussion

The increase in the area occupied by juvenile fish in riverbed depressions is primarily due to a decrease in flow velocity characteristics. As water levels drop, in our case study, the proportion of areas with average flow velocities less than 0.75 m s^{-1} significantly increases, while areas with flow velocities of 0.75–1.3 m s^{-1} almost completely disappear. The developed low-velocities zones in the riverbed depressions are similar in dynamic characteristics to resident (littoral) habitats, located in the riverbed at a distance from the coastline and at great depths; nevertheless, there is a concentration of juvenile fish here.

This feature is linked to juvenile fish preferring calm areas with low velocities [39]. For example, fish with small, elongated body shapes (representatives of the Cyprinidae family and Nemacheilidae), prefer zones of reduced speeds in the range of 0.15–0.51 m s^{-1} [40]. Additionally, it was established that the area occupied by fish decreases as flow speed increases in the central part of the riverbed (midstream), and fish tend to accumulate and migrate mainly near the riverbank [41]. This phenomenon is likely related to flow velocity characteristics exceeding critical velocities for fish during high discharge. Similarly, in our case, at higher water levels, flow velocities are higher, leading to a larger proportion of high-velocity areas; as flow discharge decreases, either these high-velocity areas diminish or are simply not recorded. Moreover, there is a shift in preferred velocity areas from the shoreline towards the river's thalweg.

Considering the velocity characteristics and distribution of juvenile fish, it becomes clear that one of the key factors ensuring the formation of juvenile fish aggregations in riverbed depressions is the hydrodynamic characteristics of these sections. This is confirmed by a direct correlation between the area occupied by juvenile carp and perch and the sizes of zones with flow speeds of 0.50, 0.55, 0.65, 0.75, and 0.90 m s^{-1} (2 m below water surface) during flood vs. low water.

Simultaneously, the formation of fish aggregations in sections with lower velocities-may occur due to loss of orientation among juveniles at minimal velocities [42] and also due to the need for rest when moving into faster areas [41]. This bioenergetic strategy allows fish to avoid being swept away by currents due to muscle fatigue under heavy currents, e.g., with an increase of flow velocity, glycogen content in fish livers decreases while lactic acid levels in their blood and muscles significantly rise [27]. Ultimately, dynamics related to hydrodynamic factors cause significant metabolic changes in fish [27,43], leading to changes in behavior expressed through shifts in preferred flow zones and redistribution across riverbed depression areas. Spatial segregation among different fish species—specifically differences in selecting zones within riverbed depression areas—results from morphological characteristics inherent to these families since it has been shown [44,45] that even closely related species exhibit varying swimming capabilities. It has been established [46] that morphological differences affect their ability to perform precise maneuvers and powerful accelerations, while influencing energy expenditure during prolonged swimming. Lifestyle differences, metabolic rate variations, and morphological traits contribute to differing swimming capabilities and resting requirements among various fish species [46–49]. Furthermore, it has been noted [25] that critical speed metrics vary significantly depending on family and species classification.

The morphology of elongated-bodied fish like juvenile roach and bleak represents their adaptive response aimed at reducing hydrodynamic resistance and energy expenditure while navigating through currents [23]. Thus, a greater occupancy area among carp species can be explained by their superior ability to withstand water currents since this capability correlates with morphological traits associated with swimming activity [45]. Although a strong relationship between swimming ability and body size has been demonstrated [50,51], body length remains a significant variable explaining critical speed magnitude [25]. In our case, we compare occupancy within riverbed depression areas for juvenile perch and carp exhibiting similar size characteristics, hence we attribute distribution differences to pronounced morphological distinctions [25] which determine swimming ability and association with specific hydrodynamic environments [52,53] despite identical body length metrics among species [43]. It has been shown [25,49] that species with more elongated body shapes can minimize energy loss due to resistance at maximum swimming speeds.

The observed flow circulations within riverbed depression areas may disorient juvenile fish [53], potentially serving as an additional factor influencing aggregation formation. Moreover, close proximity between significantly differing speed zones typical for lentic and lotic aquatic systems within researched river sections promotes fish aggregations; many authors note [54,55] that higher species diversity and density are recorded during summer studies at inflow zones where streams enter lakes.

An analysis of swimming abilities—including determining critical speed as well as optimal and maximum swimming speeds—for different freshwater fishes, including roach and perch, revealed [56] that trout (*Oncorhynchus mykiss* Walbaum), roach, and carp (*Cyprinus carpio* Linnaeus) exhibited the highest swimming capabilities and endurance. The highest recorded critical speed values—1 m s^{-1} —were noted for large trout individuals capable of extensive movements. For roach under sizes <10 cm, critical speed values ranged from 0.5 to 0.6 m s^{-1}, while for perch measuring around 10 cm, critical speed reached approximately 0.81 m s^{-1} [56]. In summarizing existing data on critical speeds, it was also established [26] that for roach, perch, and zander, these metrics hold high values compared to other freshwater species. Slightly lower critical speed values for roach and perch obtained under laboratory conditions [56] compared to preferred high-velocity zones recorded here cannot be entirely applied to individuals within natural habitats [57]; however, data obtained by [56] provide an overall understanding regarding swimming capabilities across various species.

It should also be noted that horizontal distribution among fishes within riverbed depression areas changed alongside water level dynamics, while juvenile preferences towards zones characterized by flow speeds around 0.75 m s^{-1} remained consistent. Notably, along with shifts in juvenile aggregation localization from flood towards low water, overall patterns regarding the positioning of more rapid sections [58] as well as dynamic axes within flows [59] along the outer banks (concave shore) of studied meanders remained unchanged.

Overall, distribution among fishes within lotic systems can be attributed to various abiotic factors, where we consider hydrodynamic factors as one key element influencing distribution patterns. This was previously demonstrated within a small tributary of the Irtysh River [60], indicating how hydrodynamic effects create environmental heterogeneity affecting both direct distribution patterns as well as indirectly through modifications on physical habitat characteristics. Nevertheless, distribution features among fishes within flows at family levels can primarily be explained through the pronounced morphological differences observed [61].

It should be noted that previous studies analyzing microhabitat selection among various fish species indicated reliable connections between variables, characterizing body morphology alongside habitat conditions including flow velocities for both cyprinids and percids, establishing significant relationships at $p < 0.05$. It becomes apparent that distribution patterns were aligned towards specific flow zones exhibiting average flows between ranges of approximately 0.45–0.70 m s^{-1} and lower flows around 0.25–0.50 m s^{-1} during floods and low flow, respectively. Since maintaining position within faster flowing zones requires greater energy expenditures, such areas are energetically unfavorable from a bioenergetics perspective.

In shallow water conditions, in particular in lakes with a thermocline, the presence of the "guiding effect" phenomenon of the redistribution of fish under the vessel's hull is recognized [62]. This is caused by the simultaneous influence of vessel noise and low-frequency noise of parts of the trawl (fishing gear) [63] as well as the environment, primarily the influence of temperature, which can be enhanced by the halocline and bathymetric characteristics of the reservoir. We believe that in a lotic environment, as well as in the absence of trawl fishing during our acoustic surveys, the influence of these factors can be leveled out and reduced to minimum values due to the constant mixing of water masses by intense multidirectional currents. In addition, it is noted that the presence of predators can have a much greater effect on the distribution of schools of fish than the noise of the vessel and parts of fishing gear [63].

At the same time, it is known that hydroacoustic studies performed using high-frequency echosounders with an operating frequency of 455 kHz and a scanning beam provide the least error in the result [62]. However, in our study, with the goal to determine distribution patterns of fish aggregations of certain taxonomic groups, we used a dual-beam echosounder with operating frequencies of 50 and 200 kHz and beam opening angles of 46 and 140, respectively. The technique we use for the remote identification of fish by the shape of the swim bladder is based on the operation of an echosounder with the specified characteristics [19].

Our study revealed that the main abiotic factors are hydrodynamic ones—the intensity of currents and their direction. The effect of current velocities is confirmed by the distribution of fish in certain speed zones, and the number of juvenile fish in the riverbed is influenced by the dynamics of the area of suitable habitats, determined by the water level during flood and low flow. The Irtysh River Basin is inhabited by around 15 million people in China, Kazakhstan, and Russia, thus the river faces multiple anthropogenic stressors [64]. Among those stressors, pollution [65–67], hydrological changes due to hydropower use [68], river-borne cargo, and irrigation [69], as well as fisheries and poaching, [70,71] are relevant. Ecosystem services generated by fish populations are an important aspect with regard to socioecology [72]. In this context, our study contributes to the knowledge about fish habitats in the lower Irtysh River and recommends the protection of those important habitat sections to ensure biodiversity conservation.

5. Conclusions

This study contributes to the understanding of the multifunctional role of riverbed depressions along river meanders. The local distribution of juvenile fish is related to their swimming abilities, the morphology of the channel and the use of low-velocity zones as recruitment areas.

The dynamics of the water level, accompanied by changes in areas and volumes of habitats in the floodplain–channel complex along the lower reaches of Irtysh River, include both an expansion and reduction of habitats, along with a change in flow velocities. The transition of juveniles from lentic to lotic water bodies is accompanied by a change in the characteristics of their habitats, first of all, regarding hydrodynamics—a change in the conditions from «still waters» to «flowing waters». The absence of low-velocity sections in the main channel would lead to a decrease in the survival rate of juvenile fish, which are evolutionarily adapted to such habitats in the floodplain–channel system. Thus, the importance of the presence of such sections for the functioning of ecosystems of floodplain–channel complexes is emphasized.

The studied river sections, which have a multifunctional biological role, should be provided with appropriate protection against changes in the morphology and relief of the riverbed. If anthropogenic transformation of the riverbed is necessary, water managers need to ensure the availability of the characteristics of velocity zones and their dynamics in the meandering sections of the river, in order to ensure fish recruitment in the floodplain–river complex.

The data obtained on the distribution of juvenile cyprinids and percids in the river meanders demonstrate the need for low-velocity sections and zones of hydrodynamic "shade", providing the juveniles with fundamental habitat conditions in the summer-autumn low-flow period, that contribute to the conservation of energy and fat reserves to successfully overcome the conditions during the subsequent winter period.

Supplementary Materials: The following supporting information can be downloaded at: https://www.mdpi.com/article/10.3390/d17010068/s1, The video files S1 and S2 show fragments of echograms with recorded depths, coordinates and targets (fish), obtained during hydroacoustic surveys during the flood period on 30 May 2022. S3 provides a scheme of the fishing gear and methodology of control fishing.

Author Contributions: Conceptualization, A.A.C. and M.S.; methodology, A.A.C. and E.I.P.; formal analysis, A.A.C. and E.I.P.; investigation, A.A.C. and E.I.P.; resources, A.A.C.; writing—original draft preparation, A.A.C., E.I.P., and M.S.; project administration, A.A.C.; funding acquisition, A.A.C. All authors have read and agreed to the published version of the manuscript.

Funding: This work was carried out with the support of the fundamental scientific research topic "Patterns of distribution and migration of fish in lotic and limnic water bodies of the Ob-Irtysh basin" (Ministry of Science and Higher Education of the Russian Federation, № 1022040700418-3-1.6.21).

Data Availability Statement: Raw data related to this study are available upon request from A.A.C.

Acknowledgments: The authors express their gratitude to the management of the Tobolsk Complex Scientific Station of the Ural Branch of the Russian Academy of Sciences (TCSS UrB RAS) for the technical support provided, as well as to the staff of the Laboratory of Ecology of Hydrobionts (TCSS UrB RAS) for the technical and physical assistance provided in collecting material for conducting the study and preparing the article. And we thank two anonymous reviewers for their valuable comments and constructive suggestions.

Conflicts of Interest: The authors declare no conflicts of interest.

ferences

Muška, M.; Tušer, M.; Frouzová, J.; Mrkvička, T.; Ricard, D.; Sed'a, J.; Morelli, F.; Kubečka, J. Real-time distribution of pelagic fish: Combining hydroacoustics, GIS and spatial modelling at a fine spatial scale. *Sci. Rep.* **2018**, *8*, 5381. [CrossRef] [PubMed]
Sarre, A.; Demarcq, H.; Keenlyside, N.; Krakstad, J.-O.; El Ayoubi, S.; Jeyid, A.M.; Faye, S.; Mbaye, A.; Sidibeh, M.; Brehmer, P. Climate change impacts on small pelagic fish distribution in Northwest Africa: Trends, shifts, and risk for food security. *Sci. Rep.* **2024**, *14*, 12684. [CrossRef] [PubMed]
Rinke, K.; Huber, A.M.R.; Kempke, S.; Eder, M.; Wolf, T.; Probst, W.N.; Rothhaupta, K.-O. Lake-wide distributions of temperature, phytoplankton, zooplankton, and fish in the pelagic zone of a large lake. *Limnol. Oceanogr.* **2009**, *54*, 1306–1322. [CrossRef]
Parker, G.A.; Smith, J.M. Optimality theory in evolutionary biology. *Nature* **1990**, *348*, 27–33. [CrossRef]

5. Pinel-Alloul, P. Spatial heterogeneity as a multiscale characteristic of zooplankton community. *Hydrobiologia* **1995**, *300–301*, 17- [CrossRef]
6. Yanygina, L.; Burmistrova, O.; Kotovshchikov, A.; Shirinina, M.; Schletterer, M. Why do phyto- and zooplankton exhibit differ patterns of seasonal dynamics in the large Ob river-floodplain system (West Siberia, Russia)? *Hydrobiologia* **2024**, *in pr* [CrossRef]
7. Jeppesen, E.; Pekcan-Hekim, Z.; Lauridsen, T.L.; Søndergaard, M.; Jensen, J.P. Habitat distribution of fish in late summer: Chan along a nutrient gradient in Danish lakes. *Ecol. Freshw. Fish* **2006**, *15*, 180–190. [CrossRef]
8. Buisson, L.; Blanc, L.; Grenouillet, G. Modelling stream fish species distribution in a river network: The relative effects temperature versus physical factors. *Ecol. Freshw. Fish* **2008**, *17*, 244–257. [CrossRef]
9. Nannini, M.A.; Goodrich, J.; Dettmers, J.M.; Soluk, D.A.; Wahl, D.H. Larval and early juvenile fish dynamics in main channel a backwater lake habitats of the Illinois River ecosystem. *Ecol. Freshw. Fish* **2012**, *21*, 499–509. [CrossRef]
10. Tucker, T.R.; DeBruyne, R.L.; Roseman, E.F.; Larson, D.; McNaught, A.S. Assessment of larval fish assemblages and nurs habitat in the St. Clair River delta. *J. Great Lakes Res.* **2019**, *45*, 762–776. [CrossRef]
11. Bickel, T.O.; Closs, G.P. Fish distribution and diet in relation to the invasive macrophyte Lagarosiphon major in the littoral zc of Lake Dunstan, New Zealand. *Ecol. Freshw. Fish* **2008**, *17*, 10–19. [CrossRef]
12. Górski, K.; Buijse, A.D.; Winter, H.V.; De Leeuw, J.J.; Compton, T.J.; Vekhov, D.A.; Zolotarev, D.V.; Verreth, J.A.J.; Nagelker L.A.J. Geomorphology and flooding shape fish distribution in a large-scale temperate floodplain. *River Res. Appl.* **2012**, 1226–1236. [CrossRef]
13. Brown, C. Habitat-predator association and avoidance in rainbowfish (*Melanotaenia* spp.). *Ecol. Freshw. Fish* **2003**, *12*, 118–1 [CrossRef]
14. Lautala, T.; Hirvonen, H. Antipredator behaviour of naïve Arctic charr young in the presence of predator odours and conspec alarm cues. *Ecol. Freshw. Fish* **2008**, *17*, 78–85. [CrossRef]
15. Lloren, J.I.; Davidson, S.M.; Twardek, W.M.; Elvidge, C.K. Baseline activity and shoal type determine antipredator behaviors bluegill from a southern Ontario lake. *Behav. Ecol. Sociobiol.* **2019**, *73*, 57. [CrossRef]
16. Gerber, K.M.; Mather, M.E.; Smith, J.M.; Peterson, Z.J. Multiple metrics provide context for the distribution of a highly mobile f predator, the blue catfish. *Ecol. Freshw. Fish* **2018**, *28*, 141–155. [CrossRef]
17. Brehmer, P.; Do Chi, T.; Laugier, T.; Galgani, F.; Laloë, F.; Darnaude, A.M.; Fiandrino, A.; Mouillot, D. Field investigations a multi-indicators for shallow water lagoon management: Perspective for societal benefit. *Aquat. Conserv. Mar. Freshw. Ecos* **2011**, *21*, 728–742. [CrossRef]
18. Simmonds, E.J.; MacLennan, D.N. *Fisheries Acoustics*, 2nd ed.; Blackwell Science: Oxford, UK, 2005.
19. Borisenko, E.S.; Degtev, A.I.; Mochek, A.D.; Pavlov, D.S. Hydroacoustic characteristics of mass fishes of Ob-Irtish basin. *J. Ichth* **2006**, *46*, 227–234. [CrossRef]
20. Schmidt, M.B.; Tuhtan, J.A.; Schletterer, M. Hydroacoustic and Pressure Turbulence Analysis for the Assessment of Fish Preser and Behavior Upstream of a Vertical Trash Rack at a Run-of-River Hydropower Plant. *Appl. Sci.* **2018**, *8*, 1723. [CrossRef]
21. Chemagin, A.A.; Schletterer, M. Insights into Daily Dynamics of Fish Migration during Spring in the Konda River. *Diversity* **20** *15*, 1211. [CrossRef]
22. Seidl, F.; Reisenbüchler, M.; Rutschmann, P.; Yanygina, L.V.; Schletterer, M. Large-Scale Hydromorphological Characteristics The Proglacial River Katun (Ob Headwaters). *Geogr. Environ. Sustain.* **2023**, *16*, 110–120. [CrossRef]
23. Alexandre, C.M.; Quintella, B.R.; Ferreira, A.F.; Romão, F.A.; Almeida, P.R. Swimming performance and ecomorphology of Iberian barbel *Luciobarbus bocagei* (Steindachner, 1864) on permanent and temporary rivers. *Ecol. Freshw. Fish* **2013**, *23*, 244–2 [CrossRef]
24. Schletterer, M.; Reindl, R.; Thonhauser, T. Options for re-establishing river continuity, with an emphasis on the special soluti "fish lift": Examples from Austria. *Rev. Eletrônica Gestão Tecnol. Ambient. (GESTA)* **2016**, *4*, 109–128. [CrossRef]
25. Cano-Barbacil, C.; Radinger, J.; Argudo, M.; Rubio-Gracia, F.; Vila-Gispert, A.; García-Berthou, E. Key factors explaining criti swimming speed in freshwater fish: A review and statistical analysis for Iberian species. *Sci. Rep.* **2020**, *10*, 18947. [CrossRef]
26. García-Vega, A.; Ruiz-Legazpi, J.; Fuentes-Pérez, J.F.; Bravo-Córdoba, F.J.; Sanz-Ronda, F.J. Effect of thermo-velocity barriers fish: Influence of water temperature, flow velocity and body size on the volitional swimming capacity of northern straight-mou nase (*Pseudochondrostoma duriense*). *J. Fish Biol.* **2023**, *102*, 689–706. [CrossRef]
27. Zhai, C.; Li, Y.; Zhu, G.; Peng, W.; E, Q.; Zhang, Y.; Ma, B. The Effects of Water Flow Speed on Swimming Capacity and Ener Metabolism in Adult Amur Grayling (*Thymallus grubii*). *Fishes* **2024**, *9*, 272. [CrossRef]
28. Scetsiv, I.V. Environmental problems of the Irtysh at the present stage. In Proceedings of the International Scientific and Practi Conference "Problems of Management and Rational Use of Water Resources in the Irtysh River Basin", Omsk, Russia, 20–21 M 2004.
29. Chemagin, A.A. Features of hydraulics of the river sectors with riverbed wintering holes of the Irtysh in open water period. *Ves Astrakhanskogo Gos. Tekhnicheskogo Univ. Seriya Rybn. Khozyaystvo* **2018**, *3*, 60–69. [CrossRef]

Pavlov, D.S.; Mochek, A.D.; Borisenko, E.S.; Degtev, A.I.; Shakirov, R.R.; Degtev, E.A. Biological significance of the Gornoslinkinskaya Riverbed Depression in the Irtysh. *J. Ichthyol.* **2006**, *46*, S125–S133. [CrossRef]
Chemagin, A. Distribution of sturgeon in the River Irtysh. *Regul. Mech. Biosyst.* **2020**, *11*, 444–448. [CrossRef]
Operator's Manual: FURUNO LS-4100. Available online: https://www.furuno.ru/fileadmin/files/Manuals/4_Echo_sounder/LS-4100/LS-4100_OM_ENG_23710C.pdf (accessed on 1 November 2024).
Aldushin, A.; Novozhilov, O. Hydroacoustic assessment of conditions of spawning migrations of fish on the example of the Prokhladnaya River in the Kaliningrad Region. *Fisheries* **2023**, *2023*, 110–114. [CrossRef]
Yudanov, K.I.; Kalikhman, I.L.; Tesler, V.D. *Rukovodstvo po Provede-Niyu Gidroakusticheskikh S'emok [Guidelines for Hydroacoustic Surveys]*; VNIRO: Moscow, Russia, 1984; 124p. (In Russian)
Kanevski, M.; Maignan, M. *Analysis and Modelling of Spatial Environmental Data*; EPFL Press: Lausanne, Switzerland, 2004; 306p.
Patti, B.; Bonanno, A.; Basilone, G.; Goncharov, S.; Mazzola, S.; Buscaino, G.; Cuttitta, A.; Lafuente, J.G.; García, A.; Palumbo, V.; et al. Interannual fluctuations in acoustic biomass estimates and in landings of small pelagic fish populations in relation to hydrology in the Strait of Sicily. *Chem. Ecol.* **2004**, *20*, 365–375. [CrossRef]
Bonanno, A.; Goncharov, S.; Mazzola, S.; Popov, S.; Cuttitta, A.; Patti, B.; Basilone, G.; Di Nieri, A.; Patti, C.; Aronica, S.; et al. Acoustic evaluation of anchovy larvae distribution in relation to oceanography in the Cape Passero area (Strait of Sicily). *Chem. Ecol.* **2006**, *22*, 265–273. [CrossRef]
Goncharov, S.M.; Popov, S.B.; Polyakov, A.V.; Savelyeva, E.A.; Mazzola, S.; Bonanno, A.; Patti, B. Determining the accuracy of estimating the stock of aquatic organisms during geostatistical interpolation of initial data. *Rybn. Khozyaystvo* **2010**, *4*, 36–39.
Cabonce, J.; Wang, H.; Chanson, H. Ventilated Corner Baffles to Assist Upstream Passage of Small-Bodied Fish in Box Culverts. *J. Irrig. Drain. Eng.* **2018**, *144*, 04018020. [CrossRef]
Chen, X.; Liu, S.; Wang, Y.; Hao, Y.; Li, K.; Wang, H.; Liang, R. Restoration of a fish-attracting flow field downstream of a dam based on the swimming ability of endemic fishes: A case study in the upper Yangtze River basin. *J. Environ. Manag.* **2023**, *345*, 118694. [CrossRef]
Liao, L.; Chen, M.; An, R.; Li, J.; Tang, X.; Yan, Z. Identifying three-dimensional swimming corridors for fish to match their swimming characteristics under different hydropower plant operations: Optimization of entrance location for fish-passing facilities. *Sci. Total Environ.* **2022**, *822*, 153599. [CrossRef]
Pavlov, D.S.; Mikheev, V.N.; Kostin, V.V. Migrations of Fish Juveniles in Dammed Rivers: The Role of Ecological Barriers. *J. Ichthyol.* **2019**, *59*, 234–245. [CrossRef]
Caputo, F.; de Oliveira, M.F.M.; Denadai, B.S.; Greco, C.C. Fatores intrínsecos do custo energético da locomoção durante a natação. *Rev. Bras. Med. Esporte* **2006**, *12*, 399–404. [CrossRef]
Jones, P.E.; Svendsen, J.C.; Börger, L.; Champneys, T.; Consuegra, S.; Jones, J.A.H.; Garcia de Leaniz, C. One size does not fit all: Inter- and intraspecific variation in the swimming performance of contrasting freshwater fish. *Conserv. Physiol.* **2020**, *8*, coaa126. [CrossRef]
Souza, R.C.R.; Pompeu, P.S. Ecological separation by ecomorphology and swimming performance between two congeneric fish species. *Zoologia* **2020**, *37*, 1–8. [CrossRef]
Boily, P.; Magnan, P. Relationship between individual variation in morphological characters and swimming costs in brook charr (*Salvelinus fontinalis*) and yellow perch (*Perca flavescens*). *J. Exp. Biol.* **2002**, *205*, 1031–1036. [CrossRef] [PubMed]
Dala-Corte, R.B.; De Fries, L. Inter and intraspecific variation in fish body size constrains microhabitat use in a subtropical drainage. *Environ. Biol. Fishes* **2018**, *101*, 1205–1217. [CrossRef]
Pettersen, A.K.; Marshall, D.J.; White, C.R. Understanding variation in metabolic rate. *J. Exp. Biol.* **2018**, *221*, jeb166876. [CrossRef] [PubMed]
Rubio-Gracia, F.; García-Berthou, E.; Guasch, H.; Zamora, L.; Vila-Gispert, A. Size-related effects and the influence of metabolic traits and morphology on swimming performance in fish. *Curr. Zool.* **2020**, *66*, 493–503. [CrossRef] [PubMed]
Srean, P.; Almeida, D.; Rubio-Gracia, F.; Luo, Y.; García-Berthou, E. Effects of size and sex on swimming performance and metabolism of invasive mosquitofish *Gambusia holbrooki*. *Ecol. Freshw. Fish* **2017**, *26*, 424–433. [CrossRef]
Castro, M.A.; Santos, H.A.; Sampaio, F.A.C.; Pompeu, P.S. Swimming Performance of the Small-Sized Characin *Bryconamericus stramineus* (Characiformes: Characidae). *Zoologia* **2010**, *27*, 939–944. [CrossRef]
Leavy, T.R.; Bonner, T.H. Relationships among Swimming Ability, Current Velocity Association, and Morphology for Freshwater Lotic Fishes. *N. Am. J. Fish. Manag.* **2009**, *29*, 72–83. [CrossRef]
Branco, P.; Santos, J.M.; Katopodis, C.; Pinheiro, A.; Ferreira, M.T. Pool-Type Fishways: Two Different Morpho-Ecological Cyprinid Species Facing Plunging and Streaming Flows. *PLoS ONE* **2013**, *8*, e65089. [CrossRef]
Willis, T.V.; Magnuson, J.J. Patterns in fish species composition across the interface between streams and lakes. *Can. J. Fish. Aquat. Sci.* **2020**, *57*, 1042–1052. [CrossRef]

55. Sandhya, K.M.; Lianthuamluaia, L.; Karnatak, G.; Sarkar, U.K.; Kumari, S.; Mishal, P.; Kumar, V.; Panda, D.; Ali, Y.; Naskar, M. Fish assemblage structure and spatial gradients of diversity in a large tropical reservoir, Panchet in the Ganges basin, India. *Environ. Sci. Pollut. Res.* **2019**, *26*, 18804–18813. [CrossRef]
56. Tudorache, C.; Viaene, P.; Blust, R.; Vereecken, H.; De Boeck, G. A comparison of swimming capacity and energy use in seven European freshwater fish species. *Ecol. Freshw. Fish* **2008**, *17*, 284–291. [CrossRef]
57. Haro, A.; Castro-Santos, T.; Noreika, J.; Odeh, M. Swimming performance of upstream migrant fishes in open-channel flow: A new approach to predicting passage through velocity barriers. *Can. J. Fish. Aquat. Sci.* **2004**, *61*, 1590–1601. [CrossRef]
58. Salmasi, F.; Ayaseh, A.; Dalir, A.H.; Arvanaghi, H. Flow pattern study in Beshar River and its two straight and meander reaches using CCHE2D model. *Appl. Water Sci.* **2019**, *10*, 19. [CrossRef]
59. Zhao, Z.; Yao, S.; Jiang, E.; Qu, B. Experimental study and a physical model on the geomorphic response mechanisms of meandering rivers under progressive sediment reduction. *Front. Earth Sci.* **2022**, *10*, 892926. [CrossRef]
60. Chemagin, A.A. Influence of abiotic factors on features and distribution dynamics of fish in small tributary of Irtysh River. *Vestn. Astrakhanskogo Gos. Tekhnicheskogo Univ. Seriya Rybn. Khozyaystvo* **2020**, *4*, 66–80. [CrossRef]
61. Wood, B.M.; Bain, M.B. Morphology and microhabitat use in stream fish. *Can. J. Fish. Aquat. Sci.* **1995**, *52*, 1487–1498. [CrossRef]
62. Guillard, J.; Balay, P.; Colon, M.; Brehmer, P. Survey boat effect on YOY fish schools in a pre-alpine lake: Evidence from multibeam sonar and split-beam echosounder data. *Ecol. Freshw. Fish* **2010**, *19*, 373–380. [CrossRef]
63. Misund, O.A.; Aglen, A. Swimming behaviour of fish schools in the North Sea during acoustic surveying and pelagic trawl sampling. *ICES J. Mar. Sci.* **1992**, *49*, 325–334. [CrossRef]
64. Radelyuk, I.; Zhang, L.; Assanov, D.; Maratova, G.; Tussupova, K. A state-of-the-art and future perspectives of transboundary rivers in the cold climate—A systematic review of Irtysh River. *J. Hydrol. Reg. Stud.* **2022**, *42*, 101173. [CrossRef]
65. Krupa, E.; Romanova, S.; Serikova, A.; Shakhvorostova, L. A Comprehensive Assessment of the Ecological State of the Transboundary Irtysh River (Kazakhstan, Central Asia). *Water* **2024**, *16*, 973. [CrossRef]
66. Zhalmagambetova, U.; Assanov, D.; Neftissov, A.; Biloshchytskyi, A.; Radelyuk, I. Implications of Water Quality Index and Multivariate Statistics for Improved Environmental Regulation in the Irtysh River Basin (Kazakhstan). *Water* **2024**, *16*, 2[illegible] [CrossRef]
67. Kondratyev, S.A.; Bryukhanov, A.Y.; Pozdnyakov, S.R.; Puzanov, A.V.; Zherelina, I.V.; Shmakova, M.V.; Vasilyev, E.V.; Rasulova, A.M.; Baboshkina, S.V.; Oblomkova, N.S. Assessment of the Load of Nutrients and Pollutants on the Russian Part of the Irtysh River. *Geogr. Nat. Resour.* **2024**, *45*, 31–41. [CrossRef]
68. Magritskii, D.V. Anthropogenic Impact on the Runoff of Russian Rivers Emptying into the Arctic Ocean. *Vodn. Resur.* **2008**, 3–16. [CrossRef]
69. Huang, W.; Duan, W.; Nover, D.; Sahu, N.; Chen, Y. An integrated assessment of surface water dynamics in the Irtysh River Basin during 1990–2019 and exploratory factor analyses. *J. Hydrol.* **2021**, *593*, 125905. [CrossRef]
70. Chemagin, A.A. Structure of poaching sturgeon species in the Irtysh River in 2013–2020 (according to ichthyologic examination). *Vestn. Astrakhan State Tech. University. Ser. Fish. Ind.* **2021**, *4*, 17–23. [CrossRef]
71. Volosnikov, G.I. Characteristics of the ecological danger of illegal poaching fish catches in the lower reaches of the Irtysh river. *E3S Web Conf.* **2023**, *390*, 07009. [CrossRef]
72. Holmlund, C.M.; Hammer, M. Ecosystem services generated by fish populations. *Ecol. Econ.* **1999**, *29*, 253–268. [CrossRef]

diversity

MDPI

cle

ayesian Inference of Human-Made Hazards in Networks of land Wetlands: The Case of the Aegean Archipelago

o M. Gaganis and Andreas Y. Troumbis *

Biodiversity Conservation Laboratory, Department of Environment, University of the Aegean, 81100 Mytilene, Lesvos, Greece; envm23043@env.aegean.gr
* Correspondence: atro@aegean.gr; Tel.: +30-22510-43979

Abstract: This research aims to provide guidelines using probabilistic methods to understand better the dominant characteristics of the 824 under-pressure wetlands on 75 islands within Greece and to inform future conservation efforts. Identifying the characteristics and types of anthropogenic pressures is crucial for developing effective conservation strategies. The study employs power-law modeling to validate the natural size distribution of wetlands, naïve Bayesian inference to model human impacts, and the epsilon statistic to assess wetland sensitivity to specific pressures, addressing potential sampling biases. Power-law modeling reveals a natural heavy-tailed distribution of wetland sizes, highlighting the ecological significance of larger, rarer systems. Naïve Bayesian inference indicates that agriculture and transportation are the predominant pressures affecting natural coastal wetlands. The epsilon statistic further differentiates wetland sensitivity, identifying estuaries, lagoons, and marshes as particularly vulnerable. By profiling the most vulnerable wetlands using these methods, the research provides a framework for assessing anthropogenic impacts and informing targeted conservation and management strategies to protect these vital ecosystems.

Keywords: anthropogenic pressure; epsilon statistic; power law

1. Introduction

Wetlands play a crucial role in the global ecosystem's functioning, providing essential services such as biodiversity refugia, water filtration and supply, gas and climate regulation, and flood control [1–3]. Wetlands provide critical nature contributions to people (NCPs) [4], greatly exceeding their geographic extent and physical size [5]. Yet their loss is alarming since natural wetlands' average annual loss rate is over three times faster than that of natural forests, i.e., −0.78%/year vs. −0.24%/year, respectively [6]. Since 1900 AD, we have lost approximately 64–71% of wetlands globally, making them one of the most endangered ecosystems in Earth's recent history [7].

The proximate causes and underlying forces contributing to wetland conversion and degradation are generally similar worldwide [8], but their impacts are unevenly distributed across different regions [9]; human activities, such as agricultural development, infrastructure construction, industrial operations, and urbanization, predominantly drive changes. These influences are collectively called anthropogenic threat complexes (ATCs), characterized by various sets of drivers that differ in intensity [9]. As the modernization transition advances at both national and regional levels, influenced by states' economic development programs to reshape the political economy of countries [10,11], variations in the targeted types of wetlands become evident, such as differences between inland vs. coastal wetlands [7].

These areas enrich species diversity in numerous island environments [12]. They often harbor endemic species [13] and play a crucial role in bird migration (see Figure 1) [14–16]. Most institutional weaknesses and deficiencies identified in the massive literature regarding the actual implementation of the Natura 2000 network across Europe [17–23] apply to their conservation processes. These wetlands are part of complex social–ecological systems and land use mosaics, which makes them susceptible to increased pressures from infrastructure and activities further exacerbated by tourism growth [14–27]. Past losses of Mediterranean wetlands were estimated to represent c. 50% over the 20th century, most since the 1970s [17]. In parallel, during the last 50 years, wetland-dependent species populations have declined by 81% [28].

The Mediterranean Basin is recognized as one of the world's biodiversity hotspots [29]. Greece houses a remarkable share of this diversity in the east Mediterranean due to processes and patterns involved in modern-day biota's arrival and settlement and anthropogenic pressures' history [30]. In Greece, a protected area network comprises 202 Special Protection Areas (SPAs), established under the EU Birds Directive, and 241 Special Areas for Conservation (SACs), designated under the EU Habitats Directive (total area of 4,294,960 ha). Together, these protected areas form the European Union's Natura 2000 network in the country, covering 4,294,960 ha, which accounts for 27.3% of the national territory and 6.1% of territorial waters [31]. This results in a high sufficiency index of protected area coverage of 99.07 out of 100 [32]. One might be tempted to view the Natura 2000 network as a "rosy" cornerstone of biodiversity conservation in Greece within the European framework. However, it is essential to recognize that, despite these impressive numbers at first glance, the approximately 400 protected areas, initially selected mainly by experts in taxonomy and systematics, have an average size of about 13,000 $\pm$ 21,000 ha. The smallest area measures 1.31 ha, while the largest exceeds 200,000 ha. Additionally, the selection process did not prioritize aspects such as complementarity and connectivity between areas, threats to natural features, and relationships between different human uses. As a result, these figures present significant and often complex challenges for conservation planning [33], usually leading in practice to downgrading, downsizing, and de-gazettement of protected areas [34].

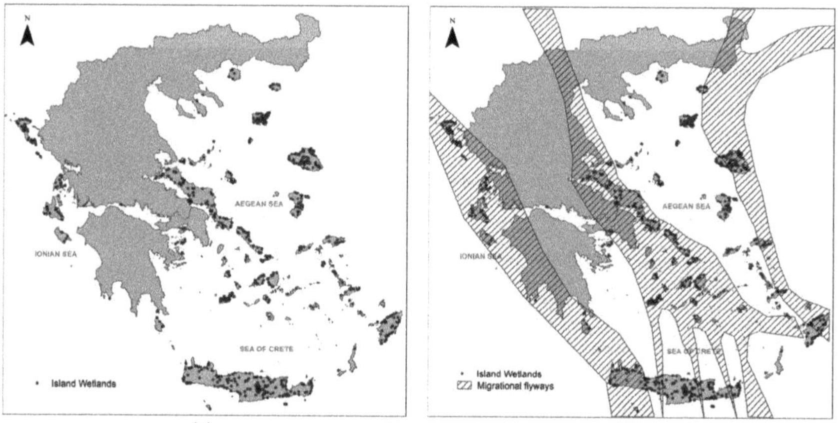

(a) (b)

Figure 1. (**a**) shows the geographical distribution of wetlands in the Greek islands, particularly in the Aegean Archipelago. (**b**) illustrates the major Balkan–Cyrenaic flyways constructed after the available literature [35]. This figure serves as a straightforward proxy representation of the significance of island wetlands in the context of a global ecological phenomenon: bird migration.

Most of these protected areas consist of wetlands with a unique "triple" insularity characteristic, which refers to the three overlapping isolation layers. This means they are:

(1) (often) protected "water bodies" (2) surrounded by irregulated land uses and (3) on islands and islets. Notable examples are (1) the 25 wetlands of Lesbos, (2) Western Milos, Antimilos, and Polyaigos, (3) the isle of Elafonissi and coastline of Crete, (4) the coastline of Samothrace, (5) Eastern Astypalea and the surrounding islets, (6) Aliki and Chortarolimni lakes on Lemnos, or (7) the islands of Oinousses.

The extensive drainage of wetlands on the Greek mainland is intricately connected to the evolution of the modern Greek state from 1830 to the present [11,28,36–39]. Island wetlands in Greece gained significant recognition after the WWF's multiannual research program "Conservation of the Island Wetlands of Greece". This initiative led to the discovery and documentation of over 800 wetlands on 75 islands in the Aegean and the Ionian Archipelagos [40]. As a result, a Presidential Decree was signed in June 2012 to provide strict protection for 380 of Greece's small (<8 hectares) natural island wetlands.

The islands of the Aegean Archipelago are particularly vulnerable due to their fragmented nature, small size, and the high density of human activities in coastal areas [40,41]. In addition to these local challenges, they face broader global issues such as climate change and biological invasions [42–44]. Although there is a growing body of research on wetland conservation within the Mediterranean region, quantitative studies focusing specifically on the Aegean Islands are still limited [45]. Most research has primarily concentrated on the fauna and flora rather than investigating the geomorphological changes and habitat alterations caused by human activities.

This contribution focuses on the biodiversity-rich wetlands of the Aegean Archipelago and addresses the challenges associated with conservation planning in these setups. These ecosystems are characterized by their triple insularity, resulting in physical boundaries that affect biodiversity patterns and processes, human usage, governance, and various threats. This triple insularity also offers conservation planners a specific set of objectives, constraints, and opportunities that can enhance the effectiveness of conservation planning strategies [46]. We propose that identifying groups of wetlands based on the interaction and co-occurrence of their ecosystemic traits, alongside the complexities of anthropogenic pressure complexes, could improve the cost efficiency of conservation operations [47]. By categorizing wetlands into groups, planners might develop general master plans for each group and then tailor these plans to address the needs of specific cases. Such a "strategy" should (1) facilitate adaptive management, allowing for updating the probability of different outcomes as new data become available, rendering more flexible and responsive conservation plans; (2) integrate various sources of data (e.g., environmental variables, pressures presence/absence); (3) help in assessing the risks associated with different conservation actions, considering both the likelihood of success and the potential consequences of failure; (4) allow the use and incorporation of prior information, which can be particularly useful when data are scarce; (5) quantify uncertainty in model predictions [46,48–53].

This study's main contribution is applying a probabilistic framework to quantify and rank the impact of different anthropogenic activities on the Aegean Islands' wetlands [31]. By systematically analyzing how proximity to urban centers, wetland size, and landscape fragmentation influence the intensity of human pressures, we seek to identify the specific characteristics that make certain wetlands more susceptible or vulnerable to degradation [54]. Such insights are essential for bridging the gap between localized ecological observations and broader regional conservation needs [43].

We employ naïve Bayesian inference to evaluate how different characteristics of wetlands—such as location, type, and size—contribute probabilistically to the likelihood of experiencing human pressures such as (1) urbanization, (2) agricultural expansion, and

(3) urban fabric development [55]. The use of Bayesian inference enables a comprehensive analysis of how multiple factors shape the resilience of wetland ecosystems under increasing human pressures, as well as the contributions of each characteristic to the observed outcomes [52,56]. Bayesian models are particularly well-suited for ecological studies because they integrate prior knowledge with new data, refining predictions as more information becomes available [53,57]. This ability to adapt makes Bayesian inference a valuable tool for wetland management, as it facilitates continuous updating and adjustment of conservation strategies based on the latest data [58]. In this regard, they align with the properties of the abovementioned conservation planning strategy. This study addresses these gaps by combining power-law modeling and Bayesian inference to analyze anthropogenic pressures on wetlands in the Aegean Archipelago. Specifically, we aim to identify pressure-sensitive wetland types and propose targeted conservation strategies to mitigate human impacts.

2. Materials and Methods

2.1. Wetland Data and Profile Building

We utilize the public domain database Ygrotopio Islands developed by the WWF (https://www.ygrotopio.gr/general/search.php?lang=en&action=map&action=export#, accessed on 17 November 2024) [59], which contains data on 824 wetlands across 75 islands. These data have been part of a national inventory of Greek wetlands since 1980. The dataset includes 37 key characteristics that describe various aspects of each wetland. These characteristics encompass the nature of the wetland (e.g., natural or artificial), water presence (seasonal, permanent, temporary/intermittent, unknown), location (inland, marine/coastal), and the area in hectares, ranging from 0.1 to 750 hectares (see Appendix A). Additionally, wetlands were classified into specific types: sand bars, seasonal freshwater ponds, underground rivers, and estuarine systems. Prevalent vegetation types were also documented, including shrubby arborescent species, wet meadow grasses, and emergent aquatic plants. These data were georeferenced to latitude and longitude coordinates, allowing for precise mapping of wetland locations within the study area. Furthermore, the database outlines each wetland's property and protection status, human activities, and intensity within each wetland, as well as its catchment area (percentage remaining untouched). The data can be accessed and linked through polygon pop-ups in WWF Greece's mapping application. The profiles built using this comprehensive set of characteristics serve as the basis for subsequent analysis of how different anthropogenic pressures impact wetland types.

2.2. Human Activity and Anthropogenic Pressure Data

In the context of islands, the proximity of many wetlands to human settlements and economic activities renders them particularly susceptible to anthropogenic pressures. Using the WWF Ygrotopio database [35], we identified and categorized significant pressures impacting each wetland [60]. Table 1 presents these categories, encompassing broader human activities such as tourism and leisure and their effects on wetland systems. These categories are valid proxies for understanding wetlands' most significant anthropogenic threats. Three categories will serve as our focal point as they provide the most essential explanatory power about human activity: agriculture and forestry (code: 10), urbanization (code: 40), and transportation and communication (code: 50)

All the collected data (Sections 2.1 and 2.2) are point-based, providing specific geographic coordinates for each island wetland and its associated human activities. This allowed for precise spatial analysis of the relationship between wetland characteristics and human pressures. The spatial analysis was conducted using GIS software (ArcGIS 10.8.1), which enabled the visualization of wetlands and the overlapping types of activities across the Aegean Islands. This approach revealed areas where wetlands are most at risk of degradation due to nearby economic activities [39].

Table 1. Categories of types and anthropogenic pressures used in the WWF Ygroscopio database for wetlands in the Aegean Islands.

Types/Pressures:
00 = Management for conservation
10 = Agriculture and forestry
20 = Fishing, hunting, and collecting
30 = Mining and material extraction
40 = Urbanization, industrialization, and similar activities
50 = Transportation and communication
60 = Leisure and tourism
70 = Pollution and other human activities
80 = Human-induced changes in hydraulic conditions
90 = Natural processes (biotic and abiotic)

2.3. *Statistical Analysis*

We present statistical methods addressing two key issues introduced earlier: (1) whether the frequency–size relationships of the Aegean Islands' wetlands deviate from heavy-tailed power laws due to anthropogenic pressures; (2) the potential patterns of co-occurrence between individual wetland traits and anthropogenic pressures applied to each of them.

2.3.1. Power-Law Distribution of Frequency–Size Relationship of Islands' Wetlands

The literature indicates that the power law and associated thresholds, representing the frequency–size distribution of wetlands, reflect the state transition dynamics of these ecosystems. They can be used to evaluate the intensity of anthropogenic disturbances [61]. The key point is that the log–log frequency–size distribution may exhibit vanishing tails and narrower spans and tends to behave as a normal or lognormal distribution [62]. However, given an island's geophysical delimitation, one should notice that such a distribution is de facto upper truncated [63]. This observation raises the hypothesis that the wetlands of the Aegean Islands might be on a path toward conversion or elimination. Alternatively, they may be experiencing anthropogenic degradation. It is important to note that the lo–log frequency–size distribution could arise from various causes, and further analysis is required to determine the relative contribution of anthropogenic factors compared to natural processes. If this is the case, anthropogenic pressure implies that the current legal protection policies are either insufficient or poorly implemented.

Mathematically, a power-law distribution follows the form $P(x) \propto x^{-a}$, where $P(x)$ is the probability of observing a wetland of size x and α is the scaling exponent that characterizes the distribution. This relationship would indicate that more extensive wetlands are exponentially less frequent than smaller ones, aligning with known patterns in natural phenomena, such as river networks, wildfires, species abundance distributions, or earthquake magnitudes [64–67]. To assess the reliability of the power-law fit, a goodness-of-fit (GoF) test is performed using the Kolmogorov–Smirnov (KS) test alongside bootstrap simulations.

2.3.2. Inferring the Relationship Between the Co-Occurrence of Wetlands and Anthropogenic Pressures

We follow the method extensively analyzed by Stephens et al. [51], adapting it to the specific characteristics of the current case by focusing on co-occurrences in space rather than time. The wetlands of the islands define a set, W, of individual units, W_i, such that

$W = \cup_i W_i = 824$. These units form the partitioned wetland ensemble of the Aegean Archipelago region, denoted as A.

We analyze two objects/observables, E and H, for each wetland and consider $n(E)$ and $n(H)$ representatives of each type. An object/observable of type E, $E_e(W_i)$ is an individual wetland in A; it is a Boolean variable, meaning it can be present or absent in wetland W_i. In other words, $e \in [1, \ldots, ne]$, meaning the number of discrete categories into which the object E has been divided. The same applies to type H, divided into h discrete categories.

The number of co-occurrences of E_e and H_h in W, within A, is given by "counting" the frequencies (dropping for simplicity e and h)

$$N(EH) = \sum_i I(E(W_i), H(W_i))$$

If the indicator function $I = 1 \Leftrightarrow E$ and H co-occur in the wetland Wi. Given $N(EH)$, the probability of a co-occurrence $P(EH) = N(EH)/W$, and the conditional probabilities $P(E \mid H) = P(EH)/N_H$ and $P(H \mid E)$. We may take $P(E \mid H)$ and $P(EH)$ as measures of the degree of co-occurrence of E and H types.

Bayesian inference was employed to model the probability of different wetland types experiencing anthropogenic pressures based on their characteristics. The Bayesian framework is particularly suitable for the present study, allowing for incorporating prior knowledge and handling uncertainties in the data [55,68]. Under the naïve assumption that all variables are independent, the core of the naïve Bayesian approach is expressed by the following equation:

$$P(H|E) = \frac{P(E|H) \cdot P(H)}{P(E)}$$

where:

- $P(H \mid E)$ is the posterior probability that a wetland is subject to a particular human pressure given its characteristics.
- $P(E \mid H)$ is the likelihood, representing the probability of observing specific wetland characteristics if the pressure is present.
- (H) is the prior probability of the wetland being under pressure before observing the evidence.
- $P(E)$ is the marginal likelihood, representing the total probability of the evidence across all hypotheses.

The model treats each characteristic—such as wetland type, size, and location—as independent predictors of the likelihood of a wetland experiencing one of the three key pressures. For example, coastal wetlands near urban centers are more likely to experience pressures related to urbanization and transport than remote inland wetlands primarily affected by agricultural practices.

In ecological modeling, especially when integrating multiple data types such as wetland characteristics and anthropogenic pressures, one of the key challenges is accounting for variations in data sample sizes across different wetlands [43,49,50]. The presence or absence of pressures in wetlands can be influenced by the size and density of observations, leading to potential biases in probabilistic models. To address this issue, we utilize the epsilon (ε) statistic, which measures the strength of association between wetland characteristics and specific human pressures while adjusting for differences in sample sizes across the dataset [69].

The epsilon statistic (ε) quantifies the deviation from a null model where the distribution of pressures is assumed to be random, providing a standardized score that can be interpreted as a measure of statistical significance. The formula for calculating ε is provided, where $\varepsilon(H_h|E_e)$ is the epsilon statistic, to evaluate whether the presence of a particular human pressure H_h (e.g., urbanization) is associated with a given wetland characteristic E_e (e.g., coastal wetland, specific vegetation):

$$\varepsilon(H_h|E_e) = \frac{N_{E_e} \cdot P(H_h) \cdot (1 - P(H_h))}{N_{H_h \cap E_e} - N_{E_e} \cdot P(H_h)}$$

1. $\varepsilon(H_h|E_e)$ is the epsilon statistic, which now evaluates whether the presence of a particular human pressure H_h (e.g., urbanization) is associated with a given wetland characteristic E_e (e.g., coastal wetland, specific vegetation).
2. $N_{H_h \cap E_e}$ represents the number of wetlands where both the human pressure H_h and the characteristic E_e are observed. This essentially counts how often both conditions coincide (i.e., the number of wetlands that both have the pressure and the characteristic), N_{E_e}.
3. N_{E_e} is the total number of wetlands where the characteristic E_e is present, regardless of whether the human pressure H_h exists.
4. $P(H_h)$ is the prior probability of a wetland being subject to human pressure H_h across all wetlands in the dataset. It is calculated as:

$$P(H_h) = \frac{\text{NHh}}{W}$$

where NHh is the number of wetlands with the pressure H_h and W is the total number of wetlands in the dataset.

The epsilon statistic ε serves as a standardized score that compares the actual occurrence of a specific wetland characteristic under pressure to what would be expected if the pressure were randomly distributed across the study area [50]. Its value can be interpreted as follows:

ε > 0: A positive epsilon value indicates that the wetland characteristic E_e (such as a coastal location or large area) is positively associated with the occurrence of the pressure H_h. This means the pressure is more likely to occur in wetlands with this characteristic than expected by chance. For example, a high positive epsilon value for coastal wetlands experiencing tourism-related pressures suggests a significant association between coastal location and the likelihood of experiencing impacts from tourism.

ε < 0: A negative epsilon value indicates a negative association between the wetland characteristic and pressure. This suggests that the pressure occurs less frequently than is expected in wetlands where the characteristic E_e is present. For instance, a negative epsilon value for inland wetlands concerning urbanization pressure may imply that these wetlands are relatively less impacted by urbanization compared to coastal wetlands.

ε = 0: An epsilon value close to zero suggests that the presence of the characteristic has no significant association with the occurrence of the pressure. This implies that the pressure distribution is similar to what would be expected under random conditions.

This study calculated the epsilon statistic for each combination of wetland characteristics (such as wetland type, size, or vegetation type) and the three primary anthropogenic pressures: agriculture, urbanization, and transportation. By comparing the epsilon ε values across different characteristics, we could rank the predictors based on their strength of association with the pressures.

3. Results

3.1. Wetlands Profiling

The studied wetlands were classified based on three key characteristics: inland vs. marine/coastal wetlands, natural vs. artificial wetlands, and the presence of water (duration/year). The classification results, shown in Figure 2, indicate that the most common type of wetland in the Aegean Islands is natural coastal wetland with permanent water. However, various other combinations are also observed, reflecting the diversity of wetland types in the region. This diversity is further illustrated in Figure 3, which presents the absolute frequencies of wetland types across the Aegean Archipelago.

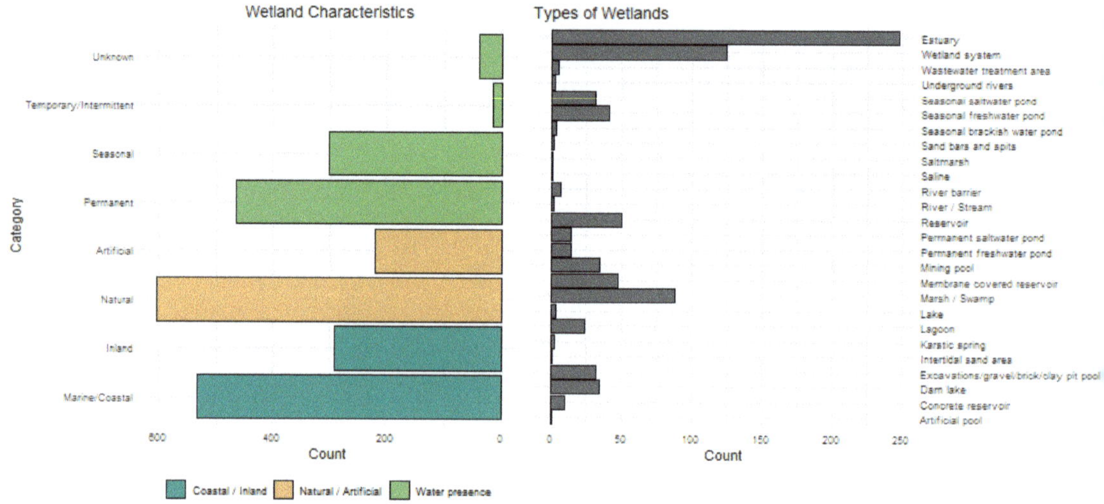

(a) (b)

Figure 2. (**a**) Major categories of wetlands in the Aegean Archipelago islands, in absolute frequencies per case. (**b**) Absolute frequencies of wetland types in the Aegean Archipelago wetlands.

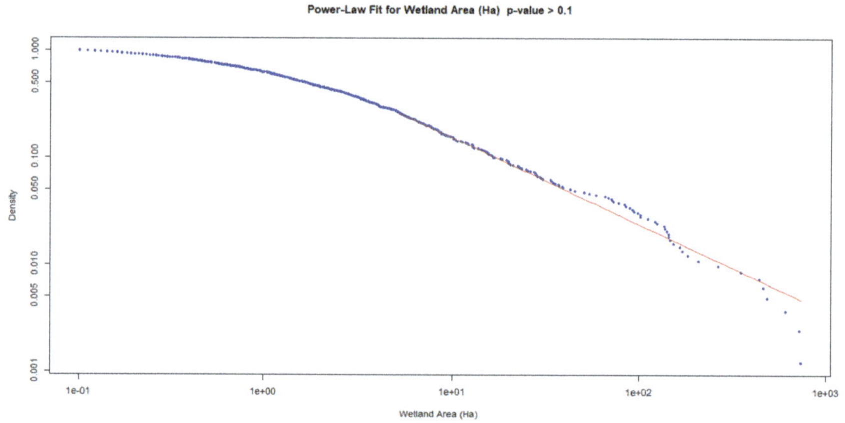

Figure 3. Aegean Archipelago wetlands(blue points) frequency–size distribution fitted to a power-law model (red line). Estimated scaling parameter (α) = 1.81 and the corresponding minimum value ($x_{\min}$) = 4.66 hectares.

These results provide an overview of the wetland types analyzed in subsequent sections and form the basis for exploring patterns and pressures affecting these ecosystems.

3.2. Power-Law Distribution and Validation

The fitting process for the distribution of wetland areas aimed to determine whether the size–frequency distribution of the Aegean Archipelago wetlands follows a power-law behavior. This behavior suggests a natural, self-organizing system commonly observed in ecological distributions. The power-law fitting yielded an estimated scaling parameter (α) of 1.81 and a corresponding minimum value (x_{min}) of 4.66 hectares (Figure 3). These values indicate that the power-law relationship holds for wetland areas larger than 4.66 hectares. This suggests that the distribution of wetland areas exhibits a heavy-tailed pattern, where larger wetlands are increasingly rare yet follow a predictable scaling law.

The fractal dimension was calculated as a key indicator of the complexity and spatial structure of the island's wetlands. It was computed using the formula $Df = 2 - \alpha = 0.189$. This low fractal dimension indicates a relatively sparse distribution of larger wetlands compared to smaller ones. This supports the hypothesis that larger wetlands are less frequent, yet their presence significantly influences the overall dynamics of the system. The p-value obtained from KS testing was 0.416, which exceeds the standard significance threshold of 0.05. This indicates that the power-law model adequately represents the distribution of the studied wetland areas, indicating no significant evidence against the hypothesis that the data follow a power-law distribution. Figure 4 illustrates the Aegean islands' wetlands network using binning relating to this power law.

Figure 4. Representation of the wetlands network in the Aegean Archipelago reflecting the power-law binning. See text for details.

3.3. Bayesian Inferences

A naïve Bayesian model was applied to assess the likelihood of different anthropogenic pressures affecting wetlands in the Aegean Islands. The focus was on three main pressures: agriculture and forestry (code: 10), urbanization (code: 40), and transportation and communication (code: 50). The analysis involved 618 samples and employed 10 predictor variables related to wetland characteristics such as location, type, and size. Model evaluation metrics were applied to assess the model's performance for each type of pressure, including accuracy, kappa, sensitivity, specificity, and confusion matrices. The results for each pressure type are presented below, emphasizing the model's predictive capabilities, key findings, and limitations.

3.3.1. Pressure 1: Agriculture and Forestry (10)

The naïve Bayes model for agriculture and forestry pressure revealed a significant class imbalance. In the dataset, 87.7% of the training samples and 87.8% of the testing samples were labeled as experiencing this type of pressure (TRUE). This imbalance indicates that most wetlands in the dataset are affected by agricultural and forestry activities, making it difficult for the model to predict the minority class (FALSE) accurately.

Model Performance:

Cross-validation was conducted using kernel smoothing, resulting in a model accuracy of 78%, compared to 59% without smoothing. Although this accuracy seems relatively high, the kappa statistic was low, approximately 0.137, suggesting that the model's predictions were only slightly better than random chance.

- Confusion matrix analysis:
    - The model successfully identified 153 instances of TRUE pressure but had difficulty predicting the minority class, with only five correct predictions for FALSE.
    - There were 27 false negatives (instances where TRUE was incorrectly predicted as FALSE) and 20 false positives (instances where FALSE was predicted but the true class was TRUE).
    - The method exhibited low sensitivity for FALSE (20%) and high specificity for TRUE (85%), indicating it performed well in identifying the more common TRUE cases but struggled to identify wetlands not impacted by agricultural and forestry activities.

An imbalance in the dataset significantly impacted the model's effectiveness in identifying wetlands affected by agricultural and forestry pressures. This imbalance resulted in a high prediction rate for the TRUE class, while many FALSE cases were overlooked, as indicated by the low sensitivity and poor precision for FALSE cases (15.6%). Although the model can provide valuable insights into which wetlands are most affected by agricultural practices, it lacks reliability in detecting those unaffected. In terms of conservation, the model is useful for pinpointing areas severely impacted by agriculture and forestry. However, its limited ability to detect the absence of such pressures could lead to an underestimation of areas that may seem unaffected but are still experiencing subtle ecological impacts.

3.3.2. Pressure 2: Urbanization (40)

The analysis of urbanization pressure revealed a more balanced dataset, with 49.35% of the training set labeled as FALSE and 50.65% labeled as TRUE. Similar proportions were found in the testing set. This balance enabled the naïve Bayes model to perform without a significant bias toward either class.

Model Performance: The cross-validated model using kernel smoothing achieved an accuracy of 54.68%, slightly better than the 52.59% accuracy without kernel smoothing. However, the kappa value was 0.0980, indicating limited predictive power and only slight agreement beyond random classification.

- Confusion matrix analysis:
  - The model accurately predicted 65 cases as FALSE and 57 as TRUE, demonstrating a relatively balanced ability to identify both classes.
  - There were 47 false negatives (TRUE cases misclassified as FALSE) and 36 false positives (FALSE cases misclassified as TRUE).
  - The sensitivity for the FALSE class was 64.36%, indicating that the model was more effective at identifying wetlands not affected by urbanization compared to those affected.
  - The specificity for the TRUE class was lower, at 54.81%, suggesting that the model had more difficulty detecting cases of urbanization pressure.

The balanced nature of the dataset allowed the naïve Bayes model to perform reasonably well. However, the low kappa value indicates the model's unreliable predictions. The results suggest that while the model can distinguish between urbanized and non-urbanized wetlands, it may overlook subtle land use changes that signal early stages of urban development. The model's ability to identify wetlands under urban pressure could be valuable for urban planning and management, especially in areas undergoing rapid development. Nonetheless, improvements to the model—potentially by incorporating additional predictors or exploring alternative modeling techniques—are necessary to enhance its reliability for more nuanced conservation decision making.

3.3.3. Pressure 3: Transportation and Communication (50)

The dataset for transportation and communication pressure was balanced, with the training set consisting of 52.1% labeled as FALSE and 47.9% labeled as TRUE. This balance created a fair foundation for the naïve Bayes model to distinguish between the two classes effectively.

Model performance:

By utilizing kernel smoothing, the model achieved an accuracy of 69.75%, an improvement over the 67.96% accuracy obtained without smoothing. The kappa statistic increased to 0.3911 with kernel smoothing, indicating a moderate agreement between the predicted and actual values.

- Confusion matrix analysis:
  - The model accurately classified 42 instances as FALSE and 28 as TRUE. However, it also exhibited many false negatives (TRUE instances classified as FALSE) and false positives (FALSE instances classified as TRUE).
  - The sensitivity for detecting TRUE cases was 60.75%, while the specificity for identifying FALSE cases was 71.43%. This indicates that the model was more effective at recognizing wetlands unaffected by transportation activities than those impacted.

The naïve Bayes model performed relatively well compared to previous assessments, with improved kappa and balanced accuracy scores. However, it still faced challenges distinguishing subtle differences in transportation-related activities that may impact wetland conditions. The sensitivity and specificity metrics suggest that while the model can effectively identify areas with transportation pressure, it might overlook some regions where such activities are less evident but still significant.

These results are valuable for evaluating transportation infrastructure's effects on coastal and inland wetlands. This assessment can help inform strategies to mitigate negative impacts, such as pollution and habitat fragmentation. Nevertheless, the model could be enhanced through additional tuning and by incorporating more detailed spatial data to improve its predictive performance regarding transportation pressures.

3.3.4. Summary of Model Results

The naïve Bayes model demonstrated varying levels of predictive accuracy and reliability across the three types of anthropogenic pressures (see Table 2). It performed best with the more balanced datasets, specifically urbanization and transportation, where the class distribution allowed for a more equitable assessment of TRUE and FALSE cases. In contrast, the model for agricultural pressure, which was affected by class imbalance, achieved a high accuracy overall but struggled to detect the minority class. This highlights the importance of addressing dataset imbalances. Low kappa values across all pressures indicated that while the model could predict the majority class, its overall reliability was limited. This suggests a need for further refinement of the model. Contribution scores produced by the naïve Bayes model showcase little fluctuations between pressures, highlighting specific wetland characteristics throughout the three. The ε value, however, provides more variety across the pressures, highlighting several different wetland characteristics with high values.

Table 2. Summary of naïve Bayes model predictive accuracy and reliability across the three types of anthropogenic pressures.

	Agriculture/Forestry		**Urbanization**		**Transportation**	
Wetland Characteristics	**Contribution**	**ε**	**Contribution**	**ε**	**Contribution**	**ε**
Inland	0.317	−4.697	0.329	−1.225	0.183	−7.899
Marine/Coastal	0.683	3.474	0.671	0.906	0.817	5.842
Artificial pool	0.001	0.374	0.000	−1.013	0.000	−0.958
Concrete reservoir	0.008	−2.672	0.017	1.223	0.013	0.135
Dam lake	0.040	−0.432	0.034	−1.107	0.025	−2.155
Excavations/gravel/brick/clay pit pool	0.032	−2.733	0.048	1.339	0.015	−3.298
Intertidal sand area	0.001	0.374	0.000	−1.013	0.000	−0.958
Karstic spring	0.004	0.648	0.002	−0.601	0.005	0.652
Lagoon	0.033	1.832	0.022	−1.290	0.041	1.843
Lake	0.006	0.748	0.007	0.973	0.000	−1.917
Marsh/Swamp	0.109	0.875	0.129	2.127	0.132	2.221
Membrane-covered reservoir	0.054	−0.992	0.038	−2.280	0.005	−5.986
Mining pool	0.039	−0.955	0.010	−4.537	0.028	−1.812
Permanent freshwater pond	0.015	−0.010	0.012	0.024	0.008	0.003
Permanent saltwater pond	0.019	0.162	0.005	0.004	0.013	0.010
Reservoir	0.051	−2.958	0.055	−0.660	0.038	−2.530
River/Stream	0.003	0.529	0.002	−0.019	0.003	0.060
River barrier	0.010	0.990	0.010	0.343	0.003	−1.779
Saline	0.001	0.374	0.002	0.987	0.003	1.043
Saltmarsh	0.001	0.374	0.000	−1.013	0.000	−0.958
Sand bars and spits	0.003	0.529	0.002	−0.019	0.005	1.476
Seasonal brackish water pond	0.006	0.748	0.000	−2.027	0.000	−1.917
Seasonal freshwater pond	0.042	−2.980	0.065	1.115	0.025	−3.598
Seasonal saltwater pond	0.037	0.692	0.019	−3.817	0.036	−0.759
Underground rivers	0.004	0.648	0.000	−1.755	0.005	0.652
Wastewater treatment area	0.006	−0.527	0.007	0.417	0.000	−2.143
Wetland system	0.155	0.881	0.211	4.521	0.256	7.485
Estuary	0.319	2.582	0.302	0.108	0.343	2.134
Permanent	0.564	−0.132	0.556	−0.335	0.574	0.314
Seasonal	0.366	0.144	0.369	0.230	0.363	−0.072
Temporary/Intermittent	0.021	0.064	0.019	−0.298	0.025	0.904
Unknown	0.050	0.015	0.055	0.695	0.038	−1.447
Artificial	0.242	−3.870	0.221	−2.688	0.129	−7.379
Natural	0.758	2.345	0.779	1.629	0.871	4.471
Shrubby_Arborescent	0.727	2.160	0.770	2.297	0.832	4.227
Wet_meadow	0.515	1.452	0.624	4.994	0.657	6.036
Emergent	0.702	3.067	0.746	2.711	0.830	5.391

4. Discussion

The results of this study offer valuable insights into the distribution and vulnerability of wetlands in the Aegean Islands, particularly concerning three key anthropogenic pressures: agriculture, urbanization, and transportation. By utilizing power-law modeling and naïve Bayesian inference, this research lays the groundwork for understanding how these pressures affect wetlands of varying types and sizes. Applying the epsilon statistic also enables a more detailed evaluation of how different wetland characteristics relate to specific pressures. Below, we discuss the implications of the findings, identify areas for improvement, and highlight the novelty of this research within the context of existing studies.

The power-law fitting demonstrated that the distribution of wetland sizes follows a natural heavy-tailed pattern, with smaller wetlands being much more frequent than larger ones [65,67]. This result supports the hypothesis that wetland areas are self-organizing and shaped by natural ecological processes, such as hydrological and geomorphological factors, rather than by artificial or random influences [70,71]. The fractal dimension of 0.189 further reinforces the idea that larger wetlands, while rare, play a critical role in the overall ecological dynamics [2,50]. The similarity to the power-law model (p-value = 0.416) supports the dataset's validity, confirming that the observed distribution reflects naturally occurring processes and enhancing confidence in its reliability [61,72].

This finding is significant since power-law distributions are often associated with robust ecological systems. In these systems, smaller components, such as smaller wetlands, comprise the majority, while larger components, like larger wetlands, play crucial ecological roles. These roles include acting as biodiversity hotspots and regulating water [1,73]. Power-law dynamics have been applied in various ecological and hydrological contexts to analyze the distribution of natural resources and environmental disturbances [74,75]. Future research could build on this by examining how human activities might disrupt this natural balance, particularly if such pressures lead to a disproportionate loss of large or small wetlands [74].

Bayesian models are particularly well-suited for ecological studies because they integrate prior knowledge with new data, refining predictions as more information becomes available [53,57]. This ability to adapt makes Bayesian inference a valuable tool for wetland management, as it facilitates continuous updating and adjustment of conservation strategies based on the latest data [58]. The results of the naïve Bayes model reveal important patterns regarding the impact of various pressures on wetlands in the Aegean Islands. However, they also highlight some limitations in the model's performance. The model achieved high accuracy rates for agricultural and transportation pressures, 78% and 69.75% respectively. However, these accuracy figures were accompanied by low kappa values, indicating that while the model effectively predicts the majority class, it struggles to accurately identify the minority class or false cases [69,74]. This limitation is particularly evident in the agriculture model, where significant class imbalance led to skewed predictions favoring wetlands under agricultural pressure, reducing the model's ability to identify wetlands unaffected by this pressure [76] accurately.

The urbanization model, which used a more balanced dataset, demonstrated moderate performance with a kappa value of 0.0980. However, its overall effectiveness was still weak. This indicates that while the model can distinguish between impacted and non-impacted wetlands, it struggles to identify subtle land use changes and the effects of early-stage urbanization [56,77]. This emphasizes the need to refine the model, especially by addressing class imbalances and potentially incorporating more complex interactions among predictor variables [78]. The results suggest that while the naïve Bayesian approach offers a useful starting point for understanding human impacts on wetlands, it could be improved by utilizing more advanced methods, such as random forests or support vector machines. These techniques better handle imbalanced data and capture non-linear relationships between variables [79,80]. Additionally, more detailed spatial and temporal data could enhance the model's predictive capabilities, particularly for identifying emerging pressures in urbanizing areas [81].

The epsilon analysis provided valuable insights into how different types of wetlands respond to various human pressures. Positive epsilon values for wetlands such as estuaries, lagoons, and coastal marine systems—particularly under agricultural and transportation pressures—indicate that these wetlands are either resilient to or significantly affected by human activities [43]. In contrast, freshwater ponds and inland wetlands showed strong negative epsilon values, highlighting their vulnerability to anthropogenic pressures [60]. These findings underscore the differing responses of various wetlands to human impacts, emphasizing the need for tailored conservation strategies [82].

The ability of epsilon to quantify associations is a novel application in wetland conservation, offering a more detailed understanding of which wetlands are at the greatest risk [31]. This approach enables the prioritization of conservation efforts based on the pressure level faced and the ecological significance of specific wetland types [83]. Future research could further enhance this method by examining temporal trends in epsilon values, providing insights into how wetland vulnerability changes as human pressures increase over time [84–86].

This research aimed to provide guidelines using probabilistic methods to understand the dominant characteristics of current under-pressure wetlands and to inform future conservation efforts. Identifying the characteristics and types of anthropogenic pressure that attract interest is invaluable for developing effective conservation strategies. By analyzing this information, we can construct a profile for each of the three pressures we have focused on, utilizing the highest contribution scores and the largest positive epsilon values.

Through the naïve Bayes analysis, the contribution scores have yielded highly similar results across all pressure types. It was found that the wetlands under pressure are predominantly natural wetlands located in coastal areas with a permanent water presence and various types of vegetation. All three examined vegetation types produced high scores (Figure 5).

This consistency throughout the pressure types may signify that these are overall targeted characteristics for agriculture, urbanization, and transport expansion. However, this conclusion is somewhat foreseeable, and to remedy this, we turn to the calculated epsilon values, as the method bypasses the problem that may be caused by sample size and variable frequency.

Positive epsilon values showcase more specific characteristics that may indicate the unique vulnerabilities of these ecosystems under different pressures. For agricultural pressure, marine/coastal wetlands have the highest epsilon value (3.47), accompanied by characteristics showcasing elevated probability through the naïve Bayes analysis—natural wetlands with emergent or shrubby arborescent vegetation.

Estuaries and lagoons exhibit significant sensitivity to agricultural activity, with epsilon values of 2.58 and 1.83, respectively. This sensitivity is likely due to their transitional nature between land and sea (Figure 6). In urbanization, wetland systems characterized as type wetlands have an epsilon value of 4.521, while marsh/swamp-type wetlands show an epsilon value of 2.127. Additionally, wetland systems demonstrate a notable epsilon value of 7.48 under transportation pressure, indicating extreme susceptibility to high-pressure levels. This is likely linked to intense human interference, such as urbanization or industrial development. Wet meadows follow closely behind with an epsilon value of 6.03, further underscoring that these systems are highly at risk.

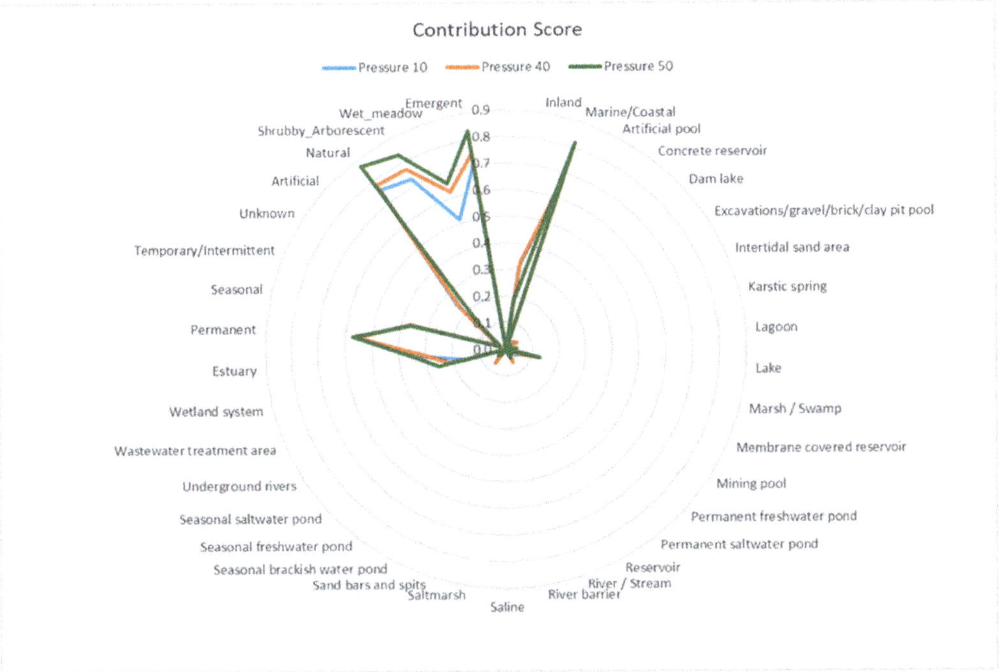

Figure 5. Spider chart that illustrates the score contributions of various wetland characteristics to the likelihood of three specific pressures: Agriculture and Forestry (indicated in blue), Urbanization, Industrialization, and Similar Activities (indicated in orange), and Transportation and Communication (indicated in green). The score contributions are relatively similar, highlighting the significance of particular characteristics concerning these pressures.

Novel Contributions and Areas for Improvement

This research makes several novel contributions to the field of wetland conservation and ecological modeling:

- Integration of Power-Law Modeling: The study effectively uses power-law distribution analysis to validate the natural distribution of wetland areas. This novel approach offers a robust statistical framework for validating wetland data and can be applied to other ecological datasets.
- Application of Naïve Bayesian Inference: The application of naïve Bayes to model the likelihood of human-induced pressures on wetlands is a relatively unexplored area. Although this method has limitations—especially in dealing with imbalanced data—it provides a simple and interpretable framework for predicting human impact based on wetland characteristics.
- Use of the Epsilon Statistic: The epsilon statistic enables a detailed assessment of wetland sensitivity to human pressures. This allows for identifying particularly vulnerable or resilient wetlands, which can help inform more targeted conservation strategies.

There are several areas for improvement in the study:

The class imbalance in the dataset, especially regarding agricultural pressures, limited the model's ability to predict the minority class accurately. Future work should investigate oversampling, undersampling, or synthetic data generation to balance the dataset and enhance model performance. While naïve Bayes provides a straightforward approach, more advanced machine learning models like random forests or gradient boosting could better capture complex interactions and improve predictive accuracy, particularly for subtle pressures such as early-stage urbanization.

Additionally, the study would benefit from higher-resolution spatial and temporal data, enabling a more dynamic understanding of how various pressures evolve and interact with wetland characteristics over time.

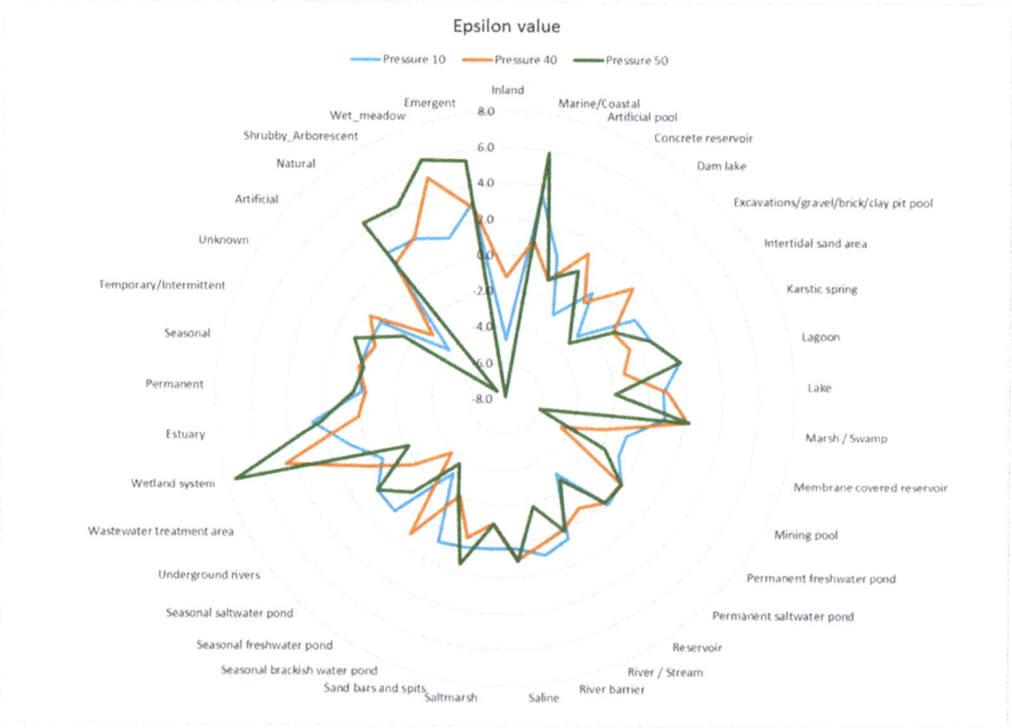

Figure 6. Spider chart that illustrates the epsilon values for each examined wetland characteristic in relation to the probability of three pressures: blue indicates pressure 10 (Agriculture and forestry), orange indicates pressure 40 (Urbanization, industrialization, and similar activities), and green indicates pressure 50 (Transportation and communication). The epsilon values show varying but comparable levels, highlighting the significant influence of specific characteristics on the likelihood of pressure presence.

5. Conclusions

This research provides important insights into the distribution of wetlands and the human pressures they face in the Aegean Islands. By utilizing power-law modeling, naïve Bayesian inference, and epsilon analysis, the study creates a comprehensive framework for evaluating wetland vulnerability and resilience. While it does identify some limitations in the model's performance, it also highlights opportunities for future research and improvement. Building on these findings, conservation efforts can be better designed to protect the most vulnerable wetlands from human-induced pressures. A conservation planning strategy based on general master plans for groups of wetlands with similar vulnerability and resilience, followed by specific tailoring to individual cases, as proposed and developed in the Introduction, could increase cost efficiency and effectiveness in such a complex archipelagic context.

Author Contributions: Conceptualization, methodology, supervision, writing—review and editing, A.Y.T.; software, validation, formal analysis, investigation, visualization, data curation, original draft preparation, C.M.G. All authors have read and agreed to the published version of the manuscript.

Funding: This research received no external funding.

Institutional Review Board Statement: Not relevant.

Data Availability Statement: The data supporting this study are publicly available and can be accessed from the Ygrotopio WWF database at https://www.ygrotopio.gr/general/search.php?lang=en&action=map&action=export (accessed on 16 October 2024).

Conflicts of Interest: The authors declare no conflicts of interest.

Appendix A

Summary of the 37 wetland characteristics and traits included in the analysis.

Characteristic	Variable Type	Possible States	Stats
Longitude	Continuous	N/A	Mean: 24.62, Std. Dev.: 1.99
Latitude	Continuous	N/A	Mean: 37.63, Std. Dev.: 1.72
Coastal/Inland	Categorical	Marine/Coastal, Inland	Marine/Coastal: 532, Inland: 291
Natural/Artificial	Binary	Natural, Artificial	Natural: 602, Artificial: 221
Wetland Type	Categorical	Estuary, Lagoon, Marsh/Swamp, Seasonal Saltwater Pond,	Estuary: 247, Wetland System: 124, Membrane-Covered Reservoir: 47, Marsh/Swamp: 87, ...
Wetland Area (Ha)	Continuous	N/A	Mean: 13.1, Std. Dev.: 56.5
Water Presence	Categorical	Seasonal, Permanent, Temporary/Intermittent, Unknown	Seasonal: 300, Permanent: 456, Temporary/Intermittent: 17, Unknown: 41
Shrubby/Arborescent	Binary	True, False	False: 244, True: 579
Wet Meadow	Binary	True, False	False: 410, True: 413
Emergent	Binary	True, False	True: 551, False: 272
10 Agricultural	Binary	True, False	True: 722, False: 101
40 Urbanization	Binary	True, False	True: 417, False: 406
50 Transportation	Binary	True, False	True: 429, False: 394

ferences

Mitsch, W.J.; Gosselink, J.G. *Wetlands*; John Wiley Sons: Hoboken, NJ, USA, 2015.
An, S.; Verhoeven, J.T. Wetland functions and ecosystem services: Implications for wetland restoration and wise use. In *Wetlands: Ecosystem Services, Restoration and Wise Use*; An, S., Verhoeven, J.T., Eds.; Springer: Cham, Switzerland, 2019; pp. 1–10.
Spieles, D.J. Wetland Construction, Restoration, and Integration: A Comparative Review. *Land* **2022**, *11*, 554. [CrossRef]
Díaz, S.; Pascual, U.; Stenseke, M.; Martín-López, B.; Watson, R.T.; Molnár, Z.; Hill, R.; Chan, K.M.A.; Baste, I.A.; Brauman, K.A.; et al. Assessing nature's contributions to people. *Science* **2018**, *359*, 270–272. [CrossRef]
Costanza, R.; De Groot, R.; Sutton, P.; Van der Ploeg, S.; Anderson, S.J.; Kubiszewski, I.; Farber, S.; Turner, R.K. Changes in the global value of ecosystem services. *Glob. Environ. Chang.* **2014**, *26*, 152–158. [CrossRef]
IPBES. *Global Assessment Report on Biodiversity and Ecosystem Services of the Intergovernmental Science-Policy Platform on Biodiversity and Ecosystem Services*; Brondizio, E.S., Settele, J., Díaz, S., Ngo, H.T., Eds.; IPBES Secretariat: Bonn, Germany, 2019.
Fluet-Chouinard, E.; Stocker, B.D.; Zhang, Z.; Malhotra, A.; Melton, J.R.; Poulter, B.; Kaplan, J.O.; Goldewijk, K.K.; Siebert, S.; Minayeva, T.; et al. Extensive global wetland loss over the past three centuries. *Nature* **2023**, *614*, 281–286. [CrossRef]
Verburg, P.H.; Vermaat, J.E.; Janse, J.H. Drivers of Wetland Conversion: A Global Meta-Analysis. *PLoS ONE* **2013**, *8*, e81292. [CrossRef]

9. Bowler, D.E.; Bjorkman, A.D.; Dornelas, M.; Myers-Smith, I.H.; Navarro, L.M.; Niamir, A.; Supp, S.R.; Waldock, C.; Winter, Vellend, M.; et al. Mapping human pressures on biodiversity across the planet uncovers anthropogenic threat complexes. *Peo Nat.* **2020**, *2*, 380–394. [CrossRef]
10. Choi, Y.R. Modernization, Development and Underdevelopment: Reclamation of Korean tidal flats, 1950s–2000s. *Ocean. Co Man.* **2014**, *102*, 426–436. [CrossRef]
11. Gaganis, C.M.; Troumbis, A.Y.; Kontos, T. Leveraging Reed Bed Burnings as Indicators of Wetland Conversion in Modern Gre *Land* **2024**, *13*, 538. [CrossRef]
12. Balian, E.V.; Lévêque, C.; Segers, H.; Martens, K.; Covich, A.P. Freshwater animal diversity assessment. *Hydrobiologia* **2008**, 5 627–637. [CrossRef]
13. Coll, M.; Piroddi, C.; Steenbeek, J.; Kaschner, K.; Ben Rais Lasram, F.; Aguzzi, J.; Ballesteros, E.; Bianchi, C.N.; Corbera, J.; Dailia T.; et al. The biodiversity of the Mediterranean Sea: Estimates, patterns, and threats. *PLoS ONE* **2010**, *5*, e11842. [CrossRef]
14. Ferrarini, A.; Gustin, M.; Celada, C. Twenty-Three Years of Land-Use Changes Induced Considerable Threats to the M Wetlands of Sardinia and Sicily (Italy) along the Mediterranean Bird Flyways. *Diversity* **2021**, *13*, 240. [CrossRef]
15. Merken, R.; Deboelpaep, E.; Teunen, J.; Saura, S.; Koedam, N. Wetland Suitability and Connectivity for Trans-Saharan Migrat Waterbirds. *PLoS ONE* **2015**, *10*, e0135445. [CrossRef]
16. Let, M.; Pal, S. Socio-ecological well-being perspectives of wetland loss scenario: A review. *J. Environ. Manag.* **2023**, *326*, 116 [CrossRef]
17. Kati, V.; Hovardas, T.; Dieterich, M.; Ibisch, P.L.; Mihok, B.; Selva, N. The challenge of implementing the European network protected areas Natura 2000. *Conserv. Biol.* **2015**, *29*, 260–270. [CrossRef] [PubMed]
18. Ferranti, F.; Turnhout, E.; Beunen, R.; Behagel, J.H. Shifting nature conservation approaches in Natura 2000 and the implicatic for the roles of stakeholders. *J. Environ. Plan. Manag.* **2014**, *57*, 1642–1657. [CrossRef]
19. Portaccio, A.; Campagnaro, T.; Sitzia, T. Birds and Natura 2000: A review of the scientific literature. *Bird Conserv. Int.* **2023**, *33*, [CrossRef]
20. Blicharska, M.; Orlikowska, E.H.; Roberge, J.M.; Grodzinska-Jurczak, M. Contribution of social science to large scale biodivers conservation: A review of research about the Natura 2000 network. *Biol. Conserv.* **2016**, *199*, 110–122. [CrossRef]
21. Blondet, M.; de Koning, J.; Borrass, L.; Ferranti, F.; Geitzenauer, M.; Weiss, G.; Turnhout, E.; Winkel, G. Participation in t implementation of Natura 2000: A comparative study of six EU member states. *Land Use Policy* **2017**, *66*, 346–355. [CrossRef]
22. Tsiafouli, M.A.; Apostolopoulou, E.; Mazaris, A.D.; Kallimanis, A.S.; Drakou, E.G.; Pantis, J.D. Human Activities in Natura 2(Sites: A Highly Diversified Conservation Network. *Environ. Manag.* **2013**, *51*, 1025–1033. [CrossRef] [PubMed]
23. Troumbis, A.Y. Imbalances in attitudes of European citizens towards biodiversity: Did the communication of the Europe Biodiversity Strategy work? *J. Nat. Conserv.* **2021**, *63*, 126041. [CrossRef]
24. Newton, A.; Icely, J.; Cristina, S.; Perillo, G.M.E.; Turner, R.E.; Ashan, D.; Cragg, S.; Luo, Y.; Tu, C.; Li, Y.; et al. Anthropoger Direct Pressures on Coastal Wetlands. *Front. Ecol. Evol.* **2020**, *8*, 144. [CrossRef]
25. Taylor, N.G.; Grillas, P.; Al Hreisha, H.; Balkız, Ö.; Borie, M.; Boutron, O.; Catita, A.; Champagnon, J.; Cherif, S.; Çiçek, K.; et The future for Mediterranean wetlands: 50 key issues and 50 important conservation research questions. *Reg. Environ. Cha* **2021**, *21*, 33. [CrossRef]
26. Perennou, C.; Beltrame, C.; Gelmami, A.; Tomas Vives, P.; Caessteker, P. Existing areas and past changes of wetland extent in t Mediterranean region: An overview. *Ecol. Mediterr.* **2012**, *38*, 53–66. [CrossRef]
27. WWF. Island Wetlands. Available online: https://www.wwf.gr/en/our_work/nature/terrestrial/protected_areas/islan wetlands/ (accessed on 26 November 2024).
28. Mpeza, P.E. Management and Sustainability of Greek Wetlands. In *Protected Area Management—Recent Advances*; IntechOp Houston, TX, USA, 2022.
29. Myers, N.; Mittermeier, R.A.; Mittermeier, C.G.; da Fonseca, G.A.B.; Kent, J. Biodiversity hotspots for conservation prioriti *Nature* **2000**, *403*, 853–858. [CrossRef] [PubMed]
30. Blondel, J.; Aronson, J. *Biology and Wildlife of the Mediterranean Region*; Oxford University Press: Oxford, UK, 1999; pp. 1–315.
31. Despoina Vokou, D.; Dimitrakopoulos, P.G.; Jones, N.; Damialis, A.; Monokrousos, N.; Pantis, J.D.; Mazaris, A.D.; The Natu 2000 Committee (2010–2013) Members. Ten years of co-management in Greek protected areas: An evaluation. *Biodivers. Conse* **2014**, *23*, 2833–2855. [CrossRef]
32. Gaston, K.J.; Jackson, S.F.; Nagy, A.; Cantú-Salazar, L.; Johnson, M. Protected Areas in Europe: Principle and Practice. *Ann. N. Acad. Sci.* **2008**, *1134*, 97–119. [CrossRef]
33. Troumbis, A.Y. Biodiversity Conservation Versus Ecological Modernization in Post-crisis Greece: Environmental Syndemics a the Biodiversity Degradation Syndrome. In *The Geography of Greece*; Darques, R., Sidiropoulos, G., Kalabokidis, K., Eds.; Wor Regional Geography Book Series; Springer: Cham, Switzerland, 2024. [CrossRef]
34. Mascia, M.B.; Pailler, S. Protected area downgrading, downsizing, and degazettement (PADDD) and its conservation implicatic *Conserv. Lett.* **2011**, *4*, 9–20. [CrossRef]

WWF. Oikoscopio/Ygrotopio Islands. 2004. Available online: https://www.ygrotopio.gr/general/search.php?lang=en&action=map (accessed on 1 August 2024).
Suratman, M.N. (Ed.) *Protected Area Management—Recent Advances*; IntechOpen: Houston, TX, USA, 2022; Volume 368. [CrossRef]
Gerakis, A.; Kalburtji, K. Agricultural activities affecting the functions and values of Ramsar wetland sites of Greece. *Agric. Ecosyst. Environ.* **1998**, *70*, 119–128. [CrossRef]
Zalidis, G.C.; Mantzavelas, A.L.; Gourvelou, E. Environmental impacts on Greek wetlands. *Wetlands* **1997**, *17*, 339–345. [CrossRef]
Zafeiriou, E.; Andrea, V.; Tampakis, S.; Karanikola, P. Wetlands Management in Northern Greece: An Empirical Survey. *Water* **2020**, *12*, 3181. [CrossRef]
Christopoulou, O.G.; Tsachalidis, E. Conservation Policies for Protected Areas (Wetlands) in Greece: A Survey of Local Residents' Attitude. *Water Air Soil Pollut. Focus* **2004**, *4*, 445–457. [CrossRef]
Catsadorakis, G.; Paragamian, K. *Inventory of the Wetlands of the Aegean Islands: Identity, Ecological Status and Threats*; World Wide Fund for Nature-WWF: Athens, Greece, 2007.
Vogiatzakis, I.N.; Mannion, A.; Sarris, D. Mediterranean island biodiversity and climate change: The last 10,000 years and the future. *Biodivers. Conserv.* **2016**, *25*, 2597–2627. [CrossRef]
Chatziioannou, I.; Bakogiannis, E.; Kyriakidis, C.; Alvarez-Icaza, L. A Prospective Study for the Mitigation of the Climate Change Effects: The Case of the North Aegean Region of Greece. *Sustainability* **2020**, *12*, 10420. [CrossRef]
Gritti, E.S.; Smith, B.; Sykes, M.T. Vulnerability of Mediterranean basin ecosystems to climate change and invasion by exotic plant species. *J. Biogeogr.* **2006**, *33*, 145–157. [CrossRef]
Strayer, D.L. Alien species in fresh waters: Ecological effects, interactions with other stressors, and prospects for the future. *Freshw. Biol.* **2010**, *55*, 152–174. [CrossRef]
Koutseri, I.; Vogiatzakis, I.N.; Tzanopoulos, J.; Sgardelis, S. Sustainable management in Aegean island wetlands: A case study of the Drepano lagoon. *J. Environ. Manag.* **2008**, *88*, 799–811.
Pressey, R.L.; Bottrill, M.C. Approaches to landscape- and seascape-scale conservation planning: Convergence, contrasts and challenges. *Oryx* **2009**, *43*, 464–475. [CrossRef]
Arponen, A.; Cabeza, M.; Eklund, J.; Kujala, H.; Lehtomäki, J. Costs of Integrating Economics and Conservation Planning. *Conserv. Biol.* **2010**, *24*, 1198–1204. [CrossRef]
Wade, P.R. Bayesian methods in conservation biology. *Conserv. Biol.* **2000**, *14*, 1308–1316. [CrossRef]
Stephens, P.A.; Buskirk, S.W.; del Rio, C.M. Inference in ecology and evolution. *Trends Ecol. Evol.* **2007**, *22*, 192–197. [CrossRef]
González-Salazar, C.; Stephens, C.R.; Marquet, P.A. Comparing the relative contributions of biotic and abiotic factors as mediators of species' distributions. *Ecol. Model.* **2013**, *248*, 57–70. [CrossRef]
Stephens, C.R.; Sánchez-Cordero, V.; Salazar, C.G. Bayesian Inference of Ecological Interactions from Spatial Data. *Entropy* **2017**, *19*, 547. [CrossRef]
Conn, P.B.; Johnson, D.S.; Williams, P.J.; Melin, S.R.; Hooten, M.B. A guide to Bayesian model checking for ecologists. *Ecol. Monogr.* **2018**, *88*, 526–542. [CrossRef]
Sienkiewicz, A.; Laska, G. Bayesian belief networks: A potential tool for conservation planning of endangered plant species populations. *J. Plant Ecol.* **2023**, *16*, rtac071. [CrossRef]
Guisan, A.; Lehmann, A.; Ferrier, S.; Austin, M.; Overton, J.M.C.; Aspinall, R.; Hastie, T. Making better biogeographical predictions of species' distributions. *J. Appl. Ecol.* **2006**, *43*, 386–392. [CrossRef]
Pearson, R.G.; Dawson, T.P. Predicting the impacts of climate change on the distribution of species: Are bioclimate envelope models useful? *Glob. Ecol. Biogeogr.* **2003**, *12*, 361–371. [CrossRef]
Zambrano, J.; Garzon-Lopez, C.X.; Yeager, L.; Fortunel, C.; Cordeiro, N.J.; Beckman, N.G. The effects of habitat loss and fragmentation on plant functional traits and functional diversity: What do we know so far? *Oecologia* **2019**, *191*, 505–518. [CrossRef]
Clark, J.S.; Ferraz, G.A.; Oguge, N.; Hays, H.; DiCostanzo, J. Hierarchical Bayes for structured, variable populations: From recapture data to life-history prediction. *Ecology* **2005**, *86*, 2232–2244. [CrossRef]
McCarthy, M.A.; Possingham, H.P. Active adaptive management for conservation. *Conserv. Biol.* **2007**, *21*, 956–963. [CrossRef]
Handrinos, G.; Akriotis, T. *The Birds of Greece (Helm Field Guides)*; Bloomsbury Publishing: London, UK, 1997.
Finlayson, C.M.; Gardner, R.C. Ten key issues from the Global Wetland Outlook for decision makers. *Mar. Frewhwater Res.* **2021**, *72*, 301–310. [CrossRef]
Serran, J.N.; Creed, I.F.; Ameli, A.A.; Aldred, D.A. Estimating rates of wetland loss using power-law functions. *Wetlands* **2018**, *38*, 109–120. [CrossRef]
Xu, F.; Zhou, Z.; Fagherazzi, S.; D'alpaos, A.; Townend, I.; Zhao, K.; Xie, W.; Guo, L.; Wang, X.; Peng, Z.; et al. Anomalous scaling of branching tidal networks in global coastal wetlands and mudflats. *Nat. Commun.* **2024**, *15*, 9700. [CrossRef]
Maschberger, T.; Kroupa, P. Estimators for the Exponent and Upper Limit, and Goodness-of-Fit Tests for (Truncated) Power-Law Distributions. *Mon. Not. R. Astron. Soc.* **2009**, *395*, 931–942. [CrossRef]

65. Newman, M. Power Laws, Pareto Distributions and Zipf's Law. *Contemp. Phys.* **2005**, *46*, 323–351. [CrossRef]
66. White, E.P.; Enquist, B.J.; Green, J.L. On estimating the exponent of power-law frequency distributions. *Ecology* **2008**, *89*, [CrossRef] [PubMed]
67. Clauset, A.; Shalizi, C.R.; Newman, M.E.J. Power-Law Distributions in Empirical Data. *SIAM Rev.* **2009**, *51*, 661–703. [CrossRef]
68. Troumbis, A.Y.; Gaganis, C.M.; Sideropoulos, H. Probabilistic Wildfire Risk Assessment and Modernization Transitions: The C of Greece. *Fire* **2023**, *6*, 158. [CrossRef]
69. Ellison, A.M. Bayesian inference in ecology. *Ecol. Lett.* **2004**, *7*, 509–520. [CrossRef]
70. Mace, G.M.; Collar, N.J.; Gaston, K.J.; Hilton-Taylor, C.; Akçakaya, H.R.; Leader-Williams, N.; Milner-Gulland, E.J.; Stu S.N. Quantification of Extinction Risk: IUCN's System for Classifying Threatened Species. *Conserv. Biol.* **2008**, *22*, 1424–1 [CrossRef] [PubMed]
71. Mandelbrot, B.B. *The Fractal Geometry of Nature*; W.H. Freeman & Co. Ltd.: New York, NY, USA, 1982.
72. Sole, R.V.; Bascompte, J. Are critical phenomena relevant to large-scale evolution? *Proc. R. Soc. Biol. Sci.* **1996**, *263*, 161–1 [CrossRef]
73. Bak, P.; Tang, C.; Wiesenfeld, K. Self-Organized Criticality: An Explanation of the 1/f Noise. *Phys. Rev. Lett.* **1987**, *59*, 381– [CrossRef]
74. Morris, P.A.; Hill, N.M.; Reekie, E.G.; Hewlin, H.L. Lakeshore diversity and rarity relationships along interacting disturba gradients: Catchment area, wave action and depth. *Biol. Conserv.* **2002**, *106*, 79–90. [CrossRef]
75. Marquet, P.A.; Quiñones, R.A.; Abades, S.; Labra, F.; Tognelli, M.; Arim, M.; Rivadeneira, M. Scaling and power-laws in ecolog systems. *J. Exp. Biol.* **2005**, *208*, 1749–1769. [CrossRef] [PubMed]
76. Hand, D.J.; Yu, K.M. Idiot's Bayes—Not so stupid after all? *Int. Stat. Rev.* **2001**, *69*, 385–398. [CrossRef]
77. With, K.A.; Crist, T.O. Critical thresholds in species responses to landscape structure. *Ecology* **1995**, *76*, 2446–2459. [CrossRef]
78. Guisan, A.; Thuiller, W. Predicting species distribution: Offering more than simple habitat models. *Ecol. Lett.* **2005**, *8*, 993–1 [CrossRef]
79. Elith, J.; Leathwick, J.R. Species Distribution Models: Ecological Explanation and Prediction Across Space and Time. Ann Review of Ecology. *Evol. Syst.* **2009**, *40*, 677–697. [CrossRef]
80. Breiman, L. Statistical modeling: The two cultures. *Stat. Sci.* **2001**, *16*, 199–215. [CrossRef]
81. Cutler, D.R.; Edwards, T.C.; Beard, K.H.; Cutler, A.; Hess, K.T. Random forests for classification in ecology. *Ecology* **2007**, 2783–2792. [CrossRef]
82. Richardson, C.J.; King, R.S.; Qian, S.S.; Vaithiyanathan, P.; Qualls, R.G.; Stow, C.A. Estimating ecological thresholds for phospho in the Everglades. *Environ. Sci. Technol.* **2008**, *41*, 8084–8091. [CrossRef]
83. Ramsar Convention Secretariat. Sustainable Development for a Viable Future: The Pivotal Role of Wetlands. Available onli https://www.ramsar.org/news/sustainable-development-viable-future-pivotal-role-wetlands (accessed on 15 October 202
84. Sutherland, W.J.; Aveling, R.; Brooks, T.M.; Clout, M.; Dicks, L.V.; Fellman, L.; Fleishman, E.; Gibbons, D.W.; Keim, B.; Lickor F.; et al. A horizon scan of global conservation issues for 2014. *Trends Ecol. Evol.* **2014**, *29*, 15–22. [CrossRef]
85. Erwin, K.L. Wetlands and global climate change: The role of wetland restoration in a changing world. *Wetl. Ecol. Manag.* **2009**, 71–84. [CrossRef]
86. Chasmer, L.; Mahoney, C.; Millard, K.; Nelson, K.; Peters, D.; Merchant, M.; Hopkinson, C.; Brisco, B.; Niemann, O.; Montgom J.; et al. Remote Sensing of Boreal Wetlands 2: Methods for Evaluating Boreal Wetland Ecosystem State and Drivers of Char *Remote Sens.* **2020**, *12*, 1321. [CrossRef]

diversity

MDPI

cle

xploring the Relationship between Ecosystem Services and ıstainable Development Goals for Ecological Conservation: Case Study in the Hehuang Valley of Qinghai-Tibet Plateau

ie Wei [1,†], Ke Wang [1,†], Yu Ma [1], Qingxiang Meng [1], Yi Yang [2,*] and Mengxue Liu [3]

[1] College of Resources and Environmental Sciences, Henan Agricultural University, Zhengzhou 450046, China; hjwei@henau.edu.cn (H.W.); wangke010603@163.com (K.W.); mayu99@henau.edu.cn (Y.M.); qxmeng@henau.edu.cn (Q.M.)
[2] College of Forestry, Henan Agricultural University, Zhengzhou 450046, China
[3] Lhasa Plateau Ecosystem Research Station, Key Laboratory of Ecosystem Network Observation and Modeling, Institute of Geographic Sciences and Natural Resources Research, Chinese Academy of Sciences, Beijing 100101, China; mengxueliu@mail.bnu.edu.cn
* Correspondence: yangyi023@henau.edu.cn
† These authors contributed equally to this work.

Abstract: With the increase in human activities and the acceleration of urbanization, over-exploitation of natural resources has led to a decline in ecosystem services (ESs), subsequently affecting the achievement of sustainable development goals (SDGs). As the key ecological zone of Qinghai-Tibet Plateau, the stability and enhancement of ESs in the Hehuang Valley are crucial for achieving SDGs and biodiversity conservation. This study quantifies nine SDGs for the Hehuang Valley in the last twenty years. Four ecological models were utilized to compute key ESs: net primary productivity (NPP), water yield, soil retention, and sand fixation. Panel data were analyzed using a coupling coordination model to quantify the relationship between ESs and sustainable development level (SDL) in each county. Additionally, the Geographically and Temporally Weighted Regression (GTWR) model was employed to examine the correlation between ESs and SDL. The results indicate the following: (1) During the period, NPP and water yield first increased and then decreased. The capacity for soil retention and sand fixation showed an overall increase, highlighting substantial variability among counties in their ability to deliver these ESs. (2) The SDL of counties in the Hehuang Valley increased, with Xining City showing slightly higher SDL than other counties. (3) The overall coupling coordination degree among NPP, water yield, soil retention, sand fixation, and SDL in the Hehuang Valley exhibited an upward trend in the last twenty years. SDL demonstrated the highest coordination degree with NPP, followed by soil retention, water yield, and sand fixation. (4) Most counties in the Hehuang Valley exhibited a lag in SDL relative to NPP, water yield, and soil retention in the last twenty years. In the early stage, sand fixation and SDL were primarily lagging in SDL, while in the late stages, sand fixation lagged behind SDL. (5) During the period, there was an increasing negative correlation observed between the four ESs and SDL. The positive contribution of NPP and sand fixation in some counties gradually shifted to a negative effect, and the negative effect of water yield and soil retention on SDL intensified. The impact of human activities on ecosystem function hindered local SDL. This study offers scientific theoretical backing and practical recommendations for promoting SDL and biodiversity conservation in the Hehuang Valley.

Keywords: ecosystem services; sustainable development goals; coupling coordination model; GTWR model; Hehuang Valley

1. Introduction

An ecosystem is a complex system influenced by various natural, economic, and social factors, and it undergoes dynamic changes over time [1]. The connotation of ecosystem services (ESs) is integral to the framework of sustainable development goals (SDGs) [2], exerting a significant influence on societal and economic sustainability. ESs function as a crucial link between natural contributions and human needs across various timeframes and geographic scales [3]. These services, which are essential functions of ecosystems, facilitate connections between natural and societal systems, playing a vital role in human well-being and progress. They contribute to the overall economic value of the earth [4] and serve as important indicators of ecological health in regions. In 2015, the United Nations' 193 member states adopted the 2030 Agenda for Sustainable Development, encompassing 17 SDGs and 169 associated targets [5]. These goals include various social, economic, and environmental objectives, each interacting with ESs in different ways [6]. The interactions and influencing factors among the SDGs are complex [7]. Many scholars have analyzed the synergies and trade-offs between SDGs [8], the prioritization of SDGs [9], and the factors influencing these goals [10]. Their findings suggest that achieving the SDGs requires proactive management of production activities to sustain livelihoods, create new job opportunities, build climate change resilience, and harmonize with environmental changes to provide robust ESs [11].

The attainment of SDGs depends on more than just social and economic progress; it also hinges on integrating ecological factors into sustainable development decisions [12]. ESs are pivotal in shaping the SDGs [13], and the primary aim of implementing ecological conservation measures is to promote regional sustainability. It is crucial to evaluate the contribution of these measures to sustainable development. The protection, restoration, and sustainable provision of ESs are crucial for attaining the SDGs [5]. Connecting ESs with the SDGs fosters synergistic development across various aspects of sustainable development [14]. ESs can serve as direct indicators of ecosystem functions and quantitatively demonstrate the direct or indirect contributions of ecosystems to SDGs from an ecological perspective [15]. For instance, provisioning services contribute to the attainment of SDG2 (Zero Hunger), SDG3 (Good Health and Well-being), and SDG6 (Clean Water and Sanitation), among others. ESs are closely related to SDG15 (Life on Land), SDG13 (Climate Action), SDG14 (Life Below Water), and SDG6, indirectly promoting the realization of 13 other goals [16]. Zhang et al. [17] conducted an analysis using SDG 15.3.1 and carbon storage as metrics to assess the impact of land use changes on ecosystem service values, providing valuable insights for improving land degradation. Wood et al. [18] identified the importance of ESs in achieving the 2030 SDGs by integrating the perspectives of researchers through the "ecosystem service-goal" contribution framework. The results of their survey indicated that 16 ESs were considered to contribute to 41 targets within 12 SDGs. Other researchers have conducted global analyses on the relationships between multiple ESs provided by urban landscapes and SDGs, finding that 17 different ESs are linked to 12 SDGs [19].

Currently, numerous scholars have utilized various methodologies to examine the complex relationships and impacts between ESs and sustainable development level (SDL). Hussain et al. [20] employed an interpretive structural model to investigate the factors influencing sustainable tourism development and their inherent logical connections. Xu et al. [12] applied the geographical detector method to analyze the cumulative effects of ESs and urbanization on achieving SDGs from a spatial perspective. Liu et al. [7] utilized a panel data model to analyze the contribution of protected areas to SDGs, thereby providing a theoretical foundation for the management of protected areas and regional SDL. Yin et al. [21], through downscaled localized SDG indicators, regression methods, and mechanism analysis, identified the contributions of ecosystem carbon sequestration services to SDGs, revealing significant progress in resource and environmental sustainability goals on the Loess Plateau. Additionally, some scholars conducted expert surveys to scrutinize the global-scale relationships between distinct ESs and SDGs [22].

The coupling coordination model comprehensively considers the interactions and coordination mechanisms among different factors, quantitatively assessing their contributions [23]. It can also adapt variables to actual conditions to accommodate changes across various temporal and spatial scales [24]. Consequently, it facilitates a more comprehensive and accurate understanding of the coupling relationship between ESs and SDGs, offering optimization and coordination recommendations to achieve equilibrium and mutual benefit. For instance, He et al. [25] scrutinized the coupling coordination relationship between water resource and eco-environment in China. Yang et al. [26] employed the coupling coordination index to reveal the temporal-spatial coupling relationship between ESs and human well-being, thereby promoting the SDL of urban agglomerations. Yang et al. [27] assessed the SDL status of land ecology in Shanxi Province using ecological footprint and coupling coordination degree (CCD) as indicators. By integrating the two indicators into a mathematical model using the TOPSIS multi-criteria decision-making approach, they obtained comprehensive scores for the SDL of land ecology in various cities in Shanxi Province. This outlines the temporal-spatial changes in SDL and ESs.

Although the coupling coordination model quantifies the relationship between ESs and SDGs, it does not indicate the correlation between the two, which is a more specific type of relationship [28]. The explanation of how ES trends affect the attainment of SDGs and the regional variability in this impact remains incomplete [29]. Recognizing the relationship between ESs and SDGs is vital for policymaking and pinpointing current management requirements [30]. Thus, employing the Geographically and Temporally Weighted Regression (GTWR) model to analyze this relationship can yield valuable insights into its extent.

The Hehuang Valley, situated at the transition zone between the Qinghai-Tibet Plateau and the Loess Plateau, occupies a unique geographical position and serves as a "sensitive area" for ecology and climate in Asia [31]. It is a relatively low-altitude region of the Qinghai-Tibet Plateau, suitable for human habitation, and is also economically significant in Qinghai Province [32]. As population and socio-economic growth continue, the interconnection between humans and nature becomes increasingly tight [33]. The pursuit of a better quality of life and improved well-being is a pressing need for people. However, intensified human activities and environmental degradation pose serious threats to ESs. Evaluating the results at smaller administrative units, particularly at the county level, is more targeted and can effectively assess SDGs, thereby promoting their implementation [30]. Additionally, in line with ongoing urbanization, Qinghai Province plans to establish the Xining-Haidong metropolitan area, with a focus on the economic development of Xining and Haidong. The unique geography and environment of the Hehuang Valley, along with its future development potential, will significantly impact its ecosystems and ecological environment [34]. Therefore, exploring the coupling relationship between ESs and SDGs in the Hehuang Valley can provide scientific support for policymaking and resource management, thereby promoting the effective coupling and synergistic development of ESs and SDGs.

With this context in mind, this study aims to achieve the following objectives: (1) Create a unique evaluation framework for SDL and ESs tailored to local conditions using available data, and assess the SDL and ESs status of the Hehuang Valley quantitatively; (2) Investigate the interconnection between key ESs and SDL by coupling model; (3) Employ the GTWR model to uncover the temporal and spatial relationships between various types of ESs and SDL.

2. Materials and Methods

2.1. Study Area

Situated at the transition zone between the Qinghai-Tibet Plateau and the Loess Plateau, the Hehuang Valley (35°~38° N, 100°~103° E) serves as a significant agricultural region of the Qinghai-Tibet Plateau [35] (Figure 1). It functions as the political, economic, and cultural center of Qinghai Province [36]. Spanning an area of approximately 35,273.77 km^2, the valley plays a crucial role in ensuring food security for Qinghai Province and the wider Qinghai-Tibet Plateau [35]. Despite occupying only 5% of the province's total area, it accommodates approximately 72.77% of the province's population and 60% of its arable land. The valley consists of broad river valleys and well-developed terraces, with elevations ranging from 1650 to 2400 m, making it suitable for agricultural practices. Wheat, corn, soybeans, rapeseed, and other crops are cultivated in this region [37]. The Hehuang Valley includes 17 counties and districts, including Chengdong, Chengzhong, Chengxi, and Chengbei districts et al., in Xining City; Ledu District and Ping'an District et al., in Haidong City; Jianzha County and Tongren County in Huangnan; Menyuan County in Haibei; and Guide County in Hainan.

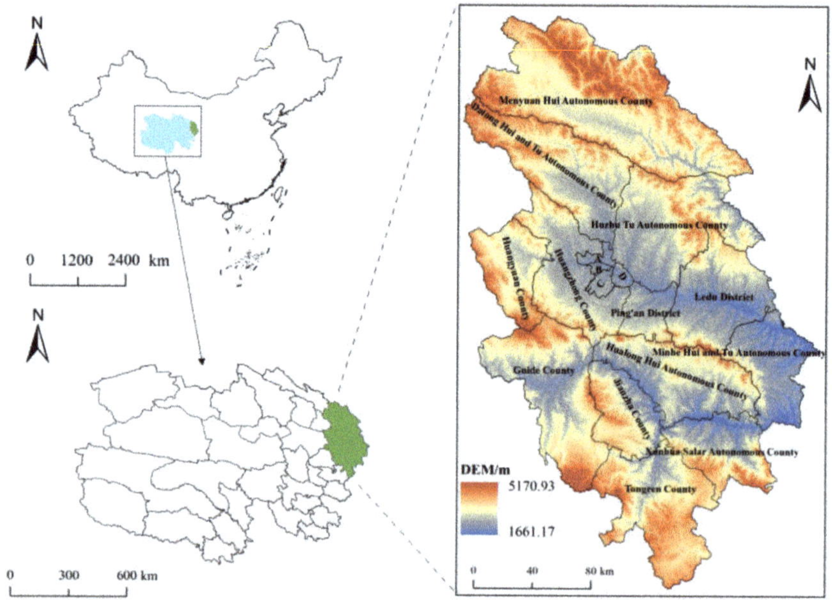

Figure 1. Location of the Hehuang Valley. Note: A represents Chengbei District; B represents Chengxi District; C represents Chengzhong District; D represents Chengdong District.

In 2020, the population of the Hehuang Valley was 1.38 million. The per capita regional GDP reached 64,537 yuan, and the per capita total retail sales of consumer goods was 13,282 yuan. The total grain and total meat output in the Hehuang Valley were 1.038 million tons and 271,000 tons, respectively, representing increases of 47.3% and 58.5% compared to the year 2000 [34]. The valley's geomorphological types include river valley areas, shallow mountainous areas, and hilly areas, with annual precipitation ranging from 166.4 to 646.6 mm, mostly occurring from May to October, coinciding with the warm season [38]. The region experiences a typical highland continental climate with distinct seasonal variations. The Hehuang Valley has been significantly uplifted due to the block movements of the Qilian Mountains. The parallel ridge-valley system of the Yemanan Mountains—Shule South Mountains (Shule Mountains), Datong Mountains, and Dabanshan Mountains—and the upper Danghe River Valley—Hala Lake, Qinghai Lake, and Huangshui Valley—extend into this area. Additionally, the remaining ridges of Dabanshan, Laji Mountain, and Xiqing Mountain enclose the Yellow River and Huangshui Valleys, forming the "three mountains enclosing two valleys" geomorphological framework of the region [34]. Grassland and forestland were the main land use types of Hehuang Valley and the construction land has increased significantly in the context of urbanization (Figure 2).

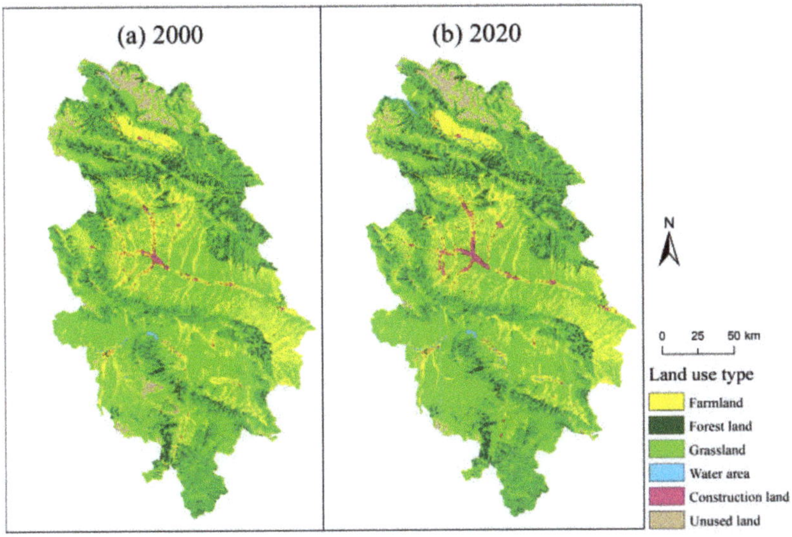

Figure 2. Land Use Status of the Hehuang Valley in 2000 (**a**) and 2020 (**b**).

2.2. Data Sources

The data employed in this study consist mainly of natural geographic information, socio-economic statistics, and land use-related data. Specific data usage and sources are detailed in Table 1. Additional data such as regional GDP, fixed asset investment, retail sales, average years of education, mortality rate, employment rate, broadband access, fixed telephone users, local fiscal expenditure, number of industrial enterprises, hospital bed numbers, public library collections, and sports venues are sourced from the Qinghai Statistical Yearbook, China County Statistical Yearbook, National Economic and Social Development Statistical Bulletin, Xining Statistical Yearbook, China City Statistical Yearbook, and China Health Statistics Yearbook.

Table 1. Data Sources and Their Uses.

Data Name	Data Format	Data Source	Data Use
DEM Elevation Data	Raster data with 30 m resolution	http://www.gscloud.cn/search, accessed on 10 February 2023	Basic parameter input for soil retention and wind erosion model
Land Use Remote Sensing Data	Raster data with 30 m resolution	http://www.resdc.cn/, accessed on 22 December 2023	Basic parameter input for NPP, water yield, and soil retention models
MOD13Q1	Raster data with 250 m resolution	https://www.nasa.gov/, accessed on 11 January 2024	Obtain Normalized Difference Vegetation Index (NDVI) and vegetation coverage data
·lobal Land Cover Data (China subset)	Raster data with 100 m resolution	http://bdc.casnw.net/yyzc/sj/250299.shtml, accessed on 13 December 2023	Obtain vegetation-type data for the study area
ERA5-Land Wind Speed Data	List data	www.ecmwf.int, accessed on 10 December 2023	Obtain wind factor and cumulative time distribution raster maps for various wind speed levels

Table 1. *Cont.*

Data Name	Data Format	Data Source	Data Use
Soil Moisture Data	Raster data with 1000 m resolution	http://bdc.casnw.net/yyzc/sj/250299.shtml, accessed on 1 December 2023	Topsoil moisture factor (0–10 cm depth range)
Monthly Precipitation Data	List data	http://data.cma.cn/, accessed on 18 November 2023	Obtain rainfall erosion factor and annual average rainfall raster maps
Soil Texture, Organic Matter, Depth	Raster data with 1000 m resolution	http://bdc.casnw.net/yyzc/sj/250299.shtml, accessed on 12 December 2023	Basic parameter input for water yield and soil retention models
Monthly Temperature, Precipitation, Radiation Data	List data	http://data.cma.cn/, accessed on 16 November 2023	Obtain monthly average temperature, radiation raster data and annual potential evaporation data
Annual Meat, Grain Production, and Population Data	Statistical data	Qinghai Statistical Yearbook, China County Statistical Yearbook	Obtain grain and meat production and county population data
Road Network Data	Vector data	http://www.resdc.cn/, accessed on 8 December 2023	Obtain road and railway data for 1995, 2012, and 2020
2020 Standard Map of China	Vector data	Ministry of Natural Resources (https://www.mnr.gov.cn/, accessed on 25 December 2023)	Obtain map with approval number GS(2020)4619 and county vector data
PM2.5 Data	Statistical data	https://quotsoft.net/air/, accessed on 10 September 2023	Obtain PM2.5 data for various periods across counties

2.3. Technical Route

The research methodology and approach in this study are outlined in Figure 3. Firstly, utilizing the Carnegie-Ames-Stanford Approach (CASA), the Integrated Valuation of ESs and Tradeoffs (InVEST) model's water yield module, the Revised Universal Soil Loss Equation (RUSLE), and a wind erosion model, four key ESs were quantified: net primary productivity (NPP), water yield, soil retention, and sand fixation. Secondly, SDL was assessed across three dimensions: foundational, developmental acceleration, and environmental aspects. The entropy weight method and linear weighting method were applied to calculate these dimensions and the overall SDL for 17 counties in the Hehuang Valley for the years 2000, 2010, and 2020. Thirdly, a coupling coordination model was employed to determine the degree of coordination and coupling between SDL and ESs. Additionally, a relative development model was utilized to analyze the development level relative to ESs and SDL. Finally, the Geographically and Temporally Weighted Regression (GTWR) model was utilized to compute correlation coefficients between SDL and ESs. This approach helped to elucidate their temporal and spatial interaction mechanisms.

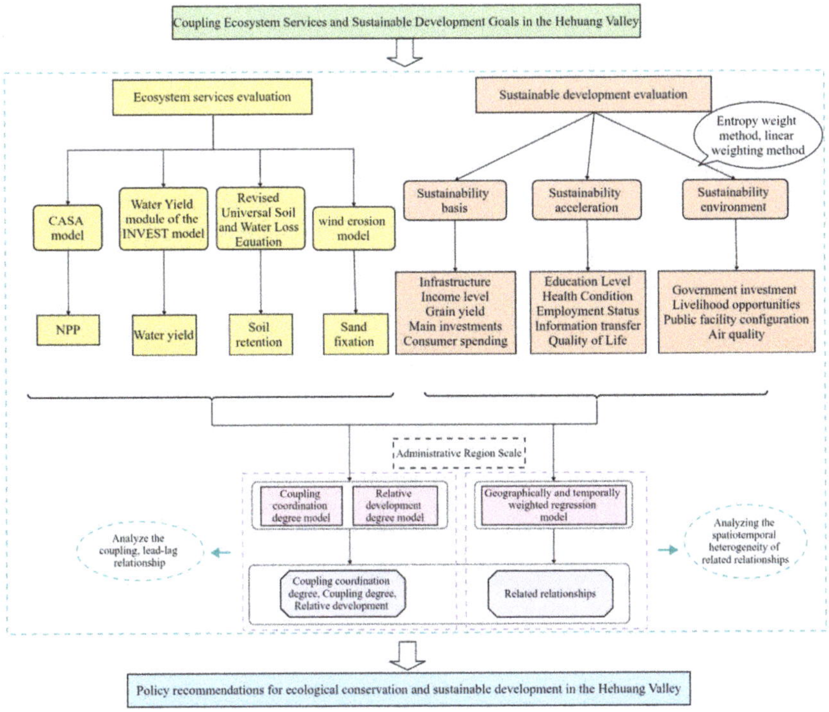

Figure 3. Technical Route of the Study.

2.4. Ecosystem Service Assessment Methods

2.4.1. Selection Basis for Ecosystem Service Indicators

Based on the Millennium Ecosystem Assessment, there are more than 20 ESs, which are divided into provisioning services, regulating services, supporting services, and cultural services. Considering the feasibility of the method, the importance of ESs, and the local physical geographical features, this study selects four ESs—NPP, water yield, soil retention, and sand fixation—to examine changes in ESs in the Hehuang Valley. The detailed illustrations are as follows. NPP, as an indicator of the capacity of vegetation communities to produce organic matter in their natural environment, is of particular importance in an agricultural area like the Hehuang Valley. Analyzing the temporal and spatial variations of NPP will provide insights into the carbon cycle and its support capacity for local agricultural production. The Hehuang Valley, being located in arid and semi-arid regions, has a high demand for water. Therefore, investigating changes in water yield will enable the analysis of the ecosystem's ability to supply water resources, which can guide the formulation of effective water resource protection policies. The complex terrain combined with a dry climate and concentrated precipitation in the Hehuang Valley results in severe soil erosion. The area's location as a transition zone makes it susceptible to natural disasters such as sandstorms, which have a remarkable impact on the local ecosystem and the livelihoods of residents. Studying the temporal and spatial changes in sand fixation will provide valuable insights into the ecosystem's ability to defend against wind and sand disasters, thereby supporting sand disaster prevention and ecological environment improvement efforts. Moreover, four ESs all belong to the regulating services or supporting services, but no provisioning services or cultural services are chosen. Firstly, because provisioning services and cultural services are terminal services, the relationship between them and SDGs is relatively direct and clear. However, regulating services and supporting services are basic functions that affect the quality of provisioning services and cultural services, and the relationship between them and SDGs is relatively complex. Current research has not yet clarified this, which is a knowledge gap. Secondly, indicators related to provisioning services and cultural services are reflected in the sustainable development indicator system, such as grain output, education, etc.

2.4.2. Measurement of Net Primary Productivity, Water Yield, and Soil Retention

NPP was estimated by utilizing the CASA model based on the principle of light energy utilization. The detailed model and parameter selection were based on Zhu et al. [39]. Water yield was assessed by the Water Yield module of the InVEST model [40] based on the principle of water balance. This assessment takes into account factors such as topography, vegetation type, meteorology, and soil texture to achieve spatialization of water yield. The RUSLE is used to quantify soil erosion in the Hehuang Valley. Soil retention is determined by the difference between potential soil erosion and actual soil erosion [41,42].

2.4.3. Measurement of Sand Fixation

This study utilizes the wind erosion model established by Li et al. [43], which includes wind erosion modulus calculation models for three types of underlying surfaces: cultivated land, grassland/shrubland, and sandy land. The wind erosion model assesses the sand fixation capability of the ecosystem by subtracting the actual wind erosion modulus from the potential wind erosion modulus.

2.5. Sustainable Development Assessment Methods

Based on the SDGs, relevant research [44,45], and local conditions, the sustainable development indicator system was established considering three dimensions: sustainable development foundation, sustainable development acceleration, and sustainable development environment. This study ultimately selected nine SDGs and 17 influencing indicators to analyze the SDL and construct the SDL evaluation indicator system (Table 2).

Table 2. SDL Evaluation Indicator System for the Hehuang Valley.

Dimension	Factor	Indicator Level	Indicator	Weight	SDG Indicator
Sustainability Foundation	Capital Status	Infrastructure	Road network density	0.024	Sustainable Cities and Communities (SDG 11)
		Income Level	Regional GDP	0.095	No Poverty (SDG 1)
		Grain Output	Total grain production	0.065	Zero Hunger (SDG 2)
	Capital Change	Major Investment	Total fixed asset investment	0.113	Sustainable Cities and Communities (SDG 11)
		Consumer Spending	Total retail sales of consumer goods	0.132	Responsible Consumptio and Production (SDG 12
Sustainability Acceleration	Quality Status	Education Level	Average years of schooling	0.012	Quality Education (SDG
		Health Status	Mortality rate	0.015	Good Health and Well-being (SDG 3)
	Skill Level	Employment Status	Employment rate	0.008	Decent Work and Economic Growth (SDG
	Living Standard	Information Transmission	Broadband access	0.125	Industry, Innovation, an Infrastructure (SDG 9)
			Fixed Telephone Users	0.063	Industry, Innovation, an Infrastructure (SDG 9)
		Quality of Life	Per capita total meat production	0.029	Zero Hunger (SDG 2)

Table 2. *Cont.*

imension	Factor	Indicator Level	Indicator	Weight	SDG Indicator
stainability nvironment	Soft Environment	Government Investment	Local fiscal general budget expenditure	0.067	Sustainable Cities and Communities (SDG 11)
		Livelihood Opportunities	Number of industrial enterprises above designated size	0.034	Industry, Innovation, and Infrastructure (SDG 9)
	Hard Environment	Public Facility Allocation	Number of hospital beds	0.061	Good Health and Well-being (SDG 3)
			Total collection of public libraries	0.120	Quality Education (SDG 4)
			Number of sports venues	0.021	Good Health and Well-being (SDG 3)
		Air Quality	Average PM2.5	0.015	Climate Action (SDG 13)

The sustainable development foundation includes capital status and capital changes [30,46,47], involving SDG1, SDG2, SDG11, and SDG12. For capital status and changes, five indicators were selected to represent the foundational conditions required for SDL: infrastructure, income levels, grain output, major investments, and consumer spending. Sustainable acceleration [30,46,47] involves quality conditions, skill levels, and living standards, representing the achievement of SDG2, SDG3, SDG4, SDG8, and SDG9. Five indicators were selected to represent the acceleration conditions required for SDL: education levels, health status, employment status, information dissemination, and quality of life. As the main actors in SDL, people determine the direction and pace of sustainable development. A well-educated population facilitates faster and better SDL [30,48]. The sustainable development environment was classified into two categories: soft and hard environments, which represent the processes involved in achieving SDG3, SDG4, SDG9, SDG11, and SDG13. Four key indicators were selected to represent the necessary environmental conditions for SDL: government investment, livelihood opportunities, public facility allocation, and air quality.

To ensure comparability across different indicators, data normalization was necessary, thereby distributing the data within the range of [0, 1]. The objective weights of the indicators were then calculated using the entropy weight method. Based on the SDL Evaluation Indicator System established in Table 2 for the Hehuang Valley, in combination with the entropy weight method, we computed the sustainability foundation, sustainability acceleration, sustainability environment, and overall SDL evaluation indices for the years 2000, 2010, and 2020.

2.6. Quantification of the Coupling Relationship between ESs and SDL

The coupling degree between NPP, water yield, soil retention, sand fixation, and SDL reflects the extent of their interaction and the developmental changes between them. The CCD indicates the level of coordination between the two systems. A higher CCD indicates a more harmonious relationship between SDL and ESs. The calculation formulas are as follows:

$$T = aG_i + bF(x) \tag{1}$$

$$C = \left\{G_i \times F(x) / [G_i \times F(x)/2]^2\right\}^k \tag{2}$$

$$D = \sqrt{T \times C} \tag{3}$$

where $F(x)$ is the SDL index, and G_i (i = 1, 2, 3, 4) are the standardized values of NPP, water yield, soil retention, and sand fixation, respectively. T is the comprehensive evaluation value of SDL and ESs, C is the coupling degree, and D is the CCD, ranging between [0, 1]. Since both systems are equally important, $a = b = 0.5$ and there are $k = 2$ system layers. Referring to related research [49], the CCD is classified into five types: severe imbalance (0.0–0.2), moderate imbalance (0.2–0.4), basic coordination (0.4–0.6), moderate coordination (0.6–0.8), and high coordination (0.8–1.0), as shown in Table 3.

Table 3. Classification Standards for Coupling Degree and CCD between ESs and SDL in the Hehuang Valley.

Coupling Degree *C* Value	Coupling Type	CCD *D* Value	Coordination Type
$0 \leq C \leq 0.3$	Low-level coupling	$0 \leq D < 0.2$	Severe imbalance
$0.3 < C \leq 0.5$	Antagonistic phase	$0.2 \leq D < 0.4$	Moderate imbalance
$0.5 < C \leq 0.8$	Running-in phase	$0.4 \leq D < 0.6$	Basic coordination
$0.8 < C \leq 1$	High-level coupling	$0.6 \leq D < 0.8$	Moderate coordination
——	——	$0.8 \leq D < 1$	High coordination

To further analyze the relationship between ESs and SDL in each county, the relative development degree of both factors was calculated. Based on previous research findings [34], all counties in the study area were classified into three types: ESs-lagging type, synchronous development type of ESs and SDL, and SDL-lagging type. The specific classification standards can be found in Table 4.

Table 4. Classification Standards for Relative Development Types of ESs and SDL in the Hehuang Valley.

Type	Relative Development Degree	Specific Classification
Relative Development Types	$0 < \beta = Gi/F(x) < 0.85$	ESs-lagging type
	$0.85 < \beta = Gi/F(x) < 1.25$	Synchronous development type
	$\beta = Gi/F(x) > 1.25$	SDL-lagging type

Note: G_i (i = 1, 2, 3, 4) are the standardized values of NPP, water yield, soil retention, and sand fixation, respectively; $F(x)$ is the SDL index.

2.7. GTWR Regression Model

The coupling coordination model and the relative development model identified the coupling coordination and leading-lagging relationship between individual ESs and SDL. However, due to the complexity of ESs and the trade-offs/synergies within them, it is necessary to simultaneously explore the relationship between all four ESs and SDL. Therefore, we further used the GTWR model, treating the four services as independent variables and SDL as the dependent variable, to comprehensively explore the spatiotemporal relationship between ESs and SDL.

The GTWR model introduces a temporal dimension to the traditional Geographically Weighted Regression (GWR) model [50], analyzing the influence mechanisms between variables from both temporal and spatial perspectives, thereby revealing the heterogeneity of variables over time and space. Thus, the GTWR model is employed to study the interaction between ESs and SDL. The specific formula [50] is as follows:

$$Y_{ik} = a_0(m_i, n_i, t_i) + \sum_{i=1}^{k} (m_i, n_i, t_i) X_{ik} + \varepsilon_i \quad (4)$$

where Y_{ik} is the SDL of county i at time k; $a_0(m_i, n_i, t_i)$ represents the geographical location; X_{ik} represents the ESs level of county i at time k; and ε_i is the random factor of county i at time k. This formula measures the correlation between ESs and SDL in county i at time k.

3. Results

3.1. Temporal and Spatial Changes in ESs and SDL in the Hehuang Valley

3.1.1. Changes in ESs

Figure 4 displays the temporal changes in ESs for each administrative unit in the Hehuang Valley from 2000 to 2020. Soil retention remained relatively stable, while NPP, water yield, and sand fixation exhibited more fluctuation, with both increases and decreases observed.

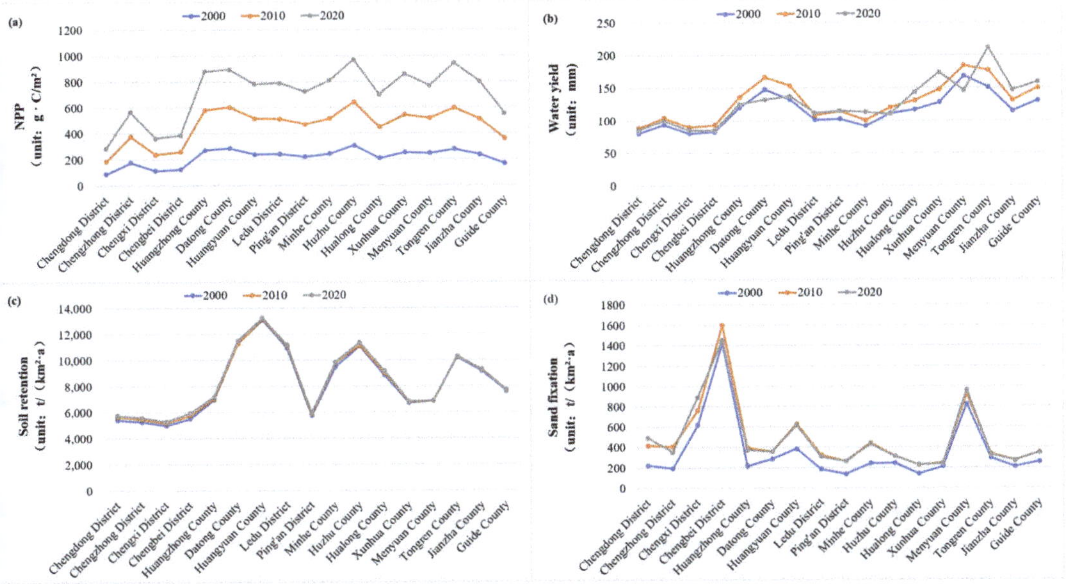

Figure 4. ES Changes in the County Scale of Hehuang Valley in 2000, 2010, and 2020, Including NPP (**a**), Water Yield (**b**), Soil Retention (**c**), and Sand Fixation (**d**).

Figure 4a reveals that between 2000 and 2010, NPP generally increased across all regions, with the most significant increase occurring in Huangzhong County, from 271.221 g·C/m^2 to 311.832 g·C/m^2, representing a net increase of 40.611 g·C/m^2. On the other hand, Chengdong District displayed the smallest increase, rising from 89.044 g·C/m^2 to 97.440 g·C/m^2, equivalent to a net increase of 8.396 g·C/m^2. However, comparing 2010 to 2020, NPP exhibited varying changes among the counties. Chengzhong District, Chengxi District, Chengbei District, Huangzhong County, Datong County, Huangyuan County, Ledu District, Ping'an District, Minhe County, Huzhu County, Hualong County, Xunhua County, and Menyuan County all experienced different degrees of decrease, with Datong County displaying the largest decrease of 30.656 g·C/m^2. In contrast, other regions showed slight increases, with Tongren County exhibiting the greatest growth from 320.044 g·C/m^2 to 343.439 g·C/m^2, representing a net increase of 23.394 g·C/m^2.

Figure 4b depicts the water yield changes, indicating that the overall change was not significant. However, when compared to 2010, several districts and counties experienced decreases in water yield, including Chengdong District, Chengzhong District, Chengxi District, Chengbei District, Huangzhong County, Datong County, Huangyuan County, Huzhu County, and Menyuan County. Of these, Menyuan County had the largest decrease, with a reduction of 38.272 mm. On the other hand, Tongren County exhibited the most substantial increase in water yield, rising from 150.749 mm to 211.542 mm, indicating a net increase of 60.793 mm.

Figure 4c illustrates the soil retention patterns, revealing no significant overall changes, as evidenced by the high overlap among the three curves. Huangyuan County consistently maintained the highest soil retention values in 2000, 2010, and 2020, with respective values of 13,131.574 t/(km^2·a), 13,185.213 t/(km^2·a), and 13,253.505 t/(km^2·a). On the other hand, Chengxi District consistently displayed the lowest soil retention values in the same years, with values of 5002.608 t/(km^2·a), 5185.912 t/(km^2·a), and 5283.454 t/(km^2·a), respectively.

Figure 4d portrays the changes in sand fixation, highlighting its instability over time. Chengbei District consistently exhibited the highest sand fixation values in 2000, 2010, and 2020, with respective values of 1425.845 t/(km^2·a), 1602.328 t/(km^2·a), and 1450.619 t/(km^2·a). Following closely was Menyuan County, with values of 823.514 t/(km^2·a), 924.560 t/(km^2·a), and 963.487 t/(km^2·a), respectively. Between 2000 and 2010, the overall sand fixation in the study area increased, primarily driven by Huangyuan County, which experienced a significant net increase of 243.346 t/(km^2·a). However, from 2010 to 2020, several districts and counties witnessed decreases in sand fixation, including Chengzhong District, Chengbei District, Huangzhong County, Datong County, Huangyuan County, Ledu District, Ping'an District, Minhe County, Huzhu County, Xunhua County, Tongren County, Jianzha County, and Guide County. Among them, Chengbei District reported the largest decrease of 151.709 t/(km^2·a). There was a notable increase in sand fixation in Chengdong District, indicating ongoing improvements in sand control in this area. Throughout the period from 2000 to 2020, Chengbei District and Menyuan County consistently exhibited the highest sand fixation values, demonstrating effective sand control in these regions.

In the grid scale, the average NPP in the Hehuang Valley increased from 246.40 g·C/m^2 to 273.68 g·C/m^2, representing a net increase of 27.28 g·C/m^2 from 2000 to 2020 (Figure 5). The average NPP values for 2000, 2010, and 2020 were 246.40 g·C/m^2, 277.23 g·C/m^2, and 273.68 g·C/m^2, respectively, indicating an initial increase followed by a decrease. The spatial distribution of NPP during this period remained largely consistent, with elevated values concentrated in the northeastern and southern regions of the Hehuang Valley. Conversely, lower NPP values were predominantly observed in the central and southwestern areas. Notably, the southern regions witnessed a significant increase in high NPP regions, while the northern regions experienced a rise in low NPP areas (Figure 5).

The average water yield in the Hehuang Valley increased from 131.09 mm to 142.27 mm, resulting in a net increase of 11.18 mm. The average water yield values for 2000, 2010, and 2020 were 131.09 mm, 146.84 mm, and 142.27 mm, respectively, indicating a similar trend of initial increase followed by a decrease. The spatial distribution of water yield exhibited a pattern centered around Xining City, with increasing values radiating outward. Higher water yield regions were mainly situated in the northern and southern parts of the Hehuang Valley, while lower water yield regions were observed in the central and northeastern regions. By 2020, the northern regions experienced a decrease in water yield compared to 2000, while the southern regions witnessed an increase. Moreover, while the overall spatial distribution of high and low water yield regions remained stable, the low-value areas expanded in 2020, primarily in the northern parts of the Hehuang Valley, compared to 2010. Additionally, the average soil retention in the Hehuang Valley increased from 8948.87 t/(km^2·a) to 9109.97 t/(km^2·a), representing a net increase of 161.10 t/(km^2·a). The average soil retention values for 2000, 2010, and 2020 were 8948.87 t/(km^2·a), 9030.82 t/(km^2·a), and 9109.97 t/(km^2·a), respectively, indicating an overall increase. The spatial distribution of soil retention exhibited a relatively scattered pattern, with high and low-value areas dispersed throughout. Higher soil retention areas were primarily concentrated in the northern-central regions, while lower soil retention areas were predominantly found in the northern regions. Overall, soil retention in the Hehuang Valley remained relatively low. The average sand fixation in the Hehuang Valley increased from 354.33 t/(km^2·a) to 457.41 t/(km^2·a), representing a net increase of 103.08 t/(km^2·a). The average sand fixation values for 2000, 2010, and 2020 were 354.33 t/(km^2·a), 456.96 t/(km^2·a), and 457.41 t/(km^2·a), respectively, indicating a gradual overall increase. The spatial distribution of sand fixation remained largely consistent over time, with high-value areas primarily located in the central and northern parts of the Hehuang Valley, while low-value areas were widespread throughout the region. The overall spatial distribution of high-value and low-value areas remained stable.

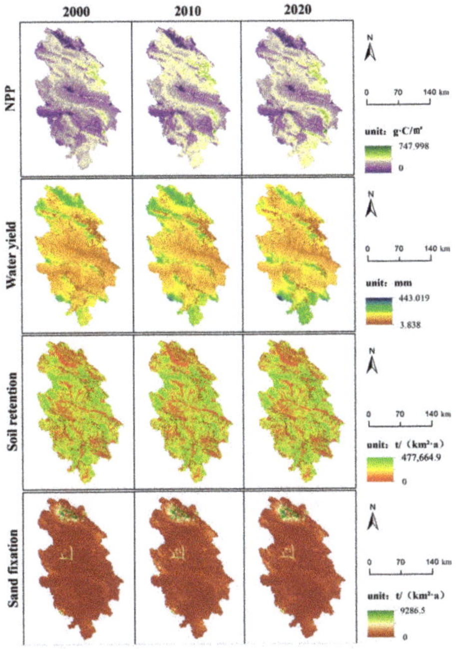

Figure 5. Spatiotemporal Distribution of ESs in the Grid Scale of Hehuang Valley from 2000 to 2020.

3.1.2. Changes in SDL

When examining the spatiotemporal changes in SDL (Figure 6), it is evident that the overall sustainability foundation and SDL increased across the 17 regions from 2000 to 2020. With the exception of Hualong County and Xunhua County, all other regions also experienced increases in sustainability acceleration scores and sustainability environment scores. High-value areas of SDL predominantly centered around the central part of the Hehuang Valley, radiating outward from Xining City. In contrast, low-value areas of SDL were primarily located in the southern part. Chengzhong District witnessed the most significant increase in sustainability foundation scores from 2010 to 2020, surpassing the increase recorded between 2000 and 2010. The scores rose from 0.062 in 2010 to 0.355 in 2020, reflecting a net increase of 0.293. Chengxi District followed closely, increasing from 0.075 in 2010 to 0.307 in 2020, resulting in a net increase of 0.232. On the other hand, Jianzhai County and Guide County observed marginal increases, both rising by 0.009. Chengxi District exhibited the most substantial increase in sustainability acceleration scores from 2010 to 2020, surpassing the increase observed between 2000 and 2010. The scores rose from 0.116 in 2010 to 0.218 in 2020, representing a net increase of 0.102. Chengbei District followed with a net increase of 0.097. While Hualong County experienced an initial increase in sustainability acceleration scores from 2000 to 2010, it subsequently declined, reducing by 0.018 in 2020 compared to 2010. Between 2000 and 2020, Chengxi District consistently achieved the highest sustainability environment scores, reaching 0.189 in 2020. From 2010 to 2020, all 17 regions witnessed increases in sustainability environment scores, although Xunhua County experienced a decrease of 0.015 between 2000 and 2010. Figure 6 illustrates the positive impact of the mentioned subsystems on the overall SDL.

Notably, from 2010 to 2020, Chengxi District consistently achieved the highest SDL scores, witnessing a remarkable increase from 0.193 in 2000 to 0.713 in 2020. Chengzhong District followed suit, while Jianzhai County recorded the lowest score in 2020, measuring at 0.107. Notably, the overall SDL of all 17 counties in the Hehuang Valley demonstrated consistent growth from 2000 to 2020, reaching their peak values in 2020.

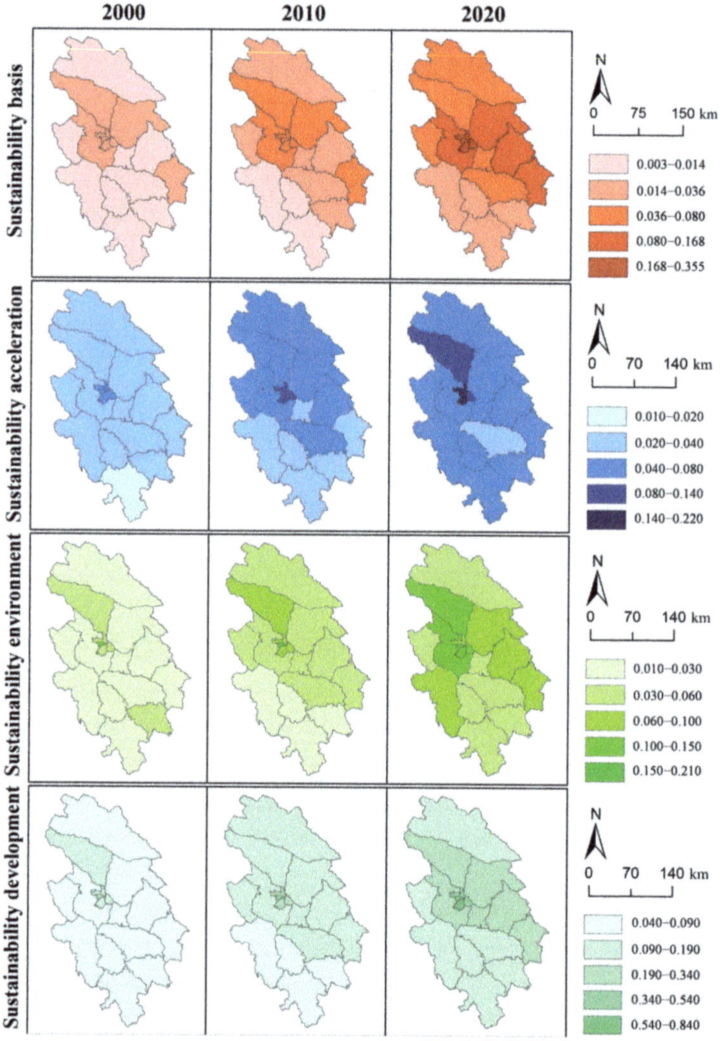

Figure 6. Spatiotemporal Distribution of SDL in the Counties of the Hehuang Valley from 2000 to 2020.

3.2. Coupling Coordination Analysis between ESs and SDL

3.2.1. Coupling Degree between ESs and SDL

Based on the spatiotemporal distribution of the coupling degree between ESs and SDL (Figure 7), it is evident that the coupling degree of all four ESs with SDL increased from 2000 to 2020.

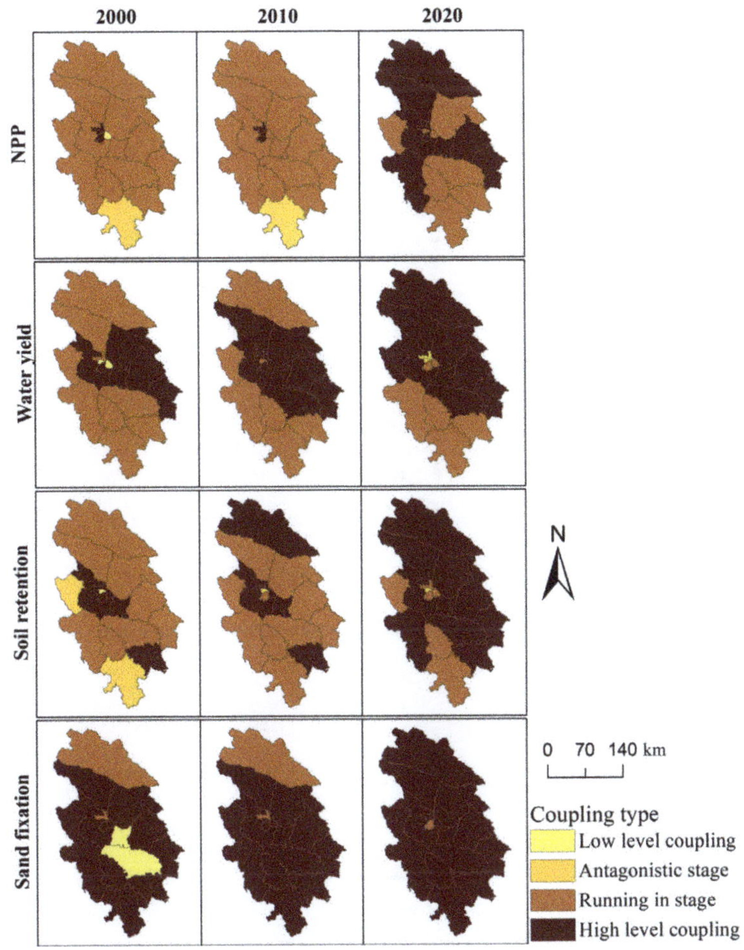

Figure 7. Spatiotemporal Distribution of the Coupling Degree between ESs and SDL in the Counties of the Hehuang Valley from 2000 to 2020.

In 2000, there was one county with a low-level coupling between NPP and SDL, and one county with a low-level coupling between soil retention and SDL. Additionally, two counties demonstrated a low-level coupling between water yield and SDL, while two other counties exhibited a low-level coupling between sand fixation and SDL. Between 2010 and 2020, all four types of ESs surpassed the antagonistic phase and progressed towards higher levels of coupling with SDL. By 2020, Ledu District and Huzhu County achieved a coupling degree value of 100% for water yield and SDL, Huangzhong County achieved the same for soil retention and SDL, and Guide County attained 100% for sand fixation and SDL. Notably, the coupling degree between water yield and SDL in Chengzhong District was 1 in 2000. However, this coupling degree continuously decreased from 2000 to 2020, suggesting a gradual divergence between water yield and SDL in this area.

3.2.2. CCD between ESs and SDL

Between 2000 and 2020, the overall CCD among NPP, water yield, soil retention, sand fixation, and SDL in the Hehuang Valley exhibited an upward trend (Figure 8). SDL demonstrated the highest coordination degree with NPP, followed by soil retention, water yield, and sand fixation.

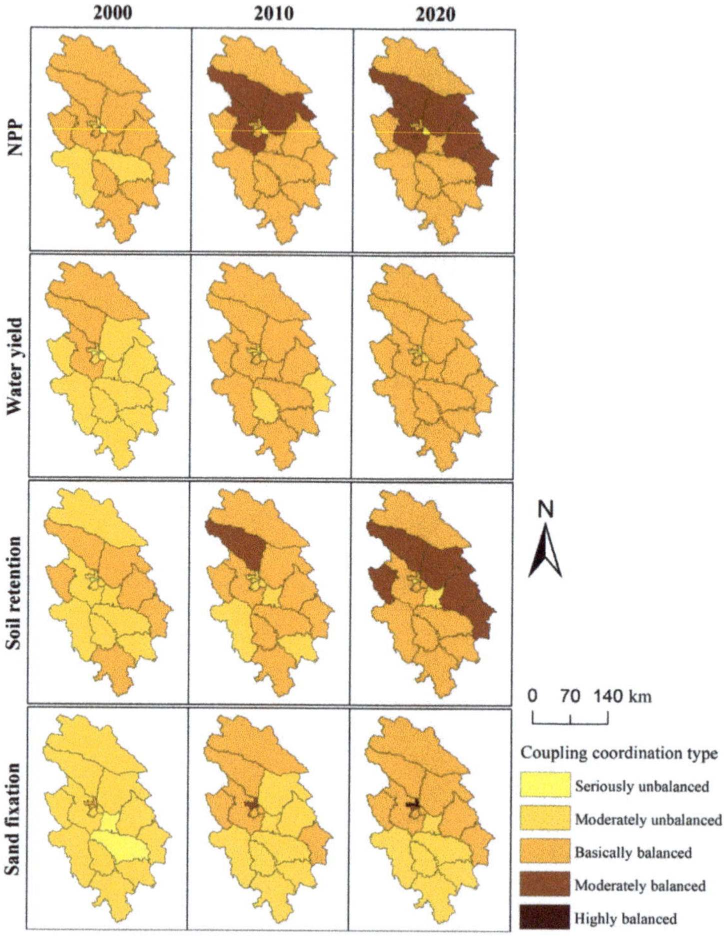

Figure 8. Spatiotemporal Distribution of the CCD between ESs and SDL in the Counties of the Hehuang Valley from 2000 to 2020.

The average CCD between NPP and SDL in the counties of the Hehuang Valley increased from 0.406 to 0.586, indicating a rise of 44.33%. The average CCDs for 2000, 2010, and 2020 were 0.406, 0.509, and 0.586, respectively, reflecting an overall increase. By 2020, the overall CCD between NPP and SDL reached basic to moderate coordination. Spatially, high-value areas were concentrated in the central and northern regions, while low-value areas were located in the northernmost and southern regions (Figure 8). Most counties achieved a state of basic coordination. From 2000 to 2020, the CCD increased in all 17 counties, with Chengdong District displaying the largest increase. The proportion of counties falling into the moderate coordination category rose from 0% to 35.29%.

From 2000 to 2020, there was a significant increase in the average CCD between water yield and SDL in the counties of the Hehuang Valley. Specifically, the degree increased from 0.321 to 0.496, representing a rise of 54.52%. The average CCDs for 2000, 2010, and 2020 were 0.321, 0.444, and 0.496, respectively, indicating an overall increase over the two-decade period. By 2020, the overall CCD between water yield and SDL reached the basic coordination stage. Spatially, only Chengbei District, Chengdong District, and Chengxi District were categorized as having moderate imbalance, while the remaining cities were in the basic coordination stage. Notably, from 2000 to 2020, the CCD increased in all 17 counties, with Chengxi District demonstrating the largest increase. Additionally, the proportion of counties in the basic coordination category saw a significant increase, rising from 17.65% in 2000 to 82.35% in 2020.

There was an increase in the average CCD between soil retention and SDL in the counties of the Hehuang Valley from 2000 to 2020. The degree rose from 0.358 to 0.516, resulting in a 44.13% increase. The average CCDs for 2000, 2010, and 2020 were 0.358, 0.439, and 0.516, respectively, showing an overall upward trend. By 2020, the overall CCD between soil retention and SDL reached the basic coordination stage. Spatially, Chengxi District and Ping'an District were categorized as having moderate imbalance, whereas Huangyuan County, Datong County, Huzhu County, Ledu District, and Minhe County were in the moderate coordination stage. The remaining cities were in the basic coordination stage. Similar to the water yield analysis, the CCD for soil retention increased in all 17 counties, with Chengxi District experiencing the largest increase. Furthermore, the proportion of counties in the basic coordination or higher category rose significantly, increasing from 35.29% in 2000 to 88.24% in 2020.

From 2000 to 2020, there was a remarkable increase in the average CCD between sand fixation and SDL in the counties of the Hehuang Valley. Specifically, the average CCD rose from 0.276 in 2000 to 0.467 in 2020, representing a 69.20% increase. The overall trend showed a consistent increase, with the average CCDs in 2010 and 2020 being 0.402 and 0.467, respectively. By 2020, half of the regions achieved a basic coordination level between sand fixation and SDL. Regarding the spatial distribution, Chengbei District was categorized as being in the high coordination stage, while Chengxi District was in the moderate coordination stage. Basic coordination was predominantly concentrated in the central and northern regions, whereas moderate imbalance was concentrated in the southern regions. It is noteworthy that the CCD increased in all 17 counties from 2000 to 2020, with Ping'an District showing the largest increase. Furthermore, the proportion of counties in the basic coordination or higher category rose significantly from 0% to 64.71%.

3.2.3. Relative Development Degree of ESs and SDL

From 2000 to 2020, the relative development types between NPP and SDL in the counties of the Hehuang Valley were predominantly SDL-lagging types (Figure 9). The SDL-lagging counties were widespread, with all counties except Chengzhong District, Chengxi District, Chengdong District, and Chengbei District belonging to this type. The number of NPP-lagging counties increased from 2 to 4.

The relative development types between water yield and SDL were also mainly SDL-lagging types (Figure 9). In 2000, all counties except those in Xining City belonged to the SDL-lagging type. By 2010, Minhe County transitioned from the SDL-lagging type to the synchronized development type, while Chengzhong District shifted from the synchronized development type to the water yield-lagging type. Compared to 2010, in 2020, Huangzhong County, Huzhu County, and Ledu District transitioned from the SDL-lagging type to the synchronized development type.

The relative development types between soil retention and SDL were predominantly SDL-lagging types (Figure 9). In 2000, all counties except those in Xining City belonged to the SDL-lagging type. By 2010, Ping'an District transitioned from the SDL-lagging type to the synchronized development type. Compared to 2000, in 2020, Huangzhong County transitioned from the SDL-lagging type to the synchronized development type, while Ping'an District transitioned from the SDL-lagging type to the soil retention-lagging type.

The relative development types between sand fixation and SDL showed significant changes. From 2000 to 2010, the predominant type was SDL-lagging; from 2010 to 2020, it shifted to sand fixation-lagging (Figure 9). From 2000 to 2020, the proportion of counties in the SDL-lagging type decreased from 35.29% to 17.65%, while the sand fixation-lagging type increased from 41.18% to 70.59%. The number of synchronized development counties decreased from 5 to 2.

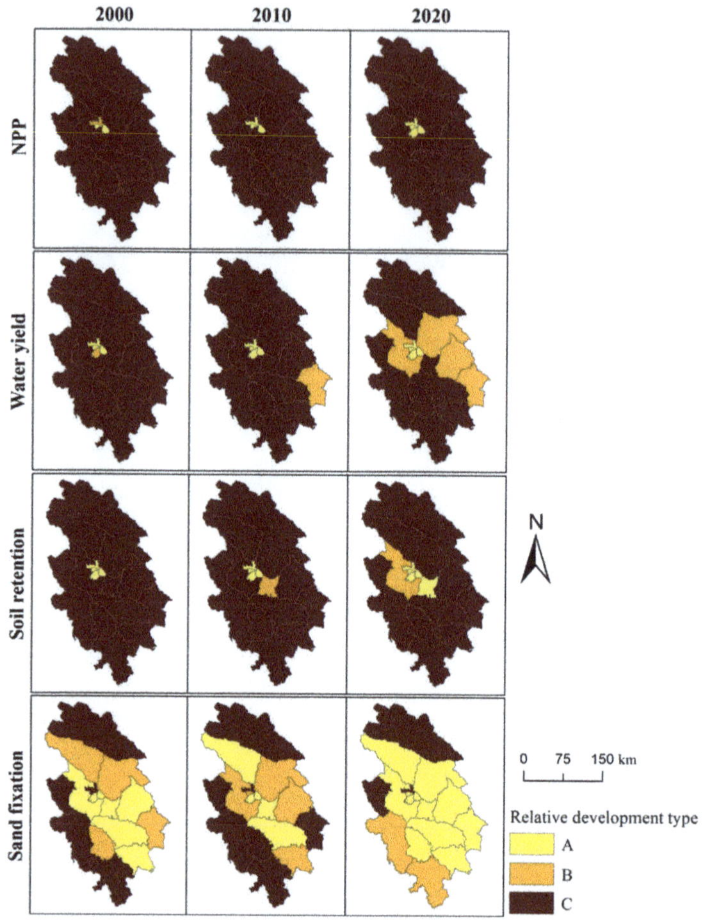

Figure 9. Spatiotemporal Distribution of the Relative Development Degree between ESs and SDL in the Counties of the Hehuang Valley from 2000 to 2020. Note: (A) Represents ESs-lagging type; (B) Represents synchronized development of ESs and SDL; (C) Represents SDL-lagging type.

3.3. Spatiotemporal Relationships between ESs and SDL

Multicollinearity among independent variables can affect regression analysis results [51]. Therefore, SPSS 26.0 software was used for multicollinearity diagnostics among the variables. The tests (Table 5) showed that the variance inflation factors (VIF) for all variables were less than 10 and the tolerances were greater than 0.1, indicating no multicollinearity.

Table 5. Multicollinearity Test.

Explanatory Variable	NPP	Water Yield	Soil Retention	Sand Fixation
VIF	3.012	1.830	1.766	1.251
Tolerance	0.332	0.546	0.566	0.799

Using ArcGIS 10.8.1 software, standardized values of the four ESs were used as explanatory variables, and the total SDL score was used as the dependent variable. GTWR models, GWR models, temporally weighted regression (TWR) models, and ordinary least squares (OLS) models were analyzed. The results (Table 6) indicate that the GTWR model generally had a higher R^2, lower AICc, and lower RSS [52].

Table 6. Comparison of GTWR, GWR, TWR, and OLS Models.

	GTWR	GWR	TWR	OLS
R^2	0.851	0.378	0.761	0.278
AICc	−57.095	−48.807	−89.523	−61.354
RSS	0.155	0.648	0.249	0.737
Adjusted R^2	0.838	0.324	0.740	——

Hence, the GTWR model was employed in this study to explore the correlation between the four ESs and SDL in 2000, 2010, and 2020, visualizing the spatial heterogeneity between ESs and SDL. This model is highly explanatory and effectively elucidates the relationship between ESs and SDL. Lower AICc values indicate better model fit [53].

Based on the parameter results of the GTWR model, correlation coefficients between ESs and SDL in the counties of the Hehuang Valley for 2000, 2010, and 2020 were obtained. These coefficients were categorized into six types: high negative, medium negative, low negative, low positive, medium positive, and high positive. Overall, the correlation between the four ESs and SDL has shown varying degrees of shift towards negative correlation over time (Figure 10).

As illustrated in Figure 10, from 2000 to 2020, there were significant changes in the correlation between NPP and SDL in the central and northern regions of our site. The negative correlation areas expanded, and the intensity of the negative correlation strengthened. By 2020, Menyuan County exhibited a high negative correlation. The central-southern regions, such as Guide County, Jianzha County, Tongren County, and Xunhua County showed medium positive correlations.

The correlation between water yield and SDL exhibited a continuous negative trend, with the degree of negative correlation intensifying. In 2000, except for Menyuan County, 16 regions showed medium negative correlations. By 2020, 12 regions exhibited high negative correlations, accounting for 70.59% of the Hehuang Valley.

The correlation between soil retention and SDL improved in Minhe County, shifting from a medium negative to a low negative correlation. However, the negative correlation intensified in Datong County and Huzhu County. From 2010 to 2020, the negative correlation between Minhe County and Menyuan County intensified, although Menyuan County maintained a low positive correlation. The remaining regions exhibited medium negative correlations.

The relationship between sand fixation and SDL in the 17 study areas in the Hehuang Valley exhibited varying degrees of positive correlation. Specifically, Tongren County, Minhe County, and Xunhua County showed high positive correlations in 2010. Significant efforts were made in the Hehuang Valley to enhance the environment and ensure sustainable sand control, thereby contributing to SDL. Between 2010 and 2020, the positive correlation in the central-southern regions increased, while in the central-northern regions, the positive correlation weakened. In fact, some areas even shifted from medium positive to medium negative correlation, such as Huangyuan County. Moreover, the number of regions with medium negative correlation rose from zero in 2000 to three in 2020.

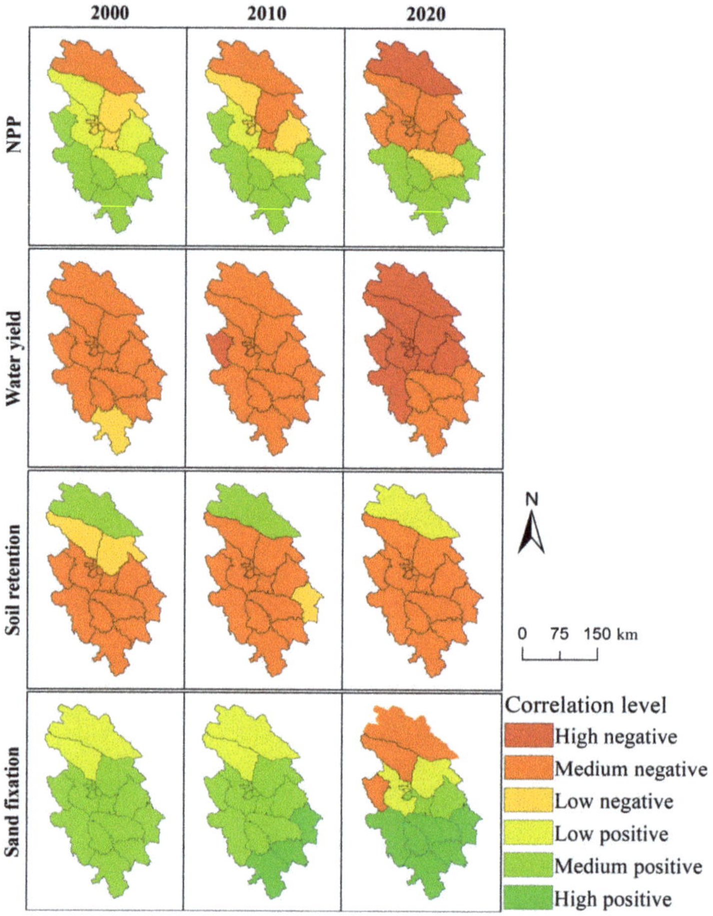

Figure 10. Spatiotemporal Distribution of Correlation Levels between ESs and SDL in the Counties of the Hehuang Valley from 2000 to 2020.

4. Discussion

4.1. Factors Influencing the Coupling Relationship between ESs and SDL

In this study, the coupling coordination model was employed to analyze the spatiotemporal changes in the coupling relationship between ESs and SDL in the Hehuang Valley from 2000 to 2020. The CCD between NPP, water yield, soil retention, and SDL generally increased. This can be attributed mainly to the continuous improvements in NPP, water yield, soil retention, sand fixation, and SDL throughout the study period. The heightened CCD between NPP and SDL can be primarily attributed to the implementation of various ecological protection policies and measures in the Hehuang Valley region. Since 2000, the Qinghai Forestry Bureau has issued pertinent policies and laws, including the Grain for Green Project, afforestation plans, and the Natural Forest Protection Project. As of 2020, Qinghai Province successfully completed 2886.7 km^2 of land greening, while forest pest control covered an area of 2139.7 km^2. These efforts have ensured the safety and healthy growth of forest and grassland resources. Furthermore, the forest and grassland resource protection system has been consistently enhanced, with 36,780 km^2 of natural forests and 39,771.1 km^2 of national public welfare forests being effectively managed.

From 2000 to 2014, Qinghai Province implemented various ecological restoration projects, including the conversion of farmland into forests and grasslands. As a result, the province increased its forest area by 6666.7 km^2, raising the forest coverage rate from 3.1% in 1999 to 6.1% in 2014. In addition, cultivation of steep slopes over 15 degrees was stopped in an area of 755.3 km^2, and soil erosion was effectively managed in an area of 7333.3 km^2, which successfully controlled soil erosion and improved soil retention. The increased precipitation in the upper reaches of the Yellow River and the source area of the Yangtze River led to an increase in river flow. Between 2003 and 2010, the average flow in the upper reaches of the Yellow River increased by 117.2 m^3/s, while the average flow in the Yangtze River source area increased by 149.4 m^3/s compared to the period of 1991–2002. The increase in precipitation, combined with the implementation of water pollution prevention regulations, contributed to a continuous rise in water yield. Moreover, the implementation of projects such as converting farmland to forests and grasslands significantly improved the structure of land use. With the introduction of mechanized operations and adjustments in crop structure, the grain yield per unit area steadily increased. Consequently, from 2005 to 2014, the total grain output in Qinghai Province increased from 932,600 tons to 1,023,700 tons, achieving steady growth and significantly enhancing income, thereby promoting local SDL. As a result, the CCD between water yield, soil retention, and SDL also steadily improved. The CCD between sand fixation and SDL underwent significant changes, transitioning from widespread imbalance in 2000 to basic coordination in some counties by 2020. This indicates an increase in the CCD. In 2020, Qinghai Province completed 1230 km^2 of sand control tasks, effectively protecting 5824 km^2 of desertified land and continuously improving the condition of desertified land in the Hehuang Valley.

4.2. Changes in the Spatiotemporal Relationship between ESs and SDL

The correlation between ESs and SDL is determined by the environmental quality and development direction of different regions [54,55]. A positive correlation implies that ecosystems contribute to SDL, while the capital investment generated by development can also enhance ecosystems. On the other hand, a negative correlation indicates that ecosystem degradation hampers SDL, which is often caused by human activities such as urbanization, industrialization, and farmland expansion [30].

From 2000 to 2020, there has been a gradual deterioration in the overall correlation between NPP, water yield, soil retention, and SDL, with a notable increase in negative correlation. Despite the implementation of policies aimed at protecting the ecological environment and promoting SDL during this period, urbanization has resulted in an expansion of construction land and transportation networks, disrupting the natural material cycles of the environment and causing damage to ecosystem functions [56]. Additionally, the predominant land use types in the region are grasslands and forests, which have slow growth rates, thus limiting the potential for improvement and hindering SDL in the area.

During the same period, the correlation between sand fixation and SDL exhibited contrasting trends in different regions. The northern region showed a negative correlation, while the southern region demonstrated a positive correlation. The northern region, which borders the Loess Plateau, is characterized by fragile ecosystems and has long been plagued by severe soil erosion [57]. Consequently, the northern region experiences severe wind erosion, which is detrimental to SDL. In contrast, the southeastern part of the valley benefits from moisture from the east [34], resulting in higher air humidity. The implementation of projects such as the conversion of farmland to forests and grasslands has increased vegetation coverage, yielding positive outcomes in terms of sand fixation in the southern region and fostering SDL.

4.3. Policies and Recommendations for Ecological Conservation

By comparing the relative development degrees of various ESs and SDL in 2020, the leading-lagging relationship between ESs and SDL for each county in the Hehuang Valley can be determined, as shown in Table 7. Based on the coupling coordination relationship and relative development types of ESs and SDL in each county, the counties are classified into the following five categories (Table 7): Comprehensive SDL Improvement Areas, NPP-Water Yield-Soil Retention SDL Improvement Areas, Water Yield Optimization Areas, Sand Fixation Optimization Areas, and Comprehensive Ecological Function Optimization Areas. Based on these classifications, specific policy recommendations are proposed.

Table 7. Relative Development Degree between ESs and SDL and Classification in the Hehuang Valley.

County	NPP-SDGs	Water Yield-SDGs	Soil Retention-SDGs	Sand Fixation-SDGs	Classification
Chengdong District	0.079	0.078	0.201	0.543	Comprehensive Ecological Function Lagging Type
Chengzhong District	0.610	0.224	0.108	0.221	Comprehensive Ecological Function Lagging Type
Chengxi District	0.183	0.046	0.048	0.724	Comprehensive Ecological Function Lagging Type
Chengbei District	0.297	0.067	0.238	1.884	Water Yield Lagging Type
Huangzhong County	2.947	1.229	0.956	0.580	Sand Fixation Lagging Type
Datong County	2.810	1.414	2.825	0.528	Sand Fixation Lagging Type
Huangyuan County	4.414	2.789	6.425	2.086	Comprehensive SDL Lagging Type
Ledu District	3.220	1.045	3.303	0.500	Sand Fixation Lagging Type
Ping'an District	3.229	1.323	0.584	0.416	Sand Fixation Lagging Type
Minhe County	2.802	0.881	2.088	0.700	Sand Fixation Lagging Type
Huzhu County	4.171	1.011	3.471	0.524	Sand Fixation Lagging Type
Hualong County	4.633	3.502	3.688	0.444	Sand Fixation Lagging Type
Xunhua County	6.966	5.586	1.753	0.547	Sand Fixation Lagging Type
Menyuan County	3.698	2.977	1.375	3.377	Comprehensive SDL Lagging Type
Tongren County	8.574	8.574	5.486	1.083	SDL Lagging in NPP-Water Yield-Soil Retention
Jianzha County	7.285	4.698	4.834	0.813	Sand Fixation Lagging Type
Guide County	3.169	4.532	2.481	1.058	SDL Lagging in NPP-Water Yield-Soil Retention

In the comprehensive SDL Improvement Areas (e.g., Huangyuan County, Menyuan County) and NPP-Water Yield-Soil Retention SDL Improvement Areas (e.g., Tongren County, Guide County), due to unique geography and environment, as well as limitations like weak infrastructure and significant poverty, the SDL in these counties are lower compared to urban areas. Emphasize county-level development to drive SDL through economic growth. Strengthen infrastructure construction, including transportation, communication, water, and electricity, to improve production and living conditions in surrounding areas. Leverage the economic influence of urban centers to transfer resources to nearby counties, achieving synchronized development and mutual benefits. Increase policy support, conduct poverty alleviation efforts, and provide basic guarantees for food, housing, and education for impoverished populations. Promote employment and improve educational conditions, as higher education levels are critical for sustainable development. Encourage industrial transformation and upgrading by introducing new technologies and cultivating new industries, optimizing and upgrading the industrial structure in surrounding areas, and aligning it with urban industries to enhance sustainable development capacity.

Water Yield Optimization Areas (e.g., Chengbei District) and Comprehensive Ecological Function Optimization Areas (e.g., Chengdong District, Chengzhong District, Chengxi District) both belong to the Urban areas, which are generally classified as ESs-lagging due to their relatively abundant resources and infrastructure, which accelerate sustainable development but also cause some environmental degradation. Protect and restore the local ecological environment by implementing strict ecological protection policies. Restore and reconstruct damaged ecosystems to enhance stability. Formulate stringent ecological protection policies to ensure that sustainable development aligns with environmental carrying capacity. Optimize industrial structure and layout to avoid overexploitation and disorderly expansion, ensuring coordination between ecosystems and sustainable development. Strengthen pollution regulation and management, ensure standard emissions, reduce environmental damage, and promote the development of low-energy, low-emission green industries to improve regional environmental quality.

Sand Fixation Optimization Areas (e.g., Huangzhong County, Datong County, Ledu District) should implement large-scale ecological projects such as afforestation to increase ground vegetation cover and enhance soil and water conservation and water source maintenance capabilities. Strengthen wetland protection to reduce the impact of sandstorms on the Hehuang Valley. Innovate in technology by developing sand fixation techniques suitable for the Hehuang Valley, improving sand control efficiency. Enforce policies for mountain closure for forest and grass cultivation, build soil and water conservation forests and artificial grasslands to reduce vegetation destruction on mountains. Strengthen soil erosion control to enhance soil retention capacity. Encourage enterprises to adopt green production methods to reduce pollution and resource waste, promoting the harmonious development of the economy and the ecological environment. Increase ecological education and awareness, improving public understanding of the importance of sand fixation and encouraging active public participation in ecological protection activities.

4.4. Limitations and Future Prospects

This study utilized three indicators—CCD, coupling degree, and relative development degree—to analyze the coupling relationship and leading-lagging dynamics between four types of ESs and SDL in the Hehuang Valley in 2020. The spatiotemporal GTWR model was also employed to explore the spatiotemporal heterogeneity between the two. However, there are some limitations to this study. The evaluation indicator system for SDL primarily focuses on social and economic indicators, with fewer ecological indicators. Additionally, the study used the entropy weight method to calculate the weights of the sustainable development indicators. Future research should consider expanding the dataset and incorporating expert opinions to determine weights in a way that combines both objective and subjective perspectives, enabling a more comprehensive and in-depth study. Despite these limitations, the research methods and results of this study provide valuable insights into the coupling relationship between ESs and SDL. The findings demonstrate that the ESs, SDL, and the CCD between the two in the Hehuang Valley are continuously improving, indicating the effective implementation of governance policies in the region. These findings serve as a reference for future ecological conservation and SDL in the Hehuang Valley. Moreover, the study's focus on the 17 counties of the Hehuang Valley as the study area and small administrative units as the research units contributes to the accuracy of the research. This approach allows for the formulation of solutions tailored to local characteristics, increasing their feasibility. Additionally, the study simultaneously analyzed the coupling relationship from both temporal and spatial perspectives, providing a comprehensive understanding of the intrinsic connections and interaction mechanisms between ESs and SDL. This provides strong support for the coordinated development of ESs and SDL.

5. Conclusions

The achievement of SDGs is not only about improving individual quality of life but is also closely linked to the health and development of the earth's entire ecosystem. By assessing the SDGs at the small administrative unit level and combining them with local geographical conditions and development status, it is possible to precisely identify and address existing issues, thereby facilitating substantial progress in sustainable development. In this context, this study explores the spatiotemporal heterogeneity and interactions between four ESs and SDL in the counties of the Hehuang Valley for the years 2000, 2010, and 2020, based on a constructed sustainable development assessment framework and utilizing the coupling model, relative development model, and GTWR model. The findings of the study are as follows:

(1) The SDL in the counties of the Hehuang Valley increased from 2000 to 2020, with Xining City exhibiting slightly higher levels compared to other counties. NPP and water yield increased from 2000 to 2010, but both decreased from 2010 to 2020. However, soil retention and sand fixation showed overall increases from 2000 to 2020. There was significant heterogeneity in the ability of the counties to provide ESs.

(2) The overall CCD between NPP, water yield, soil retention, sand fixation, and SDL increased from 2000 to 2020. The coordination between SDL and NPP was the highest, followed by soil retention and water yield, with sand fixation having the lowest coordination but the largest increase at 69.20%.

(3) From 2000 to 2020, most counties in the Hehuang Valley exhibited a relative development type where NPP, water yield, and soil retention lagged behind SDL. The relative development type for sand fixation and SDL varied significantly. From 2000 to 2010, sand fixation and SDL were mainly of the SDL-lagging type, whereas from 2010 to 2020, they were primarily of the sand fixation-lagging type.

(4) The negative correlation between the four ESs and SDL intensified from 2000 to 2020. In some counties, the positive contribution of NPP and sand fixation to SDL gradually turned negative, and the negative impact of water yield and soil retention on SDL continuously increased. The Impact of human activities on ecosystem function hindered local sustainable development.

Author Contributions: Conceptualization, Y.Y.; Formal analysis, Y.M.; Funding acquisition, H.W.; Methodology, K.W.; Visualization, M.L.; Writing—original draft, H.W. and K.W.; Writing—review & editing, Q.M. and Y.Y. All authors have read and agreed to the published version of the manuscript.

Funding: This research was funded by the Second Tibetan Plateau Scientific Expedition and Research Program (2019QZKK0608) and the Key Project of Henan Provincial Science and Technology R&D Plan Joint Fund (225200810045).

Data Availability Statement: The data presented in this study are available on request from the corresponding author.

Conflicts of Interest: The authors declare no conflicts of interest.

References

1. Yackinous, W.S. *Understanding Complex Ecosystem Dynamics: A Systems and Engineering Perspective*; Academic Press: Cambrid MA, USA, 2015.
2. Liu, J.; Mooney, H.; Hull, V.; Davis, S.J.; Gaskell, J.; Hertel, T.; Lubchenco, J.; Seto, K.C.; Gleick, P.; Kremen, C.; et al. Syste integration for global sustainability. *Science* **2015**, *347*, 1258832. [CrossRef] [PubMed]
3. Comberti, C.; Thornton, T.; de Echeverria, V.W.; Patterson, T. Ecosystem Services or Services to Ecosystems? Valuing Cultivat and Reciprocal Relationships between Humans and Ecosystems. *Glob. Environ. Chang.* **2015**, *34*, 247–262. [CrossRef]
4. Mononen, L.; Auvinen, A.P.; Ahokumpu, A.L.; Rönkä, M.; Aarras, N.; Tolvanen, H.; Kamppinen, K.; Viirret, E.; Kumpula, Vihervaara, P. National ecosystem service indicators: Measures of social–ecological sustainability. *Ecol. Indic.* **2014**, *61*, 27– [CrossRef]
5. Dai, E.R.; Zhao, Z.X.; Jia, L.Z.; Jiang, X.T. Contribution of ecosystem services improvement on achieving Sustainable Developm Goals under ecological engineering projects on the Qinghai–Tibet Plateau. *Ecol. Eng.* **2024**, *199*, 12. [CrossRef]

Dasgupta, R.; Hashimoto, S.; Gundimeda, H. Biodiversity/ecosystem services scenario exercises from the Asia–Pacific: Typology, archetypes and implications for sustainable development goals (SDGs). *Sustain. Sci.* **2019**, *14*, 241–257. [CrossRef]
Liu, Y.; Lü, Y.; Fu, B.; Zhang, X. Landscape pattern and ecosystem services are critical for protected areas' contributions to sustainable development goals at regional scale. *Sci. Total Environ.* **2023**, *881*, 163535. [CrossRef]
Liu, X.; Yuan, M. Assessing progress towards achieving the transport dimension of the SDGs in China. *Sci. Total Environ.* **2023**, *858*, 159752. [CrossRef]
Wu, X.; Liu, J.; Fu, B.; Wang, S.; Wei, Y.; Li, Y. Bundling regions for promoting Sustainable Development Goals. *Environ. Res. Lett.* **2022**, *17*, 044021. [CrossRef]
Liu, X.; Zhao, W.; Liu, Y.; Hua, T.; Hu, X.; Cherubini, F. Contributions of ecological programs to sustainable development goals in Linzhi, over the Tibetan Plateau: A mental map perspective. *Ecol. Eng.* **2020**, *176*, 106532. [CrossRef]
Griggs, D.; Nilsson, M.; Stevance, A.; McCollum, D. *A Guide to SDG Interactions: From Science to Implementation*; International Council for Science: Paris, France, 2017.
Xu, Z.; Peng, J.; Liu, Y.; Qiu, S.; Zhang, H.; Dong, J. Exploring the combined impact of ecosystem services and urbanization on SDGs realization. *Appl. Geogr.* **2023**, *153*, 102907. [CrossRef]
Smith, P.; Adams, J.; Beerling, D.J.; Beringer, T.; Calvin, K.V.; Fuss, S.; Griscom, B.; Hagemann, N.; Kammann, C.; Kraxner, F.; et al. Land-Management Options for Greenhouse Gas Removal and Their Impacts on Ecosystem Services and the Sustainable Development Goals. *Annu. Rev. Environ. Resour.* **2019**, *44*, 255–286. [CrossRef]
Mandle, L.; Shields-Estrada, A.; Chaplin-Kramer, R.; Mitchell, M.G.E.; Bremer, L.L.; Gourevitch, J.D.; Hawthorne, P.; Johnson, J.A.; Robinson, B.E.; Smith, J.R.; et al. Increasing decision relevance of ecosystem service science. *Nat. Sustain.* **2021**, *4*, 161–169. [CrossRef]
Costanza, R.; de Groot, R.; Braat, L.; Kubiszewski, I.; Fioramonti, L.; Sutton, P.; Farber, S.; Grasso, M. Twenty Years of Ecosystem Services: How Far Have We Come and How Far Do We Still Need to Go? *Ecosyst. Serv.* **2017**, *28*, 1–16. [CrossRef]
Zuo, L.; Liu, G.; Zhao, J.; Li, J.; Zheng, S.; Wang, J.; Su, X. Effects of ecosystem services on sustainable development goals in western Sichuan. *Acta Ecol. Sin.* **2024**, *44*, 4203–4216. (In Chinese)
Zhang, L.; Zhao, J.; Chen, G.; Lin, Y.; Liu, F.; Peng, S. Impact of Land-use Change on the Value of Ecosystem Services Based on SDG 15.3.1. *J. Soil Water Conserv.* **2024**, *38*, 268–277. (In Chinese)
Wood, S.L.R.; Jones, S.K.; Johnson, J.A.; Brauman, K.A.; Chaplin-Kramer, R.; Fremier, A.; Girvetz, E.; Gordon, L.J.; Kappel, C.V.; Mandle, L.; et al. Distilling the role of ecosystem services in the sustainable development goals. *Ecosyst. Serv.* **2018**, *29*, 70–82. [CrossRef]
Rozas-Vásquez, D.; Spyra, M.; Jorquera, F.; Molina, S.; Caló, N.C. Ecosystem Services Supply from Peri-Urban Landscapes and Their Contribution to the Sustainable Development Goals: A Global Perspective. *Land* **2022**, *11*, 2006. [CrossRef]
Hussain, K.; Sun, H.; Ramzan, M.; Mahmood, S.; Zubair Saeed, M. Interpretive Structural Modeling of Barriers to Sustainable Tourism Development: A Developing Economy Perspective. *Sustainability* **2024**, *16*, 5442. [CrossRef]
Yin, C.; Zhao, W.; Ye, J.; Muroki, M.; Pereira, P. Ecosystem carbon sequestration service supports the Sustainable Development Goals progress. *J. Environ. Manag.* **2023**, *330*, 117155. [CrossRef]
Yang, S.; Zhao, W.; Liu, Y.; Cherubini, F.; Fu, B.; Pereira, P. Prioritizing sustainable development goals and linking them to ecosystem services: A global expert's knowledge evaluation. *Geogr. Sustain.* **2020**, *1*, 321–330. [CrossRef]
Fang, X.; Shi, X.; Phillips, T.K.; Du, P.; Gao, W. The Coupling Coordinated Development of Urban Environment Towards Sustainable Urbanization: An Empirical Study of Shandong Peninsula, China. *Ecol. Indic.* **2021**, *129*, 107864. [CrossRef]
Wei, H.; Zheng, J.; Xue, D.; Dong, X.; Liu, M.; Zhang, Y. Identifying the Relationship between Livelihoods and Land Ecosystem Services Using a Coupled Model: A Case Study in the "One River and Two Tributaries" Region of Tibet. *Land* **2022**, *11*, 1377. [CrossRef]
He, L.; Du, X.; Zhao, J.; Chen, H. Exploring the coupling coordination relationship of water resources, socio-economy and eco-environment in China. *Sci. Total Environ.* **2024**, *918*, 170705. [CrossRef] [PubMed]
Yang, J.; Ma, X.; Zhao, X.; Li, W. Spatiotemporal of the Coupling Relationship between Ecosystem Services and Human Well-Being in Guanzhong Plain Urban Agglomeration. *Int. J. Environ. Res. Public Health* **2022**, *19*, 12535. [CrossRef]
Yang, N. *Evaluation and Research on Sustainable Development of Land Ecology in Shanxi Province*; Harbin Institute of Technology: Harbin, China, 2020. (In Chinese)
Rezaee, Z.; Aliabadi, S.; Dorestani, A.; Rezaee, N.J. Application of Time Series Models in Business Research: Correlation, Association, Causation. *Sustainability* **2020**, *12*, 4833. [CrossRef]
Reyers, B.; Selig, E.R. Global targets that reveal the social–ecological interdependencies of sustainable development. *Nat. Ecol. Evol.* **2020**, *4*, 1011–1019. [CrossRef]
Yang, Z.; Zhan, J.Y.; Wang, C.; Twumasi-Ankrah, M.J. Coupling coordination analysis and spatiotemporal heterogeneity between sustainable development and ecosystem services in Shanxi Province, China. *Sci. Total Environ.* **2022**, *836*, 155625. [CrossRef]
Zhu, J.; Gong, J.; Li, J. Spatiotemporal change of habitat quality in ecologically sensitive areas of eastern Qinghai-Tibet Plateau: A case study of the Hehuang Valley, Qinghai Province. *Resour. Sci.* **2020**, *42*, 991–1003. [CrossRef]
Li, X.; Xin, Z.; Yang, J.; Liu, J. The spatiotemporal changes and influencing factors of vegetation NDVI in the Hehuang Valley of Qinghai Province from 2000 to 2020. *J. Soil Water Conserv.* **2023**, *13*, 79–90.
Vince, G. An epoch debate. *Science* **2011**, *334*, 32–37. [CrossRef]

34. Xue, D. Research on the Spatiotemporal Evolution of Land Use Functions and Its Coupling Relationship with Human Well-Bei A Case Study of Hehuang Valley. Master's Thesis, The Henan Agricultural University, Zhengzhou, China, 2023. (In Chinese
35. Chen, R.; Zhou, Q.; Liu, F.; Zhang, Y.; Chen, Q.; Chen, Y. Contribution of climate change to food yield in Yellow River-Huangs River Valley. *Hubei Agric. Sci.* **2018**, *57*, 114–119. (In Chinese)
36. Li, H.-H.; Huang, Y.-M.; Guo, W.; Hou, H.-Y.; Fan, M.-Y.; Qi, X.-P.; Jia, P.-H.; Guo, Q.-B. Influence of land use and land co patterns on water quality at different spatio-temporal scales in Hehuang Valley. *Environ. Sci.* **2022**, *43*, 4042–4053. (In Chines
37. Hou, G.; Xu, C.; Li, X.; Qi, B. Spatial Pattern of Water Conservation Function in Hehuang Valley. *Ecol. Sci.* **2021**, *40*, 169–1 (In Chinese)
38. Wang, P.; Li, R.; Liu, D.; Wu, Y. Dynamic characteristics and responses of ecosystem services under land use/land cover char scenarios in the Huangshui River Basin, China. *Ecol. Indic.* **2022**, *144*, 109539. [CrossRef]
39. Zhu, W.Q.; Pan, Y.Z.; Zhang, J.S. Estimation of net primary productivity of Chinese terrestrial Vegetation based on remote sensi *J. Plant Ecol.* **2007**, *31*, 413–424.
40. Sharp, R.; Douglass, J.; Wolny, S.; Arkema, K.; Bernhardt, J.; Bierbower, W.; Chaumont, N.; Denu, D.; Fisher, D.; Glowinski, et al. *InVEST 3.10.2. User's Guide*; The Natural Capital Project, Stanford University: Stanford, CA, USA, 2020.
41. Chen, S.X.; Yang, X.H.; Xiao, L.L.; Cai, H.Y. Study of Soil Erosion in the Southern Hilly Area of China Based on RUSLE Mo *Resour. Sci.* **2014**, *36*, 10. (In Chinese)
42. Yi, K.; Wang, S.; Wang, X.; Yao, H. The Characteristics of Spatial–temporal Differentiation of Soil Erosion Based on RUSLE Mo A Case Study of Chaoyang City, Liaoning Province. *Sci. Geogr. Sin.* **2015**, *35*, 365–372.
43. Li, G.Z.; Zou, X.Y.; Cheng, H. Method of wind erosion sampling survey in China. *Sci. Soil Water Conserv.* **2013**, *11*, 17.
44. Qiu, J.Q.; Yu, D.Y.; Huang, T. Influential paths of ecosystem services on human well-being in the context of the sustaina development goals. *Sci. Total Environ.* **2022**, *852*, 158443. [CrossRef]
45. Ma, Y.; Ai, X. Evaluation of Sustainable Urbanization Development in Jilin Province Based on the 2030 Sustainable Developm Goals. *Sci. Geogr. Sin.* **2019**, *39*, 487–495.
46. Kwatra, S.; Kumar, A.; Sharma, P. A Critical Review of Studies Related to Construction and Computation of Sustaina Development Indices. *Ecol. Indic.* **2020**, *112*, 106061. [CrossRef]
47. Meadows, D. Indicators and information systems for sustainable development. In *The Earthscan Reader in Sustainable Ci* 1st ed.; Satterthwaite, D., Ed.; Routledge: London, UK, 1999; pp. 1–30.
48. Li, S.; Yu, D.; Li, X. Exploring the impacts of ecosystem services on human well-being in Qinghai Province under the framew of the sustainable development goals. *J. Environ. Manag.* **2023**, *345*, 118880. [CrossRef] [PubMed]
49. Wei, H.; Xue, D.; Huang, J.; Liu, M.; Li, L. Identification of Coupling Relationship between Ecosystem Services and Urbanizati for Supporting Ecological Management: A Case Study on Areas along the Yellow River of Henan Province. *Remote Sens.* **20** *14*, 2277. [CrossRef]
50. Fotheringham, A.S.; Crespo, R.; Yao, J. Geographical and temporal weighted regression (GTWR). *Geogr. Anal.* **2015**, *47*, 431– [CrossRef]
51. Kim, J.H. Multicollinearity and misleading statistical results. *Korean J. Anesthesiol.* **2019**, *72*, 558–569. [CrossRef]
52. Kumar, S.; Lal, R.; Liu, D. A geographically weighted regression kriging approach for mapping soil organic carbon stock. *Geoder* **2012**, *189*, 627–634. [CrossRef]
53. Kletting, P.; Glatting, G. Model selection for time-activity curves: The corrected akaike information criterion and the f-test. *Med. Phys.* **2009**, *19*, 200–206. [CrossRef]
54. An, Q.; Yuan, X.; Zhang, X.; Yang, Y.; Chen, J.; An, J. Spatio-Temporal Interaction and Constraint Effects between Ecosyst Services and Human Activity Intensity in Shaanxi Province, China. *Ecol. Indic.* **2024**, *160*, 111937. [CrossRef]
55. Li, W.; Wang, Y.; Xie, S.; Cheng, X. Coupling coordination analysis and spatiotemporal heterogeneity between urbanization a ecosystem health in Chongqing municipality, China. *Sci. Total Environ.* **2021**, *791*, 148311. [CrossRef]
56. Mo, W.; Wang, Y.; Zhang, Y.; Zhuang, D. Impacts of road network expansion on landscape ecological risk in a megacity, Chi A case study of Beijing. *Sci. Total Environ.* **2017**, *574*, 1000–1011. [CrossRef]
57. Sun, W.; Song, X.; Mu, X.; Gao, P.; Wang, F.; Zhao, G. Spatiotemporal vegetation cover variations associated with climate char and ecological restoration in the Loess Plateau. *Agric. For. Meteorol.* **2015**, *209*, 87–99. [CrossRef]

diversity

MDPI

ew

uman Communities in Protected Natural Areas .d Biodiversity Conservation

abeth Olmos-Martínez [1], Heidi Leticia Romero-Schmidt [2], María del Carmen Blázquez [2,*], ılo Arias-González [2,*] and Alfredo Ortega-Rubio [2]

1 Department of Administrative Economic Sciences, Universidad Autónoma de Occidente, Mazatlán C.P. 82149, Mexico; eolmosm.udo@gmail.com
2 Centro de Investigaciones Biológicas del Noroeste, La Paz C.P. 23096, Mexico; heidi04@cibnor.mx (H.L.R.-S.); aortega@cibnor.mx (A.O.-R.)
* Correspondence: blazquez@cibnor.mx (M.d.C.B.); bio.gonzalez.camilo@gmail.com (C.A.-G.)

Abstract: Using socioecological concepts and within a historical biodiversity conservation context, this research study reviews the main interactions between human communities and protected natural areas (PNAs) to describe their different stages over time and assess the implications arising from climate change. The review suggests that both society and governments have raised awareness and interest regarding the importance of biodiversity conservation. The interactions of human communities in these areas have had different effects on biodiversity use, management, and conservation. Local communities have historically developed traditional uses of natural resources that allow them to remain over time and conserve them. Thus, the interest in PNAs as a conservation instrument has increased worldwide. Regarding climate change, this study found evidence indicating that PNAs may act as buffer barriers to prevent biodiversity loss and mitigate the impacts of extreme events; nevertheless, a great variation can be expected. The magnitude of the impacts on human communities, levels of vulnerability, and resilience capacity of PNAs facing climate change (CC) rely on many factors, such as location, extent, composition, management, and ecology of a given protected area. Therefore, the new scenarios that CC may bring are challenging current systematic conservation planning and traditional management methods of the natural resources that are vital for people. Finally, the authors suggest that society is increasingly aware of PNAs as one of the best tools humans have to prevent biodiversity loss, and potentially buffer the effects of CC. These increases in social awareness of biodiversity conservation importance and PNAs are gratifying and spread optimism about the future that next generations may face.

Keywords: biodiversity; conservation; protected natural areas; human communities; climate change

1. Introduction

The mission of natural heritage conservation through different modalities has its main instrument in the creation and management of protected natural areas (PNAs) based on a conservation culture of natural ecosystems and sustainable development [1]. Protected areas are well received throughout the world as environmental policy tools characterized by the preservation and protection of diverse ecosystems, where the original environment has not been essentially altered, producing a series of increasingly valued environmental services [2].

Academic, social, and legislative systems are conscious of their obligation to protect natural spaces, as well as the systems and cycles that are completely dependent on natural resources and that have supported life throughout millions of years of evolution [3,4].

The potential for economic, social, and political wellbeing that a PNA offers depends on the strategies implemented for natural biodiversity management and conservation [5]. Therefore, this study analyzes the interactions between human communities and biodiversity with emphasis on climate change impact and resilience in protected natural areas.

2. Methods

This literature review article does not have the typical "background section" [6], because it is not an introductory substantiation for a primary research study. Instead, this article has a predominant objective: to synthesize the literature in Socioecology, without analyzing or even collecting primary data [7].

The article started by formulating the research question: What is the relationship between socioecology and biodiversity conservation in natural protected areas? The second step was to define the objective of this study, which is to create a solid starting point for all members of the community interested in socioecology and biodiversity conservation. Then, the authors proceeded to search extant literature, for suitability of the material to be considered in this review [8], which was based entirely on the research background of each co-author.

3. The History of Biodiversity Conservation

In the current context of climate change and the post-COVID-19 pandemic, people in the developed world are looking anew at rural environments and natural ecosystems where human societies have had an historically balanced long-term relationship with nature. Populations trapped in crowded cities are showing more interest in nature, not just looking for beautiful landscapes but also places where they can escape to on holidays. As a result of this worldwide crisis, the idea of having a better quality of life, even urban life, through the conservation of natural ecosystems and nature-based solutions has become common [4,9].

Media coverage now favors this viewpoint of better life quality. For example, people now value the need for healthy food, organic cosmetics, space for leisure, clothes, and furniture made from ethically produced materials. Nowadays, sustainable consumption practices are becoming trendy in everyday urban life, if not common. For instance, regional suppliers, local markets, organic farms, sustainable fisheries, urban gardens, and allotments are examples of new ways to fulfill desires and return to harmony with nature in a more rational way of living [10].

Additionally, a renewed interest in the natural world is growing. Apart from the obvious and now more frequent catastrophic climate events (fires, droughts, heavy rains, and floods), the perception is that society is equally interested in topics such as pollinator decline, ocean pollution, and air quality. Specialized conservationists have had a particular resonance with the public, such as Carl Safina, whose books are global best-sellers.

Where is all this public interest coming from? Ecological studies and conservation efforts are not new. As early as 1869, the British naturalist Alfred Russel Wallace wrote in his book *The Malay Archipelago* about the value of biodiversity itself. He noticed the perfect equilibrium of the species on the islands and reflected on how their beauty is wasted in the "dark and gloomy woods, with no intelligent eye to gaze upon their loveliness, but when "civilized man" reaches the islands, he will certainly upset the balance of nature and make the birds extinct" [11] (p. 486).

The tradition of setting aside wilderness –its beauty and equilibrium, away from human actions–was the ideal of American conservationist pioneers, Henry David Thoreau, John Muir, and Aldo Leopold. However, the idea of preserving landscapes, flora, and fauna alone developed into one of conservation in conjunction with human activities in the 20th century [12].

More than half a century ago, the science of conservation biology started in 1985 when a group of researchers met in La Jolla, San Diego, CA, USA, concerned about increased loss of species and the accelerated reduction in tropical forest [13]. The term biodiversity first appeared at the National Forum on Biodiversity held by the National Academy of Sciences (NAS) and Smithsonian Institute in September 1986. The focus of the forum was habitat destruction and species extinction, and the results were published with the title *Biodiversity* [14]. The concept of biodiversity, or biological diversity, relates to evolutionary biology and ecology. The key elements are individual species, but biodiversity also considers the role of the evolutionary process in species emergence and extinction, the causes of biodiversity loss, habitat destruction, invasive species, pollution, and overexploitation. In the Rio Convention, the definition of biological diversity covered three levels: species diversity, genetic diversity, and ecosystem diversity [15], but what ways have been found to preserve biodiversity whilst sharing the same space?

The Biological Dynamics of Forest Fragments Project demonstrated that to maintain species diversity, the variety of habitats was more important than the size of an area [16]. For that reason, the importance of the protected areas was clearly to be big enough and with sufficient habitat diversity suitable for keystone species. These type of species are those that maintain ecosystem cohesion over time and ensure population viability [17].

MacArthur and Wilson [18] developed the theory of island biogeography and Simberloff and Abele [19] used it to propose that creating several small but connected NPAs could eventually contain and preserve more species than a bigger but more isolated reserve. Shaffer [20] developed the concept of minimum viable population; meanwhile, Newmar [21] demonstrated that in fragmented isolated habitats, small populations became extinct more quickly. All these elements together were used as key tools in designing wildlife reserves capable of maintaining and preserving biodiversity.

Nevertheless, and despite the creation of natural protected areas, numerous conferences, meetings, and thousands of published articles on the importance of biodiversity conservation remained confined to the academic world. Even broad political efforts, such as the United Nations Earth Summit in Rio de Janeiro [15] or the Paris Agreement [22], seemed to hold little interest for the public.

Because lifestyle is now in danger, as early as 2007, George Beals Schaller pointed out that conservation problems are also social and economic as well as scientific, since humans benefit from wildlife [23]. In order to assign an intrinsic value to biodiversity, one looks at services that can be obtained from it. Ecosystem services have been defined as the conditions and processes through which natural ecosystems sustain and fulfill human life [24]. The services that species of microorganisms, animals, and plants provide and that are required to sustain human life are uncountable. The systems and natural cycles are product of billions of years of evolution. Whether humans appreciate it or not, all living beings depend entirely on ecosystems because both knowledge and ability are required to substitute them. For millennia, ecosystem services have been almost unnoticed, because they were not disrupted, although nowadays few places on Earth remain untouched (chemically, physically, or biologically). However, society likely values ecosystem services more as human impacts on the environment intensify, and the cost and limits of technological substitution (when possible) become more difficult or expensive [24].

Because humans have evolved and lived embedded in nature until recent times, the relationship between biophysical and sociocultural components of the ecosystems are historically interdependent, and a biocultural approach is now recognized as critical in conservation and restoration projects [25]. In fact, 11 of the 17 Sustainable Development Goals of the United Nations are directly related to biodiversity conservation [26,27].

Since the mid-20th century, society has found that an effective method to improve biodiversity conservation is to involve local communities in the government of protected natural areas [12], which could either be by collaborative management (helping authorities) or co-management (jointly with authorities) [28]. Both methods incorporate traditional knowledge and current necessities of the inhabitants and can lead to more long-term, sustainable results. Collaborative governance is a way to eliminate imbalance between strong governmental politics and weak communities who live closely with wildlife. At the end, life quality improves both local communities and society [29–31].

Unfortunately, the richest countries in biodiversity are not always the most developed; for that reason, in 2014, the Nagoya Protocol, signed by many countries, established a global policy to improve biodiversity conservation, including sharing the benefits arising from using genetic resources fairly and equitably. Despite being signatories, many governments, companies, and institutions still do not adhere to the treaty. They still retain numerous patents, plant breeder rights, genetic resources, and plant varieties. However, the international community is trying to correct that and working to ensure that the profits of any natural exploitation or genetic resources stay in each nation. On top of that, in the current post-2020 Global Biodiversity Framework (GBF), the Aichi Targets about biodiversity loss have not been met, so society and researchers must work together to achieve them [31–34].

In the last two decades, involving people in biodiversity conservation through a process called citizen science has become a powerful tool. In this process, the public is involved in scientific projects in different ways. The easiest and most common one is to ask lay people for their help in gathering data for scientific projects following a specific protocol. In other cases, nonscientists can be engaged in the process of decision-making about policy issues that have technical or scientific components; alternatively, the scientific community becomes part of the policy decision-making process [35]. These projects are sometimes global, involving people from every continent, such as the example below about pollination [36]: (https://www.beeproject.science/croppollination.html, accessed on 10 April 2022).

This project was created to compile crop pollination data from published studies across the world and help predict pollination services. It also permits sharing data since the database has open access online, so scientists and institutions that want to contribute to new pollination datasets can add them easily.

In conclusion, a renewed desire (or necessity) is perceived to reinforce the reciprocal relationship with nature, and return to traditional crop methods instead of industrialized agriculture, small farms instead of macro livestock production, fields free of chemicals instead of insecticides, rational logging instead of deforestation, and a biocultural approach to restoration [37]. All these practices are perfect on a small scale, but none of them are possible on a major scale if consumption habits do not change. The amount of energy and resources and ecological cost of crowded urban societies have become unsustainable and new models, energies, productive forms, and imaginative solutions for waste and recycling are imperative.

4. Protected Natural Areas

In the evolution of humanity, the passage from nomadic hunter-gatherers to sedentary farmers has had significant effects on ecosystems and the environment. These changes were caused by tree felling and forest clearing, using wood for construction and as fuel, competing for land and water, and evicting and consuming wild animals in the same area [38]. At the same time, some plants and animals were intensively exploited, which led to the appearance of the first domesticated species [39,40].

However, ancient cultures, such as Egyptians, Vikings, Persians, Romans, and Indians, have protected natural spaces for several centuries.

In pre-Columbian Mesoamerica and in almost every ancient culture, people worshiped the sun, stars, moon, Earth, water, many animals, and almost all vegetation. All these elements of nature were conceived as divine, and cultures expressed their own realities linked to a cosmogony of which they were part and had to respect to maintain harmony with nature [41].

Incas used agricultural techniques, such as the construction of platforms or terraces to avoid soil erosion and take advantage of the slopes and hills [42]. Mayans linked their development with the tropical forest, basing their agricultural, horticultural, and forestry practices on protected pluriculturalism, called Pet-koot (round fence) [43].

In Mexico in the 15th century, the Mexica King Nezahualcóyotl established the first botanical garden in Texcotzingo, prohibiting hunting and imposing limits on obtaining firewood to protect the forests of his dominions in Texcoco. In 1428, he fenced the Chapultepec forest, enriched flora, planted the famous ahuehuete trees, and implanted a rich fauna, thus initiating its protection [44].

In 1450, Moctezuma Ilhuicamina established the Oaxtepec Garden, and in 1465, he took over Bosque de Chapultepec, a tradition that was followed by his successors. Moctezuma

Xocoyotzin established gardens at the hillsides of Popocatépetl volcano and Atlixco in Puebla and maintained the one in Oaxtepec, Morelos, which had been operating as a protected area for more than 75 years when the Spaniards arrived [45].

During the Middle Ages, royalty promoted laws for forests and wildlife protection. Nevertheless, after the emergence of industrial cities in the mid-1700s, the exploitation of mineral coal polluted waters, soil, and atmosphere, which had an impact on policies and actions for the conservation of natural areas [41].

When Europeans discovered North America in the 15th and 16th centuries, they found it populated by diverse groups of indigenous people who showed a deep respect for the land and animals. This continent had abundant and seemingly inexhaustible stocks of wood, fertile soil, wildlife, fresh water, minerals, and other resources, so the colonizers felt that it should be conquered, opened, disassembled, and used as quickly as possible [46,47].

During the conquest and colonization by the Spaniards, the demand for construction materials and fuels increased, paying for forest clearance and making alterations to ecosystems. To stop such deforestation, King Carlos V of Spain decreed in 1530 that Bosque de Chapultepec and the hill surrounding it be owned by the City of Mexico, making this area the first protected forest of the colonial era.

Throughout the expansion of Europeans in North America from 1607 to 1832, forests and wildlife were degraded and depleted at an alarming rate, so the task of protecting was passed down to subsequent generations. However, in the United States, federal action on the conservation of forest resources and wildlife did not properly begin until 1872 with the creation of Yellowstone National Park [47,48]. In the same decade, "Desierto de los Leones" was the first Mexican protected area, created in 1876 with the purpose of protecting the springs that supplied water to Mexico City, and decreed in 1917 as a national park [49]. These parks were the first of their kind in North America and set a precedent for preservation of biodiversity and cultural history (Figure 1).

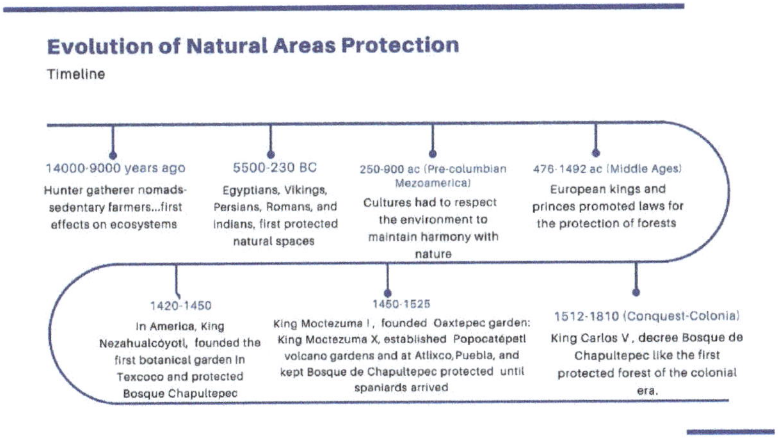

Figure 1. Timeline of the evolution of Natural Protected Areas until the 20th Century.

The definition of a protected area for the International Union for Conservation of Nature (IUCN) is "a clearly defined geographical space, recognized, dedicated, and managed, through legal or other effective means, to achieve the long-term conservation of nature with associated ecosystem services and cultural values" [50]. Although Yellowstone has been recognized as the first national park in the world, it was not the first protected natural area by a government. The oldest national parks on each continent have their own stories of origin and are home to evidence the world's complex and changing relationship to public conservation [51]. Some examples of protected areas worldwide are described below, with a timeline of the first PNAs summarized in Figure 2.

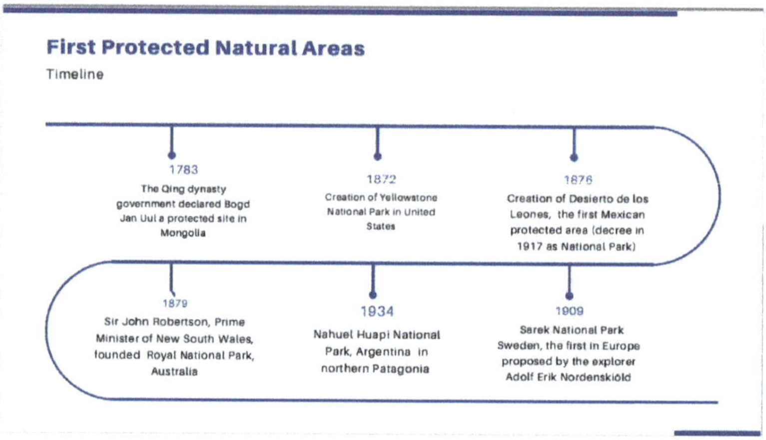

Figure 2. Timeline of some of the first Protected Natural Areas.

4.1. Bogd Jan Uul Protected Area, Mongolia

Since the 14th century, Bogd Jan has been considered one of the holiest mountains in Mongolia. The Qing dynasty government declared it a protected site in 1783. Since then, the coniferous forests, endangered musk deer, and other wildlife of Bogd Jan have survived political upheaval that lasted until the 1990s. Finally, in 1996 the United Nations Educational, Scientific, and Cultural Organization (UNESCO) declared it a biosphere reserve [52].

4.2. Yellowstone National Park, United States of America

The creation of the protected area by President Ulysses S. Grant in 1872 was a great success for early conservation initiatives and influenced global growth of national parks. As was the case in many national parks, poaching and illegal logging plagued Yellowstone from the 1880s to 1890s, until Congress passed tougher protective measures [48,49].

4.3. Royal National Park, Australia

This environment of more than 15,000 ha is home to rare birds, clifftops, and lagoons. Centuries earlier, it was home to the Gweagal Aboriginal community. In 1879, Sir John Robertson, Prime Minister of New South Wales, founded the park through a series of land reforms [52].

4.4. Nahuel Huapi National Park, Argentina

More than 800,000 ha of vast slopes and glaciers are found in the depths of northern Patagonia. In the 1880s, the explorer Francisco Pascasio Moreno surveyed the terrain for the Argentine government that contributed to the inclusion of the area under federal jurisdiction. In 1903, he donated part of the land to the government with the intention of making it a national park, which was achieved in 1934 [51].

4.5. Sarek National Park, Sweden

With more than 100 glaciers and thousands of elks, Sarek National Park is the jewel of Sweden's diverse flora and fauna. The explorer Adolf Erik Nordenskiöld proposed the foundation of national parks, which influenced the decision of the Swedish parliament to create the first one in Europe in 1909. The indigenous Lapp or Saami people lived in the area before Sarek obtained park status [53].

The greatest conservation success story in the 21st century is the exponential growth of protected areas, a primary defense mechanism against Earth's biodiversity loss [54]. Protected areas play a key role in preserving the benefits or "ecosystem services" that nature brings to people and includes the provision of food and fresh water, regulation of floods and droughts, nutrient cycling, and recreational opportunities [55].

The initial goal was to keep nature safe from people. While it remains true today, new targets were added over time as protected wildlife areas often have often become tourist attractions and important economic mainstays since the beginning of the 1970s. Since the Convention on Biological Diversity was opened for signing in 1992, its 168 member states have nearly doubled the size of their protected areas in an effort to meet the goal of that treaty: each nation committed to have at least 17% of its land area and 10% of its oceans protected by 2020. As of 2019 the nations of the world have legally designated more than 202,000 protected lands covering 19.8 million square kilometers, or 15% of the Earth's surface [54,56,57].

Globally, a third of all protected areas are under intense human pressure. They face pressure from agriculture, encroachment on human settlements, roads, light pollution, railways, and infrastructure development on waterways. Only 10% of the lands are completely free from human activity, but they are in remote areas of high-latitude nations, such as Russia and Canada [58]. Protected areas are globally underfunded, and their mismanagement opens the door to many human impacts, including poaching, illegal logging and fishing, and habitat destruction [59].

Although a rapid progress in developing protected areas and a great increase in protected land and ocean percentages have been reported [56,60,61], such efforts are not ecologically representative nor do they optimally protect biodiversity [62–65]. Quantitative targets tend to overshadow qualitative ones. For example, within the Aichi Target 11 established by the United Nations Convention on Biological Diversity, 17% terrestrial/10% marine protected areas have tended to be the primary goal, rather than the qualitative elements of ecological representation, equity and effectiveness management, and integration into wider landscapes [15,61,66].

Similarly to terrestrial PNAs, Marine Protected Areas (MPAs) can be directed at conserving overall biodiversity or some portions of it (e.g., key species, top predators, commercial fish species, endangered species, communities, or ecosystems). If the main goal is biodiversity conservation, today's approach is generally the development of "ecologically representative" protected areas [67–69].

A well-conserved protected area is the result of collaboration between law enforcement departments, policy makers, local communities, and other stakeholders. Management objectives of the protected areas, in addition to the main one of biodiversity conservation, now also include social and economic ones. Reconciling goals with the needs of local livelihoods requires building partnerships and alliances with local businesses and communities [54].

5. Human Communities in Protected Natural Areas

Traditionally, the communities that have lived within protected natural areas or in their area of influence depend on the use of available natural resources and this condition defines the characteristics of the population [70]. The study of human communities in PNAs has expanded into more than just the interaction of nature and society. From a sociological perspective, all these interactions become models that integrate social and natural systems for better human development through sustainable and environmental management, conservation, and governance, among other disciplines [71,72].

Thus, Salas-Zapata et al. [73] assume that social (culture, society, economy, and politics) and ecological systems converge in such a way that they interact, couple, dominate, and cause impacts and disturbances among them. This section shows some examples of the interactions between both kinds of systems grouped in different interacting categories: coupled, one dominating the other, or mutual disturbance.

Parathian [74] studied the impact-based conservation of two communities of Tikuna in the Colombian Amazon. Those communities have adopted some transcultural beliefs and shown innovation and resilience, which allowed them to cope with climate and environmental change and generate conservation processes and initiatives that influence local livelihoods; the study concluded that indigenous populations are malleable groups with a practical understanding and vital dynamics for identifying solutions to socio-ecological problems.

In other cases, community perception is an important way of evaluating conservation performance. In this case, the community members work on projects for effective biodiversity protection and the wellbeing of the people who live within or near protected areas. For instance, in Tarangire National Park in Tanzania and in Mole National Park in Ghana, watching wildlife and enjoying nature tourism not only have a positive impact on livelihoods and community development, and especially in governance matters, but also on national and international tourism [75].

Natural and human systems sometimes do not seem to influence each other. Harris et al. [76] studied interactions and links among users and mammals that inhabit a protected area, for example, direct (poaching) and indirect (domestic animals) human activities to the fauna of the largest complex of protected areas in West Africa (W-Arly-Pendjari). They found that although human pressure is dominant, it did not influence species richness or composition, which led them to label it a coupled human-dynamic system.

In other cases, forcing regulations in protected natural areas are necessary as a strategy for biodiversity protection. A study from Adamik and Berros [77] analyzed one of such situations on the Argentine coast (Santa Fe, Entre Ríos, Corrientes, and Misiones). In the case of Santa Fe and Misiones, the authors indicate that there is a sustainable exploitation of natural resources based on existing regulations. Although the legislation allows for the creation of PNAs in the four territories, their classification and specific treatment result on different management categories. However, these authors warn that in territories where few protected areas exist, natural environments are underrepresented. On the contrary, where many protected areas are established, precarious control by the authorities is observed, which causes impacts and disturbances.

In another example Olmos-Martínez et al. [70] studied the effect of implementing a protected natural area on the wellbeing of rural communities in Baja California Sur, México. In this case, although the creation of the protected area has hampered access to natural resources, it has also increased the sense of wellbeing of the population based on social and economic indicators in the Sierra La Laguna Biosphere Reserve.

Karanth et al. [78] carried out a study on voluntary community resettlements of the protected areas of India. The authors studied the strategy of people exclusion to minimize the anthropogenic threats of wildlife. The project is challenging for the government since an estimated 4.3 million people share spaces with megafauna within protected areas, indicating that resettlement policies focus on paying in cash or a compensation package for land for families relocating voluntarily. These authors also found that 89% wanted to move to get better education, health care, roads, less human conflict, and better wildlife; thus, households that depend on wage labor decided to move and smallholders looked for better opportunities in agriculture.

In addition, research on the analysis of conservation policies in urban contexts leads to challenges faced by decision makers and society in general. In that sense De la Mora [79] shows two convergent conservation policies in Mexico, which are about Environmental Services and PNAs as instruments of public policy and the interactions from their implementation for the conservation of urban socio-ecosystems. The author alluded that urban and peri-urban PNAs should enjoy greater political visibility in urban and environmental decision-making territorial processes for their permanence. Thus, the conservation policy of environmental services can lead to the social and political reassessment of their strategic importance, which would contribute to urban planning (housing, hydraulic sector, forest areas, among others).

On the other hand, Mutanga et al. [80] conducted a study on community perception of wildlife conservation and tourism in adjacent communities within four protected areas in Zimbabwe. These authors found that one community had neutral perceptions on wildlife conservation, but the other three had positive perceptions on the same issue. However, all four communities had negative tourism perceptions because they have not been involved, and natural resource benefits are not shared fairly among stakeholders.

Related to the above, Olmos-Martínez et al. [81] alluded to this situation in their study on anthropogenic pressure in the PNA Archipelago Espíritu Santo, in Baja California Sur, Mexico. The authors concluded that tourism exerts greater pressure than other activities, such as fishing, because of its high demand and growing activity. Although the type of tourism practiced is eco- and adventure tourism, their activities produce pollution, flora and fauna alteration, exotic species introduction, and fishing sector problems.

However, Olmos-Martínez et al. [82] found a successful case related to the use of natural resources and ecotourism in the PNA "Tortuguero El Verde Camacho" sanctuary, in Mazatlán, Sinaloa, Mexico. Based on the collaboration between the society, the government, and the organization within an ecotourism cooperative society, the community has generated socioeconomic development that bolsters ecotourism services with an emphasis on the conservation, preservation, and management of the coastal wetland and endangered species, such as the olive ridley turtle. In addition, tourism in Mexico is established as a permitted activity in the PNA management program since it allows raising awareness of biodiversity protection, which also represents an economic alternative that benefits local communities and users [3].

Some studies of PNAs are related to evaluating vulnerability as a management element; this is the case in the analysis by Verdugo et al. [83], which calculated the environmental vulnerability of the marine part of Bahía de Loreto National Park in Baja California Sur, Mexico, based on environmental indicators using the exposure-sensitivity-capacity method. Their results show that the greatest vulnerability is found in the coastal zone in front of the Peninsula and in Isla del Carmen where ecotourism is practiced, as well as in Isla Catarina due to the pressure from fisheries and, therefore, interactions in social and ecological systems.

Andrade and Rhodes [84] mention that PNA management, in some cases, fully integrates important factors, such as social, cultural, and political factors. However, they have caused adverse social impacts in some cases on local communities that altered their traditional ways of life, limited their control and access to natural resources, and generated conflicts between administrators and local communities. The authors analyzed the key factors that lead to better compliance with conservation strategies and observed that local community participation is the only variable that reacted significantly with the level of compliance with the PNA policies—the greater the participation is, the greater the compliance is.

In addition to the above and because tourism has a social representation approach, this activity allows for the inclusion of interactions, history, and culture. Sarr et al. [85] reviewed these interactions in La Langue de Barbarie National Park in Northern Senegal. Members of the community found evidence of the failure of tourism programs and projects with community participation given the perception of the same communities, which implies division in groups with different interests, to such a degree that inhabitants consider emigration from the site.

Likewise, Olaniyi, et al. [86] state that biological diversity in Nigeria is highly degraded and far from the Sustainable Development Goals. This situation leads to a major problem since PNAs constitute the food base for survival of the host communities in that area. Thus, the authors propose community participation policies to improve rural livelihoods.

In addition, Fay [87] studied the process of merging Dwesa-Cwebe Nature Reserves with Pafuri Triangle reserves in South Africa. This author addresses that since PNA joint management has emerged, the communities have been involved in formal negotiations with government officials and groups of non-governmental organizations in search of mutual benefit. However, they have found that mutual gains and distributive approach have reduced the power of communities in decision-making.

Finally, Olmos-Martínez et al. [2] studied the perception of the population living in the PNAs of Baja California Sur, Mexico on climate change (CC) and environmental changes in the area where they live. The study was carried out from different topics (environment, coastal areas, soil, water, society, fishing, agriculture, livestock, and tourism). The results indicated that the local population has empirical knowledge about the effects of climate change, both economically and socially. This knowledge is a tool for decision makers that allows recognizing human behavior in social, economic, and environmental changes that may contribute to mitigation and adaptation strategies in facing the phenomenon.

In summary, human communities have a dynamic and adaptive relationship with nature within PNAs. Beyond being an instrument of environmental policy, PNA management, in some cases, fully integrates important social, cultural, and political factors.

6. Impacts and Resilience to Climate Change in Protected Natural Areas

Protected natural areas are widely recognized as one of the most effective ways to preserve biodiversity in both marine and terrestrial habitats [55,88], protect water supplies, and many other environmental and ecosystem services [89,90]. For example, there is evidence that supports conservation actions have been successful at buffering species from historical drivers of population declines, such as land-use change, pollution, hunting, deforestation, fragmentation, and habitat degradation [88,89,91–93]. However, locals and authorities in PNAs cannot similarly buffer species and natural ecosystems from the effects of anthropogenic CC, so their role in mitigating impacts has been questioned [60,94].

Some of the major limitations for ensuring biodiversity conservation under CC are the static nature of PNAs, their spatial and ecological bias, and low coverage [95–100]. Increases in greenhouse gas (GHG) emissions, linked to fossil fuel combustion and land-use changes, have distorted the chemical composition of the Eart's atmosphere, oceans, and freshwaters, and in turn, the global climate system, influencing global temperatures, precipitation patterns, and intensity and frequency of extreme weather events [101,102]. The observed impacts of CC across terrestrial and aquatic biomes are now affecting every ecosystem around the world [101,103–105] and most biological and ecological processes on Earth [106].

The perception of the PNA role of in relation to CC began in the late 1980s. Studies have pointed out paleoecological evidence of increasingly strong and widespread geographical species range shifts as a response to global climate changes [107–109]. A variety of aquatic and terrestrial species are already responding with elevation, depth (in the ocean), and latitudinal shifts in their distribution ranges, tracking shifting isotherms [16–18]. Therefore, changes in the relationships between species range limits (dynamic), protected area boundaries (fixed) [110–113], and land-use change [60,94,96,97] should be expected. These statements have been integrated into much of the literature on CC potential impacts on biodiversity management and nature reserve systems [60,94,113–116].

Peters and Darling [111] defined two important hypotheses concerning the expected process of CC impacts on nature reserves and natural habitats. The first one is that under changing land-use and climate conditions, the static nature of PNA boundaries may fail to encompass the species climate ranges they were intended to protect. The second one of species responses to CC concerns the potential species range movements along altitudinal gradients [111]. Under this hypothesis, increases in local temperature would move the species climate habitats upslope, and each successive altitudinal range would be replaced by the species habitat occupying the zone directly below. The expected CC impacts in mountainous PNAs would include the loss of the coldest climate habitats at the mountain peaks and linear migration of all remaining habitats upslope. The management deduction deriving from this altitudinal-response theory is that PNAs showing large altitudinal ranges and topographic relief offer the largest range of climate habitats. Therefore, they allow the greatest amount of internal species movement under changing climates. In contrast, climate shifts in PNAs having a limited altitudinal range would be expected to force species to migrate outside the reserve area boundaries—a conceptual model that has been used to analyze altitudinal ranges of global NPA networks in initial assessments [96,97,114].

Multispecies approaches examining potential CC impacts on species representation levels in PNA networks have been developed in several parts of the world [95,97,113]. The results suggest that range shifts cause declines in species representation in some reserves, but considering the assumptions about species dispersal abilities, they might result in increases in representation in other reserves [60,96,97]. The net effect of movements on individual reserve representation and across whole networks depends greatly on dispersal abilities and intervening land uses that may facilitate, complicate, or impede dispersal, affecting the PNA individual reserve and/or network ecological representation [60,113,117]. In terms of ecological representation, the issues or species of concern depend on the objectives of the PNA network and the particular characteristics of the geographic location [97,118–120].

System planning based entirely or partially on biotic elements may be the most vulnerable to CC impacts because species distribution, abundance, and habitat are largely determined by climate [121]. Theoretically, CC causes species range redistribution [105,106,122], so new communities and ecosystems emerge with novel species complements and assemblages that may not meet existing ecological representation targets [113,123]. The impact magnitude on ecological representation within PNAs depends on the location, composition, and ecology of a given one (or networks of NPAs). Therefore, the new scenarios that CC may bring are challenging current systematic conservation planning and management methods, such as the use of 'ecological representation' as a tool to identify the effectiveness of individual or networks of PNAs [91,97,102,124,125].

An alternative approach suggests that a classification framework of "enduring features", such as topographic relief and the origin and texture of parent materials and soil, should be the basis for representation of natural regions in a PNA network [118,124,126]. The assumption of this alternative is that these natural features are more stable in their distribution than biotic elements and yet reflect biological features. For example, a hierarchical classification approach for marine conservation that uses long-term and recurrent geophysical and oceanographic features has been advocated [127,128]. Other approaches have used a combination of biological and physical attributes to create classification hierarchies to be used in designing representative MPAs [118,128]. In this sense, the key to resilience in a complex adaptive system lies within heterogeneity maintenance—the essential variation that enables adaptation [129]. An ecologically resilient ecosystem can better absorb (climate-induced) disturbance, reorganize while undergoing change, and retain the same or similar function in a rapidly changing climate than a system that is not as resilient. Consequently, a PNA may encompass a greater range of physical characteristics and altitude, as well as a variety of habitat types, climates, and soils, and likely be more resilient to CC impacts [130–133].

The great diversity of PNAs in the world suggests that CC impacts are not uniform. Considering the relative climate sensitivity on different regions of the planet [101,134,135], differentiated impacts on individual areas or networks would be expected. Within the most natural and human systems, the current climate and its variability is a frequently limiting factor, but not the only one that determines the overall condition of these systems [134,135]. For example, projected population changes in different countries of the world, as well as the subsequent change in land-use and air and water quality, continue to put pressure on biodiversity, natural resources, and PNA effectiveness. Therefore, CC is seen as an additional factor that can affect the protected areas in different ways.

Warren et al. [135] examined the potential regional CC impacts at different degrees of temperature rise, identifying vulnerable sectors in different regions of the world. The authors assessed climate impacts projected to the end of the 21st century on water stress, agriculture, coastal flooding, human health, energy demand, and ecosystems (including biodiversity and global vegetation). A valuable review of the field on protected areas and CC [96] underscored five broad categories of management recommendation for protected areas facing CC: redundant reserves; larger reserves; buffer zones around protected areas; landscape connectivity; and management of existing threats and disturbance regimes.

Another relevant publication by Hanna [97] summarizes progress related to the design of protected area networks, connectivity and corridors, individual protected areas, mobile reserves, and alternative management measures when protection alone is not enough.

The expected variety of impacts on PNAs arising from CC suggests that managers need to adopt response options to alleviate the negative ones or take advantage of opportunities arising from CC effects, tailored to the specific needs and characteristics of individual reserves [97,136–140]. Although all PNAs of the world are vulnerable to CC impacts to some degree, those located in different regions of the planet may exhibit differential sensitivity [134–136]. Organisms, populations, species, and ecological communities do not respond to approximated global averages; instead, regional changes are more relevant in the context of ecological response to CC [103,134,141]. The geographic asymmetry in warming trends of global temperatures and changes in precipitation patterns contribute to heterogeneity in ecological dynamics across systems and in the range of impacts that PNAs face in different regions of the world [101,134,135,142].

In that scenario, managers need to adopt response options to alleviate the negative impacts or take advantage of opportunities that address them, tailoring to the PNA needs and characteristics [97,118,126,143]. At this point, information availability is critical on the potential CC impacts on these areas and their interaction with other factors in each region that influence the integrity and effectiveness of conservation strategies [102,134].

Future CC impacts outside and within PNAs are projected to become more pronounced in the next decades with variable relative effects depending on the climate scenario and geographic region [101,134,135]. Changes in species ranges can negatively influence the capacity to safeguard species, increase local species turnover, and substantially increase the risk of extinction [112,113,115,144,145]. For example, some communities, such as coral reefs, are particularly vulnerable to CC, especially to the effects of sea warming and ocean acidification [146,147]. Increasing temperatures lead to increased levels of coral bleaching and mortality, and biodiversity depletion, even in the Great Barrier Reef World Heritage Area [148], which could have serious consequences for the Reef's biodiversity, ecology, and dependent recreational use and economic activities [146,147]. Overall, coral reefs are projected to decline from 10% to 30% of former cover at 1.5 °C warming and to less than 1% of former cover at 2 °C warming. Therefore, scenarios show that limiting global warming to well below 2 °C plays a critical role in reducing adverse impacts on coral reefs, ecosystem services, PNAs, natural resources, and their contributions to people [66,149].

A consistent conceptual vulnerability model that can be applied to any system has emerged in CC literature. This model is understood as function of nature, magnitude, and speed of the climate variation to which the system is exposed, its sensitivity to climate and weather conditions induced by anthropogenic CC, and the adaptive capacity of that system to cope, adapt, or recover from the effects of those conditions [150–153]. Exposure is the extent of CC likely to be experienced by a species or system, in this case PNAs, which is an external feature that depends on the rate and magnitude of physical changes. Some examples are temperature, precipitation, sea level rise, flood frequency, CO_2 concentrations, pH in the region where the system of assessment (PNAs) is located, climate variability, and extreme events. Sensitivity is the degree to which the system persistence, performance, or regeneration is dependent on prevalent climate, particularly climate variables that are likely to suffer a change soon. In the case of species, sensitivity depends on factors such as ecophysiology, life history, and microhabitat preferences. For instance, more sensitive PNAs are likely to show a greater degree of change with even small changes in climate variables. Highly sensitive protected areas could therefore be those located in the highest mountains, Artic, Tundra, or protecting coral reefs, where small changes in climate variables and patterns causes great impacts [135,143].

Finally, adaptive capacity refers to the ability of a species to cope with CC, depending on intrinsic factors, such as phenotypic plasticity, genetic diversity, evolutionary rates, dispersal, and colonization ability. In the case of PNAs, adaptive capacity can be seen as resilience of the reserve to recover to a starting condition after being impacted by a disturbance caused by CC [132,154]. The capacity of a system to adapt to CC is also highly variable among regions, countries, and socioeconomic groups. For example, countries with limited economic resources, poor infrastructure and skills, unstable and weak institutions, and unequal distributions of power and access to resources have limited adaptive capacity and thus highly vulnerable to CC [151,155–157].

México is considered one of the most vulnerable countries to CC impacts [158]. The country joined the United Nations Framework Convention on Climate Change (UNFCCC), committing to work together with other countries to stabilize GHG. The Climate Change Intersectoral Commission (CCIC) was created in 2005 to improve collaboration between federal and regional agencies, minimize conflicts between sectors and regions, and maximize benefits of the political integration. Its mission is to accomplish the commitments acquired and be in charge of the coordination, tracing, and evaluation of the National Strategy for Climate Change (NSCC) approved in 2007. Since that time, CC was incorporated as a priority in the National Development Plan (NDP) 2007–2012, establishing objectives, strategies, and priorities that different sectors (transport, health, energy, and environment, etc.) are intended to follow. This decision resulted in the creation of the Special Program of Climate Change (SPCC) 2009–2012, whose main goal was to guide mitigation and adaptation strategies at sectorial and regional levels [158]. Mitigation strategies from the SPCC are focused on encouraging the use of renewable technologies, promoting energetic efficiency, reducing greenhouse gas emissions in productive processes, and preserving carbon sinks, such as forests and soils. On the other hand, adaptation strategies are oriented to assess vulnerability from different sectors to strengthen their adaptation capacities. For example, with the purpose of forest preservation and their related ecosystem services (carbon storage, biodiversity conservation, water balance, etc.), the SPCC incorporates the strategies from the Natural Protected Areas Program (NPAP) created in 1996, Forest Development Program (PRODEFOR) in 1997, Forest Ecosystem Conservation and Restoration Program (PROCOREF), and others currently approved [158].

In 2010, the National Commission of Natural Protected Areas (CONANP) published the first version of the Climate Change Strategy from Natural Protected Areas (ECCAP), positioning the PNAs as natural solutions to face CC. The first version of the ECCAP was built upon two main substantive axes: mitigation and adaptation, with three support components (knowledge, communication, and capacity development), and a transversality component (public policy and bonding). Recently, the second version of the ECCAP (Climate Change Strategy from Natural Protected Areas: A Call to Resilience of México 2015–2020), sees mitigation and adaptation actions as two faces of the same coin, and underscore the role of PNAs to achieve the Sustainable Development Goals (SDG) and targets established in the United Nations Framework Convention on Climate Change [159].

The ECCAP 2015–2020 is structured by four components: institutional arrangements, land gestion in a climate change context, knowledge for decision-making, and communication and social participation. Those components are intended to increase resilience of socio-ecosystems based on the following seven principles: (1) maintaining and increasing diversity and redundancy; (2) managing connectivity; (3) managing slow change variable; (4) promoting the notion of complex adaptive systems; (5) encouraging knowledge and learning; (6) increasing participation; and (7) promoting a polycentric governance. Therefore, the ECCAP 2015–2020 is intended to be a dynamic instrument that guides the actions and decision-making of the CONANP at local, regional, and national levels to reduce the vulnerability of the socio-ecosystem to CC, contribute to ecosystem service conservation, and to provide and promote carbon capture and storage in PNAs [159].

Studies have demonstrated that PNAs have been managed for a much wider range of values than just biodiversity conservation. Society expects now that these areas also provide sustainable resource use, carbon sequestration, mitigation, and adaptation to CC, as well as vital ecosystem services, spiritual values, and support for local communities [55,89,160]. Indeed, PNAs are now widely recognized as natural solutions to face CC impacts and are considered within the ecosystem-based adaptation strategies to cope with future changes [161]. However, CC impacts on these areas may have strong unexpected effects on communities and people´s wellbeing by altering the natural resources and ecosystem services they rely on [160,161].

Nowadays, the potential CC impacts on PNAs come from the convergence of a better understanding of the physical science behind climate change [101] and a recognition of the multiple uncertainties that CC brings to conservation planning [119,162–165]. Additionally, the accumulative evidence from paleoecological observations, phenological and microevolutionary responses, field experiments, global circulation models, improved modeling tools, reserve selection algorithms, and fine-scale climate projections [141,166–168] allows for improvements in the way that PNAs are established and managed in order to reach global conservation targets in a changing world.

7. Concluding Remarks

This research study reviews the main interactions between human communities in protected natural areas and natural biodiversity conservation. To place these interactions in context, the diversity conservation evolution concept was analyzed, as well as the stages of the public interests alongside this evolution.

Protection of natural diversity has evolved over many years from old civilizations to modern day. In the 20th century, human activities began to be related to the importance of conservation in the search for a balance, giving rise to the science of conservation biology, which brought about the appearance of the term biodiversity in 1986, encompassing three levels: species, genetics, and ecosystems. This step was the gateway to the first United Nations Earth Summit with the Rio Convention on biological diversity in 1992 and the signing of international treaties with a biocultural approach, recognizing them as fundamental axes in conservation and restoration projects. These axes were recognized in different conventions and treaties, such as the Biological Diversity Convention where the Aichi goals were signed in 2010, the Nagoya Protocol in 2014, and the Sustainable Development Goals approved by the United Nations Assembly in 2015.

Currently, there has been a renewed interest in reinforcing the reciprocal relationship between nature and human activities from the international community [9,56,61]. To this end, at international level, one of the environmental policy tools most committed to biodiversity conservation is the creation of PNAs throughout the world. Since the Convention on Biological Diversity in 1992, 168 member states have nearly doubled their protected areas to meet the goal of at least 17% of land area and 10% of oceans. It should be noted that the objectives related to PNA management, in addition to biodiversity conservation, also have social and economic objectives because of the intense interaction of human communities with nature.

Moreover, the evolution of PNAs worldwide was reviewed with emphasis on Mexico, indicating that public attention regarding the PNA benefits has also increased. However, the relationship between protected areas and human settlements is complex, so they are either coupled, one dominates, and/or they cause themselves mutual impacts and disturbances. Thus, in cases starting from responsible PNA management, multiple innovative solutions were generated for their coupling and or decreasing impacts, which evidences the need to continue reviewing management policies for biodiversity conservation in the PNAs.

Concerning benefit awareness evolution of the interactions among local people inhabiting PNAs and biodiversity conservation, the social perception is that ancient local communities have had a positive effect on biodiversity conservation [75,80,169]. In this review, the evidence suggests that ancient local communities have developed sustainable traditional usages of Natural Resources that benefit not only human population, but also biodiversity conservation.

To conclude, the great PNA diversity worldwide suggests that CC impacts are not uniform. Considering the relative climate sensitivity on different regions of the planet, the impacts should be differentiated in the PNAs. System planning based on biotic elements may be the most vulnerable to CC impacts because species distribution, abundance, and habitat are largely determined by climate. The impact magnitude on the ecological representation within PNAs depends on the location, composition, and ecology of a given protected area. Therefore, the new scenarios that CC may bring are challenging current systematic conservation planning and management methods of the natural resources.

Finally, the authors suggest that society is increasingly aware of PNAs as one of the best tools humans have to prevent biodiversity loss, and potentially buffer the effects of CC. These increases in social awareness of biodiversity conservation importance and PNAs are gratifying and spread optimism about the future that next generations may face. Research limitations have been identified related to future research on biodiversity conservation. Some of them are: detailed analyses of the different international efforts, monitoring, and follow-up; analyses at continent level or by level of development of the countries; and literature reviews on social involvement in biodiversity conservation projects with a gender perspective.

The state of knowledge, limitations, and uncertainties on the potential CC impacts on PNAs and their resilience capacity come from the convergence of better understanding the physical science behind CC, different levels of vulnerability, and the acceptance of multiple challenges and uncertainties that future scenarios bring to conservation planning and to the wellbeing of human communities.

Author Contributions: Conceptualization, A.O.-R.; writing—original draft preparation, E.O.-M., H.L.R.-S., M.d.C.B. and C.A.-G.; writing—review and editing E.O.-M., M.d.C.B. and C.A.-G. All authors have read and agreed to the published version of the manuscript.

Funding: The study was part of the activities developed with the economic support of CONACYT (Centro de Investigaciones Biologicas del Noroeste) Basic Science Project 251919; Red Temática de Investigación Áreas Naturales Protegidas (RENANP) of the Mexican Council for Science and Technology (CONACYT) and CIBNOR.

Institutional Review Board Statement: Not applicable.

Informed Consent Statement: Not applicable.

Data Availability Statement: Not applicable.

Acknowledgments: The Authors thank the anonymous reviewers for their time and effort devoted and whose many useful suggestions improved this paper with their accurate and sound revisions and Diana D. de Fischer for English edition.

Conflicts of Interest: The authors declare no conflict of interest. The funders had no role in the design of the study; in the collection, analyses, or interpretation of data; in the writing of the manuscript or in the decision to publish the results.

·erences

Guerrero, E. México. El Paraíso de Los Pinos, Robles y Cactus. In *Las Áreas Protegidas de América Latina: Situación Actual y Perspectivas Para el Futuro*; Elbers, J., Ed.; UICN: Quito, Ecuador, 2011; pp. 69–78.

Olmos-Martínez, E.; González-Ávila, M.E.; Contreras-Loera, M.R. Percepción de La Población Frente Al Cambio Climático En Áreas Naturales Protegidas de Baja California Sur, México. *Polis Rev. Latinoam.* **2013**, *12*, 459–481. [CrossRef]

Arizpe, O.; Olmos-Martínez, E.; Ibáñez, R.M.; Armenta Martínez, L.F. Áreas Naturales Protegidas y Turismo Sustentable En Baja California Sur. In *Turismo, Desarrollo Económico y Sustentabilidad en Baja California Sur*; Juárez Mancilla, J., Cruz Chávez, P.R., Torres García, A.F., Cruz Chávez, G.R., Eds.; Universidad Autónoma de Sinaloa: Culiacan, Mexico, 2018.

Acreman, M.; Smith, A.; Charters, L.; Tickner, D.; Opperman, J.; Acreman, S.; Edwards, F.; Sayers, P.; Chivava, F. Evidence for the Effectiveness of Nature-Based Solutions to Water Issues in Africa. *Environ. Res. Lett.* **2021**, *16*, 063007. [CrossRef]

Rascón Palacio, E. Áreas Protegidas: Aproximación A Su Proyección Socio-Económica Y Política En Centroamérica. *Delos Desarro. Local Sosten.* **2010**, *3*, 13.

Sylvester, A.; Tate, M.; Johnstone, D. Beyond Synthesis: Re-Presenting Heterogeneous Research Literature. *Behav. Inf. Technol.* **2013**, *32*, 1199–1215. [CrossRef]

Green, B.N.; Johnson, C.D.; Adams, A. Writing Narrative Literature Reviews for Peer-Reviewed Journals: Secretsof the Trade. *J. Chiropr. Med.* **2006**, *5*, 101–117. [CrossRef]

Cooper, H.M. Organizing Knowledge Syntheses: A Taxonomy of Literature Reviews. *Knowl. Soc.* **1988**, *1*, 104–126. [CrossRef]

9. Herrmann-pillath, C.; Hiedanpää, J.; Soini, K. The Co-Evolutionary Approach to Nature-Based Solutions: A Concept Framework. *Nat.-Based Solut.* **2022**, *2*, 100011. [CrossRef]
10. Cárdenas, M.L.; Wilde, V.; Hagen-Zanker, A.; Seifert-Dähnn, I.; Hutchins, M.G.; Loiselle, S. The Circular Benefits of Participati in Nature-Based Solutions. *Sustainability* **2021**, *13*, 4344. [CrossRef]
11. Wallace, A.R. *Archipiélago Malayo*; Consejo Nacional para la Cultura y las Artes, Cien del Mundo: Mexico City, Mexico, 1869.
12. Ansell, C.; Gash, A. Collaborative Governance in Theory and Practice. *J. Public Adm. Res. Theory* **2008**, *18*, 543–571. [CrossRef
13. Soulé, M.E. (Ed.) *Conservation Biology: The Science of Scarcity and Diversity*; Sinauer Associates: Sunderland, MA, USA, 1986.
14. Wilson, E.O. (Ed.) *Biodiversity*; National Academy Press: Washington, DC, USA, 1988.
15. Cooper, H.D.; Noonan-Mooney, K. Convention on Biological Diversity. In *Encyclopedia of Biodiversity*, 2nd ed.; Levin, S.A., E Elsevier Academic Press. Secretariat of the Convention on Biological Diversity: Montreal, QC, Canada, 2013; pp. 306–3 [CrossRef]
16. Bierregaard, R.O.; Gascon, C.; Lovejoy, T.E.; Mesquita, R. (Eds.) *Lessons from Amazonia: The Ecology and Conservation of a Fragmen Forest*; Yale University Press: New Haven, CT, USA, 2001.
17. Soulé, M.E.; Simberloff, D. What Do Genetics and Ecology Tell Us about the Design of Nature Reserves? *Biol. Conserv.* **1986**, 19–40. [CrossRef]
18. Macarthur, R.H.; Wilson, E.O. *The Theory of Island Biogeography*; Princeton University Press: Princeton, NJ, USA, 1967.
19. Simberloff, D.S.; Abele, L.G. Island Biogeography Theory and Conservation Practice. *Science* **1976**, *191*, 285–286. [CrossR [PubMed]
20. Shaffer, M.L. Minimum Population Sizes for Species Conservation. *Bioscience* **1981**, *31*, 131–134. [CrossRef]
21. Newmark, W.D. A Land-Bridge Island Perspective on Mammalian Extinctions in Western North American Parks. *Nature* **19** *325*, 430–432. [CrossRef] [PubMed]
22. UNFCCC. Adoption of the Paris Agreement. Available online: https://unfccc.int/sites/default/files/english_paris_agreeme pdf (accessed on 26 March 2022).
23. Schaller, G.B. *A Naturalist and Other Beasts: Tales from a Life in the Field*; Sierra Club Books: San Francisco, CA, USA, 2007.
24. Daily, G. Introduction: What Are Ecosystem Services. In *Nature's Services: Societal Dependence on Natural Ecosystems*; Daily, G., E Island Press: Washington, DC, USA; Covelo, CA, USA, 1997; pp. 1–10.
25. Chang, K.; Winter, K.B.; Lincoln, N.K. Hawai'i in Focus: Navigating Pathways in Global Biocultural Leadership. *Sustainabi* **2019**, *11*, 283. [CrossRef]
26. Opoku, A. Biodiversity and the Built Environment: Implications for the Sustainable Development Goals (SDGs). *Resour. Cons Recycl.* **2019**, *141*, 1–7. [CrossRef]
27. UNDP. Sustainable Development Goals. Available online: https://www.undp.org/sustainable-development-goals (accessed 27 March 2022).
28. Borrini-Feyerabend, G. *Collaborative Management of Protected Areas: Tailoring the Approach to the Context*; Issues in Social Poli IUCN: Gland, Switzerland, 1996.
29. Ullah, I.; Kim, D.Y. A Model of Collaborative Governance for Community-Based Trophy-Hunting Programs in Developi Countries. *Perspect. Ecol. Conserv.* **2020**, *18*, 145–160. [CrossRef]
30. Biodiversity Collaborative Group. *Report of the Biodiversity Collaborative Group*; Biodiversity (Land and Freshwater) Stakeholc Trust: Wellington, New Zeland, 2018.
31. Xu, H.; Cao, Y.; Yu, D.; Cao, M.; He, Y.; Gill, M.; Pereira, H.M. Ensuring Effective Implementation of the Post-2020 Glol Biodiversity Targets. *Nat. Ecol. Evol.* **2021**, *5*, 411–418. [CrossRef]
32. Williams, B.A.; Watson, J.E.M.; Butchart, S.H.M.; Ward, M.; Brooks, T.M.; Butt, N.; Bolam, F.C.; Stuart, S.N.; Mair, McGowan, P.J.K.; et al. A Robust Goal Is Needed for Species in the Post-2020 Global Biodiversity Framework. *Conse Lett.* **2020**, *14*, e12778. [CrossRef]
33. Nicholson, E.; Watermeyer, K.E.; Rowland, J.A.; Sato, C.F.; Stevenson, S.L.; Andrade, A.; Brooks, T.M.; Burgess, N.D.; Cheng, S Grantham, H.S.; et al. Scientific Foundations for an Ecosystem Goal, Milestones and Indicators for the Post-2020 Glol Biodiversity Framework. *Nat. Ecol. Evol.* **2021**, *5*, 1338–1349. [CrossRef]
34. Nature. Biodiversity Faces Its Make-or-Break Year. *Nature* **2022**, *601*, 1. [CrossRef]
35. Nunes, N.; Björner, E.; Hilding-Hamann, K.E. Guidelines for Citizen Engagement and the Co-Creation of Nature-Based Solutio Living Knowledge in the URBiNAT Project. *Sustainability* **2021**, *13*, 13378. [CrossRef]
36. Beeproject.Science. Available online: https://www.beeproject.science/croppollination.html (accessed on 27 March 2022).
37. Winter, K.B.; Ticktin, T.; Quazi, S.A. Biocultural Restoration in Hawai'i Also Achieves Core Conservation Goals. *Ecol. Soc.* **20** *25*, 18. [CrossRef]
38. Ibañez, J.J.; Muñiz, J. De Cazadores Recolectores Nómadas a Agricultores y Ganaderos Sedentarios. Available online: htt //fundacionpalarq.com/de-cazadores-recolectores-nomadas-a-agricultores-y-ganaderos-sedentarios/ (accessed on 27 Mar 2022).
39. Ahmad, H.I.; Ahmad, M.J.; Jabbir, F.; Ahmar, S.; Ahmad, N.; Elokil, A.A.; Chen, J. The Domestication Makeup: Evoluti Survival, and Challenges. *Front. Ecol. Evol.* **2020**, *8*, 103. [CrossRef]

Groß, D.; Piezonka, H.; Corradini, E.; Schmölcke, U.; Zanon, M.; Dörfler, W.; Dreibrodt, S.; Feeser, I.; Krüger, S.; Lübke, H.; et al. Adaptations and Transformations of Hunter-Gatherers in Forest Environments: New Archaeological and Anthropological Insights. *Holocene* **2019**, *29*, 1531–1544. [CrossRef]
Luminet, J.-P. Creation, Chaos, Time: From Myth to Modern Cosmology. *Cosmology* **2016**, *24*, 501–515.
Meghji, S. *The Innovative Technology That Powered the Inca*; BBC: London, UK, 2021.
De la Maza Elvira, R. Una Historia de Las Áreas Naturales Protegidas En México. *Gac. Ecol.* **1995**, *51*, 15–68.
Givens, H. Nezahualcoyotl: History and Historiography. 2014. Available online: https://hannahgivens.wordpress.com/2014/05/23/nezahualcoyotl-history-and-historiography/ (accessed on 1 March 2022).
Berdan, F.F. *The Aztecs. Lost Civilizations*, 1st ed.; Reaktion Books Ltd.: London, UK, 2021.
Castañeda, R.J. Las Áreas Naturales Protegidas de México: De Su Origen Precoz a Su Consolidación Tardía. *Scr. Nova Rev. Electrón. Geogr. Cienc. Soc.* **2006**, *10*, 1138–9788.
Vásquez Sánchez, M.Á. Conservación de La Naturaleza y Áreas Naturales Protegidas En Territorios de Los Pueblos Originarios de La Frontera Sur de México. *Soc. Ambient.* **2017**, *15*, 117–130. [CrossRef]
Bahia de Aguiar, P.C.; dos Santos Moreau, A.M.S.; de Oliveira Fontes, E. Protected Natural Areas: A Brief History of the Emergence of National Parks and Extractive Reserves. *Rev. Geogr. Am. Cent.* **2013**, *50*, 195–213.
Chavarría, Y.O.; Martínez García, A.L.; Ortíz, C.E.; Goyenechea, I. Evolución En La Selección de Áreas Protegidas En El Continente Americano: El Caso de Estados Unidos, México y Costa Rica. *Boletín Científico del Inst. Cienc. Básicas e Ing.* **2019**, *7*, 47–53.
IUCN. International Union for Conservation of Nature and United Nations Environment Programme. "Protected Areas: About". Available online: https://www.iucn.org/theme/protected-areas/about (accessed on 1 April 2022).
Sheail, J. *Nature's Spectacle, The World's First National Parks and Protected Places*, 1st ed.; Taylor & Francis: London, UK, 2010. [CrossRef]
Dixon, W. Los 10 Parques Nacionales Más Antiguos Del Mundo. Available online: https://www.yourtripagent.com/1723-10-oldest-national-parks-in-world (accessed on 27 March 2022).
Djossa, C. These Are the World's First National Parks. Available online: https://www.nationalgeographic.com/travel/national-parks/article/worlds-first-protected-lands-conservation-yellowstone (accessed on 27 March 2022).
Rahoof, K.K.A. Protected Areas: The Past, Present, and Future of Conservation. Available online: https://earth.org/protected-areas-the-past-present-and-future-of-conservation/ (accessed on 27 March 2022).
Secretariat of the Convention on Biological Diversity. *Protected Areas in Todays' World: Their Values and Benefits for the Welfare of the Planet*; Secretariat of the Convention on Biological Diversity: Montreal, QC, Canada, 2008; Volume 43.
Prip, C. The Convention on Biological Diversity as a Legal Framework for Safeguarding Ecosystem Services. *Ecosyst. Serv.* **2018**, *29*, 199–204. [CrossRef]
Convention on Biological Diversity. *Decision UNEP/CDB/COP/DEC/X/2 Adopted by the Conference of the Parties to the Convention on Biological Diversity at Its Tenth Meeting. Decision X/2. Strategic Plan for Biodiversity 2011–2020 and the Aichi Biodiversity Targets*; Secretariat of the Convention on Biological Diversity: Montreal, QC, Canada, 2010.
Jones, K.R.; Venter, O.; Fuller, R.A.; Allan, J.R.; Maxwell, S.L.; Negret, P.J.; Watson, J.E.M. One-Third of Global Protected Land Is under Intense Human Pressure. *Science* **2018**, *360*, 788–791. [CrossRef] [PubMed]
Pullin, A.S.; Bangpan, M.; Dalrymple, S.; Dickson, K.; Haddaway, N.R.; Healey, J.R.; Hauari, H.; Hockley, N.; Jones, J.P.G.; Knight, T.; et al. Human Well-Being Impacts of Terrestrial Protected Areas. *Environ. Evid.* **2013**, *2*, 19. [CrossRef]
Hannah, L.; Midgley, G.; Andelman, S.; Araújo, M.; Hughes, G.; Martinez-Meyer, E.; Pearson, R.; Williams, P. Protected Area Needs in a Changing Climate. *Front. Ecol. Environ.* **2007**, *5*, 131–138. [CrossRef]
Kumar, A. CBD's Global Biodiversity Outlook 5: Final Assessment of Aichi Biodiversity Targets and Beyond. In *Asian Biotechnology Development Review*; Chaturvedi, S., Ed.; Elsevier: New Delhi, India, 2020; Volume 22, pp. 5–20.
Forero-Medina, G.; Joppa, L. Representation of Global and National Conservation Priorities by Colombia's Protected Area Network. *PLoS ONE* **2010**, *5*, e13210. [CrossRef]
Pimm, S.L.; Jenkins, C.N.; Abell, R.; Brooks, T.M.; Gittleman, J.L.; Joppa, L.N.; Raven, P.H.; Roberts, C.M.; Sexton, J.O. The Biodiversity of Species and Their Rates of Extinction, Distribution, and Protection. *Science* **2014**, *344*, 1246752. [CrossRef]
Pimm, S.L.; Jenkins, C.N.; Li, B.V. How to Protect Half of Earth to Ensure It Protects Sufficient Biodiversity. *Sci. Adv.* **2018**, *4*, eaat2616. [CrossRef]
Sayre, R.; Karagulle, D.; Frye, C.; Boucher, T.; Wolff, N.H.; Breyer, S.; Wright, D.; Martin, M.; Butler, K.; Van Graafeiland, K.; et al. An Assessment of the Representation of Ecosystems in Global Protected Areas Using New Maps of World Climate Regions and World Ecosystems. *Glob. Ecol. Conserv.* **2020**, *21*, e00860. [CrossRef]
Pörtner, H.O.; Scholes, R.J.; Agard, J.; Archer, E.; Arneth, A.; Bai, X.; Barnes, D.; Burrows, M.; Chan, L.; Cheung, W.L.; et al. *Scientific Outcome of the IPBES-IPCC Co-Sponsored Workshop Report on Biodiversity and Climate Change*; IPBES Secretariat: Bonn, Germany, 2021. [CrossRef]
Cantú-Salazar, L.; Gaston, K.J. Species Richness and Representation in Protected Areas of the Western Hemisphere: Discrepancies between Checklists and Range Maps. *Divers. Distrib.* **2013**, *19*, 782–793. [CrossRef]
Hazen, H.D.; Anthamatten, P.J. Representation of Ecological Regions by Protected Areas at the Global Scale. *Phys. Geogr.* **2004**, *25*, 499–512. [CrossRef]

69. Polak, T.; Watson, J.E.M.; Bennett, J.R.; Possingham, H.P.; Fuller, R.A.; Carwardine, J. Balancing Ecosystem and Threate Species Representation in Protected Areas and Immplications for Nations Achieving Global Conservation Goals. *Conserv. I* **2016**, *9*, 438–445. [CrossRef]
70. Olmos-Martínez, E.; Rodríguez Rodríguez, G.; Salas, S.; Ortega-Rubio, A. Efecto de La Implementación de Un Área Proteg Sobre El Bienestar de Comunidades Rurales de Baja California Sur. In *Las Áreas Naturales Protegidas y la Investigación Científic México. (pp. 249–282)*; Ortega–Rubio, A., Pinkus-Rendón, M.J., Espitia-Moreno, I.C., Eds.; Centro de Investigaciones Biológi del Noroeste S. C., La Paz, Universidad Autónoma de Yucatán, Mérida, Yucatán y Universidad Michoacana de San Nicolás Hidalgo: Morelia, Mexico, 2015; p. 572.
71. Olmos-Martínez, E.; Ortega-Rubio, A. Socio-Ecological Studies in Natural Protected Areas. Linking Community Developm and Conservation in Mexico. In *Socio-Ecological Studies in Natural Protected Areas*; Ortega-Rubio, A., Ed.; Springer: La Paz, Mex 2020; p. 625. [CrossRef]
72. Challenger, A.; Bocco, G.; Equihua, M.; Lazos Chavero, E.; Maass, M. La Aplicación Del Concepto Del Sistema Socioecológ Alcances, Posibilidades y Limitaciones En La Gestión Ambiental de México. *Investig. Ambient. Cienc. Política Pública* **2014**, *6*, 1-
73. Salas-Zapata, W.A.; Ríos-Osorio, L.A.; Álvarez-Del Castillo, J. Conceptual Bases for a Classification of Socioecological System Sustainability Research. *Rev. Lasallista Investig.* **2011**, *8*, 136–142.
74. Parathian, H. Understanding Cosmopolitan Communities in Protected Areas: A Case Study from the Colombian Ama *Conserv. Soc.* **2019**, *17*, 26–37. [CrossRef]
75. Abukari, H.; Mwalyosi, R.B. Local Communities' Perceptions about the Impact of Protected Areas on Livelihoods and Commu Development. *Glob. Ecol. Conserv.* **2020**, *22*, e00909. [CrossRef]
76. Harris, N.C.; Mills, K.L.; Harissou, Y.; Hema, E.M.; Gnoumou, I.T.; VanZoeren, J.; Abdel-Nasser, Y.I.; Doamba, B. First Cam Survey in Burkina Faso and Niger Reveals Human Pressures on Mammal Communities within the Largest Protected A Complex in West Africa. *Conserv. Lett.* **2019**, *12*, e12667. [CrossRef]
77. Adamik, S.; Berros, M.V. Áreas Naturales Protegidas En El Litoral Argentino: Un Análisis Comparativo De Las Regulacio Vigentes. *Derechos En Acción* **2021**, *19*, 410–436. [CrossRef]
78. Karanth, K.K.; Kudalkar, S.; Jain, S. Re-Building Communities: Voluntary Resettlement from Protected Areas in India. *Front. E Evol.* **2018**, *6*, 183. [CrossRef]
79. Mora-De la Mora, G.D.L. Sociopolitical Approach for the Analysis of Conservation Policies in Urban Contexts: Betw Environmental Services and Protected Natural Areas. *Perf. Latinoam.* **2019**, *27*, 53. [CrossRef]
80. Mutanga, C.N.; Vengesayi, S.; Gandiwa, E.; Muboko, N. Community Perceptions of Wildlife Conservation and Tourism: A C Study of Communities Adjacent to Four Protected Areas in Zimbabwe. *Trop. Conserv. Sci.* **2015**, *8*, 564–582. [CrossRef]
81. Olmos-Martínez, E.; Arizpe, O.; Maldonado-Alcudia, C.; Roldán-Clarà, B. Conservation of Biodiversity vs Tourism and Fishin the Archipelago Espiritu Santo in the Gulf of California. In *Mexican Natural Resources Management and Biodiversity Conserva Recent Case Studies*; Ortega-Rubio, A., Ed.; Springer International Publishing AG: Cham, Switzerland, 2018. [CrossRef]
82. Olmos-Martínez, E.; Ibarra-Michel, J.P.; Velarde-Valdez, M. Socio-Ecological Effects of Government and Community Collabora Work to Local Development in a Natural Protected Area. In *Socio-Ecological Studies in Natural Protected Areas*; Ortega-Rubio, Ed.; Springer International Publishing AG: Cham, Switzerland, 2020; pp. 511–535. [CrossRef]
83. Verdugo, M.D.; Arizpe-Covarrubias, O.; Olmos Martinez, E. Evaluación de La Vulnerabilidad Como Elemento de Manejo Áreas Naturales Protegidas: Parque Nacional Bahía de Loreto, México. In *Estudios Recientes Sobre Economía Ambiental y Agrícol México*; Hernández Trejo, V., Valdivia Alcalá, R., Cruz Chávez, P.R., Cruz Chávez, G.R., Eds.; Universidad Autónoma de B California Sur, Universidad Autónoma Chapingo: La Paz, Mexico, 2019; pp. 325–344.
84. Andrade, G.S.M.; Rhodes, J.R. Protected Areas and Local Communities: An Inevitable Partnership toward Successful Conser tion Strategies? *Ecol. Soc.* **2012**, *17*, 17. [CrossRef]
85. Sarr, B.; González-Hernández, M.M.; Boza-Chirino, J.; de León, J. Understanding Communities' Disaffection to Participate Tourism in Protected Areas: A Social Representational Approach. *Sustainability* **2020**, *12*, 3677. [CrossRef]
86. Olaniyi, O.E.; Akinsorotan, O.A.; Zakaria, M.; Martins, C.O.; Adebola, S.I.; Oyelowo, O.J. Taking the Edge off Host Communi Dependence on Protected Areas in Nigeria. *IOP Conf. Ser. Earth Environ. Sci.* **2019**, *269*, 13. [CrossRef]
87. Fay, D.A. Mutual Gains and Distributive Ideologies in South Africa: Theorizing Negotiations between Communities and Protec Areas. *Hum. Ecol.* **2008**, *35*, 81–95. [CrossRef]
88. Saxon, E. Noah's Parks: A Partial Antidote to the Anthropocene Extinction Event. *Biodiversity* **2008**, *9*, 5–10. [CrossRef]
89. Stolton, S.; Dudley, N.; Çokçalışkan, B.A.; Hunter, D.; Ivanić, K.-Z.; Kanga, E.; Kettunen, M.; Kumagai, Y.; Maxted, N.; Senio et al. Values and Benefits of Protected Areas. In *Protected Area Governance and Management*; Worboys, G.L., Lockwood, Kothari, A., Feary, S., Pulsford, I., Eds.; ANU Press: Canberra, Australia, 2015; pp. 145–168.
90. Lovejoy, T.E. Protected Areas: A Prism for a Changing World. *Trends Ecol. Evol.* **2006**, *21*, 329–333. [CrossRef]
91. Watson, J.E.M.; Dudley, N.; Segan, D.B.; Hockings, M. The Performance and Potential of Protected Areas. *Nature* **2014**, *515*, 67- [CrossRef]
92. IPBES. *Summary for Policymakers of the Global Assessment Report on Biodiversity and Ecosystem Services of the Intergovernme Science-Policy Platform on Biodiversity and Ecosystem Services*; Díaz, S., Settele, J., Brondízio, E.S., Ngo, H.T., Guèze, M., Agard Arneth, A., Balvanera, P., Brauman, K.A., Butchart, S.H.M., et al., Eds.; IPBES Secretariat: Bonn, Germany, 2019.

Geldmann, J.; Coad, L.; Barnes, M.; Craigie, I.D.; Hockings, M.; Knights, K.; Leverington, F.; Cuadros, I.C.; Zamora, C.; Woodley, S.; et al. Changes in Protected Area Management Effectiveness over Time: A Global Analysis. *Biol. Conserv.* **2015**, *191*, 692–699. [CrossRef]
Hannah, L.; Midgley, G.F.; Lovejoy, T.; Bond, W.J.; Bush, M.; Lovett, J.C.; Scott, D.; Woodward, F.I. Conservation of Biodiversity in a Changing Climate. *Conserv. Biol.* **2002**, *16*, 264–268. [CrossRef]
Monzón, J.; Moyer-Horner, L.; Palamar, M.B. Climate Change and Species Range Dynamics in Protected Areas. *Bioscience* **2011**, *61*, 752–761. [CrossRef]
Halpin, P.N. Global Climate Change and Natural-Area Protection: Management Responses and Research Directions. *Ecol. Appl.* **1997**, *7*, 828–843. [CrossRef]
Hannah, L. Protected Areas and Climate Change. *Ann. N. Y. Acad. Sci.* **2008**, *1134*, 201–212. [CrossRef] [PubMed]
Joppa, L.N.; Pfaff, A. High and Far: Biases in the Location of Protected Areas. *PLoS ONE* **2010**, *4*, e8273. [CrossRef] [PubMed]
Brooks, T.M.; Bakarr, M.I.; Boucher, T.; Da Fonseca GA, B.; Hilton-Taylor, C.; Hoekstra, J.M.; Moritz, T.; Olivieri, S.; Parrish, J.; Pressey, R.L.; et al. Coverage Provided by the Global Protected-Area System: Is It Enough? *Bioscience* **2004**, *54*, 1081–1091. [CrossRef]
Rodrigues, A.S.L.; Akçakaya, H.R.; Andelman, S.J.; Bakarr, M.I.; Boitani, L.; Brooks, T.M.; Chanson, J.S.; Fishpool, L.D.C.; Da Fonseca, G.A.B.; Gaston, K.J.; et al. Global Gap Analysis: Priority Regions for Expanding the Global Protected-Area Network. *Bioscience* **2004**, *54*, 1092–1100. [CrossRef]
IPCC. Summary for Policymakers. In *Climate Change 2021: The Physical Science Basis. Contribution of Working Group I to the Sixth Assessment Report of the Intergovernmental Panel on Climate Change*; Masson-Delmotte, V., Zhai, P., Pirani, A., Connors, S.L., Péan, C., Berger, S., Caud, N., Chen, Y., Goldfarb, L., Gomis, M.I., et al., Eds.; Cambridge University Press: Cambridge, UK, 2021.
IPCC. *Climate Change and Land: An IPCC Special Report on Climate Change, Desertification, Land Degradation, Sustainable Land Management, Food Security, and Greenhouse Gas Fluxes in Terrestrial Ecosystems*; Shukla, P.R., Skea, J., Buendia, E.C., Masson-Delmotte, V., Pörtner, H.-O., Roberts, D.C., Zhai, P., Slade, R., Connors, S., van Diemen, R., et al., Eds.; Intergovernmental Panel on Climate Change: Geneva, Switzerland, 2019.
Parmesan, C. Ecological and Evolutionary Responses to Recent Climate Change. *Annu. Rev. Ecol. Evol. Syst.* **2006**, *37*, 637–669. [CrossRef]
Root, T.L.; Price, J.T.; Hall, K.R.; Schneider, S.H.; Rosenzweig, C.; Pounds, J.A. Fingerprints of Global Warming on Wild Animals and Plants. *Nature* **2003**, *421*, 57–60. [CrossRef]
Walther, G.R.; Post, E.; Convey, P.; Menzel, A.; Parmesan, C.; Beebee, T.J.; Fromentin, J.M.; Hoegh-Guldberg, O.; Bairlein, F. Ecological Responses to Recent Climate Change. *Nature* **2002**, *416*, 389–395. [CrossRef]
Scheffers, B.R.; De Meester, L.; Bridge, T.C.L.; Hoffmann, A.A.; Pandolfi, J.M.; Corlett, R.T.; Butchart, S.H.M.; Pearce-Kelly, P.; Kovacs, K.M.; Dudgeon, D.; et al. The Broad Footprint of Climate Change from Genes to Biomes to People. *Science* **2016**, *354*, aaf7671. [CrossRef]
Mason, S.C.; Palmer, G.; Fox, R.; Gillings, S.; Hill, J.K.; Thomas, C.D.; Oliver, T.H. Geographical Range Margins of Many Taxonomic Groups Continue to Shift Polewards. *Biol. J. Linn. Soc.* **2015**, *115*, 586–597. [CrossRef]
Thomas, C.D. Climate, Climate Change and Range Boundaries. *Divers. Distrib.* **2010**, *16*, 488–495. [CrossRef]
Kelly, A.E.; Goulden, M.L. Rapid Shifts in Plant Distribution with Recent Climate Change. *Proc. Natl. Acad. Sci. USA* **2008**, *105*, 11823–11826. [CrossRef] [PubMed]
Peters, R.L.; Myers, J.P. Preserving Biodiversity in a Changing Climate. *Issues Sci. Technol.* **1991**, *8*, 66–72.
Peters, R.L.; Darling, J.D.S. The Greenhouse Effect and Nature Reserves. Global Warming Would Diminish Biological Diversity by Causing Extinctions among Reserve Species. *Bioscience* **1985**, *35*, 707–717. [CrossRef]
Alagador, D.; Cerdeira, J.O.; Araújo, M.B. Shifting Protected Areas: Scheduling Spatial Priorities under Climate Change. *J. Appl. Ecol.* **2014**, *51*, 703–713. [CrossRef]
Araújo, M.B.; Cabeza, M.; Thuiller, W.; Hannah, L.; Williams, P.H. Would Climate Change Drive Species out of Reserves? An Assessment of Existing Reserve-Selection Methods. *Glob. Chang. Biol.* **2004**, *10*, 1618–1626. [CrossRef]
Lawler, J.J.; Ruesch, A.S.; Olden, J.D.; Mcrae, B.H. Projected Climate-Driven Faunal Movement Routes. *Ecol. Lett.* **2013**, *16*, 1014–1022. [CrossRef]
Shoo, L.P.; Williams, S.E.; Hero, J.M. Detecting Climate Change Induced Range Shifts: Where and How Should We Be Looking? *Austral Ecol.* **2006**, *31*, 22–29. [CrossRef]
Brooker, R.W.; Travis, J.M.J.; Clark, E.J.; Dytham, C. Modelling Species' Range Shifts in a Changing Climate: The Impacts of Biotic Interactions, Dispersal Distance and the Rate of Climate Change. *J. Theor. Biol.* **2007**, *245*, 59–65. [CrossRef]
Pearson, R.G. Climate Change and the Migration Capacity of Species. *Trends Ecol. Evol.* **2006**, *21*, 111–113. [CrossRef]
Soto, C.G. The Potential Impacts of Global Climate Change on Marine Protected Areas. *Rev. Fish Biol. Fish.* **2002**, *11*, 181–195. [CrossRef]
Watson, J.E.M.; Rao, M.; Kang, A.L.; Xie, Y. Climate Change Adaptation Planning for Biodiversity Conservation: A Review. *Adv. Clim. Chang. Res.* **2012**, *3*, 1–11. [CrossRef]
Popescu, V.D.; Rozylowicz, L.; Cogălniceanu, D.; Niculae, I.M.; Cucu, A.L. Moving into Protected Areas? Setting Conservation Priorities for Romanian Reptiles and Amphibians at Risk from Climate Change. *PLoS ONE* **2013**, *8*, e79330. [CrossRef] [PubMed]

121. Lemieux, C.J.; Beechey, T.J.; Gray, P.A. Prospects for Canada's Protected Areas in an Era of Rapid Climate Change. *Land Use Po* **2011**, *28*, 928–941. [CrossRef]
122. Heller, N.E.; Zavaleta, E.S. Biodiversity Management in the Face of Climate Change: A Review of 22 Years of Recommendatic *Biol. Conserv.* **2009**, *142*, 14–32. [CrossRef]
123. Araújo, M.B.; Rahbek, C. How Does Climate Change Affect Biodiversity? *Science* **2006**, *313*, 1396–1397. [CrossRef]
124. Lemieux, C.J.; Scott, D.J. Climate Change, Biodiversity Conservation and Protected Area Planning in Canada. *Can. Geogr.* **20** *49*, 384–397. [CrossRef]
125. Abrahms, B.; Di Pietro, D.; Graffis, A.; Hollander, A. Managing Biodiversity under Climate Change: Challenges, Framewor and Tools for Adaptation. *Biodivers. Conserv.* **2017**, *26*, 2277–2293. [CrossRef]
126. Barr, S.L.; Larson, B.M.H.; Beechey, T.J.; Scott, D.J. Assessing Climate Change Adaptation Progress in Canada's Protected Ar *Can. Geogr.* **2021**, *65*, 152–165. [CrossRef]
127. Day, J.C.; Roff, J.C. *Planning for Representative Marine Protected Areas: A Framework for Canada's Oceans*; Laughren, J., Ed.; Wo Wildlife Fund Canada: Toronto, ON, Canada, 2000.
128. Roff, J.C.; Taylor, M.E. National Frameworks for Marine Conservation, a Hierarchical Geophysical Approach. *Aquat. Cons Mar. Freshw. Ecosyst.* **2000**, *10*, 209–223. [CrossRef]
129. Levin, S.A. Ecosystems and the Biosphere as Complex Adaptive Systems. *Ecosystems* **1998**, *1*, 431–436. [CrossRef]
130. Thomas, C.D.; Gillingham, P.K. The Performance of Protected Areas for Biodiversity under Climate Change. *Biol. J. Linn. S* **2015**, *115*, 718–730. [CrossRef]
131. Lawrence, A.; Hoffmann, S.; Beierkuhnlein, C. Topographic Diversity as an Indicator for Resilience of Terrestrial Protected Ar against Climate Change. *Glob. Ecol. Conserv.* **2021**, *25*, e01445. [CrossRef]
132. Walker, B.; Holling, C.S.; Carpenter, S.R.; Kinzig, A. Resilience, Adaptability and Transformability in Social–Ecological Syste *Ecol. Soc.* **2004**, *9*. Available online: http://www.ecologyandsociety.org/vol9/iss2/art5 (accessed on 7 April 2022). [CrossRe
133. Carroll, C.; Roberts, D.R.; Michalak, J.L.; Lawler, J.J.; Nielsen, S.E.; Stralberg, D.; Haann, A.; Mcrae, B.H.; Wang, T. Scale-Depend Complementarity of Climatic Velocity and Environmental Diversity for Identifying Priority Areas for Conservation under Clim Change. *Glob. Chang. Biol.* **2017**, *23*, 4508–4520. [CrossRef]
134. IPCC. *The Regional Impacts of Climate Change—An Assessment of Vulnerability*, 1st ed.; Watson, R.T., Zinyoweera, M.C., Moss, R. Eds.; Cambridge University Press: Cambridge, UK, 1998; Volume 88. [CrossRef]
135. Warren, R.; Arnell, N.; Nicholls, R.; Levy, P.; Price, J. *Understanding the Regional Impacts of Climate Change. Research Report Prepa for the Stern Review on the Economics of Climate Change*; Tyndall Centre: Norwich, UK, 2006.
136. Ackerly, D.D.; Loarie, S.R.; Cornwell, W.K.; Weiss, S.B.; Hamilton, H.; Branciforte, R.; Kraft, N.J.B. The Geography of Clim Change: Implications for Conservation Biogeography. *Divers. Distrib.* **2010**, *16*, 476–487. [CrossRef]
137. Mackey, B.G.; Watson, J.E.M.; Hope, G.; Gilmore, S. Climate Change, Biodiversity Conservation, and the Role of Protected Are An Australian Perspective. *Biodiversity* **2008**, *9*, 11–18. [CrossRef]
138. Nwe, T.; Zomer, R.J.; Corlett, R.T. Projected Impacts of Climate Change on the Protected Areas of Myanmar. *Climate* **2020**, *8*, [CrossRef]
139. Groves, C.R.; Game, E.T.; Anderson, M.G.; Cross, M.; Enquist, C.; Ferdaña, Z.; Girvetz, E.; Gondor, A.; Hall, K.R.; Higgins, J.; et Incorporating Climate Change into Systematic Conservation Planning. *Biodivers. Conserv.* **2012**, *21*, 1651–1671. [CrossRef]
140. van Kerkhoff, L.; Munera, C.; Dudley, N.; Guevara, O.; Wyborn, C.; Figueroa, C.; Dunlop, M.; Hoyos, M.A.; Castiblanco, Becerra, L. Towards Future-Oriented Conservation: Managing Protected Areas in an Era of Climate Change. *Ambio* **2019**, 699–713. [CrossRef]
141. McCarty, J.P. Consecuencias Biológicas de Cambios Climáticos Recientes. *Conserv. Biol.* **2001**, *15*, 320–331. [CrossRef]
142. Bellard, C.; Bertelsmeier, C.; Leadley, P.; Thuiller, W.; Courchamp, F. Impacts of Climate Change on the Future of Biodivers *Ecol. Lett.* **2012**, *15*, 365–377. [CrossRef] [PubMed]
143. Fischman, R.L.; Meretsky, V.J.; Babko, A.; Kennedy, M.; Liu, L.; Robinson, M.; Wambugu, S. Planning for Adaptation to Clim Change: Lessons from the US National Wildlife Refuge System. *Bioscience* **2014**, *64*, 993–1005. [CrossRef]
144. Urban, M.C. Accelerating Extinction Risk from Climate Change. *Science* **2015**, *348*, 571–573. [CrossRef]
145. Thomas, C.D.; Cameron, A.; Green, R.E.; Bakkenes, M.; Beaumont, L.J.; Collingham, Y.C.; Erasmus, B.F.N.; Ferreira De Siqueira, Grainger, A.; Hannah, L.; et al. Extinction Risk from Climate Change. *Nature* **2004**, *427*, 145–148. [CrossRef]
146. West, J.M.; Salm, R.V. Resistance and Resilience to Coral Bleaching: Implications for Coral Reef Conservation and Manageme *Conserv. Biol.* **2003**, *17*, 956–967. [CrossRef]
147. Keller, B.D.; Gleason, D.F.; McLeod, E.; Woodley, C.M.; Airamé, S.; Causey, B.D.; Friedlander, A.M.; Grober-Dunsmore, Johnson, J.E.; Miller, S.L.; et al. Climate Change, Coral Reef Ecosystems, and Management Options for Marine Protected Are *Environ. Manag.* **2009**, *44*, 1069–1088. [CrossRef]
148. Done, T.; Whetton, P.; Jones, R.; Berkelmans, R.; Lough, J.; Skirving, W.; Wooldridge, S. *Global Climate Change and Coral Bleachi on the Great Barrier Reef. Final Report to the State of Queensland Greenhouse Taskforce through the Department of Natural Resources a Mines*; AIMS: Townsville, Australia, 2003.
149. IPBES. *Global Assessment Report of the Intergovernmental Science-Policy Platform on Biodiversity and Ecosystem Services*; Brondízio, E Settele, J., Díaz, S., Ngo, H.T., Eds.; IPBES Secretariat: Bonn, Germany, 2019.

Foden, W.B.; Garcia, R.A.; Platts, P.J.; Carr, J.A.; Hoffmann, A.A.; Visconti, P. *Selecting and Evaluating CCVA Approaches and Methods*; Wiley: Hoboken, NJ, USA, 2016.
Smit, B.; Pilifosova, O.; Burton, I.; Challenger, B.; Huq, S.; Klein, R.J.; Yohe, G. Adaptation to Climate Change in the Context of Sustainable Development and Equity. In *Climate Change 2001: Impacts, Adaptation, and Vulnerability. Working Group II. IPCC*; McCarthy, J., Canziani, O., Leary, N., Dokken, D., White, K., Eds.; Cambridge University Press: Cambridge, UK, 2001; Volume 8, pp. 879–912.
Still, S.M.; Frances, A.L.; Treher, A.C.; Oliver, L. Using Two Climate Change Vulnerability Assessment Methods to Prioritize and Manage Rare Plants: A Case Study. *Nat. Areas J.* **2015**, *35*, 106–121. [CrossRef]
Williams, S.E.; Shoo, L.P.; Isaac, J.L.; Hoffmann, A.A.; Langham, G. Towards an Integrated Framework for Assessing the Vulnerability of Species to Climate Change. *PLoS Biol.* **2008**, *6*, 2621–2626. [CrossRef]
Hansen, L.J.; Biringer, J.L.; Hoffman, J.R. *Buying Time: A User ´s Manual for Building Resistance and Resilience to Climate Change in Natural Systems*; Hansen, L.J., Biringer, J.L., Hoffman, J.R., Eds.; WWF: Gland, Switzerland, 2003.
Smit, B.; Wandel, J. Adaptation, Adaptive Capacity and Vulnerability. *Glob. Environ. Chang.* **2006**, *16*, 282–292. [CrossRef]
Tanner-McAllister, S.L.; Rhodes, J.; Hockings, M. Managing for Climate Change on Protected Areas: An Adaptive Management Decision Making Framework. *J. Environ. Manag.* **2017**, *204*, 510–518. [CrossRef] [PubMed]
Adger, W.N.; Huq, S.; Brown, K.; Conwaya, D.; Hulmea, M. Adaptation to Climate Change in the Developing World. *Prog. Dev. Stud.* **2003**, *3*, 179–195. [CrossRef]
Sosa-Rodríguez, F. Política Del Cambio Climático En México: Avances, Obstáculos y Retos. Real. Datos y Espac. *Rev. Int. Estadística y Geogr.* **2015**, *6*, 4–23.
Comisión Nacional de Áreas Naturales Protegidas. *Estrategia de Cambio Climático Desde Las Áreas Naturales Protegidas: Una Convocatoria Para la Resiliencia de México (2015–2020)*, 1st ed.; Secretaria de Medio Ambiente y Recursos Naturales: Ciudad de Mméxico, Mexico, 2015.
Coleman, M.A.; Palmer-Brodie, A.; Kelaher, B.P. Conservation Benefits of a Network of Marine Reserves and Partially Protected Areas. *Biol. Conserv.* **2013**, *167*, 257–264. [CrossRef]
Dudley, N.; Stolton, S.; Belokurov, A.; Krueger, L.; Lopoukhine, N.; MacKinnon, K.; Sandwith, T.; Sekhran, N. (Eds.) *Natural Solutions: Protected Areas Helping People to Cope with Climate Change*, 1st ed.; IUCN-WCPA, TNC, UNDP, WCS: Gland, Switzerland; The World Bank: Washington, DC, USA; WWF: New York, NY, USA, 2010. [CrossRef]
Mckellar, R.; Midgley, G.F.; Yates, C.J.; Abbott, I.; GIOIA, P.; Le Maitre, D. The Need to Develop a Coherent Research Approach for Climate Change Vulnerability Impact Assessment and Adaptation in High-Biodiversity Terrestrial Ecosystems. *Austral Ecol.* **2010**, *35*, 371–373. [CrossRef]
Baron, J.S.; Gunderson, L.; Allen, C.D.; Fleishman, E.; McKenzie, D.; Meyerson, L.A.; Oropeza, J.; Stephenson, N. Options for National Parks and Reserves for Adapting to Climate Change. *Environ. Manag.* **2009**, *44*, 1033–1042. [CrossRef]
Diniz-Filho, J.A.F.; Bini, L.M.; Rangel, T.F.; Loyola, R.D.; Hof, C.; Nogués-Bravo, D.; Araújo, M.B. Partitioning and Mapping Uncertainties in Ensembles of Forecasts of Species Turnover under Climate Change. *Ecography* **2009**, *32*, 897–906. [CrossRef]
Peterson, G.D.; Cumming, G.S.; Carpenter, S.R. Scenario Planning: A Tool for Conservation in an Uncertain World. *Conserv. Biol.* **2003**, *17*, 358–366. [CrossRef]
Dawson, T.P.; Jackson, S.T.; House, J.I.; Prentice, I.C.; Mace, G.M. Beyond Predictions: Biodiversity Conservation in a Changing Climate. *Science* **2011**, *332*, 53–58. [CrossRef]
Navarro-racines, C.; Tarapues, J.; Thornton, P.; Jarvis, A.; Ramirez-Villegas, J. High-Resolution and Bias-Corrected CMIP5 Projections for Climate Change Impact Assessments. *Sci. Data* **2020**, *7*, 7. [CrossRef]
Stein, B.A.; Staudt, A.; Cross, M.S.; Dubois, N.S.; Enquist, C.; Griffis, R.; Hansen, L.J.; Hellmann, J.J.; Lawler, J.J.; Nelson, E.J.; et al. Preparing for and Managing Change: Climate Adaptation for Biodiversity and Ecosystems. *Front. Ecol. Environ.* **2013**, *11*, 502–510. [CrossRef]
Bowen-Jones, E.; Entwistle, A. Identifying Appropriate Flagship Species: The Importance of Culture and Local Contexts. *Oryx* **2002**, *36*, 189–195. [CrossRef]

diversity

MDPI

Article

Ethnobotanical Knowledge and the Distribution of Pesticidal Plants in Northern Tanzania: A Multi-Ethnic Perspective

Immaculate K. Constantine [1,2,*], Richard A. Giliba [2] and Angela G. Mkindi [2]

[1] National Herbarium of Tanzania (NHT), Tanzania Plant Health and Pesticides Authority (TPHPA), Ngaramtoni Area, Arusha P.O. Box 3024, Tanzania
[2] School of Life Science and Bioengineering, The Nelson Mandela African Institution of Science and Technology (NM-AIST), Tengeru, Arusha P.O. Box 447, Tanzania; richard.giliba@nm-aist.ac.tz (R.A.G.); angela.mkindi@nm-aist.ac.tz (A.G.M.)
* Correspondence: constantinei@nm-aist.ac.tz

Abstract

Pesticidal plants are vital for pest management in sub-Saharan Africa, yet knowledge about them is culturally and geographically unevenly documented. This study examined ethnobotanical knowledge and the distribution of pesticidal plants among six ethnic groups (Maasai, Chagga, Iraqw, Pare, Nguu, Zigua) in northern Tanzania. Data related to ethnobotanical knowledge were collected from questionnaire surveys involving 266 participants, 24 focus group discussions, 26 key informant interviews, complemented by field verifications across the six ethnic groups. Pesticidal plant coordinates were accessed from herbarium voucher specimens from The National Herbarium of Tanzania. Chi-square tests of independence assessed associations between ethnic groups and knowledge transmission pathways. Penalized logistic regression analysis was conducted to assess the influence of demographic factors on reported knowledge of pesticidal plants. Spatial overlay was conducted to examine the distribution of pesticidal plant species occurrences across agroecological zones and rainfall gradients. The results revealed a significant association between ethnic group and the source of pesticidal plant knowledge. Across all ethnic groups, knowledge was predominantly acquired through family/community traditional sources, with the highest frequencies recorded among the Maasai, Iraqw, and Zigua. Knowledge is mainly transmitted orally, particularly among the Maasai, Iraqw, and Zigua. A total of one hundred and six distinct species were recorded across the six ethnic groups surveyed, with *Tephrosia vogelii* and *Solanum incanum* being the most frequently cited. Leaves were the most frequently used plant part across all ethnic groups, with notably high usage among the Chagga, Iraqw, and Maasai. Perceptions of the declining population of pesticidal plants were the highest among the Maasai. Spatial mapping revealed pesticidal plant hotspots in the Northern Rift and Volcanic Highlands agroecological zones, and they fall within zones receiving moderate to relatively high rainfall. The findings highlight that ethnobotanical knowledge of pesticidal plants in northern Tanzania is strongly shaped by ethnic affiliation, oral knowledge transmission, and localized ecological availability, with clear spatial hotspots aligned to specific agroecological zones and high-rainfall areas.

Keywords: pesticidal plants; indigenous knowledge; ethnic groups; cultural diversity; agroecological zones; northern Tanzania

 https://doi.org/10.3390/d18050

1. Introduction

In sub-Saharan Africa, agricultural production faces persistent challenges from pest infestations that threaten food security, livelihoods, and ecological balance [1,2]. Although synthetic pesticides are widely used for pest control [1,3–5], they are often inaccessible to smallholder farmers and are costly, and their application raises major concerns related to environmental degradation, pest resistance, and adverse effects on human and animal health [6–9]. In common with numerous agricultural regions, farmers have historically utilized indigenous plants with pesticidal attributes to safeguard their crops and mitigate losses. Additionally, there is a growing global interest in sustainable pest management practices rooted in traditional ecological knowledge systems [3].

Among many rural communities in Africa, the use of pesticidal plants is a long-established practice embedded within local knowledge and cultural traditions [2–4]. Ethnobotanical studies have extensively documented using plant-derived substances for pest control across diverse indigenous societies [10]. These pesticidal plants, which contain bioactive compounds with insecticidal, fungicidal, bactericidal, herbicidal, or nematocidal properties, serve as effective and environmentally friendly alternatives to chemical pesticides [1,8,11]. Commonly referred to as botanical pesticides, these substances are typically accessible, cost-effective, and biodegradable, making them especially valuable for resource-constrained smallholder farmers [11].

The knowledge associated with these plants, ranging from their identification to harvesting, preparation, and application, is intrinsically linked to cultural knowledge and ethnic identity. As such, understanding local knowledge creates the basis for advancing scientific research and promoting sustainable agricultural practices [12,13]. However, such knowledge remains under-documented and faces increasing erosion due to the expansion of modern agricultural systems, synthetic pesticide use, and the Westernization of younger generations [4,14]. Cultural diversity, reflected in differences in traditions, languages, and ecological practices among ethnic groups, influences how communities interact with natural resources [15]. These cultural systems also shape local or indigenous ecological knowledge transmitted orally through generations and represent a rich reservoir for sustainable development and ecological adaptation [15,16].

Within this context, ethnobotanical knowledge is understood as a biocultural system in which biological and cultural diversity are closely interconnected, with plant use and management shaped by culturally specific knowledge developed through long-term interactions between communities and their environments [17]. Within this biocultural diversity framework, variations in pesticidal plant knowledge among ethnic groups reflect differences in livelihoods, belief systems, and historical relationships with local ecosystems [17,18]. Traditional ecological knowledge (TEK) further conceptualizes the use of pesticidal plants as cumulative knowledge and practices transmitted across generations through direct environmental engagement, particularly within agro-pastoral and smallholder farming systems [19]. These knowledge systems are dynamic and subject to both erosion and adaptation, influenced by modernization, land-use change, and increased reliance on synthetic pesticides [20].

In this biocultural and TEK-based framework, the significant plant and ethnic biodiversity of Tanzania provides an ideal setting for investigating the relationship between culture and pest control methods. Previous research has highlighted the use of pesticidal plants by agro-pastoral communities in regions like Mbulu, yet much of this knowledge remains fragmented or at risk of being lost due to oral modes of transmission, declining intergenerational knowledge transfer, and increased reliance on synthetic pesticides [7,21]. In the northern regions of Tanzania, home to ethnic groups such as the Maasai, Chagga, Iraqw, Pare, Nguu, and Zigua, indigenous knowledge systems continue to support eco-

https://doi.org/10.3390/d18050253

logical farming, even as they face threats from socio-economic transitions and modern agricultural inputs [15,22]. The gradual erosion of this oral heritage, compounded by declining intergenerational transfer and language shifts, highlights the urgency of the documentation and protection of this knowledge [23]. Given these dynamics, this study aims to examine ethnobotanical knowledge of pesticidal plants among different ethnic groups in northern Tanzania by identifying the predominant sources of knowledge; assessing the influence of age, gender, education level, and ethnicity on knowledge possession; and evaluating the patterns of plant part use and access. Specifically, this study seeks to analyze ethnic variation in harvesting and utilization practices, document the local perceptions of pesticidal plant population trends as indicators of conservation status and usage pressure, and map the distribution of commonly used pesticidal plant species occurrences across agroecological zones and rainfall gradients.

2. Materials and Methods

2.1. Study Area

This study was conducted in northern Tanzania, which geographically spans latitudes 0.5° S to 11.5° S and longitudes 29.5° E to 40.5° E (Figure 1). The landscape encompasses a range of agroecological zones, including the Volcanic Highlands, Northern Rift Valley, Eastern Plateaux and Mountain Blocks, Central Plateau, and Undifferentiated Rocky Terrain, each supporting distinct vegetation types and ethnobotanical practices. The elevation ranges from 600 to over 5000 m above sea level. Climatic conditions vary widely: rainfall ranges between 500 and 1400 mm, and cooler temperatures (10–20 °C) prevail in highlands, while lowland zones are relatively warmer and drier. This diversity in altitude and climate fosters high biodiversity, including various pesticidal plant species [24]. Vegetation types across these zones reflect ecological gradients, such as the Volcanic Highlands being dominated by montane forests, bamboo thickets, and highland grasslands. The Northern Rift Valley features dry savannahs, open woodlands, and acacia-dominated grasslands, with riparian vegetation along seasonal rivers. The Eastern Plateaux and Mountain Blocks comprise miombo woodlands, bushlands, and scattered montane forest patches. The Central Plateau is characterized by semi-arid grasslands, dry woodlands, and scattered shrubs, whereas Undifferentiated Rocky Terrain supports sparse vegetation composed of hardy shrubs, succulents, and scattered trees. This heterogeneity provides a rich source of plant species traditionally used by local communities for pest management. Agriculture and livestock keeping are the dominant economic activities, with communities growing maize, beans, sorghum, bananas, vegetables, and coffee while also engaging in cattle rearing [21]. Ethnically, the region is home to the Maasai, Iraqw, Chagga, Pare, Zigua, and Barabaig communities, each with unique ethnobotanical knowledge and agricultural–cultural practices. An ethnobotanical survey was conducted across six districts: Longido (inhabited by Maasai), Monduli (inhabited by Maasai), Same (inhabited by Pare), Hai (inhabited by Chagga), Kiteto (inhabited by Maasai, Nguu, Zigua), and Mbulu (inhabited by Iraqw), located in the Arusha, Kilimanjaro, and Manyara regions.

https://doi.org/10.3390/d180502

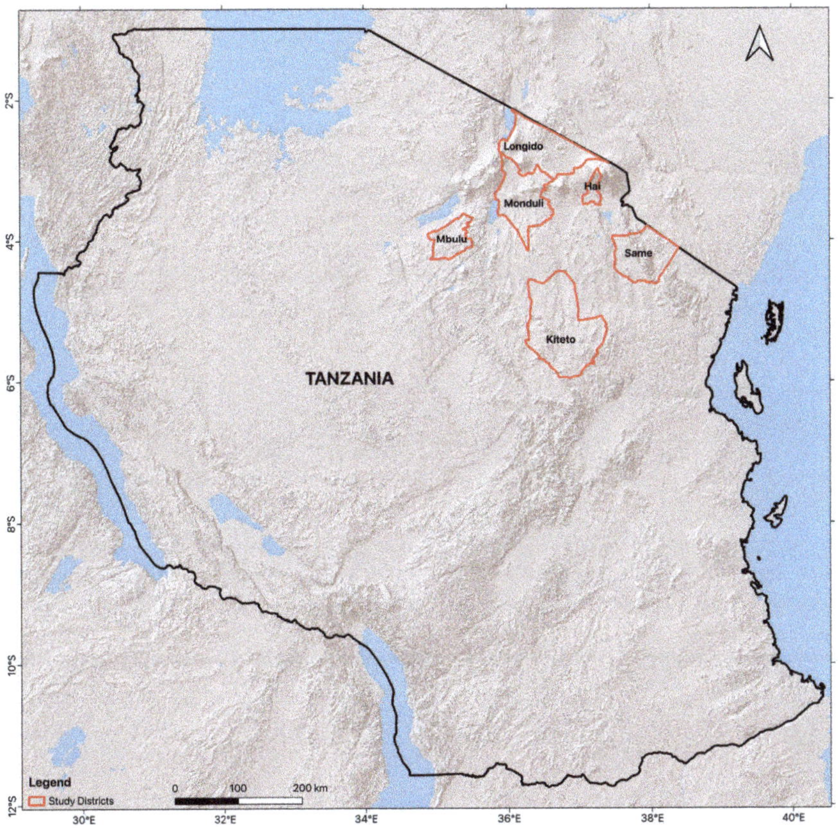

Figure 1. The location of study districts in northern Tanzania where ethnobotanical data on pesticidal plants were collected. The highlighted districts, Hai, Same, Monduli, Longido, Mbulu, and Kiteto, represent diverse ecological zones and are predominantly inhabited by distinct ethnic groups, including the Chagga, Pare, Maasai, Iraqw, Nguu, and Zigua.

2.2. Study Design

A cross-sectional ethnobotanical study design was employed using purposive and snowball sampling techniques. Districts and villages were purposively selected based on ecological diversity, ethnic representation, and the presence of traditional pesticidal plant use. Initial participants from agricultural extension officers, traditional healers, experienced farmers, and community elders were purposively selected for interviews based on their recognized expertise in the use of pesticidal plants. Subsequently, snowball sampling was used to identify additional knowledgeable respondents.

2.3. Data Collection

2.3.1. Ethnobotanical Interviews

Data were collected between October and December 2024 using questionnaire surveys (S4) (n = 266), focus group discussions (FGDs; n = 24 groups), key informant interviews (KIIs; n = 26), and field verifications across the six ethnic groups. Data collected included plant local names, plant parts used, preparation and application methods, perceived effectiveness, sources of knowledge, access pathways, and perceived population trends in pesticidal plants. Data were recorded digitally using Kobo Toolbox (https://www.kobotoolbox.org/). Questionnaires were prepared in English, translated into Swahili, and

https://doi.org/10.3390/d18050253

further translated into local languages where necessary to ensure comprehension. All participants provided oral informed consent prior to the interviews and their participation in the study.

2.3.2. Species Verification and Occurence Data

Plant species reported during interviews were identified and collected in the field with the assistance of a botanist and local collaborators to prepare herbarium specimens. Collected specimens were cross-referenced with those held at the National Herbarium of Tanzania (NHT) and the Flora of Tropical East Africa (FTEA) to verify taxonomic identity. Scientific names were further confirmed using the online database Plants of the World Online (POWO, https://powo.science.kew.org/). Species occurence coordinates data were obtained from the geo-referenced herbarium specimens.

2.4. Data Analysis

Descriptive statistics were used to summarize respondent demographics, and plant species used as pesticidal plants.

Chi-square or Fisher tests of independence were used to assess associations between ethnic groups and (i) knowledge transmission pathways, (ii) plant parts used, (iii) sources of plant materials, and (iv) perceptions of pesticidal plant population trends. Penalized logistic regression analysis was conducted to assess the influence of demographic factors (age, gender, education and ethnic) on reported knowledge of pesticidal plants. Analyses were conducted in R version 4.3 at a 95% confidence level. QGIS (https://qgis.org/) was used to visualize the occurrence patterns of the pesticidal plant species across agroecological zones and rainfall gradients. Thematic analysis was used to analyze the qualitative data obtained from key informant interviews and focus group discussions using the approach in [25].

3. Results

3.1. The Demographic Characteristics of the Respondents

Two hundred and sixty-six individuals were interviewed to assess ethnobotanical knowledge and the use of pesticidal plants across various ethnic groups. The distribution among ethnic groups was 36.1% Maasai (96), 22.6% Iraqw (60), 21.4% Chagga (57), 10.2% Pare (27), 5.6% Zigua (15), and 4.1% Nguu (11). Gender distribution comprised 139 males (52.3%) and 127 females (47.7%). Educational attainment was as follows: primary, 67.7%; secondary, 19.1%; no formal education, 8.3%; and tertiary, 4.9%. Age distribution was 9.4% 14–23 years, 28.2% 24–39 years, 30.5% 40–53 years, 19.9% 54–67 years, and 12% 68–80 years. Most respondents reported multiple roles, with the majority being farmers at 85.7% (228), followed by herdsmen at 29.3% (78), business persons at 3.4% (9), those with no job at 1.5% (4), executive officers at 1.1% (3), teachers at 0.8% (2), students at 0.8% (2), and livestock officers at 0.4% (1).

3.2. Sources and Determinants of Knowledge on Pesticidal Plants

The sources of knowledge on pesticidal plants varied significantly across ethnic groups ($\chi^2 = 144.46$, df = 10, $p < 0.001$; Figure 2). Family/community tradition emerged as the dominant source of knowledge among Maasai, Iraqw, Nguu, and Zigua respondents, with nearly all participants in these groups ($\geq$98%) reporting reliance on intergenerational transmission. This quantitative pattern aligns closely with qualitative evidence from key informants. As one Maasai elder explained, "*We learned about these plants from our fathers and grandfathers. It is part of our tradition. We don't write it, we live it.*" This statement reinforces the statistical finding that knowledge in these communities is primarily orally transmitted and embedded within cultural practice rather than formal institutions.

https://doi.org/10.3390/d18050

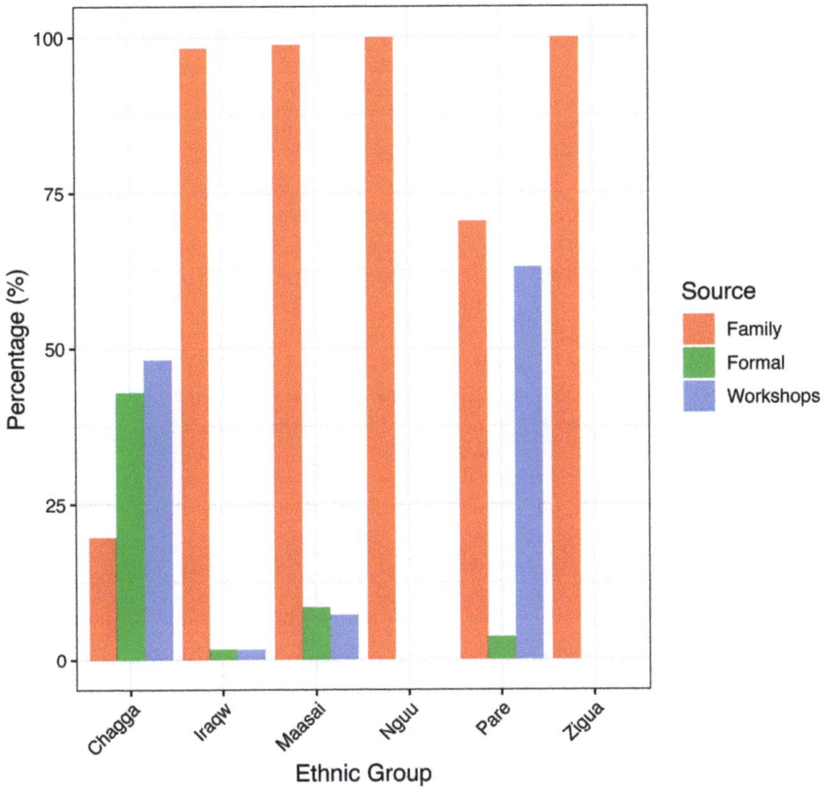

Figure 2. Sources of knowledge on pesticidal plants across ethnic groups (n = 246). Significant variation was observed among groups. Intergenerational transmission through family and community traditions predominated among Maasai, Iraqw, Nguu, and Zigua respondents, whereas Chagga and Pare communities reported greater reliance on formal education and workshops, reflecting integration of traditional and institutional knowledge systems.

In contrast, Chagga respondents reported substantially higher reliance on formal education (42.9%) and workshops or training (48.2%), indicating a more institutionalized knowledge pathway. This divergence is consistent with focus group discussions from Hai District, where participants noted that "*agricultural officers and NGOs have occasionally provided training on pesticidal plant use, especially for vegetable farming*." The quantitative data therefore reflect the greater exposure of Chagga communities to extension services and organized training programs, suggesting the integration of traditional and formal knowledge systems.

Among Pare respondents, both workshops or training (63.0%) and family/community tradition (70.4%) were frequently reported, indicating a hybrid knowledge system where traditional transmission coexists with formal agricultural outreach. This mixed pattern suggests that Pare communities may be transitioning toward more structured knowledge dissemination mechanisms while retaining cultural foundations. Notably, respondents from the Nguu and Zigua ethnic groups reported exclusive reliance on family and community-based traditional knowledge, with no reported formal education or workshop-based training related to pesticidal plants. Furthermore, qualitative insights from Kiteto indicated the intergenerational erosion of knowledge, as one elder observed, "*In these days, many of the youth have never even seen some of the pesticidal plants we grew up using. They*

https://doi.org/10.3390/d18050253

now trust what they buy in shops over what they can grow or find." This perception suggests emerging shifts in knowledge preference and a possible decline in traditional knowledge transmission among younger generations, despite the high overall reliance on tradition shown in the quantitative data.

Penalized logistic regression analysis (Table 1) showed that age and gender were not significant predictors of possessing knowledge on pesticidal plants ($p > 0.05$), indicating that knowledge distribution does not differ significantly across age groups or between males and females. Education level, however, revealed mixed effects. Respondents with primary education were significantly more likely to report knowledge compared to those with no formal education, whereas secondary and tertiary education were not statistically significant predictors. The qualitative findings provide important contextual explanations for these patterns. During focus group discussions, several participants emphasized that individuals with only primary education often remain closely involved in farming and traditional livelihood activities, which enhances exposure to plant-based pest control practices. As one elder explained, "*Those who stopped at primary school are still in the fields with us. They know the plants because they use them.*"

Table 1. The penalized logistic regression results showing the effect of age, gender, ethnic group, and education level on knowledge of pesticidal plants. The model uses "female" as the reference category for gender, "Maasai" for ethnic group, and "no formal education" for education level.

	Odds Ratio	Low CI	Upper CI	*p*-Valu
(Intercept)	0.976	0.1	8.752	0.98
Age	1.014	0.972	1.064	0.53
Femaleas Referencecategory				
Male	1.946	0.6	6.596	0.26
Did not attend school as Referencecategory				
Primary education	5.712	1.239	29.266	0.026
Secondary education	0.977	0.201	4.66	0.97
Tertiary education	3.789	0.242	574.54	0.37
Maasai as Reference category				
Chagga	14.457	1.487	1943.433	0.017
Iraqw	5.476	1.159	53.538	0.03
Nguu	2.342	0.166	349.48	0.57
Pare	7.344	0.685	1012.147	0.1
Zigua	7.837	0.744	1082.426	0.09

* $p < 0.05$.

In contrast, younger and more formally educated participants were frequently described as having limited practical experience with pesticidal plants. An FGD participant in Hai District noted, "*Many of the youth who went to secondary school or college do not know these plants. They rely more on what they buy in shops than what we used traditionally.*" Similarly, another elder remarked the following: "*Modern education teaches chemicals and modern farming. It does not teach the plants we used before.*" These narratives help explain why higher levels of formal education did not correspond to increased ethnobotanical knowledge in the quantitative analysis. Instead, the findings suggest that experiential learning and continued engagement in agricultural practice may be more influential determinants of knowledge than formal schooling beyond the primary level.

Ethnic group emerged as a significant determinant of reported knowledge on pesticidal plants. Using Maasai respondents as the reference category, Chagga and Iraqw respondents were significantly more likely to report knowledge of pesticidal plants, indicating meaningful variation in the familiarity and use of botanical pest control practices

https://doi.org/10.3390/d180502

across ethnic groups. These differences appear to reflect variations in farming systems, crop specialization, and exposure to agricultural extension services.

The qualitative findings support this interpretation. In predominantly Chagga areas, participants described the active use of plant-based pesticides, particularly in vegetable farming systems. One participant explained, "*We were trained on how to prepare plant extracts for controlling pests in tomatoes and cabbages. It helps us reduce chemical use.*" Similarly, an Iraqw farmer noted, "*We use some plants to control storage pests and field insects. Some we learned from elders, others from agricultural officers.*" These accounts align with the significantly higher likelihood of pesticidal plant knowledge among Chagga and Iraqw respondents, suggesting that both traditional transmission and formal agricultural outreach contribute to knowledge retention in these communities.

Although Pare and Zigua respondents showed elevated odds relative to Maasai respondents, these differences were not statistically significant. Discussions suggested that pesticidal plant use in these communities may be present but less systematically practiced or less widely transmitted. In contrast, several Maasai participants indicated that while traditional plant knowledge exists, it is more strongly associated with veterinary or medicinal uses than crop pest control. As one elder stated, "*We know many plants for livestock and medicine, but for crop pests we often buy chemicals now.*" Nguu respondents did not differ significantly from Maasai respondents, and qualitative responses similarly reflected reliance on traditional knowledge systems with limited structured training on pesticidal plant preparation. Taken together, the quantitative and qualitative evidence suggests that ethnic differences in pesticidal plant knowledge are shaped by agricultural specialization, extension exposure, and the extent to which botanical pest control remains embedded within local farming practices.

3.3. Plant Species Used as a Pesticidal Plants

One hundred and six pesticidal plant taxa were recorded across the six ethnic groups surveyed, with a cumulative frequency of 688 mentions (S1). Among the most frequently cited species (Table 2), *Tephrosia vogelii* Hook.f. had the highest relative frequency of citation (RFC = 0.293) with 72 mentions, primarily from the Chagga (48) and Pare (23). Other widely cited taxa included *Tithonia diversifolia* (Hemsl.) A.Gray (57 citations, RFC = 0.232), *Croton dichogamus* Pax (54, RFC = 0.22), *Tagetes minuta* L. (37, RFC = 0.15), and *Solanum incanum* L. (36, RFC = 0.146). Those with over 20 mentions across multiple ethnic groups included *Tithonia diversifolia* (Hemsl.) A.Gray, *Tagetes minuta* L., *Azadirachta indica* A.Juss, *Capsicum frutescens* L., *Lantana camara* L., and *Aloe volkensii* Engl., although their distribution varied substantially among communities (χ^2 = 562.34, df = 40, $p < 0.001$, Table 2).

The qualitative findings provide context for these citation patterns. During focus group discussions, Chagga elders emphasized the importance of *Tephrosia vogelii*, describing it as follows: "*The plant our fathers always used for bean insects.*" They further explained that the leaves were harvested from farms and dried in shaded conditions to enhance potency before application. Iraqw participants similarly highlighted the role of *Croton dichogamus*, noting that "*our ancestors kept it near the farm boundaries for controlling pests and protecting stored grain*". In contrast, Maasai informants described the use of *Solanum incanum* primarily in livestock management, particularly for protection against ectoparasites, reflecting differences in livelihood specialization across communities.

https://doi.org/10.3390/d18050253

Table 2. Frequency of citation (FC) and relative frequency of citation (RFC) of major pesticidal plant taxa recorded across six ethnic groups in northern Tanzania.

Taxa	Chagga	Iraqw	Maasai	Nguu	Pare	Zigua	FC	RFC
Tephrosia vogelii Hook.f.	48	0	1	0	23	0	72	**0.293**
Tithonia diversifolia (Hemsl.) A.Gray	41	0	0	0	16	0	57	**0.232**
Croton dichogamus Pax	0	54	0	0	0	0	54	**0.22**
Tagetes minuta L.	6	20	1	0	4	6	37	**0.15**
Solanum incanum L.	1	12	19	0	4	0	36	**0.146**
Azadirachta indica A.Juss	3	3	6	5	13	5	35	**0.142**
Capsicum frutescens L.	4	1	15	0	11	0	31	**0.126**
Lantana camara L.	29	0	0	0	1	0	30	**0.122**
Aloe volkensii Engl.	1	0	13	1	11	1	27	**0.11**
Total	**133**	**90**	**55**	**6**	**83**	**12**	**379**	
$\chi^2 = 562.34$, df = 40, $p < 0.001$								

3.4. Pesticidal Plant Part Preferences

A Chi-square test was conducted to determine whether the use of pesticidal plant parts differed among the six surveyed ethnic groups (Chagga, Iraqw, Maasai, Nguu, Pare, and Zigua). The results revealed variation between ethnic affiliation and the plant part used ($\chi^2 = 70.754$, df = 25, $p < 0.001$; Table 3). Leaves were the most frequently used plant part across all ethnic groups, with particularly high citation among Chagga (56 mentions), Iraqw (53), and Maasai (52) respondents. In contrast, bark, seeds, roots, and whole plants were cited less frequently overall but displayed marked ethnic variation. The Maasai exhibited the greatest diversity in plant part utilization, reporting relatively high use of bark (20 citations), seeds (11), roots (8), and whole plants (12). Conversely, Zigua and Nguu respondents demonstrated a narrower pattern of use, relying predominantly on leaves and reporting minimal use of roots or seeds.

Table 3. Frequency of plant parts used for pesticidal purposes across six ethnic groups in study area.

Ethnic Group	Whole Plant	Leaves	Bark	Seeds	Roots	Fruits	Total
Chagga	1	56	2	0	1	3	**63**
Iraqw	5	53	1	4	1	1	**65**
Maasai	12	52	20	11	8	5	**108**
Nguu	1	4	0	1	1	0	**7**
Pare	2	25	0	1	2	5	**35**
Zigua	0	14	0	0	0	0	**14**
Total	**21**	**204**	**23**	**17**	**13**	**14**	**292**
$\chi^2 = 70.754$, df = 25, $p < 0.001$							

The qualitative findings from FGDs provide additional insight into these patterns. A Maasai elder explained, "*Bark and roots are strong medicines for stubborn pests on cattle, but we know not to take too much from one tree.*" This statement highlights both the perceived potency and awareness of sustainable harvesting practices. Similarly, Chagga respondents emphasized the regenerative advantage of leaf harvesting, with one participant noting, "*We only pick leaves as they grow back quickly, and the plant lives on.*" These narratives reinforce the quantitative findings, suggesting that plant part selection is influenced not only by efficacy but also by sustainability considerations and livelihood needs.

To further explore practical applications, fifteen plant species with the highest frequency of citation were selected for detailed analysis. These species were examined in terms of plant parts used, preparation methods, modes of administration, targeted pests,

https://doi.org/10.3390/d18050

perceived efficacy, and reported limitations (S2). This deeper characterization provides insight into how pesticidal plant knowledge translates into applied pest management practices across communities.

3.5. Access and Sourcing Pathways of Pesticidal Plants

The sourcing of pesticidal plants varied across ethnic groups (χ^2 = 19.957; df = 10; *p*-value = 0.02966), with the wild being the predominant source overall (Figure 3). Of the total 280 reported sourcing instances, 207 (73.9%) originated from wild habitats, 65 (23.2%) from home gardens, and only 8 (2.9%) from markets. Across all ethnic groups, wild collection dominated. The Maasai reported the highest number of wild-sourced plants (69), followed by the Iraqw (53) and Chagga (46). Home garden sourcing was also notable among the Maasai (21) and Chagga (23). Pare respondents reported moderate reliance on both wild (22) and home gardens (11).

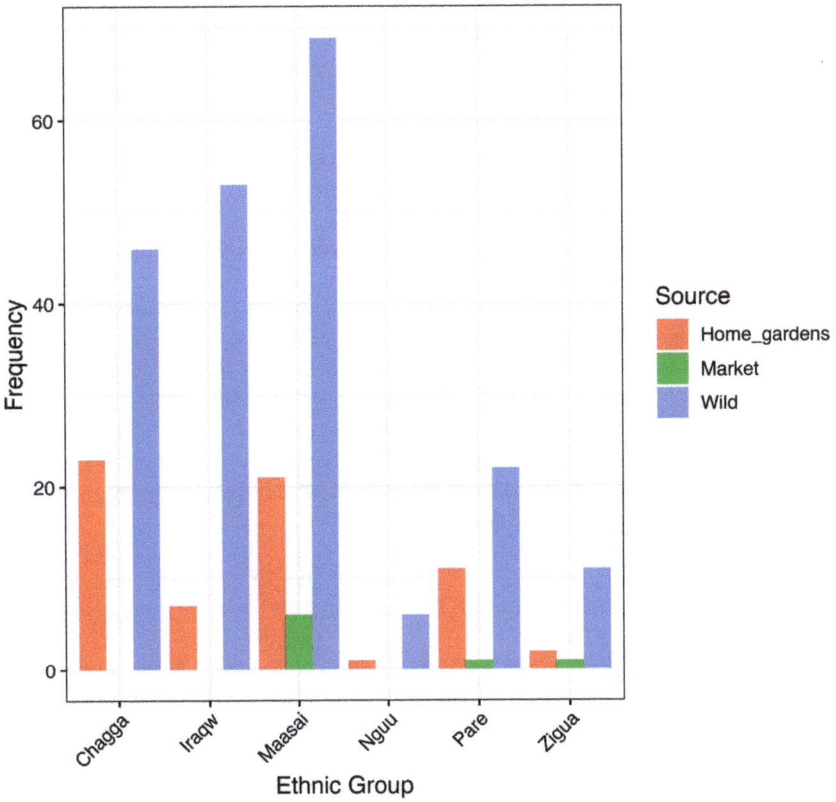

Figure 3. Percentage distribution of pesticidal plant sourcing pathways by ethnic group, showing variation in reliance on wild collection, home gardens, and markets.

Market acquisition was minimal across all groups, with only eight total mentions. The Maasai recorded the highest number of market-sourced plants (6), while Pare and Zigua each reported one instance. Chagga, Iraqw, and Nguu respondents reported no market-based sourcing. The qualitative findings from FGDs reinforce these patterns. Several participants emphasized the continued importance of wild landscapes as primary sources of pesticidal plants. A Maasai elder explained, "*Most of these plants grow in the bush. We collect them when we need them; we do not usually buy them.*" Similarly, an Iraqw participant

https://doi.org/10.3390/d18050253

noted, "*These plants are around our farms and forests. We know where to find them.*" In contrast, Chagga respondents highlighted the increasing role of home gardens for commonly used species. One participant stated, "*Some plants like Tephrosia and Tithonia we grow near the house so that they are easy to access.*" Market sourcing was rarely mentioned, and when it was, it was generally associated with species that are either scarce locally or already commercialized. As one Pare respondent observed, "*Sometimes we buy neem from the market, but most other plants we collect ourselves.*" To sum up, the quantitative and qualitative findings indicate that pesticidal plant use remains strongly embedded within local ecological systems, with limited dependence on formal markets.

3.6. The Perceptions of the Pesticidal Plants' Population Status

The results showed an association between ethnic affiliation and the perceptions of pesticidal plant population trends ($\chi^2 = 28.554$, df = 10, $p < 0.001$; Table 4).

Table 4. Perceptions of pesticidal plant population trends (declining, increasing, or stable) among six ethnic groups.

Population of Pesticidal Plants	Maasai	Chagga	Iraqw	Nguu	Pare	Zigua	Total
Declining	39	11	12	2	6	7	77
Increasing	37	41	39	3	14	6	**140**
Remained stable	7	4	8	2	7	1	**29**
Total	**83**	**56**	**59**	**7**	**27**	**14**	**246**
$\chi^2 = 28.554$, df = 10, $p < 0.001471$							

Across all respondents, the majority perceived pesticidal plant populations to be increasing (140 out of 246 responses), with this perception most frequently reported by the Chagga (41 responses) and Iraqw (39). In contrast, the Maasai reported the highest number of perceptions indicating declining plant populations (39 responses). Key informant interviews supported this concern, with one elder noting, "*Some plants we used to find easily are now disappearing because the land is changing and more people are using chemicals*". The perceptions of stable populations were comparatively rare across all ethnic groups, with responses ranging from one to eight per group.

3.7. Distribution of Pesticidal Plant Occurrences Across Agroecological Zones and Rainfall Gradients

The distribution of pesticidal plant occurrences across the study area shows a clear ecological pattern associated with agroecological zones and rainfall gradients. Most records are concentrated within the Northern Rift Zone and Volcanic Highlands as well as the Eastern Plateaux and Mountain Blocks, while relatively fewer occurrences are observed in the Central Plateau and Undifferentiated Rocky Terrain (Figure 4). Districts such as Same and Monduli exhibit pronounced clustering, particularly in volcanic and montane landscapes. These areas are characterized by fertile soils, favorable microclimates, and high plant biodiversity, which may support a greater diversity of species producing bioactive secondary metabolites with pesticidal properties. Overlaying occurrence data with annual rainfall further reveals that most pesticidal plant records fall within zones receiving moderate to relatively high rainfall (approximately 800–1200 mm annually, Figure 5). Areas with lower rainfall show comparatively fewer occurrences. This suggests that moisture availability may play a role in supporting plant diversity and biomass productivity, thereby increasing the likelihood of encountering pesticidal species. Generally, the findings indicate that agroecological conditions, rainfall variability and volcanic soil fertility collectively influence the spatial distribution of pesticidal plants in the region.

https://doi.org/10.3390/d18050

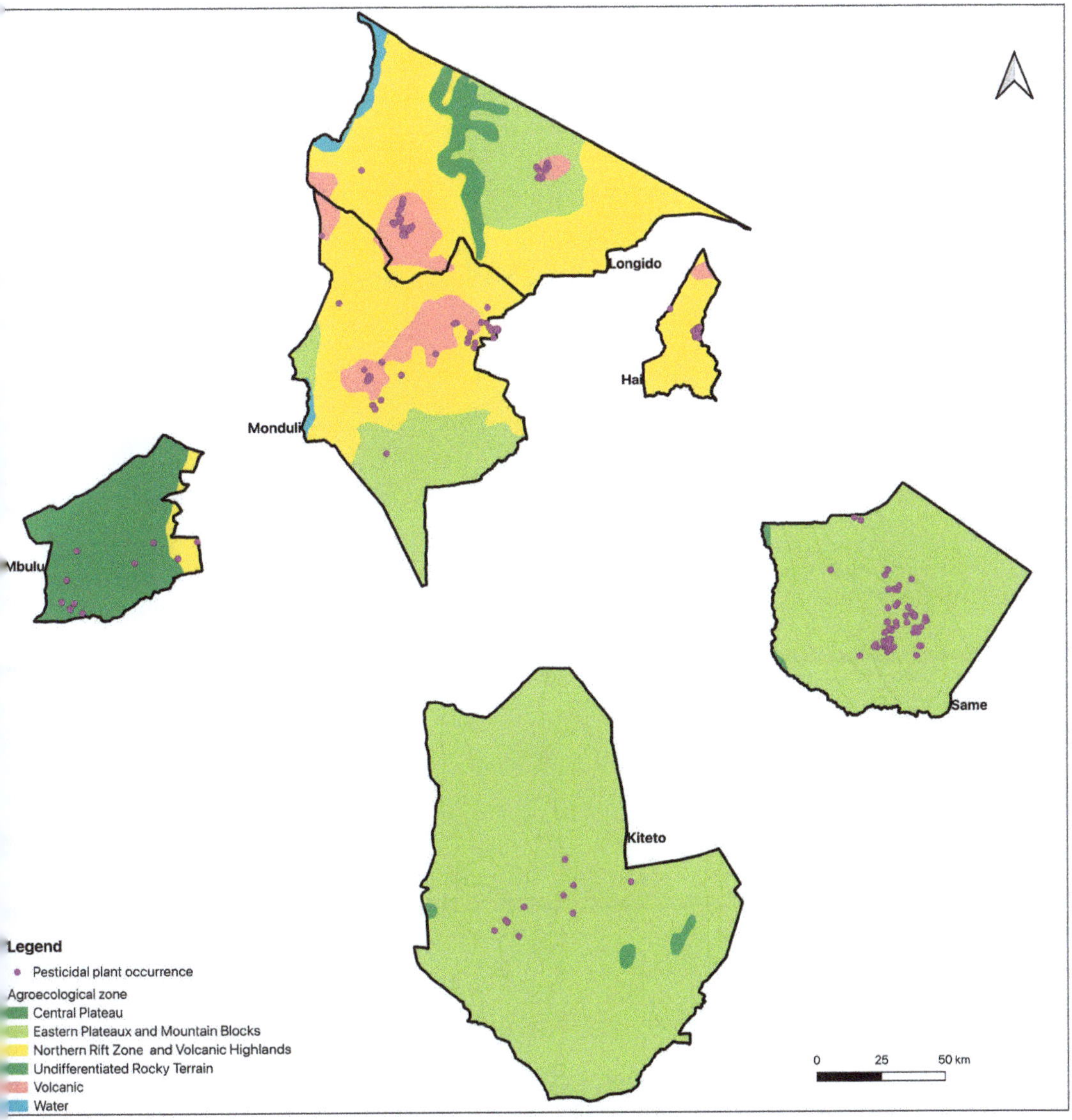

Figure 4. Occurrence of commonly used pesticidal plant species across agroecological zones in northern Tanzania. Species coordinates used to produce this map are found in S3.

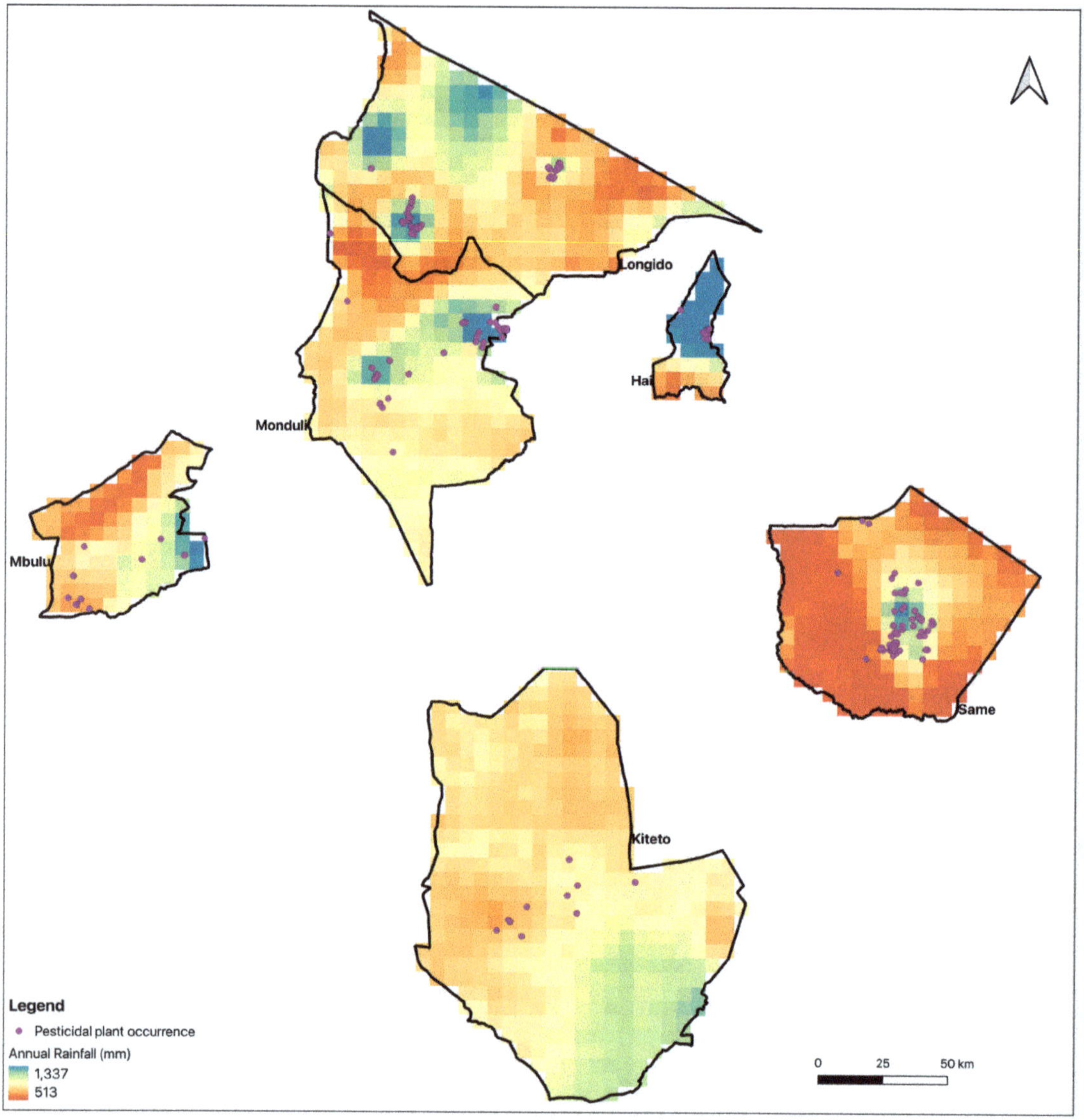

Figure 5. Spatial distribution of commonly used pesticidal plant species across rainfall gradient in northern Tanzania.

4. Discussion

4.1. Knowledge Transmission Pathways

The dominance of oral knowledge transmission among the Maasai, Iraqw, and Zigua reflects deeply rooted cultural systems in which pesticidal plant knowledge is inherited through family/community structures. Focus group discussions (FGDs) emphasized intergenerational learning through observation, participation, and daily practice, with elders identified as the primary custodians of knowledge. KIIs also pointed to limited formal education and unequal access to extension services as key factors reinforcing dependence on oral knowledge among these groups. These insights from FGDs and KIIs confirmed that such knowledge is rarely written and is instead learned through practice and observation, reinforcing the findings from previous ethnobotanical studies on other African countries [2,16,26]. Reliance on this oral transmission alone leaves pesticidal plant knowledge vulnerable to erosion over time, particularly in the context of socio-economic change, modernization, and shifting livelihood practices [20].

The hybrid knowledge systems observed among the Chagga and Pare appear to result from increased interaction with extension services and NGO-led trainings, a pattern repeatedly emphasized by FGD participants and agricultural officers' similar trends reported by [22]. These interactions enhance local knowledge and practices but may gradually alter traditional knowledge structures, potentially affecting how orally transmitted knowledge is maintained within the communities [19]. The absence of formal knowledge exposure among the Nguu highlights structural inequities in extension access, suggesting the need for culturally inclusive outreach approaches led by extension services, NGOs, and local community organizations.

4.2. Demographic Predictors of Knowledge Possession

The findings indicate that ethnic affiliation and primary education level significantly influenced knowledge reporting, whereas age and gender were not statistically significant predictors. Ethnic group emerged as the strongest determinant of pesticidal plant knowledge. Compared to Maasai respondents, Chagga and Iraqw respondents were significantly more likely to report knowledge of pesticidal plants. This pattern likely reflects differences in agroecological settings, farming systems, and cultural knowledge transmission. The Chagga and Iraqw communities traditionally practice more intensive and permanent agricultural systems, particularly in highland and volcanic zones characterized by crop diversification. Such farming systems may promote greater experimentation with plant-based pest control methods and facilitate the retention of ethnobotanical knowledge. In contrast, Maasai communities are historically pastoralist, with relatively less emphasis on crop production, which may partly explain the lower reported likelihood of pesticidal plant knowledge in this group. However, these findings should not be interpreted as the absence of knowledge but rather as differences in domain-specific ethnobotanical specialization.

Education level also showed an important pattern. Respondents with primary education were significantly more likely to report pesticidal plant knowledge compared to those who had not attended school. This suggests that basic literacy may enhance the articulation or recall of traditional knowledge or may reflect greater interaction with agricultural extension systems. Interestingly, secondary and tertiary education were not significant predictors, and the wide confidence intervals suggest instability due to small sample sizes. In many contexts, higher formal education has been associated with erosion of traditional knowledge systems; however, the present findings do not provide strong evidence of such an effect. These supported the findings from [27], which observed that moderate education levels may facilitate traditional knowledge retention, whereas advanced formal education often correlates with knowledge loss due to curricular shifts. The qualitative evidence from FGDs further highlights the role of modern lifestyles and the allure of synthetic pesticides in eroding traditional practices, consistent with the concerns raised by [28].

Regarding gender, the findings in this study aligned with [29], which found no significant differences between men and women in medicinal plant knowledge across various cultural settings. Contrasting evidence from [30–33] highlighted that in some contexts, women, especially household caregivers, tend to hold greater ethnobotanical knowledge due to their roles in family health and food preparation. Although age was statistically not a significant predictor in this study, observational and qualitative evidence suggested its underlying influence. Focus group discussions revealed that older participants ex-

https://doi.org/10.3390/d18050253

pressed concerns about knowledge erosion, particularly among youth. Many lamented that younger generations tend to lose interest in farming practices; these intergenerational gaps reflected the broader trends of declining interest in traditional knowledge among youth, as noted by [22,28]. Such insights emphasized the importance of combined quantitative and qualitative approaches in evaluating the dynamics of knowledge transmission.

It is important to note that several predictors exhibited wide confidence intervals, reflecting sparse outcome categories and potential quasi-complete separation. Although Firth's penalized logistic regression was applied to reduce small-sample bias, caution is warranted in interpreting the magnitude of some odds ratios. Future studies with more balanced outcome categories and stratified sampling across ethnic groups would improve estimate precision. Generally, the findings highlight the importance of cultural and agroecological context in shaping ethnobotanical knowledge systems. The significant role of ethnic affiliation suggests that pesticidal plant knowledge is embedded within localized farming traditions and ecological adaptations. These insights highlight the need for culturally sensitive knowledge documentation and conservation strategies that recognize intra-regional diversity in traditional pest management practices.

4.3. Cultural Specificity in Plant Use

The pesticidal plants we documented across ethnic groups, with highly significant inter-ethnic variation in species preference ($p < 0.001$), highlight the depth of traditional ecological knowledge. Specific plant use patterns, such as *Tephrosia vogelii* among the Chagga and Pare and *Croton dichogamus* among the Iraqw, reflect both ecological adaptation and cultural selection criteria, consistent with the biocultural frameworks proposed by [34]. Such diversity in plant selection highlights the importance of localized conservation and documentation efforts. As key informants noted, many species are becoming less known or available due to changing land use and reduced reliance on traditional methods. These insights reaffirm the need for urgent ethnobotanical documentation and participatory conservation support, as noted by [35].

4.4. Harvesting Practices and Sustainability

Significant variation in plant part use was observed across ethnic groups. Leaf harvesting, favored by the Chagga and Iraqw, is more sustainable due to rapid regeneration, as supported by [36]. The broader use of roots, bark, and whole plants among the Maasai, while demonstrating diverse applications, raises sustainability concerns aligned with [37]. Community narratives stress the awareness of sustainable harvesting, such as not overharvesting bark, yet practical constraints and the lack of alternatives may lead to unsustainable practices. Tailored interventions that promote leaf-based formulations and the home cultivation of high-use species could mitigate ecological pressure while honoring cultural traditions.

4.5. Access and Sourcing Patterns

Ethnic variation in sourcing pathways indicates that access to pesticidal plants is shaped not only by ecological factors but also by social and economic dynamics. The predominance of wild-sourced plants, particularly among the Maasai and Iraqw, reflects continued reliance on natural ecosystems. However, the emergence of home gardens and market-based access among the Chagga and Pare suggests evolving strategies for plant availability. These findings mirror [38], which emphasizes that plant domestication and market integration often arise in response to environmental pressures and livelihood shifts. Supporting community-based propagation and improving access to seedlings may enhance both conservation and local livelihoods.

https://doi.org/10.3390/d180502

4.6. Perceptions of Pesticidal Plant Population Status

The significant association between ethnic affiliation and the perceptions of pesticidal plant population trends suggests that views on plant availability are shaped by localized ecological conditions, livelihood systems, and resource management practices. While the majority of respondents perceived pesticidal plant populations as increasing, this optimism was not uniformly distributed across communities. The relatively high reporting of increasing trends among Chagga and Iraqw respondents may reflect greater engagement in plant cultivation, home garden domestication, and agroforestry practices [22]. These communities are more crop-oriented, and their integration of pesticidal plants into farm systems likely enhances both visibility and perceived abundance [22].

In contrast, the Maasai reported the highest number of declining perceptions, which may be linked to their stronger reliance on wild landscapes for plant collection. Pastoral systems often depend on open rangelands where environmental degradation, land-use change, bush clearing, overgrazing, and climate variability can directly affect plant availability [39]. The qualitative evidence reinforces this interpretation, as one elder explained, "*Some plants we used to find easily are now disappearing because the land is changing and more people are using chemicals.*" This statement highlights two important pressures, environmental transformation and the growing use of synthetic pesticides. The shift toward commercial agrochemicals may reduce the active use and management of traditional species, potentially leading to neglect, the reduced transmission of knowledge, and lower conservation priority for certain taxa [4,14,21].

The relatively small proportion of respondents reporting stable populations suggests that most communities perceive ongoing ecological change rather than equilibrium. This aligns with the broader patterns of land-use intensification, population growth, and climate variability documented across northern Tanzania [40]. The perceptions of increasing populations may also reflect improved awareness, domestication efforts, and adaptation strategies, whereas the perceptions of decline likely signal the vulnerability of wild-harvested species under environmental stress.

4.7. Influence of Agroecological Zones and Rainfall Gradients on the Distribution of Pesticidal Plants

The spatial clustering of pesticidal plant occurrences within the Northern Rift Zone and Volcanic Highlands, as well as the Eastern Plateaux and Mountain Blocks, highlights the strong influence of agroecological conditions on ethnobotanical resource distribution. These regions are characterized by relatively fertile volcanic soils, heterogeneous topography, and favorable microclimatic conditions, which collectively support higher plant diversity [24]. Numerous ecological studies demonstrate that montane and volcanic landscapes often harbor greater floristic richness due to improved soil nutrient availability and moisture retention, factors that enhance biomass productivity and secondary metabolite synthesis [24,41,42]. The concentration of pesticidal plant species in these zones therefore aligns with broader biodiversity patterns observed in tropical highland ecosystems.

The association between pesticidal plant occurrences and moderate to high annual rainfall (approximately 800–1200 mm) further reinforces the importance of climatic drivers in shaping plant distribution. Rainfall availability is a fundamental determinant of vegetation productivity and species richness in savanna and montane systems [43]. Areas receiving higher precipitation typically exhibit greater plant biomass and structural complexity, increasing the probability of encountering species with bioactive secondary metabolites. Moisture availability also influences plant physiological processes, including the production of defensive compounds such as alkaloids, terpenoids, and phenolics, many of which underpin pesticidal properties [42].

https://doi.org/10.3390/d18050253

In contrast, the comparatively lower occurrence of pesticidal plant records in the Central Plateau and Undifferentiated Rocky Terrain may reflect harsher edaphic conditions, reduced soil fertility, and lower rainfall regimes. Semi-arid zones are often characterized by lower species richness and biomass productivity, limiting the diversity of plant taxa available for ethnobotanical use [39]. However, it is important to note that lower recorded occurrences may also reflect sampling intensity, land-use pressure, or differences in traditional knowledge distribution rather than purely ecological constraints.

The pronounced clustering observed in districts such as Same and Monduli suggests that both environmental suitability and human knowledge networks contribute to spatial patterns. Ethnobotanical resource distribution is shaped not only by ecological gradients but also by cultural transmission and livelihood specialization [18]. Communities residing in ecologically diverse and agriculturally productive landscapes may have greater opportunities to identify, experiment with, and domesticate pesticidal plant species.

4.8. Limitations of This Study

While this study provides important insights into the distribution and ethnobotanical knowledge of pesticidal plants, several considerations should be taken into account when interpreting the findings.. First, the occurrence records of pesticidal plants were not derived from random sampling. Instead, data were collected through purposive ethnobotanical surveys and existing herbarium records, which may introduce spatial clustering and sampling bias. Consequently, the observed distribution patterns may partly reflect sampling effort rather than true ecological abundance. Second, ethnobotanical information relied on self-reported knowledge from respondents. Such data may be influenced by recall bias, personal experience, cultural transmission patterns, or selective reporting. Informants may also emphasize well-known or culturally significant species, potentially underrepresenting less familiar but ecologically important taxa. Finally, herbarium records do not necessarily indicate current abundance or population status. Specimens document historical presence at a given location and time but may not reflect present-day distribution due to land-use change, habitat degradation, climate variability, or local extinction.

5. Conclusions

This study demonstrates that ethnobotanical knowledge and the use of pesticidal plants in northern Tanzania are strongly shaped by ethnic affiliation, cultural transmission pathways, and ecological context. Knowledge acquisition is predominantly rooted in family, community, and traditional systems, with significant variation across ethnic groups, while formal education and workshops play a secondary but notable role among some communities, particularly the Chagga. Although age and gender did not significantly influence knowledge possession, ethnicity and primary education emerged as key predictors, highlighting the importance of culturally embedded learning systems in sustaining pesticidal plant knowledge. The documentation of 106 pesticidal plant species highlights a rich and diverse ethnobotanical repertoire, with strong ethnic specificity in species selection, plant part use, and sourcing pathways. Leaves were the most commonly harvested plant part across all groups, reflecting both effectiveness and sustainability considerations, while reliance on wild habitats as the primary source raises concerns about potential overharvesting. The perceptions of declining plant populations, especially among the Maasai, signal increasing pressure on key species and point to emerging conservation challenges. Spatial analyses further revealed that commonly used pesticidal plants are unevenly distributed, with clear hotspots in the Northern Rift Zone and Volcanic Highlands and areas with high rainfall, closely aligning with areas of high ethnobotanical knowledge and use. These findings emphasize the need for integrated conservation and land-use strategies

https://doi.org/10.3390/d18050

that safeguard critical habitats, promote sustainable harvesting, and recognize the role of indigenous knowledge systems. Strengthening community-based conservation, supporting knowledge transmission, and incorporating spatially explicit information into policy and planning will be essential for sustaining pesticidal plant resources and the livelihoods that depend on them under increasing socioecological change.

Supplementary Materials: The following supporting information can be downloaded at: https://www.mdpi.com/article/10.3390/d18050253/s1. S1. Pesticidal plants checklist with their frequency of citation and relative frequency of citation. S2. Plant species used as a pesticidal plant in the study areas. S3. List of commonly used pesticidal plant species for spatial analysis. S4. Questionnaire on pesticidal plants in northern Tanzania.

Author Contributions: I.K.C., A.G.M. and R.A.G. conceived this study; I.K.C., R.A.G. and A.G.M. designed this study; I.K.C. conducted data collection; I.K.C. and R.A.G. performed data curation and analysis; I.K.C., R.A.G. and A.G.M. designed the methodology; R.A.G. and I.K.C. developed the manuscript; A.G.M. and R.A.G. reviewed the manuscript; A.G.M. acquired funding. All authors have read and agreed to the published version of the manuscript.

Funding: This research was supported by Organization for Women in Science for the Developing World (OSWD), Grant Number 4500501096.

Institutional Review Board Statement: Ethical review and approval were waived for this study due to the nature of the research. The study was conducted following community-level Free, Prior and Informed Consent (FPIC) processes, and individual informed consent was obtained from all participants before collecting any data.

Informed Consent Statement: Informed consent has been obtained from the communities to publish this paper.

Data Availability Statement: The datasets generated and analyzed during the current study are available from the corresponding author upon reasonable request.

Acknowledgments: We would like to acknowledge the local communities, district officials, agricultural and livestock officer in northern Tanzania for their cooperation and for generously sharing their indigenous knowledge and experiences on the use of pesticidal plants. Their contributions are the backbone of this study. Thanks to National Herbarium of Tanzania (NHT) for giving us this opportunity and access to herbaria specimens for this study.

Conflicts of Interest: The authors declare no conflicts of interest.

Abbreviations

The following abbreviations are used in this manuscript:

FGDs	Focus Group Discussions
KIIs	Key Informant Interviews
NGOs	Non-Governmental Organizations
NHT	National Herbarium of Tanzania
QGIS	Quantum Geographic Information System
TEK	Traditional Ecological Knowledge

ferences

Dimetry, N.Z. Different plant families as bioresource for pesticides. In *Advances in Plant Biopesticides*; Singh, D., Ed.; Springer: New Delhi, India, 2014; pp. 1–20. [CrossRef]
Mwine, J.; van Damme, P.; Kamoga, G.; Kudamba, K.; Nasuuna, M.; Jumba, F. Ethnobotanical survey of pesticidal plants used in South Uganda: Case study of Masaka district. *J. Med. Plants Res.* **2011**, *5*, 1155–1163.
Acheuk, F.; Basiouni, S.; Shehata, A.A.; Dick, K.; Hajri, H.; Lasram, S.; Yilmaz, M.; Emekci, M.; Tsiamis, G.; Spona-Friedl, M. Status and prospects of botanical biopesticides in Europe and Mediterranean countries. *Biomolecules* **2022**, *12*, 311. [CrossRef] [PubMed]

https://doi.org/10.3390/d18050253

4. Ngegba, P.M.; Cui, G.; Khalid, M.Z.; Zhong, G. Use of botanical pesticides in agriculture as an alternative to synthetic pestici *Agriculture* **2022**, *12*, 600. [CrossRef]
5. Yigezu, G.; Wakgari, M. Local and indigenous knowledge of farmers management practice against fall armyworm (*Spodopt frugiperda*) (JE Smith) (Lepidoptera: Noctuidae): A review. *J. Entomol. Zool. Stud.* **2020**, *8*, 765–770.
6. Ali, A.D.; Ior, L.D.; Dogo, G.A.; Joshua, J.I.; Gushit, J.S. Ethnobotanical survey of plants used as biopesticides by indigenc people of Plateau State, Nigeria. *Diversity* **2022**, *14*, 851. [CrossRef]
7. Mkindi, A.; Mtei, K.M.; Njau, K.N.; Ndakidemi, P.A. *The Potential of Using Indigenous Pesticidal Plants for Insect Pest Control to Sn Scale Farmers in Africa*; Scientific Research Publishing Inc.: Wuhan, China, 2015.
8. Souto, A.L.; Sylvestre, M.; Tölke, E.D.; Tavares, J.F.; Barbosa-Filho, J.M.; Cebrián-Torrejón, G. Plant-derived pesticides as alternative to pest management and sustainable agricultural production: Prospects, applications and challenges. *Molecules* **20** *26*, 4835. [CrossRef]
9. Yadav, S.K. Pesticide applications-threat to ecosystems. *J. Hum. Ecol.* **2010**, *32*, 37–45. [CrossRef]
10. Pila, D.; Maqueda, R.H. Traditional knowledge and use of plants as agricultural insecticides from a gender perspective in th rural communities of the Ecuadorian Andes. *Ethnobot. Res. Appl.* **2023**, *26*, 1–12. [CrossRef]
11. Koul, O. Plant biodiversity as a resource for natural products for insect pest management. In *Biodiversity and Insect Pests: K Issues for Sustainable Management*; John Wiley & Sons, Ltd.: Hoboken, NJ, USA, 2012; pp. 85–105. [CrossRef]
12. Abate, T.; van Huis, A.; Ampofo, J. Pest management strategies in traditional agriculture: An African perspective. *Annu. R Entomol.* **2000**, *45*, 631–659. [CrossRef]
13. Melash, A.A.; Bogale, A.A.; Mengstu, S.G.; Aberra, D.A.; Tsegay, A.; Mengistu, D.K. Sustainable management practices for dur wheat production: Analyzing specific agronomic interventions on productivity, grain micronutrient content, and quality. *Heliy* **2023**, *9*, e18733. [CrossRef]
14. Tembo, Y.; Mkindi, A.G.; Mkenda, P.A.; Mpumi, N.; Mwanauta, R.; Stevenson, P.C.; Ndakidemi, P.A.; Belmain, S.R. Pestici plant extracts improve yield and reduce insect pests on legume crops without harming beneficial arthropods. *Front. Plant S* **2018**, *9*, 1425. [CrossRef]
15. Gómez-Baggethun, E. Is there a future for indigenous and local knowledge? *J. Peasant Stud.* **2022**, *49*, 1139–1157. [CrossRef]
16. Singh, S.S.; Saha, A. Documentation of indigenous knowledge in farming system. *Int. J. Bio-Resour. Environ. Agric. Sci.* **2017** 548–553.
17. Bridgewater, P.; Rotherham, I.D. A critical perspective on the concept of biocultural diversity and its emerging role in nature a heritage conservation. *People Nat.* **2019**, *1*, 291–304. [CrossRef]
18. Pretty, J.; Adams, B.; Berkes, F.; De Athayde, S.F.; Dudley, N.; Hunn, E.; Maffi, L.; Milton, K.; Rapport, D.; Robbins, P. T intersections of biological diversity and cultural diversity: Towards integration. *Conserv. Soc.* **2009**, *7*, 100–112. [CrossRef]
19. Berkes, F.; Colding, J.; Folke, C. Rediscovery of traditional ecological knowledge as adaptive management. *Ecol. Appl.* **2000**, 1251–1262. [CrossRef]
20. Reyes-García, V.; Guèze, M.; Luz, A.C.; Paneque-Gálvez, J.; Macía, M.J.; Orta-Martínez, M.; Pino, J.; Rubio-Campillo, X. Evider of traditional knowledge loss among a contemporary indigenous society. *Evol. Hum. Behav.* **2013**, *34*, 249–257. [CrossRef]
21. Qwarse, M.; Mihale, M.J.; Sempombe, J.; Mugoyela, V.; Henry, L.; Sunghwa, F. Ethnobotanical survey of medicinal and pestici plants used by agropastoral communities in Mbulu District, Tanzania. *Tanzan. J. Sci. Technol.* **2018**, *1*, 22–35. [CrossRef]
22. Mkindi, A.; Coe, R.; Stevenson, P.; Ndakidemi, P.; Belmain, S. Qualitative cost-benefit analysis of using pesticidal plants smallholder crop protection. *Agriculture* **2021**, *11*, 1007. [CrossRef]
23. Fernández-Llamazares, Á.; Lepofsky, D.; Lertzman, K.; Armstrong, C.G.; Brondizio, E.S.; Gavin, M.C.; Lyver, P.O.B.; Nichol G.P.; Pascua, P.a.; Reo, N.J. Scientists' warning to humanity on threats to indigenous and local knowledge systems. *J. Ethnob* **2021**, *41*, 144–169. [CrossRef]
24. Nkwabi, A.K.; Bukombe, J.; Maliti, H.; Liseki, S.; Lesio, N.; Kija, H. An overview of biodiversity in tanzania and conservati efforts. In *Global Biodiversity*; Apple Academic Press: Palm Bay, FL, USA, 2018; pp. 295–340.
25. Braun, V.; Clarke, V. Using thematic analysis in psychology. *Qual. Res. Psychol.* **2006**, *3*, 77–101. [CrossRef]
26. Moloise, S.D.; Matamanda, A.R.; Bhanye, J.I. Traditional ecological knowledge and practices for ecosystem conservation a management: The case of savanna ecosystem services in Limpopo, South Africa. *Int. J. Sustain. Dev. World Ecol.* **2024**, *31*, 29– [CrossRef]
27. Rinto, R.; Iswari, R.S.; Mindyarto, B.N.; Saptono, S. Bridging the generational gap: Exploring youth understanding on ethn otanical knowledge and its integration in higher education curricula. *Ethnobot. Res. Appl.* **2023**, *26*, 1–16. [CrossRef]
28. Casagrande, A.; Ritter, M.R.; Kubo, R.R. Traditional knowledge in medicinal plants and intermedicality in urban environmer A case study in a popular community in southern Brazil. *Ethnobot. Res. Appl.* **2023**, *25*, 1–34. [CrossRef]
29. Torres-Avilez, W.; Medeiros, P.M.d.; Albuquerque, U.P. Effect of gender on the knowledge of medicinal plants: Systematic revi and meta-analysis. *Evid. Based Complement. Altern. Med.* **2016**, *2016*, 6592363. [CrossRef]

https://doi.org/10.3390/d18050

Cantalice, A.S.; Gonçalves-Souza, T.; Albuquerque, U.P. Effects of socioeconomic and human-modified landscape variables on medicinal species richness at a macroscale: The case of the Caatinga, Brazil. *J. Ethnobiol. Ethnomed.* **2025**, *21*, 5. [CrossRef]
Müller, J.G.; Boubacar, R.; Guimbo, I.D. The "how" and "why" of including gender and age in ethnobotanical research and community-based resource management. *Ambio* **2015**, *44*, 67–78. [CrossRef] [PubMed]
Teixidor-Toneu, I.; Elgadi, S.; Zine, H.; Manzanilla, V.; Ouhammou, A.; D'Ambrosio, U. Medicines in the kitchen: Gender roles shape ethnobotanical knowledge in Marrakshi households. *Foods* **2021**, *10*, 2332. [CrossRef]
Nyirenda, S.P.; Sileshi, G.W.; Belmain, S.R.; Kamanula, J.F.; Mvumi, B.M.; Sola, P.; Nyirenda, G.K.; Stevenson, P.C. Farmers' ethno-ecological knowledge of vegetable pests and pesticidal plant use in Malawi and Zambia. *Afr. J. Agric. Res.* **2011**, *6*, 1525–1537.
de Boer, H.; Lamxay, V. Plants used during pregnancy, childbirth and postpartum healthcare in Lao PDR: A comparative study of the Brou, Saek and Kry ethnic groups. *J. Ethnobiol. Ethnomed.* **2009**, *5*, 25. [CrossRef]
Phengmala, K.; Saensouk, S.; Saensouk, P.; Souladeth, P. Ethnobotanical study of medicinal plants used by Lao ethnic group in Central Laos. *Not. Bot. Horti Agrobot. Cluj Napoca* **2024**, *52*, 13633. [CrossRef]
Hamilton, A.C. Medicinal plants, conservation and livelihoods. *Biodivers. Conserv.* **2004**, *13*, 1477–1517. [CrossRef]
Schippmann, U.; Leaman, D.J.; Cunningham, A.B. Impact of cultivation and gathering of medicinal plants on biodiversity: Global trends and issues. In *Biodiversity and the Ecosystem Approach in Agriculture, Forestry and Fisheries*; FAO: Rome, Italy, 2002.
De Albuquerque, U.P.; De Medeiros, P.M.; De Almeida, A.L.S.; Monteiro, J.M.; Neto, E.M.d.F.L.; de Melo, J.G.; Dos Santos, J.P. Medicinal plants of the caatinga (semi-arid) vegetation of NE Brazil: A quantitative approach. *J. Ethnopharmacol.* **2007**, *114*, 325–354. [CrossRef]
Selemani, I.S. Indigenous knowledge and rangelands' biodiversity conservation in Tanzania: Success and failure. *Biodivers. Conserv.* **2020**, *29*, 3863–3876. [CrossRef]
Lou, H.; Banerjee, A.; Mango, J.; Khan, U.; Singh, S.K. Understanding vegetation change in northern Tanzania: Interactions between climate variability and human activity. *Phys. Chem. Earth Parts A/B/C* **2025**, *141*, 104064. [CrossRef]
Akula, R.; Ravishankar, G.A. Influence of abiotic stress signals on secondary metabolites in plants. *Plant Signal. Behav.* **2011**, *6*, 1720–1731. [CrossRef]
Isah, T. Stress and defense responses in plant secondary metabolites production. *Biol. Res.* **2019**, *52*. [CrossRef] [PubMed]
Johansson, E.; Martin, R.; Mapunda, K. Climate vulnerability of agroecological and conventional smallholders in Mvomero district, Tanzania: Using mixed-methods to uncover local experiences and motivations of farming for the future. *Front. Sustain. Food Syst.* **2024**, *8*, 1423861. [CrossRef]

https://doi.org/10.3390/d18050253

diversity

MDPI

Article

A Step Too Far: Culling a Native Australian Honeyeater, the Noisy Miner (*Manorina melanocephalis*), for 'Conservation': Biases, Contradictions, and Myth-Making

Gisela Kaplan

School of Science and Technology, University of New England, Armidale, NSW 2350, Australia; gkaplan@une.edu.au

Abstract

Birds of the Anthropocene have to adapt to changing and often very unfavourable conditions, among them habitat fragmentation or outright habitat loss. Many organisations worldwide are deeply committed to stemming the tide of extinctions of native species and maintaining biodiversity. The question is how far scientists and practitioners are willing to go to achieve conservation goals in situations that are not entirely resolved, are contradictory, or involve dubious claims about alleged causative agents. The noisy miner, *Manorina melanocephala*, has been painted as such a causative agent in the decline of small woodland birds. The noisy miner is a highly successful, flexible, and socially complex small native Australian honeyeater (woodland bird). As will be shown in a new data analysis, the noisy miner also ranks highly in cognitive abilities. Despite its status as a native species, a protracted campaign against the species has led to government policies permitting its culling in New South Wales due to its alleged 'overabundance' and 'harmful' impact on small woodland birds. As a consequence, noisy miners can now be shot legally and have been culled in their thousands in the last decade. Allegedly, these actions have been taken for conservation purposes. This paper raises significant doubts about the claims against this species, and the methods and ethics of how a native species can become the sole bearer of the ills of the Anthropocene. This paper exposes bias or misinterpreted evidence and shows how myth-making is possible in modern science and how language can purposefully mislead the public via characterisations of avian behaviour. In essence, this paper is a case study of ethical issues in science: about the degree and type of intervention, and how far we are willing to go in the name of conservation, particularly when based on spurious or contradictory evidence and at the cost of native animals.

Keywords: noisy miner; cognition; cooperation; ecology of fear; typology of woodlands; habitat degradation; victimisation; demonisation; culling as conservation

1. Introduction

Birds function in a range of social, ecological, and human-manipulated environments and must learn to adapt and survive under very specific and often dramatically changing and challenging conditions. Social organization shapes the behavioural options available to birds and profoundly influences the evolution of their social behaviour and reproductive strategies [1]. External circumstances can lead to unfavourable conditions that individual groups and species may have to overcome and solve via cooperation [2]. Species have had to deal with such issues throughout their evolution, and speciation has often depended on changes in environments, but never with so many variables and in such a short period of time as in the Anthropocene [3]. Birds may adapt by changing their behaviour, location, or even food sources to reproduce successfully, or, if none of these strategies has worked and other strategies of adaptation have so far failed in the short period of time that human encroachments on their living space have allowed, they slide towards extinction.

The decline of bird numbers is well-known, and the decline of woodland-dependent birds, especially small woodland birds [4–6], is not specific or unique to Australia but is a worldwide phenomenon. Woodland birds (of a similar weight range) are declining all over the world [7,8], although this is not limited to their category alone. Farmland and grassland bird numbers and species are also declining [9–12]. Many publications on the current state of birds centre around conservation and focus on identifying (and rectifying) some of the causes of decline. To save birds and other vertebrates, we need to know as much as we can about why their numbers are declining, and this presupposes knowledge about every aspect of their life histories, along with their basic needs, behaviour, abilities, and reproductive habits. Indeed, we have never collected so much information on every aspect of avian life as in the last 50 years [13–15].

However, conservation has moved in many different directions and actions. Some theories and practices have been very insightful, progressive, and successful, while other recent actions might be regarded as highly controversial or counter-conservationist. This paper will report on the latter: concerning just one native Australian honeyeater species, the noisy miner, *Manorina melanocephala* [16], raising many issues, including the 'kill for conservation' dictum [17], which has no redress if the culls turn out to be futile, wrong, or based on false assumptions. The noisy miner is such a test case. Particularly in the last twenty-five years or so, a truly extraordinary attitudinal shift from positive to negative opinions has occurred [18], possibly unique in recent conservation history, excepting the long-standing persecutory history of the common starling (*Sturnus vulgaris*) [19]. The shift in attitude perhaps gained momentum with the publication of the *Handbook of Australian, New Zealand and Antarctic Birds* [20].

Since then, the noisy miner has been denigrated in all sorts of ways. No effort seems to have been spared to create and reinforce the disrepute of this species that occurs by calling it 'aggressive' [21,22], 'hyper-aggressive' [23], 'despotic' [24–30], and even 'notorious' [31]. These adjectives are opinion-based, value judgments usually applied to humans, not scientific, neutral, or descriptive terms for animal characteristics.

Recent scientific publications have argued that the behaviour of noisy miners is undesirable and needs to be managed and that the birds need to be eliminated from areas in which they are deemed 'overabundant' [32,33]. The term 'overabundance' originated from (a) sightings of noisy miners congregating in large numbers at specific sites (apparently as many as 200/250 individuals), and (b) observations that the miners excluded other woodland-dependent birds from their territory [23].

The focus on this apparently 'villainous' songbird reached the point at which the media published negative views of noisy miners. In 2011, the Adelaide Advertiser newspaper reported that "Scientists want noisy miners culled" [33], and I am not aware that the public had any major objections at the time. The Australian Broadcasting Corporation (ABC) broadened the condemnation of the species to apply throughout Australia and proclaimed that the noisy miner is allegedly "Australia's most hated bird" [34,35]. I doubt that this statement is correct. Time and again the public has tended to confuse (in name at least, not necessarily in appearance) the noisy miner (60–80 g) with the common myna, *Acridotheres tristis*, an introduced, clever and highly invasive, large starling species (110–140 g) that actively competes and even kills native parrots for access to their nest holes [36]. In 2017,

https://doi.org/10.3390/d18020099

in a newsletter of a local Bird Observers Club, there appeared an article with the dubious title "Miner surgery: repairing woodland bird diversity by Noisy Minectomy" [37].

Based on the published evidence [23–30], it was considered an urgent task for the survival of avian woodland species to address the noisy miner 'problem' by acting decisively to reduce noisy miner numbers. Such action, it was believed, would be restorative and allow other woodland-dependent avian species to recover [38]. Eventually, these characterisations of the noisy miner escalated into declaring the bird a threat to the survival of avian woodland species [21,39]. The matter culminated in 2013 in a 'final determination' of a legally enshrined framework that the noisy miner was a threat and had to be suppressed [40]. Alarmingly, this officially now permitted the killing of noisy miners where and when the bird was considered 'overabundant' [40].

Once the conclusion of 'overabundance' was reached and the 'suppression' of noisy miners was put in place for reasons of the alleged positive consequences for woodland birds, a strategy was devised to shoot thousands of noisy miners (mainly between 2010 and 2020). This killing spree happened with government approval, with the full permission of university ethics committees, and the methods were not queried by journals that accepted the resulting papers for publication, since culling this native bird was permitted within new state government regulations specifically naming the noisy miner as a threat and giving the green light to cull them [40].

The official condemnation of noisy miners is a permanent legislative feature and, as I understand it, would need to be challenged and rescinded if its native status as a protected species is to be regained. The species is now an outlaw in New South Wales, the state that has probably the majority of all noisy miners extant in Australia. If the legal determination allowing culling is not removed in the near future, the policy is, in fact, an official verdict on and recipe for the legislated extinction of a native bird. Perhaps this new category of 'legislated extinction' is a novelty in conservation, although so far, the birds have fought back and recolonised in most places [41].

Moreover, there may be very different interpretations for the phenomenon of 'overabundance', and some of these will be proposed in this paper. For instance, unusual bird assemblages may be an indicator (an outcome) of ecosystem dysfunction and instability in general [42], i.e., they cannot be attributed to one species alone. Indeed, it is now recognised that whole ecosystems are degrading even more rapidly than previously thought [43]. Hence, the perceived overabundance of noisy miners, even if just in some areas, might be an outcome and symptom of ecosystem dysfunction, not based on the threatening quality of the outcast species. If this conclusion applies, the case against noisy miners becomes very weak. The judgement of the species as an intrinsic disruptor may have hinged on false premises and led to incorrect conclusions, which, nevertheless, have led and could well lead to more and more lethal outcomes for many miners.

Further, to imply or state that a single native bird alone is responsible for the decline of other woodland birds is both scientifically and logically implausible, as well as ethically questionable, thus ultimately misleading, if not entirely incorrect. One also fails to see how the decision-making process of meeting ecological instability by creating even harsher instability could be a logical solution for creating stability. Further, whether a native bird is 'too' abundant ('overabundant') and/or is a perceived risk to other native bird communities may not be all that easy to establish over the species' entire range (especially difficult in a continent as large as Australia).

Woinarski asked the broad question of whether it is right to 'kill Peter to save Paul' [44]. This question has broad theoretical and applied ramifications for how species are managed. It has sparked ongoing ethical and policy debates in conservation circles, for instance, in North American debates about two native owls, one declining, the other invading the

https://doi.org/10.3390/d18020

territory of the declining one [45]. It raises the questions of what purpose such management decisions serve and whether any such management plans consider the entire life history, needs, and possible stress levels of the concerned species, regardless of numbers (too few or too many individuals).

This paper will give a summary of what we know about the noisy miner's evolutionary history, ecology, social life, cognition, and reproductive and feeding strategies, and it will refer to the proximate and ultimate causes of this species' behaviour. This is to provide a framework for arguing that noisy miners are mostly victims, as are other birds, and indeed other vertebrates and invertebrates [46], and that they do not qualify as perpetrators of environmental damage or the decline of other species. The many reasons for the declining status of birds have been well explored worldwide and in Australia. In Australia, the specific points of concern are habitat fragmentation, dangerous agricultural practices causing sudden resource reductions [47], increased and often inappropriate fire regimes [48], and climate change, including the increasing occurrence of unprecedented and, for birds, non-survivable heatwaves [49], as well as substantial and ongoing habitat loss [50].

Another objective of this paper is to examine the risks of sometimes choosing interpretations that may seem obvious but fail to examine or overlook vital factors, which may well offer more insight into this species and into specific environmental factors. Indeed, the noisy miner case is shown here as an example of why the simplest explanations are not always the best [51] and why 'naïve' anthropomorphisms (Burghardt's term [52]) slipping into descriptions of animal behaviour may invalidate current interpretations of noisy miner behaviour [53,54]. The management plans for noisy miners are used as an example of the risks and misdirection caused by killing native species for conservation purposes.

This paper aims to expose the contradictions and how conservation of native small- to medium-sized avian woodland species, of which the noisy miner is a part, can go so spectacularly wrong and even fall prey to myth-making (details can be found in Section 7.3).

Tragically, the noisy miner may be "innocent", and the behavioural evidence against it spurious. Even if this were not so, the question is whether conservation efforts are now overstepping the mark by manipulating avian dynamics in natural environments through selective culling and determining which species shall live or die.

Ultimately, this paper raises the often-controversial ethical issues about the degree and type of intervention concerning native species in the natural world (the issue of invasive species is another debate not raised here), how far we are willing to go in the name of conservation [55–58], and how much harm may be caused by misinformation or perhaps influenced by powerful vested interests.

2. Evolution, Distribution, and Confirmed Positive Attributes of Noisy Miners

2.1. Evolution and Social Ecology of Noisy Miners

It has now been acknowledged for at least two decades that all modern songbirds (post the mass extinctions of 65 mya) first evolved in Australia [59,60]. Only much later did the first oscines, the clade formed by Sylviida, Muscicapida, and Passerida, radiate to Eurasia [61]. The evidence thus shows that Australian songbirds have a distinct indigenous and continuous evolutionary history in East Gondwana, now Australia [59–62]. The 22 lineages that are said to have survived the mass extinction [62] are purported to have their origin as far back as the late Cretaceous [63], while modern songbirds speciated after the mass extinction of 65 million years ago [63–65].

Speciation may happen for several reasons [66–68]. Successful adaptive radiation may also depend on the discovery of underused food sources, a process that led to the evolution of and shift to ancestral nectivory. This shift resulted in the exceptional diversity of a large

https://doi.org/10.3390/d18020099

songbird clade, the honeyeaters, Meliphagidae [69], containing co-occurring species that vary in size more than non-nectivorous clades [70]. Latest assessments indicate that the family of honeyeaters (Meliphagidae) consists of 195 species and 74 genera [71]. Some 34 genera are found exclusively in Australia, and some 23 genera on islands and nations around the north and east of Australia (including New Zealand), with a particularly strong presence in New Guinea, once part of the Australian landmass [72].

The Australian honeyeaters include several families: the Australian chats (Epthianura ssp); Myzomelas, the largest group of honeyeaters with 40 species, widespread through Oceania; the spinebills (genus Acanthorhynchus); the friarbirds (genus Philemon); and the wattlebirds (genus Anthochaera), including the seriously endangered regent honeyeater, *Anthochaera phrygia*, as well as the miners (Manorina). The noisy miner belongs to the latter.

The Meliphagidae have been described as "an iconic and largely Australo-Papuan group" for which "the complexity of the biogeography underlying the group's spectacular radiation, especially within Australia", has now been revealed [72]. Ecological and phenotypic diversity is particularly pronounced in the honeyeater subclade (Table 1). Marki et al. [70] pointed out, as others have done before, that the honeyeaters possess a number of unique morphological and physiological adaptations for nectarivory, including structural modifications to the renal system for more efficient balancing of fluid intake [73] and a brush-tipped protrusible tongue [74]. These adaptations helped honeyeaters exploit a novel food source (nectar) and take up an ecological niche largely unoccupied, at least by birds. Their radiation seemed to coincide broadly with the accelerating aridification of Australia [75] in the earlier part of the Oligocene (33.9–23.03 million years ago) and later during the Miocene epoch (23.03–5.03 million years ago) in the absence of any human populations, which, in Australia, began to appear only about 55–65 thousand years ago [76,77].

Table 1. Examples of weights and sizes of honeyeaters. Source [20]: **S**, small; **N**, nectar; **Fr**, fruit, berries, flowers; **Ins.**, insects/may incl. Spiders; **R**, small reptiles/amphibians; **E/B**, eggs; **B**, small birds; **Sd**, seeds; **M**, migratory; **HM**, highly mobile; **LM**, locally mobile/seasonally nomadic; **S**, sedentary. (In the food column: Lower-case letters indicate a secondary food source; capital letters indicate the main food source or food of equal value). Focal species is bolded.

Common Name	Latin Name	Weight (g)	Size (cm)	Small-Large	Food	Mobility
Scarlet honeyeater	*Myzomela sanguinolenta*	8–9	11	smallest	N/Fr/Ins.	N/S
Brown honeyeater	*Lichmera indistincta*	9–11	12–16	small	N/Insects	LM
Eastern spinebill	*Acanthorhynchus tenuirostris*	11	13–16	small	N/Ins.	LM
Bell miner	*Manorina melanophrys*	25–29	17–20	small	lerps Ins.	S
Lewin's honeyeater	*Meliphaga lewinii*	27–49	20–22	small to medium	N/Fr/Ins.	LM
Regent honeyeater	*Xanthomyza phrygia*	35–50	24–28	medium	N/Ins.	HM
Noisy miner	*Manorina melanocephalis*	40–80	24–28	medium	N/Fr/I/R	S/LM
Blue-faced honeyeater	*Entomyzon cyanotis*	105	26–32	med. to large	INS/Fr,n.+	S/LM
Little wattlebird	*Anthochaera lunulata*	45–85	27–35	large	N/Fr/Ins.	S/LM
Noisy friarbird	*Philemon corniculatus*	80–130	30–35	large	N/Fr/Ins./E/B	LM
Yellow wattlebird	*Anthochaera paradoxa*	123–168	37–45	largest	N/Fr/Ins.	S/M/LM

As the small sample of honeyeaters from smallest to largest shows (Table 1), the differences between various taxa in the same family can be substantial, not just in size

https://doi.org/10.3390/d18020

and weight range but in mobility and range of food sources. The noisy miner and the other miner species (Figure 1) are thus part of a remarkable evolutionary history and an extremely diverse group found in Australia.

Figure 1. Three miner species: (**A**) black-eared miner; (**B**) noisy miner; (**C**) yellow-throated miner (Photocredit: G. Kaplan).

Manorina, the special focus of this paper, contains four species: the bell miner (*M. melanophrys*), noisy miner (*M. melanocephala*), yellow-throated miner (*M. flavigula*), and black-eared miner (*M. melanotis*). The bell miner (*Manorina melanophrys*) is distinct from the other three by its green plumage and smaller size, but the other three are very similar (Figure 1) and can easily be confused. Although the black-eared and noisy miners live in very different habitats, the yellow-throated miner overlaps with the habitat of the noisy miner in coastal Queensland and central New South Wales.

Unlike the relatively smaller ranges of other honeyeaters, the noisy miner and the yellow-throated miner together cover the entirety of the Australian continent: noisy miners occur in a broad sweep along the Eastern Seaboard right up to the tropics and the north-eastern reaches of Australia, while the yellow-throated miner covers almost the rest of the continent, overlapping only slightly in some eastern regions with the range of the noisy miner (Figure 2).

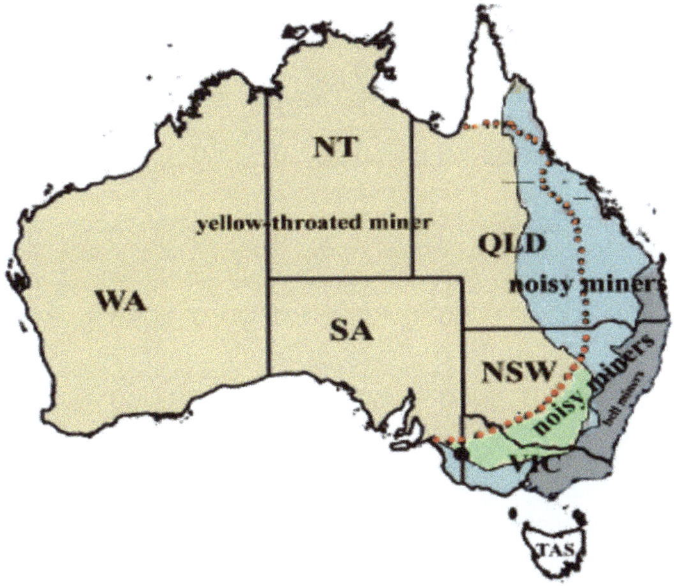

Figure 2. Distribution of miners (Manorina). **Beige**: Range of the yellow-throated miner (*M. flavigula*) across Australia, from Western Australia to the red dotted line in the east. The overlapping area with

https://doi.org/10.3390/d18020099

the noisy miner is shown in blue. Grey: Range of the bell miner (*M. melanophrys*), which is in the same territory as the noisy miner. **Green**: Potential overlaps of the yellow-throated miner and the noisy miner. The range of the bell miner and noisy miner is largely the same in NSW and part of southern Victoria. **Black dot**: Range of the black-eared miner (*M. melanotis*), a highly endangered species, now only found in the mallee region of north-western Victoria, east to Hattah-Kulkyne National Park, and through the Murray Mallee of South Australia. (WA = Western Australia; NT = Northern Territory; QLD = Queensland; SA = South Australia; NSW = New South Wales; VIC = Victoria; TAS = Tasmania).

The overlap between noisy and bell miners on the Eastern Seaboard is somewhat misleading because bell miners and noisy miners are rarely found in the same territories. However, there may be quite a few other honeyeaters that live in proximity to each other, such as blue-faced honeyeaters (*Entomyzon cyanotis)* and little wattlebirds (*Anthochaera chrysoptera*), sometimes up to five different Meliphagidae species.

The black-eared miner is the same size and more or less has the same plumage as the noisy miner, but it has a dark facial mask and an orange tinge on its beak and legs (Figure 1). The difference is one of habitat. Noisy miners live in the woodlands of eastern Australia, while their black-eared cousin is now confined to a very small plot in the Murray Mallee region in the far southwest of NSW. The Threatened Bird Index identified the key threats for the black-eared miner: (1) destruction of habitat via fire, (2) overgrazing, and (3) fragmentation of habitat [78].

2.2. *Loss of Avian Biodiversity*

The precarious situation of the black-eared miner is just one example of the trends over recent decades. Increasingly, the contest between humans and birds for the very same territories and environments has made many native mammals and birds losers on a large scale. As is well-known, urbanisation alone has often meant a significant drop in species richness [79–81]. Although great efforts have been made to halt the avian extinction rate in Australia by a myriad of organisations, often funded only by donations, the sharp decline of avian biodiversity has continued [81–85]. The Threatened Bird Index [78] showed an overall decline of 60 percent since 1985 for all bird taxa across Australia, slightly higher in Queensland and South Australia, and slightly lower in New South Wales. Australia's 'Environment Protection and Biodiversity Conservation Act 1999' (EPBC Act) [78] list was updated on 5 March 2025. One bird was added, the little tern (*Sternula albifrons*), which is now listed as 'vulnerable'.

The latest government report [78] on threatened species, as part of its ten-year Action Plan 2023–2032, has also identified 22 priority birds for conservation work in NSW, and despite the large clade of honeyeaters, only two have so far been identified to be at high risk of extinction: the black-eared miner, already mentioned, and the regent honeyeater *Anthochaera phrygia*. Other threatened species include several island birds and, on the Australian mainland, priorities include several parrots, two species of black cockatoos (Carnaby cockatoo, *Zanda latirostris*, and the red-tailed black cockatoo, *Calyptorhynchus banksii graptogyne*), and, among several other taxa, the noisy scrubbird, *Atrichornis clamosus*, and some species specific to New South Wales [78]. It is important to mention this here because none of the small woodland birds for whose decline noisy miners have been deemed responsible, and 'punished' by the elimination of thousands of them, are listed among the threatened species at this time.

Of the four passerines on the list, scrubbirds (order Passeriformes, family Atrichornithidae, size 16 cm) are found in terrestrial/rainforest habitat. The decline of the rufous scrubbird, *Atrichornis rufescens*, is believed to have been driven by the clearance of the

https://doi.org/10.3390/d18020

species' lowland habitat and the ongoing logging practices and subsequent disappearance of the understorey of the remaining eucalypt stands [78]. Of course, decline is not just measured in categories of threatened species but in declines in numbers within species and by a species' disappearance from specific sites, and this is a challenge for present-day conservation [86].

Considering noisy miners in this context, however, creates a puzzle. Their alleged 'overabundance' runs counter to general international and Australian trends chronicling the ubiquitous decline of half of all avian species on every continent, and according to the *State of the World's Birds Report* [50], only 6% of avian species worldwide are increasing in number. Apart from natural adaptations, these relatively rare examples are often due to intense conservation efforts.

Moreover, after decades of emphasising the negative attributes of noisy miners, be this in the media or in research papers, some of their ecologically important and positive characteristics are often no longer mentioned or are entirely forgotten, even though they may have a bearing on the perception and assessment of the role of noisy miners in the ecosystem.

2.3. The Overlooked Beneficial Role of Noisy Miners

The social ecology of honeyeaters has rarely been fully investigated [87], although they play an important role in the ecosystem. Noisy miners are no doubt of great importance to avian communities in the bush and in biodiverse environments. They may cohabit with several species of honeyeaters in one transect and live in stable multispecies bird communities [88].

Indeed, with their loud and persistent alarm calls, noisy miners warn the bird community about any predators in the vicinity. Moreover, they fly close to the source if the threat is terrestrial and/or stationary (e.g., as may be the case with snakes and lace monitors), thus giving an indication of the location of a potential threat. Noisy miners also collaborate with other species, chiefly Australian magpies (*Gymnorhina tibicen*) [89], in driving out predators, thereby protecting themselves and other species, including small woodland birds living in their territory (pers.observ.)

Collaboration between Australian magpies and noisy miners in instances of danger (such as spotting predators) is well-known [89,90]. I have observed noisy miners coming to the aid of a magpie pair when the magpies were unable to shift a stationary wedge-tailed eagle from a ground position in the magpies' territory. These collaborative defences of territory seem to hold in Australia's Eastern Seaboard. 'Squabbles' between magpies and noisy miners have been reported in South Australia—although of a different subspecies of magpie than in the east [91,92].

Individually, noisy miners have few, if any, physical defences. They do not have sharp claws or a pronounced beak, cryptic colouration, or venomous defences against any predator or larger birds. Towards birds their own size, they may use mobbing flights in certain situations, or, when a rare food source is discovered, they may squabble amongst each other, and the more dominant bird may give a double beak-clapping warning (pers. observ.), a signal that is understood not just among conspecifics but also heterospecifics and is usually sufficient to make one of the competing birds retreat.

Mostly, however, noisy miners behave like the sentinels of the bush. Usually first on the scene, they have the rare ability to warn others about danger or an intrusion within their territory, and they are equipped to do so because they have specific, functionally referential (and different) alarm calls that other birds have also learned to decipher, judging by the response of other birds within their territory. As Cunningham and Magrath [93] discovered, the noisy miners they studied gave different calls to airborne raptor models

https://doi.org/10.3390/d18020099

compared to terrestrial or perched raptor models and even switched from 'aerial' alarm calls to 'chur' alarm calls when a hawk had landed on the ground.

Noisy miners are among a few songbirds that use referential signals (Australian magpies also use functionally referential signals, such as their 'eagle alarm calls' [89] to indicate to conspecifics when a predator has been spotted [89]. As a deterrent to predators, noisy miners can also use a synchronised chorus (pers. observ.). Hall and Magrath [94] found in their study of duetting in magpie larks, *Grallina cyanoleuca,* that the temporal coordination of duetting signalled more than just their pair status. To competitors and possibly to predators, it was 'read' as a signal of the pair's coalition quality [94]. One might well draw the same conclusion about the highly coordinated physical and vocal performance of noisy miners that also includes duetting and displays [95]. They indeed have a rich repertoire of vocal signals used for and expressed as cooperative behaviour, particularly in the context of threats (such as avian predators, competitors, cats).

Dealing with reptiles also involves noisy miners. Noisy miners can do little about snakes such as pythons (genus) or lace monitors (*Varanus various)* that may be 2.5 m in length, but they can warn others. In Figure 3, an excerpt of a sequence of photographed events is shown (which took less than three minutes) that took place in a native bangalow palm tree (*Archontophoenix cunninghamiana*) in which a large carpet python, *Morelia spilota spilota,* had taken position. The upper horizontal fronds of this palm tree were used regularly by many different birds as a (usually brief) strategic lookout. Landing on one of these fronds would certainly have risked the life of any bird because it was located within striking distance of the snake (Figure 3).

Pythons do not seem to respond much to vocal signals and tend to have relatively poor eyesight, but via mechanoreceptors, thermo- and chemoreceptors, they are excellent at spotting any warm-blooded or moving target. Via a "cheek pit" structure on their heads, pythons can sense subtle changes in heat emanating from a living organism [96], detect odours, and respond to the slightest movement [96]. Apart from birds of prey, pythons or lace monitors are largely the reason for heterospecific collaboration between noisy miners and Australian magpies. In some groups, this heterospecific collaboration is well established. In one instance, I observed such a collaborative effort between noisy miners and magpies. In this incident, the noisy miners had alarm-called and then seemed to wait for support. First, another noisy miner arrived and, less than a minute later, a magpie landed, immediately found the snake, and then chose an effective, if rather risky, strategy of stopping in front of the snake's head and flapping its wings. The method was successful, and the snake left (see Figure 3).

This level of cross-species cooperation is a remarkable and cognitively highly advanced method of keeping an entire bird community/assembly safe. It must be noted, however, that no general agreement exists as to what determines community dynamics, in which direction such dynamics move, and whether there are clear winners and losers or whether homeostasis maintains such communities, ironically via competition. By contrast, Garcìa-Navas et al. argued that the presence of noisy miners increases the instability of bird communities [97]. However, Hubbell suggested that the total number of individuals in a community remains relatively constant or only tends to vary "as a result of stochastic forces" [98]. Such theoretical positions do not help very much in ascertaining whether noisy miners have a destabilising role to play in communities or bird assemblies, or whether instability of stochastic forces is at all related to the presence of noisy miners.

https://doi.org/10.3390/d18020

Figure 3. Deterring a snake: (**A**) a noisy miner arrives after spotting a carpet python on the palm frond below and is alarm-calling loudly; (**B**) a second noisy miner arrives, and the two birds seem to be calling for help and then waiting; (**C**) a magpie arrives promptly on the scene and, on seeing the snake, emits a short alarm call; (**D**) the magpie flaps its wings as a warning to the snake; (**E**) the magpie flies closely over the snake; (**F**) in a daring manoeuvre, the magpie lands in front of the snake's head; (**G**) the harassment is enough to make the snake slowly slither away and down from the palm tree. (Collage via Adobe Photoshop version 22.4.3; Photocredit: G. Kaplan).

One notes, however, that the many positive qualities of noisy miners remain generally unmentioned, have not been reported in detail in recent journal articles, or might have simply been regarded as immaterial. Such omissions of crucial abilities and manifest cognitive behaviour are, in fact, manipulating evidence, be this inadvertently or intentionally, if the charge of the miners' major influence on the decline of small woodland birds is to be upheld. The argument might be the other way around: manifest cooperative behaviour and collaborative efforts of noisy miners are not the signature of the disrupters of bird communities. What their ability to warn and actively engage in assisting heterospecific members suggests, however, is that their presence and actions establish or preserve *stability* rather than *disrupt* bird assemblies. This is the very opposite of what has been proposed by recent papers that have recommended and/or carried out noisy miner culls [29,38,41].

https://doi.org/10.3390/d18020099

2.4. *Cognitive Abilities and Cooperative Kinship Ties in Noisy Miners*

Cooperative behaviour, which is expressed during the breeding season and beyond, is an important strategy for survival. For vast stretches of Eurasia and the Americas, avian species showing cooperative behaviour rank below 2%, and in some pockets up to 5%, but in Australia, cooperative behaviour has been identified in almost a quarter of all avian species [2,99] (see also maps of the worldwide distribution of cooperative behaviour in Feeney et al. 2013 [100]).

Social and vocal skills and most other survival skills are important learned behaviours that are imparted by parents and the family unit. Since, in birds, over 95% of species raise offspring in bi-parental care (compared to a mere 5% of mammals [101]), the additional support by helpers in a significant number of species, including noisy miners, suggests strong social support, protection, and opportunities to acquire new skills while safe. Making important decisions throughout life depends on the ability to have learned, memorised, and successfully acted on relevant memory when finding food or encountering potentially dangerous situations [102], and this involves cognitive competence.

Using an extensive published data set of log-transformed brain and body weights of native Australian birds [103,104], I was able to extract data on the brain mass of the Meliphagidae relative to body mass and compare these data with all other recorded Australian songbirds, identifying the position of noisy miners among Meliphagidae and with the mean relative brain size of Australian songbirds in general. The results are 391 presented in Figure 4. The figure shows that some honeyeaters, including noisy miners, belong to the top 5% of honeyeaters with the largest brain (by weight relative to body weight) and among the top 10% in log-transformed measures, on par with some of the Corvoidea such as Australian magpies (in log-transformed brain size relative to body weight) (Figure 4).

Note that the noisy miner has a substantially larger brain size relative to body weight than its close relative, the yellow-throated miner (bolded in Figure 4). Darwin has often been misquoted as saying that the strongest survive (survival of the fittest), but he said in *On the Origin of the Species* that the species that survive are the ones that can best adapt and adjust to the changing environment in which they find themselves [105,106]. In terms of cognitive ability, one might expect that noisy miners might be better adapted to changing environments.

Presumably, such adaptability would be fostered by both cooperation and possibly a more complex cognitive ability that allows for problem solving, innovation, and flexibility [107–110]. The noisy miner has shown such flexibility in great measure by its adaptability to urban and changing environments, a trait that was confirmed in a recent PhD thesis [111].

In recent decades, comparative studies on brain sizes in birds have revealed very specific differences, some of which are relevant to noisy miners. For instance, it was shown in earlier research that certain brain structures differ in size between sedentary and migratory birds of the same species [112]. Cristol and colleagues showed that in migratory dark-eyed juncos (*Junco hyemalis*), the density of neurons in the hippocampus is higher than that in resident juncos [113]. Upon this finding, Sol and colleagues [114] investigated whether whole brain sizes were also different in other migratory versus sedentary birds. They found that, in passerine birds, migratory species tend to have substantially smaller brains (relative to body size) than sedentary species. They suggested that selection for smaller brains in migratory birds may arise from the energetic and developmental costs associated with a highly mobile life cycle, and perhaps with costs associated with cognitive functions that have become less necessary in migratory species [114]. Noisy miners belong to the sedentary species among honeyeaters and hence, according to Sol et al. [114], should fall into the group of songbirds with larger brains. As shown in Figure 4, they do.

https://doi.org/10.3390/d18020(

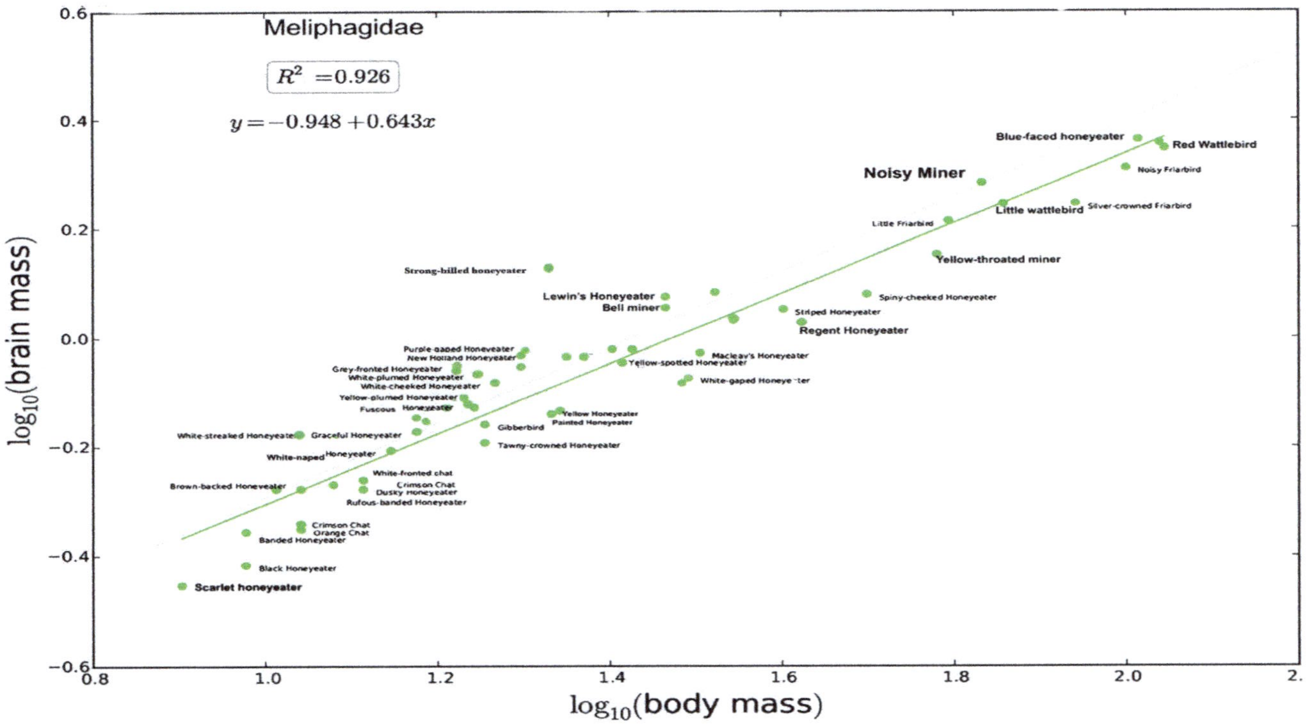

Figure 4. Brain sizes (log-transformed body versus brain mass) in Meliphagidae. The faint grey line represents the mean of all passerines, and the green line is the mean of all Meliphagidae. Only some of them are named here, and the noisy miner is marked in bold large font, showing that its brain size relative to body size is well above the mean of honeyeaters and at the mean of all passerines (the faint grey line includes honeyeaters). Note that the bell miner ranks well above the mean of passerines, while the yellow-throated miner is well below the mean of honeyeaters. Data calculated from [103,104].

https://doi.org/10.3390/d18020099

The 'big brain' theory suggests that there are several benefits and identifiable advantages that protect against extinction and stabilize populations [115]. This is achieved by behavioural plasticity, strongly associated with cognition, including the ability to live in variable habitats [115,116]. Indeed, noisy miners have adjusted admirably well to suburban and urban life. This is why every bird count conducted by citizen science identifies the noisy miner as one of the top three birds seen [117].

Another behavioural aspect associated with brain size [118], specifically with neuron density in the forebrain [118,119], and high levels of associative neuron numbers [119,120] are innovation, problem solving, and feeding innovations. Feeding innovations, whether in identifying or retrieving new food items and sources [121,122], changing nest location or materials [123], or recognition of and strategies against a predator [124], have been examined globally, and the link between cognition and innovation has been firmly established over the last two decades. These details are particularly relevant to noisy miners. Thanks to extensive research by Sulikowski and Burke [125] using wild-caught noisy miners [125–129], we know about the species' different food-searching strategies for finding either invertebrates or nectar [128]. Via the research conducted by Cunningham and Magrath [92], we also know that noisy miners have functionally referential alarm calls specifically designed to communicate about predator behaviour [92].

Note also in Figure 4 that the brain mass relative to body weight of the yellow-throated miner, while closely related to the noisy miner, is well below the brain mass relative to body weight of the noisy miner, and so is the brain mass relative to body weight of the endangered regent honeyeater (Figure 4). However, the noisy miner, bell miner, and Lewin's honeyeater, as well as a cluster of other smaller honeyeaters, share in proportionately large brain weights relative to body mass (Figure 4). Larger brain weights relative to body weight usually indicate expanded cognitive abilities, with positive effects on sociality [130] and on dealing more effectively and quickly with complex and stressful situations and conditions [131–135].

To summarise this section, noisy miners are flexible; they are feeding generalists with sophisticated resource-searching strategies [126–129], and they support each other and defend others by driving predators out (see also Figure 3). They communicate effectively in variations of duets, alarm calls, and in referential and cryptic signals [93,129], and they even recognise each other individually through acoustic means alone [135]. Indeed, their vocalisations have been called a 'complex' acoustic repertoire [136]. They are resourceful problem solvers [137,138], cooperative, not just in raising offspring but in active conspecific and even heterospecific collaboration and defence [139], and cognitively complex [140].

Indeed, noisy miners are in the top 5% of honeyeaters in (log-transformed) brain weight relative to body weight (Figure 4) and cognition and most likely among the top 10 percent of passerines with a highly developed complex cognitive skill set, something they may have in common with many of the Corvoideae, such as Australian magpies and butcherbirds, and even with some corvids [141].

In my work with noisy miners, I also found that they recognise faces and cars (in prep.), just as do common ravens, *Corvus corax* [141], American crows, *Corvus brachyrhynchos* [142,143], Australian magpies [89], northern mockingbirds, *Mimus polyglottos* [144], and possibly a whole range of other avian species, as has recently been discovered in barn swallows, *Hirundo rustica* [145,146].

There is no doubt that the noisy miner is an exceptionally clever and resourceful bird, with the ability to access a wide range of foods, live near allies, and form close ties with conspecifics and even some heterospecifics for defence. What this miner species lacks in personal prowess and size, it makes up for with vigilance, cooperation in coalitions, and a range of remarkable cognitive abilities.

https://doi.org/10.3390/d18020

In future research, it might be worth comparing common raven life histories and behaviour with that of noisy miners and their cooperative behaviour, alliances, sub-alliances, and flexibility in food intake because noisy miners, just as ravens, may have among the most complex social organisation among songbirds [147,148].

3. Vulnerabilities of Noisy Miners

3.1. Noisy Miners and Natural Predators

No matter what a small songbird's attributes might be, noisy miners are at risk from a whole range of native predators, and their breeding attempts are regularly thwarted by a wide range of predators (Figure 5). There are also 'silent' or indirect killers, rarely taken into account. These are the Australian cuckoos, secretive quasi-predators, who disrupt or eliminate nesting attempts by woodland birds, including noisy miners. Native cuckoos are plentiful on the Australian continent. Europe has one cuckoo species, Australia has twelve species, of which only one, unusually, raises its own brood (the pheasant coucal, *Centropus phasianinus*). In 1933, Alex Chisholm wrote that "Sydney itself is surely the most cuckoo-haunted city in the world, since in any normal springtime, one may see or hear nine species of cuckoos upon its fringes, and some of them occasionally near the center of the city" [149]. This may have changed, since bird populations have declined in urban environments, but a recent survey of birdlife at Bootawa Dam near Taree (mid-north coast of New South Wales) listed several cuckoo species as being present in a relatively small area [150]. Fantailed-cuckoos, *Cacomantis flavelli formis*, preferentially parasitise the nests of flycatchers (Monarchidae), fairy-wrens (Maluridae), scrubwrens (Acanthizidae), and, in particular, brown thornbills, *Acanthiza pusilla*. Guppy et al. [151] showed that small woodland birds may have their nests raided or parasitised to the extent of 90%. Such losses might increase in fragmented woodlands, even on second or third renesting attempts, because activities near nest sites are easy to spot by predators and cuckoos alike. Fragmentation and diminished understorey cover of woodlands may further dramatically reduce the effectiveness of cryptic nests [152,153], and this factor may be one important contributor to the decline of woodland birds.

There are further risks for nestlings and even adults from a variety of other avian species. At breeding times, many birds that are otherwise peaceful neighbours can turn into dangerous marauders (no matter what their main food sources are as adults) and steal eggs or nestlings from other nests to secure high-value protein for their own growing broods [154]. Eggs are probably the single most sought-after food item in the world for any vertebrate (including humans) because of their concentrated protein and other nutritious properties.

Some songbirds, such as shrikes, however, are year-round killers of birds at any stage of development and also adult birds, especially targeting small woodland birds. In Australia, examples are the pied butcherbirds, *Cracticus nigrogularis*, and grey butcherbirds, *Cracticus torquatus* (Figure 5). They are known to prey on lizards and even snakes and on adults of small as well as relatively large-sized birds year-round [155]. They, as well as kookaburras, share the same forest environment as small woodland birds (Figure 5).

https://doi.org/10.3390/d18020099

Figure 5. Shrike (**left**) and kookaburra (**right**) with prey. Australian butcherbirds tend to wedge their prey between the forks of trees. The kookaburra has caught and killed a brown honeyeater, and the head of the small woodland bird hangs down lifelessly. The bird was consumed shortly thereafter. (Photocredit: **Left**: source unknown; **Right**: Teale Britstra, appeared in an ABC radio and online program called 'Australia All Over' [156].

Apart from raptors specialising in feeding on birds, such as the Little Eagle, *Hieraaetus morphnoides* [157], and the small Australian hobby, *Falco longipennis* [157,158], larger passerines, normally with various food preferences, may feed their nestlings on protein-rich fare. Currawongs, for instance, are largely fruit eaters [159] and, as mentioned previously, one of the most important species for seed dispersal of native fruiting shrubs across Australia. However, when raising their own young, currawongs will also kill lizards, steal eggs, and kill small nestlings, while Australian ravens, *Corvus coronoides*, may kill mice, lizards, and even relatively large birds, such as pigeons, noisy miners, and starlings [160].

However, there is one important reprieve for the victims. Large birds, including pied currawongs, commonly raise only one brood per season. Nest-raiding of other bird nests tends to stop completely when their own offspring fledge and the adults and juveniles switch to fruit and berries.

Food switching is quite common in birds for several reasons, such as seasonal supplies of food, changing nutritional needs [161], or a reduction in food acquisition. The latter means that smaller-sized birds can renest, and they usually succeed in raising at least one brood per breeding season (pers. observ.), although often at a smaller clutch size [162,163]. Parasitising a nest may be more destructive because the parasitised parents often successfully fledge the cuckoo chick and may not renest. In the case of parental death, the nestlings may starve to death unless youngsters are adopted, which tends to happen only very rarely.

One study showed that as many as 87% of currawong nests had been parasitised by channel-billed cuckoos [164]. Figure 6b shows the potential causes of nest failure in noisy miners and in woodland birds in general. Many nest failures in woodland birds, especially in cup- or dome-shaped nests (noisy miners have cup-shaped nests), are due to nest predation. Fulton showed that cup- and dome nests were at the highest risk level, and in his study, he identified 94 different nest predators in total [165], a long list of potential dangers during the breeding season, particularly for woodland birds in general or noisy miners specifically.

However, one would have thought that researchers keen on noisy miner suppression would be aware of their struggle with cuckoos and other nest raiders, but this is never or rarely mentioned when noisy miners and their allegedly destructive role are mentioned.

https://doi.org/10.3390/d18020

(a)

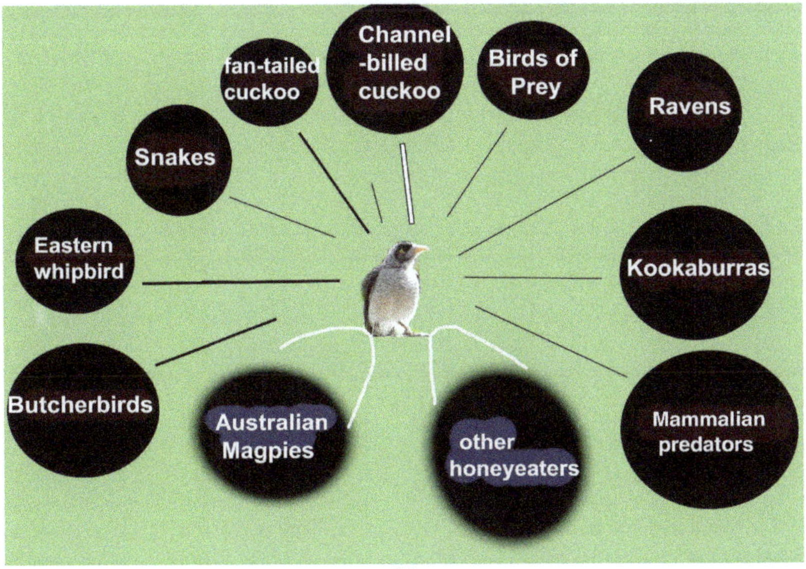

(b)

Figure 6. (**a**) Channel-billed cuckoo (migratory, from New Guinea). **Left**: Juvenile channel-billed cuckoo, the largest cuckoo in the world. **Right**: Adult noisy miner. A single channel-billed cuckoo female may parasite up to 5 nests per breeding season, one per host nest [163], preferentially choosing currawong and magpie hosts. (**b**) Aerial and other native predators and parasitic birds (red background) and straight lines, with the lower two circles (blue) with rounded lines showing potential allies. Centre image: Noisy miner. The inclusion of the eastern whipbird and the fan-tailed cuckoo is based on discoveries of their nest-raiding or parasitising [151]. Not listed are frogs and ant invasions, which are relatively rare. (Photocredit 6a + 6b: G. Kaplan).

https://doi.org/10.3390/d18020099

3.2. Difficulties in Reproduction

Overall, the impression has been that noisy miners have bullied their way into dominance, overwhelming hapless bird communities and taking over territories in large numbers. The idea of overabundance inevitably suggests that a species is particularly successful in reproduction due to having access to plentiful resources, showing foraging efficiency, and/or facing low predation risks.

Further, successful reproduction may depend on good defence of an existing clutch or successful renesting to counteract the negative effects of various reproductive failures within the same season. Small passerines in Australia tend to have larger clutches and longer breeding seasons and may breed more frequently within one season than larger species [166], that is, some species may have time to renest, suggesting that they may raise at least one surviving clutch per season. However, the success rate of surviving nestlings may vary vastly between species, regardless of intact or fragmented environments, and also within species of small woodland birds, as earlier research by Ford and colleagues found [167]. It is a biological puzzle that the reproductive data of noisy miners do not fit the description of an 'overabundant' species. Noisy miners tend to have the smallest clutch sizes of any small Australian passerine, on average, just two eggs [168]. Dow [168], who conducted one of the very few detailed studies on breeding in noisy miners, found that while adults were conscientiously and incessantly feeding offspring (at rates as high as 87 times per hour), nest success was extremely low at 15% and at best 16.7% in any given year, in a three-year study (Table 2). The study area offered a reasonably large sample size—within seven hectares, birds had been colour-ring-banded. The sex ratio favoured males by as much as 3.3: (with most males acting as helpers). In the study area used by Dow, the noisy miner population did not grow but declined over the three years from 9.0 to 7.7 birds per hectare [168], despite a very long breeding season. In fact, the breeding season for Meliphagidae is particularly long. Noisy miners seem to be able to breed at any time of year [168] but generally do so between June and January. Arnold's study of nesting success and timing/position [169] also found that nearly twice as many nests were built in winter/early spring (August/September) than later in summer (from October to January) and that early broods were significantly less likely to be preyed upon than those built later in the year [169], a view also reflected by the NSW Department of Education [170].

Table 2. Nesting success of various species (NSW only) (Source [167], excerpts only). There are 3 categories: very low (less than 20%), low (more than 30%), high (more than 50%) of nestling survival. Examples are of (a) intact or (b) fragmented environments.

Species	Latin Name	Habitat	Nesting Success (%)	Reference
		(a)-general (b) fragmented	**less than 20%**	
Scarlet Robin	*Petroica multicolor*	a	10	Robinson [171]
Rufous Whistler	*Pachycephala rufiventris*	b	13	Bridges [172]
Noisy Miner	*Manorina melanocephala*	b	15	Dow [168]
Brown Thornbill	*Acanthiza pusilla*	a	16	Bell & Ford [173]
			more than 30%	
Superb Fairy-wren	*Malurus cyaneus*	b	37	Nias [174]
New Holland Honeyeater	*Phylidonyris novaehollandiae*	a	37	Paton [175]
Red Wattlebird	*Anthochaera carunculata*	b	38	Ford [176]
			more than 50%	
Yellow-rumped Thornbill	*Acanthiza chrysorrhoa*	a	52	Ford [177]
Willie Wagtail	*Rhipidura leucophry*	a	67	Cameron [178]
Rufous Fantail	*Rhipidura rufiventris*	a	71	Cameron [178]

https://doi.org/10.3390/d18020

Table 2 clearly shows that in terms of reproductive success noisy miners rank in the lowest group of woodland nesting success [179]. In summary, the unusual combination of the exceptional qualities of noisy miners, linked with the pronounced vulnerabilities of the species, has never been fully considered or explained. However, neither aspect (smart adaptations on the one hand and vulnerabilities on the other) would seem to qualify noisy miners for the harsh assessment that, single-handedly, these small passerines are allegedly responsible for the decline of the multitude of small woodland birds in their range.

3.3. *From Protected Native Species to Outcast*

The call to cull noisy miners, as mentioned previously, reached the level of state government intervention in 2013, devising a framework that legally enshrined a statement that the noisy miner was a threat. It explicitly permitted noisy miners to be suppressed or eliminated [38]. Alarmingly, this officially permitted the killing of noisy miners where and when considered 'overabundant' [31,38]. This was supported and welcomed by researchers at major Australian universities. Approved culling was carried out at specific sites, systematically testing at each site whether the removal of noisy miners improved the conditions for other woodland birds [22–25,28,29,37]. Thousands of miners died in the process (the most detailed studies across several sites alone stated that well over 6000 miners were killed). The results ranged from mildly successful, in the sense that numbers of some small woodland species briefly increased, to unchanged conditions for other species and outright failure. As one report stated, "Before the culls, we counted 510 noisy miners across the sites, and after the culls, there were 512". This was clear evidence that culling noisy miners does not work in certain landscapes [180].

Noisy miners, dubbed "a reverse keystone species" [97,111,181], have been a strong focus of research and 'ameliorative' attention for several decades. Reverse keystone species is the name given to a species that is thought to be disruptive to other species by competing for the same resources, driving competitors out, and thereby decreasing the biodiversity and local stability of bird communities.

There are apparently sites where noisy miners appear in large flocks of hundreds and then stay, while chasing out or displacing other avian species [22,25–27].

This may not be the norm, of course. The noisy miner communities I have seen in many different locations in New South Wales consisted at best of 12–20 individuals and often fewer than this. And my observations tally closely with early fieldwork observations of noisy miner behaviour [95,139], before the most recent waves of drastic clearing of Australia's subtropical, temperate forests, and bushlands happened between 2000 and 2017, totalling some 43 million hectares [182,183].

This level of destruction of prime habitat in such a short time is almost unfathomable in terms of its impact and consequences for the entire natural world. Of this total deforestation, nearly a million hectares of forest were cleared in Queensland and New South Wales alone in a mere decade, often in hotspot areas [184,185]. I suspect that the cleared areas were precisely the favoured habitat of noisy miners.

The recent and overwhelmingly negative verdict on Australia's conservation record, expressed in statistics on deforestation (amongst the worst in the world according to the *WWF Living Planet Report 2024* [183]), makes blaming the decline of woodland birds on one species, the noisy miner, an argument ad absurdum. Surely, it is impossible to overlook the fact that small birds, especially species dependent on forests and woodlands, are the most likely group of animals to be directly harmed by the drastic and relatively sudden contraction of their habitat.

Further, from the discrepancies between my observations and those that report 'over-abundance' and have recommended culling, one may surmise that incidents of the so-called

https://doi.org/10.3390/d18020099

disruptive behaviour of noisy miners have remained sporadic and, so far, seem confined to specific localities. The noisy miners' change from living in small family groups to forming large assemblies, if only in some locations, is significant in terms of social behaviour and needs to be explained first before proceeding to kill native birds

Those who have promoted 'culling' seem not to have queried such inconsistencies, let alone explored any potential alternative reasons for the dramatic switch in social groupings and behaviour by noisy miners. At the very least, one ought to consider explanations for this abnormal and sudden alleged clustering in large groups. It is possible that some of their behaviour might well be related to the substantial curtailment of suitable habitat, which, by itself, has nothing to do with being 'despotic' [25–27]. The moment scientists switched language from scientific reporting to calling a species 'despotic' or even 'notorious' [31] suggests a noteworthy shift from science to opinion.

3.4. *The Victimisation of Noisy Miners—A Miscarriage of Justice?*

So far, the debate on the role of noisy miners has centred on a very skilled public trial of noisy miners until public opinion of this species shifted to consider noisy miners the 'most hated bird' in Australia [35], and government support for their culling was achieved surprisingly readily [21,40]. The one-sided evidence contributed to a view that noisy miners are aggressors, not, as they often are, victims of bad luck and consistent threats around them at breeding time, very similar to the predatory and/or also anthropogenic threats experienced by other small- to middle-sized woodland birds and indeed birds in general.

Furthermore, the assumption that one single native species is largely responsible for the decline of other woodland birds is unlikely. Monocausal explanations are rarely correct, considering the complexity of natural (dynamic) environments and the anthropogenic factors acting on such environments. For instance, vegetation heterogeneity is of great basic importance (known as the vegetation heterogeneity hypothesis [186]) but has rarely been critically examined in research papers on noisy miner 'overabundance' (more on this can be found in Section 6.3).

The 'aggression hypothesis', which has resulted in the extensive culling of thousands of noisy miners, may be partially misleading. Instead of aggression, one usually refers to 'resource competition' as one of the well-established drivers of evolution. The damage that large flocks can cause (through displacement and taking over food supplies) may account for less overall damage to other birds, however, than actual predation rates and environmentally detrimental factors, as found in several studies [7,12], and this may apply to noisy miners as well, perhaps specifically in fragmented habitats. Assessing resource competition has also become more complicated due to the human role as major competitors, working in practice very much against biodiversity [187,188].

The designation of noisy miners as a 'reverse keystone species' may also be problematic, as stated in the recommendations to cull noisy miners [97,111,181]. One notes that neither European nor African ornithologists use the term 'reverse keystone species' for their native avian species that flock and are known to be a problem for agriculture. For instance, the flocks of starlings (*Sturnus vulgaris*) in Europe [189,190] or of quelea (weaver family Ploceidae) in Africa [191] can be spectacularly large. However, the purpose of such displays (murmuration) of thousands of starlings, a breath-taking spectacle even if the birds cause damage, has been found to be largely an anti-predator strategy, i.e., an evolutionarily important adaptation to the many aerial predators of these species [190,191].

As far as can be ascertained, the noisy miner may be one of the very few, if not the only native songbird in the world that has been named a reverse keystone species, suggesting an extravagant exaggeration, the consequences of which are deadly for noisy miners.

https://doi.org/10.3390/d18020(

Finally, in the specific case of noisy miners, there is also an opportunity to reexamine theoretical frameworks. For instance, the role of competition and cooperation in evolution. As Heinsohn and colleagues put it so succinctly, 'cooperate or speciate' [2]. Conversely, Day and Young asked what the role of competition is in evolution. They argued, and I quote:

> "Theory and empirical evidence have shown that competition between existing species can promote evolutionary diversification through ecological character displacement and, more recently, that competition may play a role both in producing species and in driving their evolutionary divergence". [187]

Competition has, after all, a major facilitative role to play in evolutionary diversification as well as stability [192], and such theoretical positions raise doubts at least that competitiveness in some contexts of noisy miner behaviour is a convincing example of a reverse keystone species. One may therefore surmise that decisions about the impact of noisy miners (even if sporadic and confined to specific areas) were made prematurely, unilaterally, and likely based on incomplete or biased information.

4. Pitfalls in Measuring Bird Numbers and Drawing Conclusions

Since noisy miners have been identified as being 'overabundant' by the pro-culling lobby [32,33,39], it is relevant to at least identify some ways in which a false perception of 'overabundance' may arise. In Australia, as in many other countries, knowledge of bird numbers has been largely or even exclusively derived from data collected via citizen science, a large cohort of volunteers who have signed up to take notes on species and numbers of birds, as well as sightings of nests, in assigned transects. The data available from this activity are then used to draw conclusions about nesting locations, abundance, and overabundance of individual birds and species for annual or longer-term records and trends. Given the willingness and the readily available large numbers of volunteers, whether for the authoritative 'Atlases of Australian Birds' [193,194], for 'BirdLife Australia 2023' [195], or for very specific projects such as 'Big City Birds' [196], citizen science has become a tremendously important and indispensable part of gaining information about bird abundance and bird locations in Australia and worldwide [197,198].

While valuable, it is important to be aware of possible skewing of such citizen data collections. Some recent publications have analyzed the potential reasons or risks for such a survey, which may have resulted in skewed data. One main concern is the size of birds: large birds are apparently recorded more often than small birds [199]. In addition, participants in cities and towns may have made many sightings of one or two species without ever leaving their own backyard. City-tolerant avian species may appear abundant when they stay/become sedentary in specific backyards and are therefore recorded frequently, i.e., such surveys are, in fact, not necessarily designed to be capable of identifying a decline of even common birds, as a Danish study found [200]. A recent American study argued that recorded abundance and the actual annual population decline of the same species are unrelated factors [201]. These are important findings and worth noting in the context of native Australian birds deemed overabundant (noisy miners) and/or 'nuisance' native species, such as some cockatoos.

As native species, whether they be songbirds or psittacines, they should normally fall under the legal environmental protection act (Environment Protection and Biodiversity Conservation Act 1999 (EPBC Act)), but their designation as 'overabundant' makes it possible for them to be officially/legally shot in large numbers [21,39,40]. It might also suggest that perhaps the existing 'environmental protection legislation' may be too weak if loopholes and amendments can so easily override such protection. In fact, the lack of enforcement of the act and its not being 'fit for purpose' have drawn much criticism [202].

https://doi.org/10.3390/d18020099

At the very least, turning against a specific native woodland bird (noisy miner) with a long Gondwanan evolutionary history and blaming it for the decline of other native woodland species must be regarded as highly controversial if not outright questionable. Yet these are firm and ongoing determinations and decisions, rarely publicly debated, let alone contested. Other Australian native species, also believed to be 'overabundant', are sulphur-crested cockatoos (*Cacatua galerita*) and little corellas (*Cacatua sanguinea*). These cockatoo species, including galahs (*Eolophus roseicapilla*), although for different reasons, have suffered from official condemnation, partly also dubbed 'expanding' species. In their cases, either implied invasion or overabundance seems to provide sufficient reason to remove these species from the list of protected species, leaving them vulnerable to being destroyed, whether in the name of science or for the sake of agriculturalists and industrialists [203]. It is more than clear that removing any native species from legal protection against poaching, poisoning, and shooting is anti-conservationist and a substantial embarrassment in terms of the international agreements Australia has signed. Such agreements state directly that each signatory is involved in safeguarding, restoring, protecting, and supporting native wildlife.

In 2025, the Department of Primary Industries and Regions (PIRSA) of the Government of South Australia conducted a workshop for stakeholders on 3 February 2025 titled 'Pest bird impacts in South Australia' [204] with the objectives of (1) understanding and quantifying the impacts of pest birds, and (2) planning for the development of a new bait for pest bird management. One needs to read further down the document to find that the majority of so-called 'pest birds' are almost exclusively native species. The same government document (on p. 6) informs readers that the South Australian state government is working on a new *Biodiversity Bill*, under which, so the text claims, "all native species will be protected". However, on the same page there is a separate category called 'Unprotected species' (NPW Act) that may be shot or otherwise culled with or without a permit in some cases, and that the government is working towards a new and more effective bait 'solution' to deal with pest species (i.e., they mean *native* birds), even though there are alternative strategies already available that protect birds and agricultural producers alike [205]. Reframing native birds as threats to other species or as nuisances to humans offers almost a 'carte blanche' to accelerate the decline of any bird species that falls under such regulations.

5. Creating Misleading Information to Influence Public Opinion with Anti-Conservation Outcomes

5.1. Bias Against Native Birds

Australia's official actions against a native bird are not new. To give just three prominent examples, there were the infamous 'Emu Wars' against emus, blamed for the prickly pear problem, even though English settlers had introduced them for cultivation; however, due to emus enjoying and feeding on the fruit, the seeds spread and overwhelmed the countryside [206]. Instead of fining the farmers and trying to exterminate the introduced plant, a price was put on the heads of emus, and thousands were killed or their eggs collected. In Western Australia, the native Australian raven (*Corvus coronoides)* and in Tasmania, the forest raven (*Corvus tasmanicus)*, have repeatedly been classified as 'pests' in some regions [207]. Recently, there was also a concerted campaign against pied currawongs (*Strepera graculina*) because some ornithologists were certain then that currawongs were largely responsible for the decline of small woodland birds [208]. As recently as 2025, 'Birds in the Backyard', an online site [209], includes in its statement on pied currawongs that, "the currawong problem is a case of a predator getting out of control. ... [the species] is believed to have increased dramatically... pied currawongs are voracious nest predators with a strong negative influence on smaller birds in some areas" [209]. Such statements are very dubious claims. Phrasings such as 'out of control', 'increased dramatically', and

https://doi.org/10.3390/d18020

'voracious predators' may well be effective rhetorical means for achieving public support against a species but are not a good educational strategy.

Unfortunately, these claims are often unsubstantiated. They also invariably fail to mention the positive role that currawongs have in the ecosystem, referred to as ecosystem services [210,211]. Currawongs indeed have a key role to play in the ecosystem and are very important to Australian forests and native fruiting plants. As a largely frugivorous species, currawongs (Strepera spp.) are semi-nomadic and, apart from bats, are the most important seed dispersers of native berry-producing plants in Australia [212,213]. They also protect highland montane areas from stick insect (phasmid) defoliation [214]. Other vital information concerning their own experience as victims is usually also not mentioned. Currawongs are heavily parasitised, largely by two migratory species of cuckoo: the channel-billed cuckoo (*Scythrops novaehollandiae*) and the common koel (*Eudynamys orientalis*). In one specific study, it was found that 87% of currawong nests had been parasitised by channel-billed cuckoos, and this had led to a local decline in currawong numbers [214].

In the examples given above, the bias is obvious: it is achieved by using value judgements (anthropomorphising) and demonising the species, and, significantly, by leaving out vital information, as occurred in the recent literature on currawongs and now noisy miners.

While this paper has not been able to discuss the role of invasive species (incl. exotic species such as domestic cats), it is worth remembering, as Low and Booth pointed out recently, that 86% of Australian wildlife extinctions have been caused by invasive species [215]. Humans introduced some 650 vertebrate species to Australia [216], and cats alone kill over 250 million birds annually in Australia [12]. So far, there has been little visible effort and interest by governments and local councils to address and solve some of these very real threats (of unsustainable magnitude) to native birds and other native vertebrates. In this context, the pursuit of an alleged noisy miner 'problem' looks more like a strange side-show.

5.2. *Bias Expressed in Non-Scientific Language and Distortions*

The Australian Museum Fact Sheet maintains the description of noisy miners as 'bold and curious', but then, from the many facts the museum might have chosen to convey, the text proceeds to single out and describe the mobbing of hawks and kookaburras by noisy miners [217]. Kookaburras and hawks are, of course, both dangerous predators of small woodland birds, and some hawk species are obligate bird killers. Kookaburras, in fact, take noisy miner eggs, nestlings, and even adults (see Figure 5). Mobbing to deter predators is usually not described as aggression but referred to as anti-predator and/or territorial defence [218]. It is a functional behaviour that is common to most native Australian birds of different sizes and species. Such defence has evolved as a biologically essential strategy and an aspect of parental care to secure the survival of offspring anywhere in the world [219,220]. Why then single this out in a brief fact sheet as though predator defence was specific to noisy miners when so many positive things could be stated that would be biologically correct and important to convey and also support native birds.

Some of the biases obviously survive for a long time and stand uncorrected. The question remains, why would a museum, devoted to public education, single out mobbing behaviour without making it clear that anti-predator behaviour is a key element of the strategy for survival? This behavior is found in plants, in the smallest invertebrates, in birds, and in mammals. Indeed, predator–prey relationships are one dynamic driving force in shaping ecosystems, population sizes, biodiversity, and the very fabric of life. This predator–prey relationship is thought to be as old as the Cambrian period (Cambrian Explosion), some 500 million years ago, when the first mobile invertebrate predators had evolved [221]. Like every other organism, birds have survived by developing some form of

https://doi.org/10.3390/d18020099

anti-predator behaviour, but a comment like the one cited above, if it stands alone, may be read as if it were an especially negative trait of this specific species, the noisy miner.

To address inaccuracies in some of the public characterisations of native species, it is vitally important to provide balanced and objective statements about any species and to make it known publicly that all living things, here native birds, have an important role to play in the ecosystem, as mentioned previously [211]. Perhaps one should feel ethically obliged to agree that they all have a right to live.

5.3. Bias Against Avian Mobility of Native Species

Of late, there is another category used for native birds of Australia, calling them 'introduced' when found in geographical areas of Australia where they had not been recorded before [222,223], and this may even include geographically close regions within the same state. This seems to be borrowed from a description of plants, native to one area of Australia and then introduced to another (possibly with a different climate and different soil conditions), where they multiply like weeds, especially if competing plants or parasitising insects are absent in the new location. The idea that native Australian wildlife should conform to arbitrary state borders on this vast continent and that we can chase them from one area to another (with most of the movement caused by human activity and the removal of food sources and habitat) must surely be controversial and perhaps even bizarre.

The Victorian Corella Strategy 2022–2032, issued by the Victorian State Government's Department of Energy, Environment and Climate Action (wildlife.vic.gov.au), mentioned previously [203], does not use the word 'introduced' but instead reports that corellas have 'expanded their range' (see the section within the 'Corella Strategy' called 'Distribution and Populations', pp. 9ff), implying that either their population is increasing or they are now invasive. Neither may be the case. In natural environments substantially altered by anthropogenic activities (e.g., expansion of cities, logging, agricultural expansion), should one not expect that many avian species, as well as native mammals and megapods, may have been forced to move?

In general, the attitude at the government level has not encouraged conservation. Governments, whether local, state, or federal, have not always been known to set a good example for respecting and needing to protect Australia's unique flora and fauna. This has implications for public attitudes as well. Periodically, officially unprotected, i.e., targeted, native species (cockatoos, lorikeets) are found dead or dying in mass poisoning episodes of unknown sources [224,225], and on private properties, many non-official, even illegal culls are suspected to occur.

All too often, birds have been judged as not conforming to an expectation of staying put and behaving well. Given that the human species has been the most invasive on Earth (expanding its home range to every corner of every continent), one wonders how such judgements about bird movements can be justified. Of late, noisy miners are certainly not beneficiaries of any benign attitudes to birds.

5.4. Bias Created via Catch Phrases of 'Explosive Reproduction'—The Large Flock Syndrome

When large flocks of birds, including noisy miners, are seen, it is rather easy to sell to a wider public the idea that they have 'explosively' reproduced and increased in number. But such a conclusion can often be entirely incorrect: large flocks may have nothing to do with the overall increase in numbers.

Petrenko and colleagues [201] argued recently that the idea of 'explosive reproduction' may well be the wrong conclusion when seeing large flocks. The formation of large flocks of usually non-migratory birds may have several causes, one related to the 'boom-and-bust'

https://doi.org/10.3390/d18020(

phenomenon of arid Australia. The breeding strategy of Australian pelicans (*Pelicanus conspicillatus*) is one example [226], and that of nomadic red-necked avocets (*Recurvirostra novaehollandiae*) is an example of waterbirds [227]. They may fly thousands of kilometers to a fickle inland salt lake to breed, called Kati Thanda, Lake Eyre. They can only hope to breed successfully there if the lake, often nearly dry, fills to three-quarters or full capacity; the former may happen only every ten to fifteen years, whereas the latter (full capacity) has been recorded only three times in 160 years (1860–2025). This happens so rarely because it requires substantial rainfall and flooding in locations hundreds of kilometres away from the lake. Based on observations by Reid [228], in one of those very rare years at Lake Eyre, some 50,000 pelicans had raised 90,000 chicks to a successful stage of fledging. In other words, these are rare events, typical for boom-and-bust periods in outback Australia [229,230]. While 'explosive reproduction' was possible in one year, such singular events often just replenish numbers (lost through droughts) to their former levels, and this in itself does not indicate abundance, let alone 'overabundance'.

Flocking is a widespread phenomenon amongst birds worldwide, but most studied in detail are mixed-species flocks [231]. A substantial number of hypotheses have been developed to account for this behaviour [231]. One recent large-scale study of flocking established that even in single-species flocking, the overall benefit was shown to be an increase in adult survival rates [232]. Hence, flocking in noisy miners in recent decades and setting up large colonies are relatively new phenomena, and once one knows about the reproductive problems in noisy miners, this is a contradiction that will require further study and explanation in the future, beyond current management plans [233]. Importantly, if assembling in large groups fosters adult survival, and if this should apply to noisy miners, it might well have been triggered by a perceived threat. One may cite an example of an interesting inconsistency in the 2023 "Aussie Bird Count", which may explain the possible misinterpretation of the results [117]. In this count, the 'little corella' (*Cacatua sanguinea*) appears in 12th position in numbers counted, a total of 75,561 times. However, the rate of reporting is particularly low (only 9.27%), and this can only mean that many corellas were observed in large clusters but in very few locations across Australia [117]. This corresponds with my outback research experiences that corellas could only be seen in very few locations and then in flocks larger than they usually prefer to travel [234]. It is hardly a sign of overabundance when most of this vast continent had to be scanned to find corellas at all. In 2022, their presence was confined to only a few places. Yet councils and policymakers tend to be readily persuaded that the appearance of a large flock means abundance or 'overabundance'. As previously mentioned, calling noisy miners 'overabundant' [233] may not be correct either. Citizen science is often city-based, and noisy miners have adjusted well to urban life; hence, miners have been recorded as one of the top three birds seen [117]. This may be true, but unless colour-banded, noisy miners are indistinguishable from one another, and since they are territorial, the same individuals could have been counted on multiple occasions.

It is also worth remembering that systematic attempts to document which avian species are most threatened with extinction are rather recent in Australia, marked by a 1990 publication [234] and finally enshrined in the 'Endangered Species Act 1990', then replaced with the 'Endangered Species Protection Act 1992', coming into force in 1999. Consistent and reliable monitoring of specific species in Australia is still relatively selective (specific to threatened species; Figure 7) and has only increased and been consistent in the last few decades or so. One suspects that there are still substantial gaps in our knowledge of the presence and abundance of certain species [235]. Moreover, one may further surmise that the role of native avian species in the Australian ecosystem, in some cases, as well as their behaviour, may not yet be fully understood. As Rayner and colleagues argued [5], apart

https://doi.org/10.3390/d18020099

from a relative paucity of rigorous population studies and time frames that are too short, observed changes in year-to-year population size can be driven by environmental factors not included in the study [5].

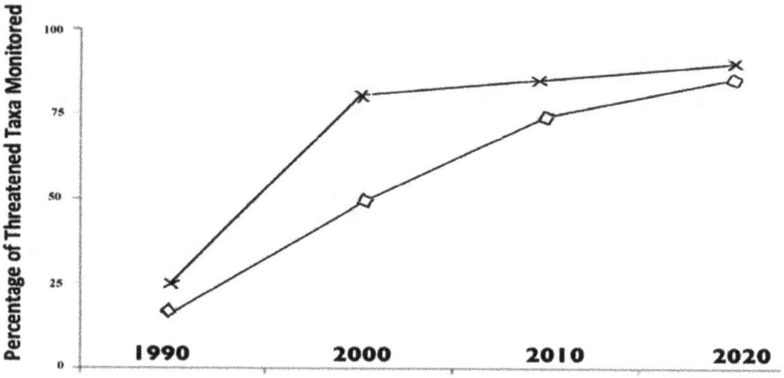

Figure 7. Percentage of endangered species monitored over time. Crosses: parrots; squares: songbirds. Excerpt and simplified from Verdon et al. [235].

Obtaining comprehensive data across all living things in Australia, whether invertebrates, fish [236], amphibians [237], birds [238], mammals [239–241], forests, or plants in general [242], is a mammoth task in a vast country/continent with a small population. Not surprisingly, there are still major inadequacies in monitoring Australian biodiversity [243,244]. Also, soil conditions have rarely been mentioned but are very important to consider for the survival of invertebrates relevant to the conservation of woodland birds that are insectivorous and terrestrial [245].

Yet, whatever the shortcomings, the available literature documents incontrovertibly major and dramatic downturns in bird numbers (an average of 44%), mammals (35%), and plants (72%) [239] in the first 25 years of the twenty-first century alone [246]. Focusing on one species, here the noisy miner, as the sole cause of all problems, obviously ignores many factors. It is important to remember that small woodland birds and larger passerines have coevolved and have lived and survived for millions of years, suggesting that checks and balances also evolved to ensure the survival of all. In a study of the nest fate of regent honeyeaters, the researchers found that daily nest survival rates showed no negative relationship with predator-community measures but rather a non-significant positive relationship between nest success and both predator abundance and predator species richness [247]. These are obviously complex relationships that may trigger different responses under different circumstances.

One would have to conclude that presenting one or two avian species as perpetrators and all others as victims is a misrepresentation and oversimplification of the dynamics of bird communities, and this also neglects the idea that human action led to the unstable and detrimental habitat conditions.

5.5. *The Problem with Bias: Ultimately a Backwards Anti-Conservation Logic*

While Australia has the highest rate of mammal extinction in the world [248], actual extinctions of birds (and those seriously endangered) have so far been largely confined to islands due to invasive predators and landscape modification [246]. Indeed, Australia has experienced radical alteration of its landscape due to the logging and felling of some 80% of pre-colonial forests, agricultural expansion, and the growth of cities, along with their increase in landscape-wide impervious surface areas from zero to about 80%, resulting in an almost immediate reduction of 50% of bird species. Climate change is also usually mentioned. And, as Wenzel et al. noted in a different context and geography, overall, open woodland birds are the biggest losers because their habitats have disappeared, been fragmented, or significantly altered in other ways [249]. One notes that noisy miners are also woodland birds, which might indicate that noisy miners, like other avian species, need help. They need help because living in constantly altered habitats or shrinking suitable habitats has and continues to turn many species into refugees in their own native environments. But this is not the conclusion drawn by lawmakers who have declared noisy miners threats to the environment and then allowed them to be eradicated and/or their numbers reduced wherever and whenever indicated [21].

In the 1800s, John Gould described noisy miners' social and individual characteristics as typically "moving in groups of four to ten individuals" and the general disposition of the species as being "restless, inquisitive, bold and noisy" [250]. 'Noisy' usually refers to alarm calls being issued if there is a perceived danger. Little had changed in this perception of noisy miners, but in the space of just a few decades, human opinion of this species has turned from a charming honeyeater to a rogue bird, from an assessment of an exquisite and clever little songbird to a destructive and dangerous bird that had to be sacrificed, often termed 'miner suppression', as though the killings could be made more palatable by using the term suppression [29,38].

This elimination of a small songbird probably represents one of the most dramatic and extraordinary public responses at government and senior academic levels to a single native bird. Beggs recently wrote a very detailed account of the changes in attitudes to noisy miners from positive to negative [18]. Suffice it to say that, ethically [55], the approach to noisy miners has been questionable and scientifically mono-causal or extremely reductionist in its explanations [251]. Such an approach, guilty due to limited evidence (and even inconsistent evidence, as shown here), reduces the chances of resolving any complex problems under investigation. This approach again raises questions of ethics, of commitment to local wildlife, and of research methods and data that can lead to intuitive but perhaps premature and certainly narrowly focused and unhelpful conclusions.

One has to be extra careful about the political fallout and by-product of blaming one native avian species for the decline of other birds and even claiming that 'habitat-clearing, fragmentation, and degradation' have created more suitable habitats for noisy miners [252,253]. This sounds very much like backwards logic against evolutionary evidence and conservation aims. No doubt, logging companies will look particularly kindly on such expert conclusions. This leaves logging companies, multinational agribusinesses, local councils, and politicians entirely without blame for the decline of biodiversity, plants, and animals in general. This is astonishing since we know that at the turn of the 21st century, approximately only 8% of woodlands in any climatic region remained in Australia [182]. Moreover, Watson and colleagues showed that there were substantial negative effects on birds due to loss of habitat and habitat complexity [254]. This, in turn, has knock-on effects, for instance, in the agricultural sector, such as the decline of insects [46,255], not to mention the devastating effects of chemicals used in agriculture, of which 42% are known to be lethal for any biota, mainly insects, including birds [256].

As many as 85% of insectivorous species foraging above ground were statistically affected by patch size and/or loss of habitat complexity, including many small woodland birds, such as Jacky winters *Microeca fascinans*, Australian robins (Petroicidae), weebills (*Smicrornis brevirostris*), golden whistlers (*Pachycephala pectoralis*), thornbills and gerygones (*Acanthizidae)*, pardalotes (Pardalotidae), as well as some honeyeaters, including noisy miners, that feed on insects from ground level but also glean insects from foliage [159,256].

It is very difficult to imagine that noisy miners were not as affected by these dramatic changes as the other small, afore-mentioned woodland birds, and even more difficult

https://doi.org/10.3390/d18020099

to believe that noisy miners specifically were the *only* species that allegedly "benefitted" from the destruction of their habitat or from the presence of white settlers, as has been claimed [253].

In fact, not all surveys of birdlife in eastern Australian woodlands have shown a decline of woodland birds or specifically highlighted a dominance of noisy miners. In a century-long study of a region in the south of New South Wales, the Tapitallee region, it was shown, for instance, that small woodland birds had thrived and noisy miners had declined over the years [88]. Some writers have argued that there is a problem. Culling programs that play off one native species against another native species are of concern to many conservationists [44,257] and one might consider the view by Batavia and colleagues pertinent to the treatment of noisy miners, and I quote, "Under these circumstances, continuing an ineffective program that perpetrates mass violence is not a right decision. It is simply wrong. Conservation is riddled with such moral conflicts." [258].

6. Environmental Factors: The Understorey and Characteristics of Woodlands

6.1. Nest Position and the Importance of the Understorey

Arnold's study [169] is one of the very few more recent research reports on noisy miners that has paid close attention not just to the timing of breeding but also to nest position. Importantly, the study found that nests in low positions (less than 2.5 m high) were more likely to be successful than nests built at heights of over 4 m. Low nests, Arnold explained, are less visible and less accessible to raptors [169].

The latter point is of special interest: researchers involved in projects designed to cull miners have repeatedly claimed that noisy miners prefer woodlands without an understorey. A fact sheet on noisy miners, made available online by the NSW Department of Education (n.d.) [170], states that "Noisy miners prefer to live in areas with eucalyptus, namely trees, and *a clear understorey*" (emphasis added) [170]. It is hard to find any evidence to show that this is, in fact, the case, but this alleged "fact" has been copied from one publication to the next, including, damagingly, by the NSW Department of Education, information that can trickle down to the entire education system [170].

Hence, the claim that noisy miners prefer no understorey is clearly contradicted by the carefully observed behaviour and circumstances in Arnold's study of the impact of timing and location on nest success. To repeat, Arnold's results show that noisy miners are more successful when nesting in a *lower understorey* [169]. An understorey is usually described as consisting of plants with heights ranging from 2 to 4 m (and possibly lower than 2 m), suggesting that contrary to the claims made for many years now, the understorey might well be of major importance for the breeding success of noisy miners and possibly of other woodland birds. Conversely, it should follow that patches without an understorey and consisting of trees with heights of 10 m or more are not suitable environments for noisy miners to successfully raise their young. One may therefore surmise that the assumption of a cleared understorey as the *preferred* habitat for noisy miners is at best on very shaky ground, if not outright incorrect.

No doubt, noisy miners have been seen to occupy open woodlands without an understorey, but does this mean that (a) they choose to be there, and (b) they prefer such an environment? It is also possible to argue that they can be found in such open spaces because they can find no other, more suitable environment. This point is discussed further in the next section.

https://doi.org/10.3390/d18020C

6.2. Ideal Environments but No Noisy Miners?

In a long-term banding project (2007–2020), bird numbers and species were recorded in a 54 ha area designated as an environmentally significant zone (the River Flat Forest) at Camden Airport, about 30 km from Sydney near the Nepean River. Although most of the large river eucalypts had been removed during WW II, there has been significant (secondary) regrowth, with an understorey and woody weeds at the time of publication [259], and the area was recorded as being home to some 50 species of birds, 20 of which were small woodland passerines, among which 9 species were honeyeaters, including the bell miner (*Manorina melanophrys*), but no noisy miners were ever recorded [259].

These discrepancies raise several concerns: Why would the behaviour of noisy miners vary so markedly at different sites? Why do they flock in large colonies in some places when the previous literature suggests that their preferred social units 'normally' consist of two to three pairs and helpers? Most reports on noisy miners describe their preferred habitat as open woodlands. The problem is that such generic descriptions may not be sufficient in any project that describes a decline or increase in species and numbers. Steve Debus's account [260] is one of the most detailed descriptions of the kind of woodland on which his study was centred. He writes that noisy miners primarily inhabit dry, open eucalypt forests dominated by spotted gum, box, and ironbark, as well as degraded woodland where the understory has been cleared, such as recently burned areas, farming and grazing areas, roadside reserves, suburban parks and gardens with trees and grass, but "without dense shrubbery", not without any shrubbery—a factor that might be of some importance. BirdLife Australia, in one of its ten booklets for the public, proposes to "restore native understorey in remnant patches or establish shrubby understorey in order "to deter noisy miners or other competitors", suggesting that noisy miners avoid the native understory but not specifying what the vegetative conditions should ideally be [261]. The comment makes little sense in light of Arnold's results, since breeding success in noisy miners was shown to be greater when nests were placed in understorey plants of 2.5–4 m height or even lower [169].

6.3. Lacking Detailed Descriptions of Woodlands/Open Woodlands/with or Without an Understorey

Descriptions of 'woodlands' and 'understorey' in the noisy miner literature are often too imprecise or contradictory, making it difficult to obtain any factual basis for the species' alleged 'preferred' environment. Therefore, it is vitally important to identify important nuances in environmental architecture, including height, structure, or density.

There are two ways to remedy the general vagueness in the description of the architecture of plants in environments, or, I believe, the discredited claim that noisy miners prefer to have no understorey [170]. The latter claim was only possible because the description and naming of plant architecture have not yet been fully standardised. Such standardisation is necessary and easy to achieve because an authoritative classification system already exists: *The Atlas of Australian Resources* (NATMAP) (vol.6) [262], which clearly defines the differences between forest and woodlands, as well as the differences between various types of woodland, understorey, and shrubbery. The definitions set out in volume 6 of the Atlas are easy to follow, and if these definitions were rigorously applied and made explicit across all projects, one might eventually find common ground and identify the specific environments in which species, such as small woodland birds and noisy miners, might find stability.

The '*Atlas of Australian Resources*' volume 6, titled 'Vegetation', provides a way to standardise exactly what a woodland is. Notably, an undisturbed woodland does have an understorey [262]. On p. 11, the Atlas presents a diagrammatic 'pictorial key to the

https://doi.org/10.3390/d18020099

structural forms of Australian vegetation'. There are drawings of six types of 'forest'. In their definition, the unifying characteristics are the closed canopy and the amount of light reaching the ground that defines a 'forest', identified as foliage cover:

(1) Closed forests fall into two groups: (a) more than 70%, and (b) 30–70%, meaning that in such environments, canopies overlap and create at best dappled light on the forest floor. There is very little left in Australia of 'tall, closed forests' (one of the very few well-known examples is the 'Daintree' in northern Queensland, which, since 1988, has been listed as part of the Wet Tropics World Heritage Area (WTWHA). Hence, all other treed areas are classified as 'woodlands'.

(2) There are five types of woodland identified. 'Open' refers to the fact that canopies do not overlap and allow more light on the forest floor: (i) tall woodlands (T2); (ii) woodlands (M2); (iii) open and (iv) low woodlands, whose canopies have 10–30% foliage cover; and (v) two types of open woodlands that provide less than 10% foliage cover. Open and low open woodlands have very sparsely distributed trees that offer less than 10% foliage cover and low native understorey vegetation [262], p. 11. Importantly, the height of trees in open woodlands is still about 20 m, and in low open woodlands it is about 10 m, but there is no understorey. Below these heights, classifications change to shrubland (4 m or less).

The woodlands in which I have observed resident noisy miners rarely belonged to the category of tall woodlands (T2) but largely to the 'woodland' category (M2), meaning that trees were 20–30 m in height and the crowns did not meet but were still relatively close together. Importantly, the T2 and M2 classifications include details of understorey heights that could reach 5–8 m (Figure 8).

However, parks established for human enjoyment, i.e., places where large numbers of noisy miners allegedly prefer to be, usually have very little, if any, understorey. In fact, these parks are artificially created, and the understorey is removed for safety reasons, i.e., to suit human needs. In natural environments, such parks may not fit into any of the categories described in the authoritative source of this Atlas.

In Figure 8, woodland profiles are shown. Figure 8D shows the parkland (woodland without an understorey) that more than one author claimed was the preferred environment of noisy miners [19]. The same statement has been repeated by the NSW Department of Education (n.d.) [170], as shown in Figure 8D. And this assumed fact is also built into description of noisy miners as a 'reverse keystone species' [181].

There is no evidence to show that a preference for an environment without an understorey for the species as a whole exists. At the culling sites, some large colonies of noisy miners have been seen in habitats with a 'cleared understorey'. The justification for culling noisy miners largely rests on both the numbers seen in one location and the assumption that they 'prefer' such a habitat and will fight to exclude others from their patch. Talking about 'preferred habitat' is risky at the best of times. As a generalisation, it is misleading, particularly when evidence exists of noisy miners breeding successfully within the understorey [168,169] (paper in prep.). To repeat, selecting the understorey for nest sites (especially at lower heights) has led to greater reproductive success than in higher trees or canopies [139].

One would therefore have to conclude that the assumption of noisy miners preferring no understorey, as so many authors have now claimed, is unresolved if not incorrect, unless noisy miners choose not to breed at all or risk even lower (and possibly unsustainable) levels of nesting success rates than those reported in the past [168] (see Table 2).

Alternatively, it is argued here that research on habitat preference (vegetation type and architecture) has simply not been described in sufficient detail and not standardised as to what understorey and 'woodlands' mean in height, density, and light. Habitat quality often determines levels of cooperation and thus also reproductive success. And since

https://doi.org/10.3390/d18020

the lives of native birds depend on human judgement, this would seem to be especially urgent. With some exceptions [221], many recent publications proposing or reporting culling events of noisy miners do not spell out details of the physical properties of the selected noisy miner habitat and sometimes avoid doing so by calling noisy miners 'edge specialists'. To end one aspect of controversy and contradiction, standardised habitat descriptions could be adopted in all papers using the categories provided by the Atlas (vol. 6, Vegetation [262]). This would help distinguish degrees of habitat complexity and enable more precise predictions of the suitability and sustainability of native bird assemblies in general and of small- to medium-sized woodland birds in particular.

Figure 8. Woodland profiles. (**A**) The amount of light penetrating through the canopy determines the category. This is an open woodland scenario: a large blackbutt tree towers above the rest but offers an open canopy. (**B**,**C**) Vistas of the substantial but open understorey, varying in height from 2 to 6 m. Note the small white bar on the lower left of (**C**), which indicates a height of 2 m. (**D**) A typical parkland landscape entirely cleared of the understorey, also called a 'woodland', which is not necessarily suitable for vulnerable woodland birds. Sources: (**A**–**C**) Photocredit: G. Kaplan; (**D**) excerpt of Lowanna Park/Chatswood, Sydney, with Red Mahogany and Sydney Red Gum, by Poytr CC BY-NC 2.0).

A very thorough study by Oldland and colleagues [263] of the environmental conditions in which noisy miners are found in Victoria suggested that the idea of 'preferred habitat' is more complex than just a question of understorey [263]. They found that the most powerful predictors of the presence of noisy miners at remnant corners were associated with deeper, more fertile soils and higher proportions of yellow gums (*Eucalyptus leucoxylon*). Yellow gums are prolific and reliable nectar producers, and nectar availability might be "more important in determining the attractiveness of a site to noisy miners than structural attributes like the presence or absence of an understorey" [263].

However, resource availability may also not be the most important factor that explains the absence or presence of noisy miners. One needs to recall that noisy miners are, in fact, more appropriately called feeding generalists than predominantly or exclusively nectivorous [125,159], as I have also observed them gleaning insects and smaller butterflies. Unfortunately, many urban-adapted birds have also taken to a variety of often harmful anthropogenic foods [264], a new avian habit that is a global problem [265–267]. Diet flexibility may be a specific adaptation to urban life, but substantial discrepancies still exist in identifying their preferred habitats in conjunction with resource availability. In a recent and very detailed review of workable and non-workable strategies for the conservation of woodland birds, Walsh and colleagues [268] showed that most conservation strategies yielded mixed results or were not successful. Condensed from hundreds of papers, mixed and unsuccessful strategies include fire control, management of noisy miners, and grazing management. Even natural regeneration and habitat protection on their own yielded mixed results.

The authors found, however, that methods of active revegetation showed consistently positive results for species richness and abundance across all bird species, including woodland birds in general and declining/threatened small woodland birds [268]. Similarly, leaving or adding large woody debris also showed consistently positive results across all birds and ground foragers.

One notes, by contrast, that almost none of the culling projects of noisy miners reported replanting activities (active revegetation). No attempts were made to reintroduce native plants, understorey, and/or ground debris to see whether, after a few years, the behaviour of noisy miners had changed, *prior* to making the decision to shoot the birds. Indeed, in only a few projects was some replanting undertaken or considered, even after the noisy miners had been removed, and then only to avoid the noisy miner resettlement. According to Walsh et al. [268], one could conclude (with hindsight) that none of the projects should have proceeded until the strategically important step of active revegetation was completed.

7. Scientific Objections and Alternative Explanations

One of the concerns underlying the biased characterisations of a native species is that the substantial findings of the proximate causes of avian behaviour were overlooked or not applied. It seems, for instance, that the only emotion that noisy miners are described as displaying is ongoing aggression, usually considered a rather extreme, if not dysfunctional, behaviour. Across all the literature against the noisy miner, it is evident that Descartes' views have not quite disappeared. He was thought to have believed that animals are genetically hardwired and are able to act purely mechanistically. In short, animals are 'automata', incapable of thinking, and, crucially, Descartes apparently believed and expressed the view that humans therefore had no moral obligation towards them. While these claims have been reinvestigated repeatedly [269], Descartes' views in the popularised version were influential well into the middle of the 20th century. In the mid-20th century, Tinbergen broke away from some of the implications of animals as 'automata' when he published two influential books on bird and animal social behaviour [270,271], for which he was later awarded a Nobel Prize. His admission that animals had emotions was modest but broke away from Descartian concepts of animals as 'automata'. He argued that animal emotions existed, were tied to motivational systems, and were basic to survival. To put it bluntly, the four fs are feed, flee, fight, and fornicate. These four basic tenets of emotions can be demonstrated and tested because each is regulated by hormones. Indeed, later research found that birds have the same or functionally equivalent hormones and neurotransmitters as humans [272], including serotonin, the dopamine system [273,274], prolactin (once thought to be unique to mammals), adrenalin, and corticosterone (the avian equivalent

https://doi.org/10.3390/d1802C

of cortisol), controlling and expressing fear. One does not necessarily expect a discussion on the function of hormones when concerned with the conservation of specific species. But should one not expect to ask why birds, here noisy miners, have switched behaviour and made a choice to congregate in large numbers and in environments they would not normally choose? First, it is worth remembering that humans, not other vertebrates, are generally believed to have a major impact on species extinction [275,276]. In other words, one would expect more detail by first trying to understand the environment from the bird's perspective.

How is this relevant to noisy miners and ongoing efforts to reduce their numbers? First, killing part of a community of a cooperative species is bound to have repercussions for their future success in breeding and even for their survival because of systematic culling. Possible motivations by noisy miners to behave as they do in some locations are questions worth asking, since the behaviour is not uniform.

It is generally agreed that fear is one of the fundamental emotions in all animals. In birds, there is a centre (both sides of the brain) designed to respond to such emotions called the amygdala or archistriatum. Activating the amygdala does two things: increases adrenalin and raises vigilance levels, enabling memory formation of the specific event that induced the fear or stress response [277,278].

Memory formation of this kind is crucial to survival. Fear can also be caused by uncertainty. Agnvall and colleagues [279] tested the relationship of fear (of humans) by examining a number of physiological and behavioural traits in junglefowl (*Gallus gallus*). They found that in fowl with low fear, basal metabolic rates were higher, feeding efficiency was greater, plasma levels of serotonin were higher, and exploratory behaviour was greater compared to birds with high levels of fear [279]. Ecology, not known to concentrate on emotions and cognition in birds, nevertheless recognises that fear is an important factor in bird behaviour, and this, I believe, might well have some explanatory power in the case of noisy miners.

7.1. *The Ecology of Fear*

Contradictory results and seemingly biased reporting and terminology have proposed many reasons as to why this is so. One problem is the characterisation of noisy miners as aggressive or hyper-aggressive. These words alone are value-loaded and generally not used in animal behaviour studies, brushing aside a century of research and fine discrimination of animal behaviour since Darwin, or at least since Tinbergen [280]. Indeed, papers on noisy miners have used the word 'aggression' and even 'hyper-aggression' rather lavishly when, perhaps, the term 'agonistic' would have been more appropriate. The latter tends to have survival-function connotations, such as nest defence, territorial defence [281,282], site-related dominance [283], defence of food resources, competition for a mate, and, if possible, driving out a predator or at least driving the predator away from offspring [284]. While styles and methods of these defences may vary between species, their functions are similar or identical and have a long evolutionary history. The question is whether the assertion of noisy miner dominance and driving out other species represents something fundamental and perhaps even alarming. It is possible and even likely that dominance behaviour may have developed in some locations in response to specific external factors.

7.1.1. Stress and Its Costs

Many external factors present in the environment (urban or natural, social or predatory) cause stress [285]. Short-term stress, such as predator–prey interactions [286], may result in actions that reduce stress [287], but in some situations, stress may become chronic, alter behaviour, and lead to poorer outcomes in social life, decision-making in general, and

https://doi.org/10.3390/d18020099

reproductive success [288–290]. Predator–prey interplay is a central theme in ecology [291], yet it seems to have played no role in the latest assessments of the behaviour of noisy miners. One needs to ask how it is that there are some noisy miners that suddenly appear in large social groups, seemingly at the exclusion of other birds, but not consistently throughout their range, and such new formations are found in some locations and not in others.

Owing to neuroscience research, it is now known that the fear and stress of life-threatening experiences can have a significant impact on the brain [292]. In humans, this is now mostly recognised and referred to as post-traumatic stress disorder (PTSD) [293], while in birds, such trauma is recognised but generally described as brain- and behaviour-altering events. In captive birds, however, PTSD has been adopted as a diagnostic tool [294]. At its core is the trauma of one or several usually life-threatening experiences that result in a number of qualitative and behavioural changes.

In 2006, Blumstein established what is known as the ecology of fear [295], and such fear is associated with identified risks, thus predominantly with predators. Notably, as Clinchy and colleagues showed, humans have now advanced to the role of 'super-predator', even more feared than larger native carnivores [296,297]. The ecology of fear in animals has been recognised as an important way to explain the total impact of predators on prey populations and bird communities, and it would be advantageous for the survival and health of avian species to understand that human presence is not an innocuous stimulus but one that, in many situations, contributes significantly to levels of fear and distress. Signs of 'hyper-aggression', as many researchers have noted in noisy miners [21–23], should immediately raise some serious concerns about noisy miners because it is atypical behaviour, out of season, and likely a consequence of ongoing stress and fear. Human presence alone and/or hostile actions against noisy miners may therefore substantially increase the birds' distress. One might expect in the future that observers watching such behaviour would show some empathetic concern for the birds' overall health and then investigate further.

Regrettably, researchers have chosen instead to interpret this in human terms as 'badly behaved bullies' and then made a case for their elimination. As Zanette and Clinchy put it so clearly:

"The ecology of fear posits that the behavioural, physiological, and neurobiological costs of avoiding predation ('fear' for short) may additionally reduce prey fecundity and survival, and the total reduction in prey numbers resulting from exposure to predators may thus far exceed that due to direct killing alone. If this is the case, then failing to consider fear as a factor risks profoundly underestimating the ecological role predators play." [298].

The distinction between functional and dysfunctional behaviour has become evident after 100 years or so of research into adaptive avian behaviour: almost all agonistic behaviour has a clear survival function [299] that is identifiable across vertebrates and indicates a long evolutionary history [220]. The 19th to 21st centuries have added a substantial number of anthropogenic stressors that did not exist before, which now often simultaneously affect the daily lives of native animals, including birds [13,300].

As mentioned previously, many studies have shown that any number of such stressors can lead to reduced foraging, increased vigilance [13,273], and reduced reproductive success, all factors that can affect behaviour and health [286–289], as they can negatively affect the functional diversity of birds [275]. The question is whether noisy miner dominance and aggression are species-typical or atypical behaviour, or have developed in some locations in response to specific external factors?

The basic emotion of fear is experienced during a stress response [301]. A study by Willcox found that free-living woodland birds in more fear-inducing habitats show a cognitive bias by making more pessimistic judgements about an ambiguous stimulus [302]. According to Willcox, location has a significant effect on fear levels, i.e., it induces different

https://doi.org/10.3390/d18020

affective states. Willcox showed that the impact of understorey cover is positive and human disturbance is negative on birds' welfare [302].

With such findings, it is at least possible to hypothesise that the lack of an understorey can induce fear because of exposure to an environment in which hiding or cryptic approaches have been made impossible and thus have exposed the birds to a greater risk of predation. Would one expect behavioural changes as a result of living in denuded environments?

There are several ways to answer this. Schmidt and Kuijper [303] suggested that the landscape of fear contains a "real death trap" because cleared environments are transparent, have no distinct refuges for prey, and make prey distribution more predictable for predators, even or especially when prey avoid the riskiest sites [303]. This is very relevant to noisy miner colonies having settled in open woodlands without an understorey. When the prey birds concentrate their foraging in fewer areas, this helps predators find prey more readily. Prey animals are aware of varying levels of predation risk 'at a spatial scale' [304]. If this is so, it is conceivable that noisy miners are also aware of varying levels of predation risk and have learned to protect themselves by forming large groups and warning other small birds of the high predatory risk. Indeed, for woodland birds, the heightened sense of exposure could well signal fear and induce chronic stress, i.e., in a negative affective state with physiological and behavioural consequences [305]. We have heard often enough that habitat loss and fragmentation threaten birds via structural changes and the ensuing reduction in resources, but as Teckentrup and colleagues [306] rightly said, this also occurs via inevitable modifications in species' interactions. They argued that an additional consequence "is perceived predation risk and hence feared, which is rarely explicitly addressed in studies on habitat modification" and, significantly, they found that fear alone had negative consequences for biodiversity [306].

The question is, why would noisy miners prevent other birds from taking residence in their territory when, in other contexts (woodland with an understorey), they live peacefully with the same species they chased away in open spaces? Clearly, such questions need answers and may require innovative research designs. We know that there is a link between cognitive bias and affective states [307]. Birds can be 'optimistic' or 'pessimistic' in their behaviour, depending on the experience and environment [308,309]. To hypothesise, it may be that other woodland birds, especially small ones, are not perceived by noisy miners as suitable recruits to help defend such an exposed environment, and hence, just to cope, the noisy miners require the recruitment of as many conspecifics as possible to defend a perceived high-risk environment. Koolhaas and colleagues suggested that stressors can result in proactive coping styles marked by higher aggressiveness, higher routine formation, and lower cue dependency [310]. Whether fear lowers or heightens response patterns, it is clear that "fear effects" can significantly alter the physiology, behaviour, and life history of prey species [311], or expressed differently, all of life becomes one of uncertainty and a perceived lack of safety.

The first response in perceived high-risk environments is usually vigilance behaviour [312]. Vigilance is an expensive form of self-preservation because the vigilant bird cannot feed while trying to look out for predators. And a simple solution, where possible, is to increase the size of the group. Vigilance for predators as a social phenomenon is one of the most studied aspects of behaviour under risk of predation [311,312]. The common observation is that individuals in a foraging group spend less time being vigilant with increasing group size, and larger flocks have a higher survival chance [313]. As shown for other classes of social animals, large-scale ecological effects can be linked back to fear. For example, the reintroduction of wolves to Yellowstone National Park increased predation risk and fear for elk, resulting in grazing changes and avoiding more wooded

https://doi.org/10.3390/d18020099

areas, which altered the vegetation in the park and literally changed the landscape to the point of stability for all animal groups [314].

It is also plausible that landscapes can engender fear when individuals feel too exposed. There may be no place to hide or no safe place to nest [315]. Urban environments, for instance, have been shown to induce fear. Behavioural observations of birds corroborate the importance of cover, not just for roosting and sleeping but also, and perhaps especially, when foraging. In a study on foraging, a video recording showed that Eurasian jays, when offered worms at an open feeder, rarely consumed the worms at the feeder but took them away to a more covered area. Carrying food to protective cover minimises exposure to predators [315], suggesting that birds perceive a high level of risk when foraging in open areas. Behavioural differences at different locations also imply that birds at 'high-fear' locations experience a negatively valenced affective state. These findings are consistent with previous research demonstrating that birds alter their foraging decisions based on their proximity to protective cover, leading Laundré et al. to argue that the concept of a 'landscape of fear' is valid and unifying in animal ecology [304].

7.1.2. Refined New Culling Techniques: An Elevated Level of Trauma for Birds

Unfortunately, the persecution of noisy miners has not ended. Just as this paper was about ready to be submitted in its final version, a paper was published that proffered a 'solution' to the noisy miner 'problem' and claimed to have found a new culling technique: culling by shooting noisy miners in the centre of the territory and letting those in the outer circle live. This was reportedly to prevent another group of noisy miners from recolonising the territory and was said to have made the surviving miners less aggressive and 'calmer' [316].

This claim of success in dealing with noisy miners and the interpretation of noisy miners' behaviour may be far off the mark. The culling technique, usually referred to as the 'doughnut tactic', may be new when applied to animals, but in the military, it is a familiar term used for eliminating the centre of a group containing 'high-value targets' and subjugating the remainder of the outer layer (hence, 'doughnut' tactic).

Stress tends not to remain with a target group but may spread from families to close allies and even throughout animal communities. As outlined in a previous section, considering the ecology of fear, the more probable interpretation would be to predict that the level of fear and stress in the surviving noisy miners was extremely high.

Also, a more likely explanation is that they are not 'calmer' [316] but traumatised to the point of 'freezing', with a poor prognosis for their own survival as a result of witnessing this event. One hopes that ethics committees in research institutes would counsel staff about the potential effects of this 'technique', be this for any purpose, let alone for conservation purposes.

No anti-predator defence that has ever evolved could have prepared miners to cope with this culling event: none of the strategies that any avian species might have perfected would have worked in this context. Had the surviving birds been tested for corticosterone levels, predictions might well be that they were highly elevated, i.e., showed signs of extreme distress, probably turning into chronic stress. These physiological changes would have been followed by behavioural signs of confusion, poor foraging, and a limited number of activities. Serotonin levels might have dropped to critical levels, leading to depression, poor digestion, and declining health. As noted previously and shown in countless studies, particularly stressful events can significantly alter the physiology, behaviour, and life history of prey species [317]. And there is a further consequence that may not have been intended but is well-known: the effects of frightening stimuli (such as predators) generally do not just remain with the targeted individual or species but tend to spread throughout the

https://doi.org/10.3390/d180200

assembly of species, that is, including any other woodland birds that might have witnessed the culling, including species that the culling event was meant to protect and help, thereby contributing to a further decline in or complete collapse of diversity [318].

7.2. Territorial Behaviour

The idea that a definition of territoriality should include 'aggression 'as a 'requirement', as Pike and colleagues recommended [305], defies long-standing documented evidence and insights. In 1956, Hinde confirmed and established the biological significance of territories and suggested that territorial defence and impulsive/emotive 'aggression' can be two very different things [281]. Indeed, a large number of species in most classes of animals, including invertebrates, have developed some form of strategy to protect at least their own breeding sites [172,174] that involves no aggression at all. Such strategies may involve deception, camouflage, or distraction displays, designed and usually successful in luring away predators. These activities are, however, largely confined to ground nesters and are common in grasslands and waterbirds worldwide. There are several kinds of these displays: feigning injury, performing impeded flight motions, feigning lameness, and playing dead. Sometimes, a combination of activities is used. Golden plovers feign injury. Oystercatchers (Haematopus ssp.) use 'impeded flight', hopping along with futile wing flutters or very slow flight. Some species mimic dying in groups, such as stilts (e.g., the pied stilt of New Zealand). Australian white-fronted chats (*Epthianura albifrons*), when performing disablement displays, encourage their neighbours to join in [319]. Stylised threat displays may involve vocalisations or parading around with wings fanned out widely. Other forms of territorial defence are purely vocal, of which duetting is the most common form. These vocal signals are usually respected, but they can also be deceitful [319]. Of course, there are also clashes and fights when nothing else works, but that is also nest defence.

Sometimes, small birds may have no choice but to flee and abandon their nest. However, not all small birds are entirely defenceless. For instance, the brown thornbill, *Acanthiza pusilla*, a bird weighing a mere 7 g, has in its repertoire aerial predator alarm calls that it can use deceptively to make potential nest raiders flee [320]. Many Australian small bush birds are not timid or anxious but quite feisty and fearless, at least when facing intruders or competition, and will mob potential predators even if they are substantially larger (including humans) [299]. Such bush birds include brown thornbills (mentioned previously); grey fantails, *Rhipidura fuliginosa* (8 g); superb fairy-wrens, *Malurus cyaneus* (10 g); rufous fantails, *Rhipidura rufifrons* (10 g); scrubwrens, Acanthizidae (12 g); willie wagtails, *Rhipidura leucophrys* (20 g); eastern yellow robins, *Eopsaltria australis* (20 g); rufous whistlers, *Pachycephala rufiventris* (25 g); sacred kingfishers, *Todirhampus sancta* ($\approx$50 g); and as many as nine species of honeyeaters, including the bell miner, *Manorina melanophrys* (10–12 g), and the noisy miner (50 g). During the breeding season, males of such small woodland birds can be seen attacking their image in windows or in rear vision mirrors on cars, indicating that any competing male would be fought.

While many confrontations among male birds may be well stylised and expressed in symbolic displays, in some cases, fights can be quite savage and, in equally matched males, may lead to severe injury or even death, which happens on rare occasions during territorial establishment or fights for a choice among capercaillie males, *Tetrao urogallus* [321]. These are some strategies for territorial maintenance, breeding success, and self-defence as part of natural selection.

Non-lethal alternatives in the management of a native bird would clearly be a more meaningful option, but to do so might involve several time-consuming and dedicated steps. One step involves appropriate plant restoration programs for natural open woodlands, including the understorey. This is a bottom-up approach to conservation. A top-down

https://doi.org/10.3390/d18020099

strategy, as argued in a recent PhD thesis by Fielding [322], considers the possible reintroduction/rewilding of top native carnivores. The author reminded us that large carnivores at the top of the food chain are also in decline but are essential for functioning and healthy ecosystems [322]. He commented on top predators that "they can trigger trophic cascades throughout an ecosystem" which, in turn, "can cause changes in behaviour at all levels": (a) a reduction in the abundance of mesoscavengers (which are a direct risk to woodland birds), and (b) decreased predation on smaller prey species, such as woodland birds, which are not the direct targets of top predators. And he believed that the reintroduction/rewilding of top carnivores is an important conservation tool for woodland bird communities [323]. As the dramatic example of the reintroduction of wolves to Yellowstone National Park so clearly showed, natural balance and dynamics can be, or become, self-regulatory once all the right elements and interactions are in place. And there is plenty of evidence to show that top predators have an important regulatory role in maintaining or restoring ecosystems [322,323]. In this chain of interdependent variables, it seems somewhat futile to single out a species, here the noisy miner, from among the multitude of living things and blame it for the failures above and below that interactive chain.

7.3. Myth-Making

Myth-making has been used rather assiduously in the pro-culling literature on noisy miners, either through omission, suggestiveness, conjecture, or other rhetorical means:

(1) Describing noisy miners using vocabulary usually reserved for value judgements of human behaviour (such as 'bullies', 'despotic' [27–29], and even 'notorious') is counterproductive [31].

Comment: Name-calling is a powerful rhetorical weapon for denigrating and/or digressing, but it is not science. Moreover, throughout the literature on noisy miner culling, the harmful role humans have played in habitat destruction (now often called the sixth mass extinction) of the Anthropocene is consistently excluded.

(2) Myth-making has been further strengthened by the claim that noisy miners have "benefitted from habitat-clearing, fragmentation, and degradation", because these have allegedly created more suitable habitats for noisy miners [253].

Comment: Since when have species benefitted from global human action? This may be a comforting thought, but it contradicts the documented evidence of both, the accelerating decline as well as the increase in extinctions of living things in the Anthropocene. In current prognoses, avian species rank highest in either category among all recorded classes of animals [[12]; Figure 1].

(3) Some writers added 'overabundance' as an ecological category, and this claim has since been further embellished. In a media publication from 2019, it was claimed (or perhaps misquoted?) that the "lead researcher [. . .] from the Australian National University [had said that] noisy miner numbers are now greater than before Europeans arrived in Australia, and on top of the loss of 80% of southern temperate woodlands, this is having a devastating impact on many other species of woodland birds." [180].

In another public document, a slightly less contentious, but possibly just as misleading reference, was made to overabundance: "since European arrival, the noisy miner, a hyper-aggressive Australian honeyeater, has become overabundant." [252], implying that the overabundance of noisy miners is a long-standing problem, but again, historical records are scarce and too recent to support such assertions at all!

Comment: The claim that "numbers [of noisy miners] are now higher than prior to or at the time of European arrival" is completely fictitious. It is also very regrettable to use such falsehood, even if 'just' as a dramatic device, seemingly to justify the killing of noisy miners.

To clarify and correct this statement: any information about species in Australia, let alone numbers, prior to European arrival, is unknown, except insofar as fossils have revealed some prehistoric presence of birds, dinosaurs, and megafauna, largely found and excavated in the 20th century. Taxonomy and palaeontology have provided the complex details of Australia's avian evolution, phylogeny, and diversification [59–67], but this is not about individual numbers of birds. Citizen science bird counts started in the 1970s, and the 'big data counts' of 'Birds in the Backyard' are in their 13th year. The long and continuous presence of Aboriginal nations over at least 55,000 years, if not more [76,324], has been largely maintained as an oral and partly pictorial culture, but they almost certainly did not count bird numbers. Europeans first claimed/invaded Australia as a permanent colony in 1788. The first ornithological organisations were formed a century later (the South Australian Ornithological Association in 1899, the Royal Australasian Ornithologists' Union in 1901, and the Bird Observer's Club in 1905). At the time of Federation in 1901 at the turn of the 20th century, Australia's white population was about 3.8 million (indigenous populations were not counted), and little was known about bird numbers, let alone their distribution across the continent, which remained relatively unexplored, as the small human population huddled mainly near the coast on the east and south, southwest of the continent.

John and Elizabeth Gould's visit to Australia between 1838 and April 1840, on which the seven volumes of Australian birds (*The Birds of Australia*, 1848) were based, produced the first meticulous collection, descriptions, and detailed illustrations of Australian birds, but it was inevitably limited in scope [304]. Their visit included Tasmania and about a 200 km radius around Sydney and Newcastle, a small section of rural New South Wales, and some forays into South Australia (Murray River, Kangaroo Island, and Mt Lofty) [325,326]. Actual systematic monitoring of some species numbers (see Figure 7), and only of select and endangered groups, started in the latter part of the 20th century [234,235].

Comment: Given these facts, how could one possibly assert or know what noisy miner numbers were like at the time of Federation, let alone 'before European arrival'? Nobody did any counting, either in the 18th or 19th centuries, and citizen science is a relatively recent phenomenon, emerging in the second half of the 20th century. Myth-making can work to legitimise whatever actions may be deemed desirable at a given time, but it diminishes science and makes a commitment to protecting native species difficult, and, of course, it is the latter that we must try to do.

8. Concluding Remarks

We have placed the noisy miner, a native bird, in an untenable position and neither solved its problems nor those of other bird assemblies. Condemning the bird's behaviour using criteria that humans find unacceptable is hardly a way forward. The proximate causes of species' decline often remain poorly understood, partly because there may be many factors involved that are difficult to analyse, and partly because such studies can be extremely time-consuming and resource-intensive. However, a very detailed study undertaken by the Royal Ornithological Society of Britain found at least seven major potential causes of avian decline [7]. Of the seven major causes, their report identified one obvious one, namely the decline of woodlands, as well as another, the decline of invertebrates as an essential food source for many woodland birds, causes that are likely to apply also to other countries, including Australia. Many other studies blamed the expansion of agriculture [8–10]. Another study from Poland wondered why two closely related species showed opposite trends: one was declining, the other increasing in numbers

https://doi.org/10.3390/d18020099

(Firecrests, *Regulus ingicapilla*, and Goldcrests, *Regulus regulus*, both very small migratory songbirds). In one case, climate change arose as a major reason for decline, and the major reason for increase was the regeneration of a forest and planting of many more of their favourite feeding and roosting trees [327]. In another recent study, Bennett et al. [328] raised the question of 'revegetation' and 'remnant' landscapes as a precondition for slowing or reversing avian decline. Hence, the matter of the decline of woodland-dependent birds in Australia may not be solved by simply culling 'aggressive' [21,23,329] noisy miners [34,57], even if the public can easily accept such an argument. The extensive culling of noisy miners, according to official edicts, was based on the species' perceived risk to other small woodland birds. The anti-noisy-miner project was discussed in this paper as a case study of ethical dilemmas in conservation, or, in this case, as taking license with lethal outcomes, on a reductionist and on what one might well perceive as a biased premise.

Moreover, the ongoing removal or reduction of noisy miners has generally not been very successful in the context of the goals set by researchers. The latest report of noisy miner culling [316], however, sounds almost self-congratulatory, as it states that the miner 'problem' has been solved. Such confidence may be premature. Using a 'doughnut technique' for targets to be shot (see Section 7.1.2) means that survivors (i.e., deliberately non-targeted birds) witnessed the carnage. This deliberate technique may have done perhaps unintended but nevertheless substantial and even long-term damage to noisy miners and even woodland bird communities as a whole. The level of intimidation and trauma from seeing one bird after another, be it their partner, kin, team member, or associate, being shot and killed, can barely be imagined. This may lead to long-term changes in behaviour, in terms of either depression (PTSD) [294] or aggression as a dysfunctional way of 'lashing out' at everything and everybody. In fact, judging by the description of noisy miner behaviour in some of the papers, this may have already happened.

This paper has suggested that the reported atypical behaviour displayed by this species may be a symptom rather than a cause of the decline of woodland birds. The reasons one species behaves entirely differently in different locations in similar environments are indeed very unusual and puzzling, and this paper recommends that more research be undertaken to understand this phenomenon.

The social organisation of noisy miners can be complicated, or one may call this extreme flexibility, rarely seen in songbirds. Noisy miners have been described as colonial, territorial, and cooperative. Some or most of these characteristics are shared by only a few other avian species, such as grey herons, *Ardea cinerea* [330]; white-fronted bee-eaters (*Merops bullockoides)* [331]; and the pied kingfisher, *Ceryle rudis rudis* [332], and these characteristics will also require some research effort to clarify how such a complex social organisation might be beneficial for the species and how effective it might be as a specific strategy for ecologically and behaviourally challenging conditions. This research is yet to be carried out. But why have noisy miners been singled out as 'nest interrupters'? This would be applicable to scores of songbirds and, of course, to the majority of birds of prey. Yet the blame has landed specifically on noisy miners, which have been deemed to cause 'avifaunal disarray' [25,26]. The claim that noisy miners have benefitted from habitat degradation [253] is an extraordinary claim that remains without any biological or ecological foundation.

The concept of the ecology of fear has thus emerged as an important concept in the Anthropocene to explain how predation risks (including pet cats and humans as super-predators [297,333]) affect animal behaviour and physiology, with potential consequences for population demography, species interactions, and ecosystem functioning [310,311]. We can help avian species by reducing their reasons for chronic fear, namely by creating usable refuges not easily accessible to predators [312] and ensuring that there are adequate and

https://doi.org/10.3390/d18020C

appropriate food resources for all endangered species, thereby increasing their ability to successfully reproduce and helping them survive:

1. Actively modifying/restoring habitat structures in favour of prey species threatened by adverse conditions.

2. Learn precisely in which habitat noisy miners maintain family groups of up to 20 in stable and peaceful populations (which requires due consideration of the understorey, as outlined in NATMAP [262]).

3. Establish why and when noisy miners form large social groups. In house sparrows, for instance, it was found that flock size is correlated with the level of perceived predation risk [313].

4. Resolve the discrepancy between poor reproductive success and cognitive abilities in noisy miners, and possibly in many other species. The potential causes of poor reproduction may, at times, be quite complex. For instance, it has been shown that removing predators does not improve annual nest and female survival, but in some instances, habitat manipulations can [168].

5. One may consider further, based on robust research, that birds experience negative emotions such as fear, stress, and frustration [317], and that they understand the world cognitively much better than had been thought even a few decades ago.

6. Active habitat restoration is likely to be a promising avenue for reducing the dynamics that create the ecology of fear in any native species, including birds. Where this is not feasible, additional conservation strategies may be required to compensate for detrimental, fear-mediated effects of land use. As Walsh and colleagues found [268], these are strategies that have demonstrated consistent success in restoring small woodland birds and biodiversity in general [268].

However, habitat restoration by planting a few trees is insufficient without considering the usefulness and compatibility of the species planted, whether in terms of structure, providing cover, or food value. It is also important not to overlook the importance of the types of understorey, with their different structures and heights, even including 'refurbishing' the ground with natural debris. Recent global research has suggested that only complex landscapes stabilise bird communities and reestablish their ecosystem services [6,314].

Having considered some of the important life history events and behavioural and cognitive strengths of the noisy miner, the conclusion, albeit tentative, is that the case of noisy miner management is a miscarriage of justice, a conservation failure, and unfinished business. Palmer and colleagues recently argued that killing for conservation is talked about in black-and-white terms and uses "demonisation to support the case that killing is necessary and morally right" and that we ought to be aware of "demonising language and imagery" [55]. These warnings may well apply to the stated attitudes toward noisy miners. This is not even just a case of killing Peter to save Paul [44], because, as has so far been shown, small woodland birds as well as most other birds continue to decline, whether noisy miners are present or not. Indeed, some may think that the culling, suppression, or local extermination of noisy miners is an anti-conservation action in terms of actual outcomes. Here, a victim of extremely stressful circumstances (and a native species that might have existed for many millions of years) is declared the perpetrator. Despite the well-documented and rapid destruction of the environment [82–84,334,335], in this particular case of noisy miners, humans and human activities seem to be excused from any wrongdoing.

The case study of the current fate of the noisy miner is a small example of a much larger problem in our relationship with nature and our environment in general. As Legge and colleagues rightly argued, if we really want recovery, it would require "society to reframe its relationship with the environment" [83].

https://doi.org/10.3390/d18020099

Funding: This research required no internal or external funding.

Institutional Review Board Statement: Not applicable.

Data Availability Statement: The original contributions presented in this study are included in the article. Further inquiries can be directed to the corresponding author.

Acknowledgments: My sincere thanks go to L.J. Rogers for the detailed comments on the draft of this paper, and special thanks to the anonymous reviewers for their valuable suggestions.

Conflicts of Interest: The author declares no conflicts of interest.

References

1. Farine, D.R. Collective action in birds. *Curr. Biol.* **2022**, *32*, R1140–R1144. [CrossRef]
2. Heinsohn, R.G.; Double, M.C. Cooperate or speciate: New theory for the distribution of passerine birds. *Trends Ecol. Evol.* **20** *19*, 55–57. [CrossRef]
3. Jørgensen, P.S.; Folke, C.; Carroll, S.P. Evolution in the Anthropocene: Informing governance and policy. *Ann. Rev. Ecol. E System* **2019**, *50*, 527–546. [CrossRef]
4. Barrett, G.; Ford, H.A.; Recher, H. Conservation of woodland birds in a fragmented rural landscape. *Pac. Conserv. Biol.* **1994** 245–256. [CrossRef]
5. Rayner, L.; Lindenmayer, D.B.; Gibbons, P.; Manning, A.D. Evaluating empirical evidence for decline in temperate woodla birds: A nationally threatened assemblage of species. *Biol. Conserv.* **2014**, *171*, 145–155. [CrossRef]
6. Belder, D.; Lindenmayer, D.; Pierson, J.; Rudder, A. Survival and home ranges of woodland birds in restoration plantings. *E Appl.* **2020**, *31*, e2268. [CrossRef]
7. Vanhinsbergh, D.; Fuller, R.J.; Noble, D. *A Review of Possible Causes of Recent Changes in Populations of Woodland Birds in Brit* BTO Research Report No 245; The British Trust for Ornithology & Joint Nature Conservation Committee: Thetford, UK, 200
8. Chollet, S.; Martin, J.L. Declining woodland birds in North America: Should we blame Bambi? *Divers. Distrib.* **2013**, *19*, 481–4 [CrossRef]
9. Rigal, S.; Dakos, V.; Alonso, H.; Auniņš, A.; Benkő, Z.; Brotons, L.; Chodkiewicz, T.; Chylarecki, P.; De Carli, E.; Del Moral et al. Farmland practices are driving bird population decline across Europe. *Proc. Natl. Acad. Sci. USA* **2023**, *120*, e22165731 [CrossRef]
10. Douglas, D.J.; Waldinger, J.; Buckmire, Z.; Gibb, K.; Medina, J.P.; Sutcliffe, L.; Beckmann, C.; Collar, N.J.; Jansen, R.; Kamp, J.; e A global review identifies agriculture as the main threat to declining grassland birds. *IBIS* **2023**, *165*, 1107–1128. [CrossRef]
11. Lauck, K.S.; Ke, A.; Olimpi, E.M.; Paredes, D.; Hood, K.; Phillips, T.; Anderegg, W.R.; Karp, D.S. Agriculture and hot temperatu interactively erode the nest success of habitat generalist birds across the United States. *Science* **2023**, *382*, 290–294. [CrossR [PubMed]
12. Kaplan, G. Human-caused high direct mortality in birds: Unsustainable trends and ameliorative actions. *Animals* **2025**, *15*, [CrossRef] [PubMed]
13. Bart, J. Monitoring the abundance of bird populations. *Auk* **2005**, *122*, 15–25. [CrossRef]
14. Greenwood, J.J. Citizens, science and bird conservation. *J. Ornithol.* **2007**, *148*, 77–124. [CrossRef]
15. Latham, J. *Supplementum Indicis Ornithologici, Sive Systematis Ornithologiae*; Leigh, J.G., Sotheby, S., Eds.; Prostat apud G. Leig et S. Sotheby: London, UK, 1801; p. 1912 xxviii, xxxiii, xxxiv, xxxv. (In Latin)
16. Wikipedia. Noisy Miner. Available online: https://en.wikipedia.org/wiki/Noisy_miner (accessed on 17 January 2026).
17. Fleming, P.J.; Ballard, G. Yes, killing is sometimes essential for conservation. *Aust. Zool.* **2019**, *40*, 5–12. [CrossRef]
18. Beggs, R. Native to Nemesis: A cultural and environmental history of the Noisy Miner 1788-2019. *Aust. Zool.* **2022**, *42*, 770–8 [CrossRef]
19. Stark, M. *Starlings: The Curious Odyssey of a Most Hated Bird*; University of Nebraska Press: Lincoln, NE, USA, 2025.
20. Higgins, P.J.; Peter, J.M.; Steele, W.K. (Eds.) *Handbook of Australian, New Zealand and Antarctic Birds, (Tyrant-Flycatchers to Cha* Oxford University Press: Melbourne, Australia, 2001; Volume 5, pp. 626–650.
21. TSSC. *Aggressive Exclusion of Birds from Potential Woodland and Forest Habitat by Over-Abundant Noisy Miners (Manor melanocephala)*; Threatened Species Scientific Committee, Department of the Environment: Canberra, Australia, 2013.
22. Dow, D.D. Indiscriminate interspecific aggression leading to almost sole occupancy of space by a single species of bird. *E Austral Ornithol.* **1977**, *77*, 115–121. [CrossRef]
23. Piper, S.D.; Catterall, C.P. A particular case and a general pattern: Hyperaggressive behaviour by one species may medi avifaunal decreases in fragmented Australian forests. *Oikos* **2003**, *101*, 602–614. [CrossRef]

https://doi.org/10.3390/d18020

Vickers, J.A.T. Demographic Shifts in Noisy Miner (*Manorina melanocephala*) Populations Following Removal. Bachelor's Thesis, School of Earth & Environmental Science, University of Wollongong, Wollongong, Australia, 2017. Available online: https://ro.uow.edu.au/thsci/145 (accessed on 20 June 2025).
Maron, M.; Grey, M.J.; Catterall, C.P.; Major, R.E.; Oliver, D.L.; Clarke, M.F.; Loyn, R.H.; MacNally, R.; Davidson, I.; Thomson, J.R. Avifaunal disarray due to a single despotic species. *Divers. Distrib.* **2012**, *19*, 1468–1479. [CrossRef]
Thomson, J.R.; Maron, M.; Grey, M.J.; Catterall, C.P.; Major, R.E.; Oliver, D.L.; Clarke, M.F.; Loyn, R.H.; Davidson, I.; Ingwersen, D.; et al. Avifaunal disarray: Quantifying models of the occurrence and ecological effects of a despotic bird species. *Divers. Distrib.* **2015**, *21*, 451–464. [CrossRef]
Barati, A.; Etezadifar, F.; McDonald, P.G. Fragmentation in eucalypt woodlands promotes nest-tree occupancy by a despotic species, the noisy miner (*Manorina melanocephala*). *Austral Ecol.* **2016**, *41*, 897–905. [CrossRef]
Mortelliti, A.; Ikin, K.; Tulloch, A.I.; Cunningham, R.; Stein, J.; Michael, D.; Lindenmayer, D.B. Surviving with a resident despot: Do revegetated patches act as refuges from the effects of the noisy miner (*Manorina melanocephala*) in a highly fragmented landscape? *Divers. Distrib.* **2016**, *22*, 770–782. [CrossRef]
Crates, R.; Terauds, A.; Rayner, L.; Stojanovic, D.; Heinsohn, R.; Wilkie, C.; Webb, M. Spatially and temporally targeted suppression of despotic noisy miners has conservation benefits for highly mobile and threatened woodland birds. *Biol. Conserv.* **2018**, *227*, 343–351. [CrossRef]
Davitt, G.; Maute, K.; Major, R.E.; McDonald, P.G.; Maron, M. Short-term response of a declining woodland bird assemblage to the removal of a despotic competitor. *Ecol. Evol.* **2018**, *8*, 4771–4780. [CrossRef] [PubMed]
Kern, J.M.; McDonald, P.G. Location and caller familiarity influence mobbing behaviour and the likely ecological impact of noisy miners around colony edges. *J. Avian Biol.* **2024**, e03231. [CrossRef]
Beggs, R.; Pierson, J.; Tulloch, A.I.; Blanchard, W.; Westgate, M.J.; Lindenmayer, D. An experimental test of a compensatory nest predation model following lethal control of an overabundant native species. *Biol. Conserv.* **2019**, *231*, 122–132.27. [CrossRef]
Grey, M.; Clarke, M.; Davidson, I.; Maron, M.; Ingwersen, D.; Tzaros, C. *The Noisy Miner: Challenges in Managing an Overabundant Species*; LaTrobe University: Melbourne, Australia, 2011.
Peddie, C. Scientists want noisy miner cull. *Adelaide Advertiser*, 10 December 2011.
Stone, L. *Noisy Miners Are One of Australia's 'Most Hated Birds'*; ABC Radio: Brisbane, Australia, 2022.
BirdLife Australia. What Bird Is That? Myna and Miner. How to Tell the Difference Between Commonly Confused Birds: Myna and Miner. Available online: https://birdlife.org.au/news/what-bird-is-that-myna-and-miner/? (accessed on 17 January 2026).
Major, R. *Miner Surgery: Repairing Woodland Bird Diversity by Noisy Minectomy*; Cumberland Bird Observers Club.: Sydney, Australia, 2017; Volume 3.
Crates, R.; Rayner, L.; Webb, M.; Stojanovic, D.; Wilkie, C.; Heinsohn, R. Sustained and delayed noisy miner suppression at an avian hotspot. *Austral Ecol.* **2020**, *45*, 636–643. [CrossRef]
Melton, C. The Value of Understorey Vegetation to Woodland and Forest Bird Communities as a Resource and Potential Deterrent of Key Threatening Process: Overabundant Native Noisy Miners (*Manorina melanocephala*). Ph.D. Thesis, University of Queensland, Brisbane, Australia, 2022.
NSWSC. *Final Determination to List Aggressive Exclusion of Birds from Woodland and Forest Habitat by Abundant Noisy Miners Manorina melanocephala (Latham 1802) as a Key Threatening Process*; New South Wales Scientific Committee: Sydney, Australia, 2013.
Beggs, R.; Tulloch, A.I.T.; Pierson, J.; Blanchard, W.; Lindenmayer, D. Patch-scale culls of an overabundant bird defeated by immediate recolonization. *Ecol. Appl.* **2019**, *29*, e01846. [CrossRef]
Birkett, S.; Rapport, D. A framework for identifying and classifying ecosystem dysfunctions. *Environmentalist* **1998**, *18*, 15–25. [CrossRef]
Willcock, S.; Cooper, G.S.; Addy, J.; Dearing, J.A. Earlier collapse of Anthropocene ecosystems driven by multiple faster and noisier drivers. *Nat. Sustain.* **2022**, *6*, 1331–1342. [CrossRef]
Woinarski, J.C.Z. Killing Peter to save Paul: An ethical and ecological basis for evaluating whether a native species should be culled for the conservation benefit of another native species. *Austral. Zool.* **2019**, *40*, 49–62. [CrossRef]
Odenbaugh, J. Should we kill one owl to save another? *Ethics Policy Environ.* **2025**, *28*, 159–177. [CrossRef]
Braby, M.F.; Yeates, D.K.; Joseph, L. Woodland birds and insect decline. *Emu-Austral Ornithol.* **2023**, *123*, 255–257. [CrossRef]
Moreau, J.; Rabdeau, J.; Badenhausser, I.; Giraudeau, M.; Sepp, T.; Crépin, M.; Gaffard, A.; Bretagnolle, V.; Monceau, K. Pesticide impacts on avian species with special reference to farmland birds: A review. *Environ. Monit. Assess.* **2022**, *194*, 790. [CrossRef]
Woinarski, J.C.Z. Fire and Australian birds: A review. In *Australia's Biodiversity: Responses to Fire: Plants, Birds and Invertebrates*; Gill, A.M., Woinarksi, J.C.Z., York, A., Eds.; Environment Australia: Canberra, Australia, 1999; pp. 55–111.
Conradie, S.R.; Woodborne, S.M.; Wolf, B.O.; Pessato, A.; Mariette, M.M.; McKechnie, A.E. Avian mortality risk during heat waves will increase greatly in arid Australia during the 21st century. *Conserv. Physiol.* **2020**, *8*, coaa048. [CrossRef]
Lees, A.C.; Haskell, L.; Allinson, T.; Bezeng, S.B.; Burfield, I.J.; Renjifo, L.M.; Rosenberg, K.V.; Viswanathan, A.; Butchart, S.H. State of the World's Birds. *Annu. Rev. Environ. Resour.* **2022**, *47*, 231–260. [CrossRef]

https://doi.org/10.3390/d18020099

51. Dyer, E.L.; Kording, K. Why the simplest explanation isn't always the best. *Proc. Natl. Acad. Sci. USA* **2023**, *120*, e23191691 [CrossRef]
52. Burghardt, G.M. Cognitive ethology and critical anthropo- morphism: A snake with two heads and hognose snakes that pl dead. In *Cognitive Ethology: The Minds of Other Animals: Essays in Honor of Donald R. Griffin*; Ristau, C.A., Ed.; Lawrence Erlba Associates: Hillsdale, NJ, USA, 1991; pp. 53–90.
53. Wynne, C.D.L. The emperor's new anthropomorphism. *Behav. Anal. Today* **2005**, *6*, 151–154. [CrossRef]
54. Wynne, C.D. The perils of anthropomorphism. *Nature* **2004**, *428*, 606. [CrossRef] [PubMed]
55. Palmer, C.; Fischer, B.; Gamborg, C.; Hampton, J.; Sandøe, P. *Wildlife Ethics. The Ethics of Wildlife Management and Conservati* UFAW Animal Welfare Series; John Wiley & Sons: Hoboken, NJ, USA; Chichester, UK, 2023.
56. Kennedy, B.P.; Boyle, N.; Fleming, P.J.; Harvey, A.M.; Jones, B.; Ramp, D.; Dixon, R.; McGreevy, P.D. Ethical Treatment of Invasi and Native Fauna in Australia: Perspectives through the One Welfare Lens. *Animals* **2022**, *12*, 1405. [CrossRef]
57. Hooper, J.; Aiello, T.; Hill, K.; Szydlowski, M.; Oxley Heaney, S. Nothing more than 'anti-cull activists': Accusations of bias a the politics of research that advocates for non-human animals. *Anim. Stud. J.* **2023**, *12*, 71–95. [CrossRef]
58. Wallach, A.D.; Batavia, C.; Nelson, M.P.; Bekoff, M.; Baker, L.; Ben-Ami, D.; Boronyak, L.; Cardilini, A.; Carmel, Y.; Celermajer, et al. Recognizing animal personhood in compassionate conservation. *Conserv. Biol.* **2020**, *34*, 1097–1106. [CrossRef]
59. Cracraft, J. Avian evolution, Gondwana biogeography and the Cretaceous-Tertiary mass extinction event. *Proc. R. Soc. B* **20** *268*, 459–469. [CrossRef]
60. Cracraft, J.; Barker, F.K. Passerine birds (Passeriformes). In *The Timetree of Life*; Blair Hedges, S., Kumar, S., Eds.; Oxford Universi Press: New York, USA, 2009; pp. 423–431.
61. Barker, F.K.; Cibois, A.; Schikler, P.; Feinstein, J.; Cracraft, J. Phylogeny and diversification of the largest avian radiation. *Pr Natl. Acad. Sci. USA* **2004**, *101*, 11040–11045. [CrossRef] [PubMed]
62. Wright, T.F.; Schirtzinger, E.E.; Matsumoto, T.; Eberhard, J.R.; Graves, G.R.; Sanchez, J.J.; Capelli, S.; Müller, H.; Scharpegge, Chambers, G.K.; et al. A multilocus molecular phylogeny of the parrots (Psittaciformes): Support for Gondwanan origin duri the Cretaceous. *Mol. Biol. Evol.* **2008**, *25*, 2141–2156. [CrossRef]
63. Cooper, A.; Penny, D. Mass survival of birds across the Cretaceous-Tertiary boundary: Molecular evidence. *Science* **1997**, *2* 1109–1113. [CrossRef] [PubMed]
64. Oliveros, C.H.; Field, D.J.; Ksepka, D.T.; Barker, F.K.; Aleixo, A.; Andersen, M.J.; Alström, P.; Benz, B.W.; Braun, E.L.; Braun, M et al. Earth history and the passerine super radiation. *Proc. Natl. Acad. Sci. USA* **2019**, *116*, 7916–7925. [CrossRef]
65. Ksepka, D.T.; Stidham, T.A.; Williamson, T.E. Early Paleocene landbird supports rapid phylogenetic and morphological divers cation of crown birds after the K–Pg mass extinction. *Proc. Natl. Acad. Sci. USA* **2017**, *114*, 8047–8052. [CrossRef]
66. McLennan, D.A.; Brooks, D.R. Complex histories of speciation and dispersal in communities: A re-analysis of some Australi bird data using BPA. *J. Biogeogr.* **2002**, *29*, 1055–1066. [CrossRef]
67. Joseph, L. Species limits in birds: Australian perspectives on interrelated challenges of allopatry, introgression of mitochondr DNA, recent speciation, and selection. *Auk* **2021**, *138*, ukab012. [CrossRef]
68. Schluter, D. *The Ecology of Adaptive Radiation*; OUP: Oxford, UK, 2000.
69. Hay, E.M.; McGee, M.D.; Chown, S.L. Geographic range size and speciation in honeyeaters. *BMC Ecol. Evol.* **2022**, *22*, [CrossRef]
70. Marki, P.Z.; Kennedy, J.D.; Cooney, C.R.; Rahbek, C.; Fjeldså, J. Adaptive radiation and the evolution of nectarivory in a lar songbird clade. *Evolution* **2019**, *73*, 1226–1240. [CrossRef]
71. Gill, F.; Donsker, D.; Rasmussen, P. (Eds.) IOC World Bird List (v 15.1). 2025. Available online: https://www.worldbirdnames.o (accessed on 17 January 2026).
72. Joseph, L.; Toon, A.; Nyári, Á.S.; Longmore, N.W.; Rowe, K.M.; Haryoko, T.; Trueman, J.; Gardner, J.L. A new synthesis of t molecular systematics and biogeography of honeyeaters (Passeriformes: Meliphagidae) highlights biogeographical and ecologi complexity of a spectacular avian radiation. *Zool. Scr.* **2014**, *43*, 235–248. [CrossRef]
73. Goldstein, D.L.; Bradshaw, S.D. Regulation of water and sodium balance in the field by Australian honeyeaters (Aves: Melipha dae). *Physiol. Zool.* **1998**, *71*, 214–225. [CrossRef] [PubMed]
74. Paton, D.C.; Collins, B.G. Bills and tongues of nectar-feeding birds: A review of morphology, function and performance, wi intercontinental comparisons. *Austr. J. Ecol.* **1989**, *14*, 473–506. [CrossRef]
75. Owen, C.L.; Marshall, D.C.; Hill, K.B.R.; Simon, C. How the Aridification of Australia Structured the Biogeography and Influenc the Diversification of a Large Lineage of Australian Cicadas. *Syst. Biol.* **2017**, *66*, 569–589. [CrossRef]
76. Clarkson, C.; Jacobs, Z.; Marwick, B.; Fullagar, R.; Wallis, L.; Smith, M.; Roberts, R.G.; Hayes, E.; Lowe, K.; Carah, X.; et al. Hum occupation of northern Australia by 65,000 years ago. *Nature* **2017**, *547*, 306–310. [CrossRef]
77. Bird, M.I.; Beaman, R.J.; Condie, S.A.; Cooper, A.; Veth, P. Palaeogeography and voyage modelling indicates early hum colonization of Australia was likely from Timor-Roti. *Quatern. Sci. Rev.* **2018**, *191*, 431–439. [CrossRef]

DCCEEW. *The 2023 Threatened Bird Index, TSX*; Department of Climate Change, Energy, the Environment and Water (DCCEEW), Australian Government: Canberra, Australia, 2023. Available online: https://www.dcceew.gov.au (accessed on 17 January 2026).
Sol, D.; González-Lagos, C.; Moreira, D.; Maspons, J.; Lapiedra, O. Urbanisation tolerance and the loss of avian diversity. *Ecol. Lett.* **2014**, *17*, 942–950. [CrossRef] [PubMed]
Banville, M.J.; Bateman, H.L.; Earl, S.R.; Warren, P.S. Decadal declines in bird abundance and diversity in urban riparian zones. *Landsc. Urban Plan.* **2017**, *159*, 48–61. [CrossRef]
Mohring, B.; Henry, P.Y.; Jiguet, F.; Malher, F.; Angelier, F. Investigating temporal and spatial correlates of the sharp decline of an urban exploiter bird in a large European city. *Urban Ecosyst.* **2021**, *24*, 501–513. [CrossRef]
Ritchie, E.; Bradshaw, C.; Dickman, C.; Hobbs, R.; Johnson, C.; Johnston, E.; Laurance, W.; Lindenmayer, D.; McCarthy, M.; Nimmo, D.; et al. Continental-scale governance and the hastening of loss of Australia's biodiversity. *Conserv. Biol.* **2013**, *27*, 1133–1135. [CrossRef]
Legge, S.; Rumpff, L.; Garnett, S.T.; Woinarski, J.C.Z. Loss of terrestrial biodiversity in Australia: Magnitude, causation, and response. *Science* **2023**, *381*, 622–631. [CrossRef]
Isbell, F.; Balvanera, P.; Mori, A.S.; He, J.S.; Bullock, J.M.; Regmi, G.R.; Seabloom, E.W.; Ferrier, S.; Sala, O.E.; Guerrero-Ramírez, N.R.; et al. Expert perspectives on global biodiversity loss and its drivers and impacts on people. *Front. Ecol. Environ.* **2023**, *21*, 94–103. [CrossRef]
Churchill, T.B.; Laidlaw, M.J.; McCall, A.H.; Hines, H.B.; Gynther, I.C.; Molyneux, J.; Teixeira, D.; Stewart, D.; Rix, M.G.; Burwell, C.J.; et al. *Bushfire Recovery 2020–2022: Priority Actions for Threatened Species in the Gondwana Rainforests of Australia World Heritage Area, South-East Queensland*; Department of Environment and Science, Queensland Government: Brisbane, Australia, 2023.
Bennett, A.F.; Haslem, A.; Garnett, S.T.; Loyn, R.H.; Woinarski, J.C.; Ehmke, G. Declining but not (yet) threatened: A challenge for avian conservation in Australia. *Emu-Austral Ornithol.* **2024**, *124*, 123–145. [CrossRef]
Andersson, E.; Barthel, S.; Ahrné, K. Measuring social–ecological dynamics behind the generation of ecosystem services. *Ecol. Appl.* **2007**, *17*, 1267–1278. [CrossRef] [PubMed]
Daly, G. Changes to the status of birds in the Tapitallee region on the south coast of New South Wales from 1928 to 2020. *Aust. Zool.* **2022**, *42*, 30–54. [CrossRef]
Kaplan, G. *Australian Magpie. Biology and Behaviour of an Unusual Songbird*, 2nd ed.; CSIRO Publishing: Melbourne, Australia, 2019.
Johnsson, R.D.; Connelly, F.; Lesku, J.A.; Roth, T.C. Australian magpies. *Curr. Biol.* **2024**, *34*, R41–R43. [CrossRef] [PubMed]
Mo, M. Agonistic behaviour from Noisy Miners. *South Aust. Ornithol.* **2023**, *47*, 2.
Mo, M. Agonistic behaviour from Noisy Miners *Manorina melanocephala* toward Australian Magpies Gymnorhina tibicen swooping humans. *South Aust.* **2024**, *47*, 72.
Cunningham, S.; Magrath, R.D. Functionally referential alarm calls in noisy miners communicate about predator behaviour. *Anim. Behav.* **2017**, *129*, 171–179. [CrossRef]
Hall, M.L.; Magrath, R.D. Temporal coordination signals coalition quality. *Curr. Biol.* **2007**, *17*, R406–R407. [CrossRef]
Dow, D.D. Displays of the Honeyeater *Manorina melanocephala*. *Zeitschr. Für Tierpsychol.* **1975**, *38*, 70–96. [CrossRef]
Zhang, X.Z.; He, J. Evolution of sensory systems in snakes: Infrared detection, chemoreception, and ecological adaptation. *Biol. Evid.* **2025**, *15*, 138–148. [CrossRef]
Garcîa-Naas, V.; Bliard, L.; Ozgul, A. Impact of a 'reverse keystone species' on the temporal dynamics of bird communities in Australia. *Biol. Conserv.* **2024**, *296*, 110698. [CrossRef]
Hubbell, S.P. Neutral theory in community ecology and the hypothesis of functional equivalence. *Funct. Ecol.* **2005**, *19*, 166–172. [CrossRef]
Cockburn, A. Prevalence of different modes of parental care in birds. *Proc. Roy. Soc. B* **2006**, *273*, 1375–1383. [CrossRef] [PubMed]
Feeney, W.E.; Medina, I.; Somveille, M.; Heinsohn, R.; Hall, M.L.; Mulder, R.A.; Stein, J.A.; Kilner, R.M.; Langmore, N.E. Brood parasitism and the evolution of cooperative breeding in birds. *Science* **2013**, *342*, 1506–1508. [CrossRef]
Clutton-Brock, T. *The Evolution of Parental Care*; Princeton University Press: Princeton, NJ, USA, 1991.
Aplin, L.; Crates, R.; Flack, A.; McGregor, P. Social learning and culture in birds: Emerging patterns and relevance to conservation. *Philos. Trans. R. Soc. B* **2025**, *380*, 20240128. [CrossRef]
Kaplan, G. Brain size and life history data of Australian landbirds. In *Bird Minds*; Appendix; CSIRO Publishing: Melbourne, Australia, 2015; pp. 239–254, 110698.
Franklin, D.C.; Garnett, S.; Luck, G.; Gutierrez-Ibanez, C.; Iwaniuk, A. Relative brain size in Australian birds. *Emu-Austral Ornithol.* **2014**, *114*, 160–170. [CrossRef]
Darwin, C. *On the Origin of Species by Natural Selection*; John Murray: London, UK, 1859.
Kaplan, G.; Rogers, L.J. Charles Darwin and animal behavior. In *Encyclopedia of Animal Behavior*; Bekoff, M., Goodall, J., Eds.; Greenwood Publishing: Westport, CT, USA, 2004; Volume 2, pp. 471–479, ISBN 0-313-32745-9.

https://doi.org/10.3390/d18020099

107. Lefebvre, L.; Reader, S.M.; Sol, D. Brains, innovations and evolution in birds and primates. *Brain Behav. Evol.* **2004**, *63*, 233–2 [CrossRef]
108. Lefebvre, L. Brains, innovations, tools and cultural transmission in birds, non-human primates, and fossil hominins. *Front. Hu Neurosci.* **2013**, *7*, 245. [CrossRef]
109. Sol, D.; Duncan, R.P.; Blackburn, T.M.; Cassey, P.; Lefebvre, L. Big brains, enhanced cognition, and response of birds to no environments. *Proc. Natl. Acad. Sci. USA* **2005**, *102*, 5460–5465. [CrossRef]
110. Iwaniuk, A.N. The evolution of cognitive brains in non-mammals. In *Evolution of the Brain, Cognition, and Emotion in Vertebra* Watanabe, S., Hofman, M.A., Shimizu, T., Eds.; Springer: Tokyo, Japan, 2017; pp. 101–124.
111. Farrow, L.F.; Hamlin, A.S.; McDonald, P. Exploring the Brain and Behaviour of a Reverse Keystone Species, the No Miner (Manorina Melanocephala). Ph.D. Thesis, University of New England, Armidale, Australia, 2024. Available onli https://hdl.handle.net/1959.11/60851 (accessed on 17 January 2026).
112. Mettke-Hoffmann, C.; Greenberg, R.S. Behavioral and cognitive adaptations to long-distance migration. In *Birds of Two Worl The Ecology and Evolution of Migration*; Greenberg, R.S., Marra, P.P., Eds.; Smithsonian Institution, John Hopkins University Pre Baltimore, MD, USA, 2005; p. 114.
113. Cristol, D.A.; Reynolds, E.B.; Leclerc, J.E.; Donner, A.H.; Farabaugh, C.S.; Ziegenfus, C.W. Migratory dark-eyed juncos, Jun hyemalis, have better spatial memory and denser hippocampal neurons than nonmigratory conspecifics. *Anim. Behav.* **2003**, 317–328. [CrossRef]
114. Sol, D.; Garcia, N.; Iwaniuk, A.; Davis, K.; Meade, A.; Boyle, W.A.; Szekely, T. Evolutionary Divergence in Brain Size betwe Migratory and Resident Birds. *PLoS ONE* **2010**, *5*, e9617. [CrossRef] [PubMed]
115. Fristoe, T.S.; Iwaniuk, A.N.; Botero, C.A. Big brains stabilize populations and facilitate colonization of variable habitats in bi *Nat. Ecol. Evol.* **2017**, *1*, 1706–1715. [CrossRef] [PubMed]
116. Ducatez, S.; Sol, D.; Sayol, F.; Lefebvre, L. Behavioural plasticity is associated with reduced extinction risk in birds. *Nat. Ecol. E* **2020**, *4*, 788–793. [CrossRef]
117. BirdLife Australia. Aussie-Bird-Count. National and State-Wide Counts and Reporting Rates. 2024. Available onli https://aussiebirdcount.org.au/ (accessed on 17 January 2026).
118. Sol, D.; Olkowicz, S.; Sayol, F.; Kocourek, M.; Zhang, Y.; Marhounová, L.; Osadnik, C.; Corssmit, E.; Garcia-Porta, J.; Martin, T et al. Neuron numbers link innovativeness with both absolute and relative brain size in birds. *Nat. Ecol. Evol.* **2022**, *6*, 1381–13 [CrossRef]
119. Herculano-Houzel, S. Number of neurons as biological correlates of cognitive capability. *Curr. Opin. Behav. Sci.* **2017**, *16*, 1 [CrossRef]
120. Ströckens, F.; Neves, K.; Kirchem, S.; Schwab, C.; Herculano-Houzel, S.; Güntürkün, O. High associative neuron numbers co drive cognitive performance in corvid species. *J. Comp. Neurol.* **2020**, *530*, 1588–1605. [CrossRef] [PubMed]
121. Lefebvre, L.; Whittle, P.; Lascaris, E.; Finkelstein, A. Feeding innovations and forebrain size in birds. *Anim. Behav.* **1997**, 549–560. [CrossRef]
122. Lefebvre, L. A global database of feeding innovations in birds. *Wilson J. Ornithol.* **2021**, *132*, 803–809. [CrossRef]
123. Overington, S.E.; Morand-Ferron, J.; Boogert, N.J.; Lefebvre, L. Technical innovations drive the relationship between innovati ness and residual brain size in birds. *Anim. Behav.* **2009**, *78*, 1001–1010. [CrossRef]
124. Ferrari, M.C.; Crane, A.L.; Chivers, D.P. Certainty and the cognitive ecology of generalization of predator recognition. *An Behav.* **2016**, *111*, 207–2011. [CrossRef]
125. Sulikowski, D.; Burke, D. Food-specific spatial memory biases in an omnivorous bird. *Biol. Lett.* **2007**, *3*, 245–248. [CrossRef]
126. Sulikowski, D.; Burke, D. Reward type influences performance and search structure of an omnivorous bird in an open-field ma *Behav. Process.* **2010**, *83*, 31–35. [CrossRef]
127. Sulikowski, D.; Burke, D. When a place is not a place: Encoding of spatial information is dependent on reward type. *Behavio* **2010**, *147*, 1461–1479. [CrossRef]
128. Sulikowski, D.; Burke, D. Movement and memory: Different cognitive strategies are used to search for resources with differe natural distributions. *Behav. Ecol. Sociobiol.* **2011**, *65*, 621–631. [CrossRef]
129. Sulikowski, D.; Burke, D. Noisy miners plan ahead: Cryptic signalling of reward location impairs search for nectar, but not invertebrates. *Anim. Behav.* **2015**, *102*, 149–155. [CrossRef]
130. Kaplan, G. Long-term attachments and complex cognition in birds and humans are linked to pre-reproductive prosociality a cooperation constructing a hypothesis. *Ann. Cogn. Sci.* **2020**, *4*, 127–142.
131. Øverli, Ø.; Sørensen, C.; Pulman, K.G.; Pottinger, T.G.; Korzan, W.; Summers, C.H.; Nilsson, G.E. Evolutionary backgrou for stress-coping styles: Relationships between physiological, behavioral, and cognitive traits in non-mammalian vertebra *Neurosci. Biobehav. Rev.* **2007**, *31*, 396–412. [CrossRef] [PubMed]
132. Sandi, C. Stress and cognition. *Wiley Interdiscip. Rev. Cogn. Sci.* **2013**, *4*, 245–261. [CrossRef]

https://doi.org/10.3390/d18020

Pravosudov, V.V. Cognitive ecology in the wild—Advances and challenges in avian cognition research. *Curr. Opin. Behav. Sci.* **2022**, *45*, 101138. [CrossRef]
Cram, D.L. Oxidative stress and cognition in ecology. *J. Zool.* **2022**, *318*, 229–240. [CrossRef]
Farrow, L.; Barati, A.; Hamlin, A.; McDonald, P. I remember you; Experimental evidence of true acoustic individual recognition in a wild passerine. *Res. Sq.* **2023**, *preprint*. [CrossRef]
Holt, J.; Barati, A.; McDonald, P.G. The complex acoustic repertoire of a highly social species, the noisy miner, Manorina melanocephala. *Emu-Austral Ornithol.* **2017**, *117*, 19–30. [CrossRef]
Audet, J.N.; Couture, M.; Lefebvre, L.; Jarvis, E.D. Problem-solving skills are predicted by technical innovations in the wild and brain size in passerines. *Nat. Ecol. Evol.* **2024**, *8*, 806–816. [CrossRef] [PubMed]
Griffin, A.S.; Diquelou, M.C. Innovative problem solving in birds: A cross-species comparison of two highly successful passerines. *Anim. Behav.* **2015**, *100*, 84–94. [CrossRef]
Arnold, K.E. Group mobbing behaviour and nest defence in a cooperatively breeding Australian bird. *Ethology* **2000**, *106*, 385–393. [CrossRef]
Boucherie, P.H.; Loretto, M.C.; Massen, J.J.; Bugnyar, T. What constitutes "social complexity" and "social intelligence" in birds? Lessons from ravens. *Behav. Ecol. Sociobiol.* **2019**, *73*, 12. [CrossRef]
Bugnyar, T. Social cognition in ravens. *Comp. Cogn. Behav. Rev.* **2013**, *8*, 1–12. [CrossRef]
Marzluff, J.M.; Walls, J.; Cornell, H.N.; Withey, J.; Craig, D.P. Lasting recognition of threatening people by wild American crows. *Anim. Behav.* **2010**, *79*, 699–707. [CrossRef]
Marzluff, J.M.; Miyaoka, R.; Minoshima, S.; Cross, D. Brain imaging reveals neuronal circuitry underlying the crow's perception of human faces. *Proc. Natl. Acad. Sci. USA* **2012**, *109*, 15912–15917. [CrossRef]
Levey, D.J.; Poulsen, J.R.; Schaeffer, A.P. Wild mockingbirds distinguish among familiar humans. *Sci. Rep.* **2023**, *13*, 10259. [CrossRef] [PubMed]
Liu, Y.; Liu, Y.; Liang, W. Breeding barn swallows recognize householders from strangers. *Anim. Cogn.* **2025**, *28*, 33. [CrossRef] [PubMed]
Niu, X.; Jiang, Z.; Peng, Y.; Huang, S.; Wang, Z.; Shi, L. Visual cognition of birds and its underlying neural mechanism: A review. *Avian Res.* **2022**, *13*, 100023. [CrossRef]
Dow, D.D. Agonistic and spacing behaviour of the noisy miner Manorina melanocephala, a communally breeding honeyeater. *IBIS* **1979**, *121*, 423–436. [CrossRef]
Barati, A. Cooperative Breeding in the Noisy Miner (*Manorina melanocephala*): The Role of Genetic Relatedness, Sex, Extra-Pair Paternity and Acoustic Signals. Ph.D. Thesis, University of New England, Armidale, Australia, 2016.
Chisholm, A.H. The cuckoo problem in Australia. *Auk* **1933**, *50*, 385–395. [CrossRef]
Carlson, A.J. Two years of bird surveys at Bootawa Dam, Bootawa, NSW. *Whistler* **2023**, *17*, 88–96.
Guppy, M.; Guppy, S.; Priddel, D.; Fullagar, P. Nest predation of woodland birds in south-east Australia: Importance of unexpected predators. *Aust. Zool.* **2014**, *37*, 105–116. [CrossRef]
Stephens, S.E.; Koons, D.N.; Rotella, J.J.; Willey, D.W. Effects of habitat fragmentation on avian nesting success: A review of the evidence at multiple spatial scales. *Biol. Conserv.* **2004**, *115*, 101–110. [CrossRef]
Mainwaring, M.C.; Reynolds, S.J.; Weidinger, K. The influence of predation on the location and design of nests. In *Nests, Eggs, and Incubation: New Ideas About Avian Reproduction*; Deeming, D.C., Reynolds, S.J., Eds.; Oxford University Press: Oxford, UK, 2015; pp. 50–64.
Hocking, P.M. Nutritional interactions with reproduction in birds. *Proc. Nutr. Soc.* **1987**, *46*, 217–225. [CrossRef]
Noske, R.A. Grey butcherbird killing an adult little lorikeet and a review of adult avian prey of butcherbirds and pied currawongs. *Corella* **2024**, *48*, 66–69.
Australian Broadcast Commission (ABC). 'Australia All Over' (13 September 2019) Segment: 'Did you Know That Kookaburras Eat Other Birds? Interview with and Photo by Teale Britstra, Calliope, Qld.
Debus, S.J.S.; Hatfield, T.S.; Ley, A.J.; Rose, A.B. Breeding Biology and Diet of the Little Eagle Hieraaetus morphnoides in the New England Region of New South Wales. *Aust. Field Ornithol.* **2007**, *24*, 137–157.
Debus, S.J.S.; Olsen, J.; Trost, S.; Judge, D. Diet of the Australian Hobby Falco longipennis breeding in Canberra, Australian Capital Territory, in 2002–2004 and 2005–2008. *Aust. Field Ornithol.* **2020**, *37*, 174–183. [CrossRef]
Barker, R.D.; Vestjens, W.J.M. *The Food of Australian Birds*; CSIRO Publishing: Melbourne, Australia, 1990; Volume 2.
Rowley, I.; Vestjens, W.J. The comparative ecology of Australian corvids. V: Food. *CSIRO Wildl. Res.* **1973**, *18*, 131–155. [CrossRef]
Blendinger, P.G.; Rojas, T.N.; Ramírez-Mejía, A.F.; Bender, I.M.; Lomáscolo, S.; Magro, J.; Núñez Montellano, M.G.; Ruggera, R.A.; Valoy, M.; Ordano, M. Nutrient balance and energy-acquisition effectiveness: Do birds adjust their fruit diet to achieve intake targets? *Funct. Ecol.* **2022**, *36*, 2649–2660. [CrossRef]
Morris, D.L.; Faaborg, J.; Washburn, B.E.; Millspaugh, J.J. Predator-induced renesting and reproductive effort in indigo buntings: More work for less pay? *Conserv. Physiol.* **2015**, *3*, cou063. [CrossRef]

https://doi.org/10.3390/d18020099

163. Langmore, N.E. Australian cuckoos and their adaptations for brood parasitism. *Avian Res.* **2013**, *4*, 86–92. [CrossRef]
164. Wood, K.A. The Channel-billed Cuckoo at Wollongong, New South Wales: Notes on migration, abundance and breeding. *Au Field Ornith.* **2004**, *21*, 89–106.
165. Fulton, G.R. Meta-analyses of nest predation in temperate Australian forests and woodlands. *Austral Ecol.* **2019**, *44*, 389–3 [CrossRef]
166. Yom-Tov, Y. The reproductive rates of Australian passerines. *Aust. Wildl. Res.* **1987**, *14*, 310–330. [CrossRef]
167. Ford, H.A.; Barrett, G.W.; Saunders, D.A.; Recher, H.F. Why have birds in the woodlands of Southern Australia declined? *B Conserv.* **1978**, *97*, 71–88. [CrossRef]
168. Dow, D.D. Breeding biology and development of the young of Manorina melanocephala, a communally breeding honeyea *Emu-Austral Ornithol.* **2016**, *78*, 207–222. [CrossRef]
169. Arnold, K.E. Strategies of the cooperatively breeding noisy miner to reduce nest predation. *Emu-Austral Ornithol.* **2000**, *1* 280–285. [CrossRef]
170. NSW Department of Education. Field of Mars Environmental Education Centre>Factsheets>Bird Facts sheets>Noisy Miner. n Available online: https://fieldofmar-e.schools.nsw.gov.au/fact-sheets/birds/noisy-miner-fact-sheet (accessed on 2 April 202
171. Robinson, D. The nesting ecology of sympatric scarlet robin Petroica multicolor and flame robin P. phoenicea populations in op eucalypt forest. *Emu-Austral Ornithol.* **1990**, *90*, 40–52. [CrossRef]
172. Bridges, L. Breeding biology of a migratory population of the rufous whistler Pachycephala rufiventris. *Emu-Austral Ornith* **1994**, *94*, 106–115. [CrossRef]
173. Bell, H.L.; Ford, H.A. A comparison of the social organisation of three syntopic species of Australian thornbills Acanthiza. *Beh Ecol. Sociobiol.* **1986**, *19*, 381–392. [CrossRef]
174. Nias, R.C. Cooperative Breeding in the Superb Fairy-Wren *Malurus cyaneus*. Ph.D. Thesis, University of New England, Armida Australia, 1987.
175. Paton, D.C. Food supply, population structure and behaviour of New Holland honeyeaters Phylidonyris novaehollandiae woodland near Horsham, Victoria. In *Birds of Eucalypt Forests and Woodlands: Ecology, Conservation, Management*; Keast, A., Rech H.F., Ford, H., Saunders, D., Eds.; Surrey Beatty and Sons: Chipping Norton, NSW, Australia, 1985; p. 219.
176. Ford, H.A. Nest site selection and breeding success in large Australian honeyeaters: Are there benefits from being differe *Emu-Austral Ornithol.* **1999**, *99*, 91–99. [CrossRef]
177. Ford, J.R. Breeding behaviour of the yellow-tailed thornbill in south-western Australia. *Emu-Austral Ornithol.* **1963**, *63*, 185–2 [CrossRef]
178. Cameron, E. Habitat usage and foraging behaviour of three fantails (Rhipidura: Pachycephalidae). In *Birds of Eucalypt Fore and Woodlands: Ecology, Conservation, Management*; Keast, A., Recher, H.F., Ford, H., Saunders, D., Eds.; Surrey Beatty and So Chipping Norton, NSW, Australia, 1985; pp. 177–191.
179. Clarke, M.F. The reproductive behaviour of the bell miner Manorina melanophrys. *Emu-Austral Ornithol.* **1988**, *88*, 88–1 [CrossRef]
180. Dielenberg, J. Culling Noisy Miners Fails in NSW Trial, Study. The Threatened Species Recovery Hub, National Environme tal Science Program, 4 March 2019. Available online: https://www.nespthreatenedspecies.edu.au/news-and-media/med releases/culling-noisy-miners-fails-in-nsw-trial-study (accessed on 17 January 2026).
181. Montague-Drake, R.M.; Lindenmayer, D.B.; Cunningham, R.B.; Stein, J.A. A reverse keystone species affects the landsca distribution of woodland avifauna: A case study using the Noisy Miner (*Manorina melanocephala*) and other Australian bir *Landsc. Ecol.* **2011**, *26*, 1383–1394. [CrossRef]
182. Ward, M.; Ashman, K.; Lindenmayer, D.; Legge, S.; Kindler, G.; Cadman, T.; Fletcher, R.; Whiterod, N.; Lintermans, M.; Zylstra The impacts of contemporary logging after 250 years of deforestation and degradation on forest-dependent threatened speci *bioRxiv* **2023**. [CrossRef]
183. WWF. *Living Planet Report 2024—A System in Peril*; WWF: Gland, Switzerland, 2024.
184. Bradshaw, C.J. Little left to lose: Deforestation and forest degradation in Australia since European colonization. *J. Plant Ecol.* **20** *5*, 109–120. [CrossRef]
185. Thomas, H.; Ward, M.S.; Simmonds, J.S.; Taylor, M.F.; Maron, M. Poor compliance and exemptions facilitate ongoing deforestati *Conserv. Biol.* **2025**, *39*, e14354.147. [CrossRef]
186. Farwell, L.S.; Gudex-Cross, D.; Anise, I.E.; Bosch, M.J.; Olah, A.M.; Radeloff, V.C.; Razenkova, E.; Rogova, N.; Silveira, E.N Smith, M.M. Satellite image texture captures vegetation heterogeneity and explains patterns of bird richness. *Remote Sens. Envir* **2021**, *253*, 112175. [CrossRef]
187. Day, T.; Young, K.A. Competitive and facilitative evolutionary diversification. *BioScience* **2004**, *54*, 101–109. [CrossRef]
188. Guillaumet, A.; Russell, I.J. Bird Communities in a Changing World: The Role of Interspecific Competition. *Diversity* **2022**, *14*, 8 [CrossRef]

https://doi.org/10.3390/d18020C

Hildenbrandt, H.; Carere, C.; Hemelrijk, C.K. Self-organized aerial displays of thousands of starlings: A model. *Behav. Ecol.* **2010**, *21*, 1349–1359. [CrossRef]
Dekker, D. The murmurations of European starlings; an anti-predator strategy and a historical misnomer. *Northwestern Nat.* **2022**, *103*, 194–197. [CrossRef]
Crook, J.H. Studies on the social behaviour of Quelea q. quelea (Linn.) in French West Africa. *Behaviour* **1960**, *16*, 1–55. [CrossRef]
Hatton, I.A.; Mazzarisi, O.; Altieri, A.; Smerlak, M. Diversity begets stability: Sublinear growth and competitive coexistence across ecosystems. *Science* **2024**, *383*, eadg8488. [CrossRef]
Blakers, M.; Davies, S.J.J.F.; Reilly, P.N. *The Atlas of Australian Birds*; Melbourne University Press: Melbourne, Australia, 1984.
Barrett, G.; Silcocks, A.; Barry, S.; Cunningham, R.; Poulter, R. *The New Atlas of Australian Birds*; Royal Ornithologists Union, Melbourne University Press: Melbourne, Australia, 2003.
BirdLife Australia. Aussie Bird Count. 2023 Results. Available online: https://aussiebirdcount.org.au (accessed on 2 May 2025).
Spotteron Citizen Science, 'Big City Birds'. Available online: https://www.spotteron.net (accessed on 23 February 2025).
De Camargo Barbosa, K.V.; Develey, P.F.; Ribeiro, M.C.; Jahn, A.E. The contribution of citizen science to research on migratory and urban birds in Brazil. *Ornithol. Res.* **2021**, *29*, 1–11. [CrossRef]
Tubelis, D.P. Spatiotemporal distribution of photographic records of Brazilian birds available in the WikiAves citizen science database. *Birds* **2023**, *4*, 28–45. [CrossRef]
Callaghan, C.T.; Poore, A.G.; Hofmann, M.; Roberts, C.J.; Pereira, H.M. Large-bodied birds are over-represented in unstructured citizen science data. *Sci. Rep.* **2021**, *11*, 19073. [CrossRef]
Kamp, J.; Oppel, S.; Heldbjerg, H.; Nyegaard, T.; Donald, P.F. Unstructured citizen science data fail to detect long-term population declines of common birds in Denmark. *Divers. Distrib.* **2016**, *22*, 1024–1035. [CrossRef]
Petrenko, J.A.; Martin, P.R.; Fanelli, R.E.; Bonier, F. Urban tolerance does not protect against population decline in North American birds. *Biol. Lett.* **2024**, *20*, 20230507. [CrossRef]
Bell-James, J. Australia Desperately Needs a Strong Federal Environmental Protection Agency. Our Chances Aren't Looking Good. The Conversation, 18 September 2024. Available online: https://theconversation.com/australia-desperately-needs-a-strong-federal-environmental-protection-agency-our-chances-arent-looking-good-239099 (accessed on 10 May 2025).
The Victorian Corella Strategy 2022–2032 9 May 2024. Available online: https://www.wildlife.vic.gov.au/__data/assets/pdf_file/0042/637989/Victorian-Corella-Strategy-2022-2032.pdf (accessed on 10 May 2025).
Department of Primary Industries and Regions (PIRSA). Pest Bird Impacts in South Australia, Stake Holder Workshop 3 February 2025, Adelaide. LGA, S.A. Available online: https://www.lga.sa.gov.au/__data/assets/pdf_file/0044/1778948/ECM_834740_v1_2024-97-Developing-a-new-bait-for-pest-birds-of-the-grains-industry-Stakeholder-Workshop-data-an.pdf (accessed on 1 March 2025).
Bird Advocacy Foundation. Cherrp-myREGION.au. Available online: https://birdadvocacyfoundation.org./ (accessed on 17 January 2026).
Johnson, M. 'Feathered Foes': Soldier Settlers and Western Australia's 'Emu War'of 1932. *J. Aust. Stud.* **2006**, *30*, 147–157. [CrossRef]
Professional Pest Manager. Native Birds as Pests in Australia, 2 September 2024. Available online: https://australian.museum/learn/animals/birds/noisy-miner/ (accessed on 17 November 2024).
Bayly, K.L.; Blumstein, D.T. Pied Currawongs and the decline of native birds. *Emu-Austral Ornithol.* **2001**, *101*, 199–204. [CrossRef]
BirdLife. Birds in Backyards-Currawongs. Available online: https://birdlife.org.au/bird-profiles/pied-currawong/ (accessed on 17 January 2026).
Whelan, C.J.; Şekercioğlu, Ç.H.; Wenny, D.G. Why birds matter: From economic ornithology to ecosystem services. *J. Ornithol.* **2015**, *156*, 227–238. [CrossRef]
Gaston, K.J. Birds and ecosystem services. *Curr. Biol.* **2022**, *32*, R1163–R1166. [CrossRef] [PubMed]
Green, R.J. Avian seed dispersal in and near subtropical rainforests. *Wildl. Res.* **1993**, *20*, 535–557. [CrossRef]
Wood, K.A. Seasonal changes in diet of pied currawongs *Strepera graculina* at Wollongong, New South Wales. *Emu-Austral Ornithol.* **1998**, *98*, 157–170. [CrossRef]
Brock, P.D.; Hasenpusch, J.W. *The Complete Field Guide to Stick and Leaf Insects of Australia*; CSIRO Publishing: Melbourne, Australia, 2009.
Low, T.; Booth, C. *GONE: Australian Animals Extinct Since the 1960s*; Invasive Species Council Inc.: Katoomba, Australia, 2023.
Whisson, D.A.; Rendall, A.R.; Gibson, M. History of Australian plant and vertebrate pests: Introductions and impacts. *Victorian Nat.* **2020**, *137*, 162–169.
Murray, M. Noisy Miner, Manorina melanocephala. Australian Museum. Online Discover & Learn/Animal Factsheets/Birds. Last updated 23/07/24. Available online: https://australian.museum/learn/animals/birds/ (accessed on 2 May 2025).

https://doi.org/10.3390/d18020099

218. Kikuchi, D.W.; Allen, W.L.; Arbuckle, K.; Aubier, T.G.; Briolat, E.S.; Burdfield-Steel, E.R.; Cheney, K.L.; Daňková, K.; Elias, Hämäläinen, L.; et al. The evolution and ecology of multiple antipredator defences. *J. Evol. Biol.* **2023**, *36*, 975–991. [CrossRef] [PubMed]
219. Bengtson, S. Origins and early evolution of predation. In *The Fossil Record of Predation*; Kowalewski, M., Kelley, P.H., Eds.; The Paleontological Society Papers 8; The Paleontological Society: Knoxville, TN, USA, 2002; pp. 289–317.
220. Yogeshpriya, S.; Rajeshwari, K.; Sivaraman, S. Predation detection and anti-predator defenses. In *Principles of Animal Behavio Aspects of Behavioural Biology and Evolution*; Rana, T., Ed.; CRC Press: Boca Raton, FL, USA, 2025; Chapter 11, pp. 103–1 [CrossRef]
221. Bicknell, R.D.; Campione, N.E.; Brock, G.A.; Paterson, J.R. Adaptive responses in Cambrian predator and prey highlight the ar race during the rise of animals. *Curr. Biol.* **2025**, *35*, 882–888. [CrossRef]
222. Blythman, M. *Guidelines for the Approved Control Technique for Introduced Corellas*; Western Australian Department of Parks Wildlife: Perth, Australia, 2017.
223. Blythman, M.; Porter, G. Movement of introduced Little Corellas Cacatua sanguinea and Long-billed Corellas C. tenuirostris south-western Western Australia. *Aust. Field Ornith.* **2020**, *37*, 48–55. [CrossRef]
224. Stock, P. 'Heartbreaking': Poisoning suspected after mass deaths of more than 200 little corellas in Newcastle. *The Guardian*, March 2025.
225. The NSW Environment Protection Authority (EPA). *EPA Investigating Suspected Corella Mass Poisoning in Newcastle 18 March 20* EPA: Parramatta, Australia, 2025.
226. Johnston, G.R. Breeding biology of Australian Pelicans (Pelecanus conspicillatus) on the coast of South Australia. *Waterbirds* **20** *39*, 300–305. [CrossRef]
227. Mott, R.; Prowse, T.A.A.; Jackson, M.V.; Sanchez-Gomez, S.; Bennet, A.T.D.; O'Connor, J.; Rogers, D.J.; Brookes, J.D.; Cassey Red-necked avocets disperse at continental scales and breed following high rainfall in distant locations. *Sci. Rep.* **2025**, *15*, 161 [CrossRef]
228. Reid, J. Australian pelican: Flexible responses to uncertainty. In *Boom and Bust. Bird Stories for a Dry Country*; Robin, L., Heinso R., Joseph, L., Eds.; BioOne Digital Library: Washington, DC, USA, 2009. [CrossRef]
229. Jordan, R.; James, A.I.; Moore, D.; Franklin, D.C. Boom and bust (or not?) among birds in an Australian semi-desert. *J. A Environ.* **2017**, *139*, 58–66. [CrossRef]
230. Pascoe, B.A.; Pavey, C.R.; Morton, S.R.; Schlesinger, C.A. Dynamics of bird assemblages in response to temporally and spatia variable resources in arid Australia. *Ecol. Evol.* **2021**, *11*, 3977–3990. [CrossRef]
231. Mangini, G.G.; Rutt, C.L.; Sridhar, H.; Buitron, G.; Muñoz, J.; Robinson, S.K.; Montaño-Centellas, F.; Zarco, A.; Fanjul, M Fernández-Arellano, G.; et al. A classification scheme for mixed-species bird flocks. *Philos. Trans. R. Soc. B* **2023**, *378*, 202201 [CrossRef]
232. Beauchamp, G. Flocking in birds increases annual adult survival in a global analysis. *Oecologia* **2021**, *197*, 387–394. [CrossRef
233. Crates, R.; McDonald, P.G.; Melton, C.B.; Maron, M.; Ingwersen, D.; Mowat, E.; Breckenridge, M.; Murphy, L.; Heinsohn Towards effective management of an overabundant native bird: The noisy miner. *Conserv. Sci. Pract.* **2023**, *5*, e12875. [CrossR
234. Brouwer, J.; Garnett, S.T. (Eds.) *Threatened Birds of Australia. An Annotated List*; Royal Australasian Ornithologists Uni Melbourn, Australia, 1990.
235. Verdon, S.J.; Davis, R.A.; Tulloch, A.; Legge, S.M.; Watson, D.M.; Woinarski, J.C.; Baker, G.B.; Driessen, J.; Geyle, H.M.; Possingha H.; et al. Trends in monitoring of Australia's threatened birds (1990–2020): Much improved but still inadequate. *Emu-Aust Ornithol.* **2024**, *124*, 21–36. [CrossRef]
236. Lintermans, M.; Geyle, H.M.; Beatty, S.; Brown, C.; Ebner, B.C.; Freeman, R.; Hammer, M.P.; Humphreys, W.F.; Kennard, M Kern, P.; et al. Big trouble for little fish: Identifying Australian freshwater fishes in imminent risk of extinction. *Pac. Conserv. B* **2020**, *26*, 365–377. [CrossRef]
237. Heatwole, H.; Rowley, J. (Eds.) *Status of Conservation and Decline of Amphibians: Australia, New Zealand, and Pacific Islands*; CSI Publishing: Melbourne, Australia, 2018.
238. Bayraktarov, E.; Ehmke, G.; Tulloch, A.I.; Chauvenet, A.L.; Avery-Gomm, S.; McRae, L.; Wintle, B.A.; O'Connor, J.; Driessen Watmuff, J.; et al. A threatened species index for Australian birds. *Conserv. Sci. Pract.* **2021**, *3*, e322. [CrossRef]
239. Woinarski, J.C.Z.; Burbidge, A.A.; Harrison, P.L. Ongoing unravelling of a continental fauna: Decline and extinction of Australi mammals since European settlement. *Proc. Natl. Acad. Sci. USA* **2015**, *112*, 4531–4540. [CrossRef] [PubMed]
240. Brodie, J.F.; Williams, S.; Garner, B. The decline of mammal functional and evolutionary diversity worldwide. *Proc. Natl. Ac Sci. USA* **2021**, *118*, e1921849118. [CrossRef]
241. Santos, J.L.; Hradsky, B.A.; Keith, D.A.; Rowe, K.C.; Senior, K.L.; Sitters, H.; Kelly, L.T. Beyond inappropriate fire regimes: synthesis of fire-driven declines of threatened mammals in Australia. *Conserv. Lett.* **2022**, *15*, e12905. [CrossRef]

Lavery, T.; Lindenmayer, D.; Blanchard, W.; Carey, A.; Cook, E.; Copley, P.; Macgregor, N.A.; Melzer, R.; Nano, C.; Prentice, L.; et al. Counting plants: The extent and adequacy of monitoring for a continental-scale list of threatened plant species. *Biol. Conserv.* **2021**, *260*, 109193. [CrossRef]

Woinarski, J.C.; Garnett, S.T.; Gillespie, G.; Legge, S.M.; Lintermans, M.; Rumpff, L. Lights at the end of the tunnel: The incidence and characteristics of recovery for Australian threatened animals. *Biol. Conserv.* **2023**, *279*, 109946. [CrossRef]

Scheele, B.C.; Legge, S.; Blanchard, W.; Garnett, S.; Geyle, H.; Gillespie, G.; Harrison, P.; Lindenmayer, D.; Lintermans, M.; Robinson, N.; et al. Continental-scale assessment reveals inadequate monitoring for threatened vertebrates in a megadiverse country. *Biol. Conserv.* **2019**, *235*, 273–278. [CrossRef]

Watson, D.M. A productivity-based explanation for woodland bird declines: Poorer soils yield less food. *Emu-Austral Ornithol.* **2011**, *111*, 10–18. [CrossRef]

Woinarski, J.C.Z.; Braby, M.F.; Burbidge, A.A.; Coates, D.; Garnett, S.T.; Fensham, R.J.; Legge, S.M.; McKenzie, N.L.; Silcock, J.L.; Murphy, B.P. Reading the black book: The number, timing, distribution and causes of listed extinctions in Australia. *Biol. Conserv.* **2016**, *239*, 108261. [CrossRef]

Gautschi, D.; Heinsohn, R.; Murphy, L.; Crates, R. Landscape-scale distribution of nest predators and its relationship with regent honeyeater nest success. *Austral Ecol.* **2021**, *46*, 1322–1334. [CrossRef]

Threatened Species Recovery Hub. A Review of Listed Extinctions in Australia. 2019. Available online: https://www.nespthreatenedspecies.edu.au/news-and-media/latest-news/a-review-of-listed-extinctions-in-australia (accessed on 17 January 2026).

Wenzel, A.; Grass, I.; Raj, V.; Nölke, N.; Subramanya, S.; Tscharntke, T. High losses of farmland birds and potential biocontrol along an urbanization gradient in a tropical megacity. *Agric. Ecosyst. Environ.* **2023**, *354*, 108571. [CrossRef]

Gould, J. *A Synopsis of the Birds of Australia and the Adjacent Islands*; 73 hand-coloured lithographic plates by and after Elizabeth Gould; Published by the Author: London, UK, 1837–1838.

Rose, S. What is wrong with reductionist explanations of behaviour? In *Novartis Foundation Symposium*; John Wiley & Sons, Ltd.: Chichester, UK, 2007; Volume 213, pp. 176–192.

Threatened Species Recovery Hub. Project 1.2.1.2. Can Culling Noisy Miners Save Declining Small Woodland Birds in Fragmented Temperate Woodlands? Research Findings Fact Sheet. National Environmental Science Programme. Available online: https://www.nespthreatenedspecies.edu.au/publications-and-tools/can-culling-noisy-miners-save-declining-small-woodland-birds-in-fragmented-temperate-woodlands (accessed on 17 January 2026).

Threatened Species Recovery Hub. Project 3.2.6. Does Culling Noisy Miners have Conservation Benefits for Small Woodland birds? Evidence from 45 Removals. Research Findings Fact Sheet. (n.d.) National Environmental Science Programme. Available online: https://www.nespthreatenedspecies.edu.au (accessed on 17 January 2026).

Watson, J.; Watson, A.; Paull, D.; Freudenberger, D. Woodland fragmentation is causing the decline of species and functional groups of birds in southeastern Australia. *Pac. Conserv. Biol.* **2002**, *8*, 261–270. [CrossRef]

Green, K.; Caley, P.; Baker, M.; Dreyer, D.; Wallace, J.; Warrant, E. Australian Bogong moths Agrotis infusa (Lepidoptera: Noctuidae), 1951–2020: Decline and crash. *Austral Entomol.* **2021**, *60*, 66–81. [CrossRef] [PubMed]

Parsons, K.C.; Mineau, P.; Renfrew, R.B. Effects of pesticide use in rice fields on birds. *Waterbirds* **2010**, *33*, 193–218. [CrossRef]

Warwick, H. *Cull of the Wild: Killing in the Name of Conservation*; Bloomsbury Publishing: London, UK, 2024.

Batavia, C.; Nelson, M.P.; Wallach, A.D. The moral residue of conservation. *Conserv. Biol.* **2020**, *34*, 1114–1121. [CrossRef]

Leishman, A.; Leishman, A. Banding Project Report No.10. *Corella* **2021**, *45*, S1–S11.

Debus, S.J. The effect of noisy miners on small bush birds: An unofficial cull and its outcome. *Pac. Conserv. Biol.* **2008**, *14*, 185–190201. [CrossRef]

BirdLife Australia. Shrubby-Understorey Specialists. A Guide to Functional Groups. Available online: https://birdlife.org.au/?srsltid=AfmBOoqxqqRqKI8DrNQvgXRAENr3A63M8X2b3z6JcwkZ3ZdHWwVjAFTv (accessed on 8 April 2025).

NATMAP; Bullen, F.; Carnahan, J.; Deveson, T. *Atlas of Australian Resources Third Series, Volume 6: Vegetation*; Commonwealth of Australia (Geoscience Australia); GA Publication: Canberra, Australia, 1990.

Oldland, J.M.; Taylor, R.S.; Clarke, M.F. Habitat preferences of the noisy miner (*Manorina melanocephala*)—A propensity for prime real estate? *Austral Ecol.* **2009**, *34*, 306–316. [CrossRef]

Fountain, J.; McDonald, P.G. Do differences in the availability of anthropogenic food resources influence the observed levels of agonistic behaviour in Noisy Miners (*Manorina melanocephala*)? *Emu-Austral Ornith.* **2022**, *122*, 61–70. [CrossRef]

Ottoni, I.; de Oliveira, F.F.; Young, R.J. Estimating the diet of urban birds: The problems of anthropogenic food and food digestibility. *Appl. Anim. Behav. Sci.* **2009**, *117*, 42–46. [CrossRef]

Burger, J.; Gochfeld, M. *Habitat, Population Dynamics, and Metal Levels in Colonial Waterbirds: A Food Chain Approach*; CRC Press: Boca Raton, FL, USA, 2016; p. 254.

Fehlmann, G.; Martin, J.M.; Safi, K.; Aplin, L.M. Wild Sulphur-crested Cockatoos match human activity rhythms to access food in the urban environment. *Urban Ecosyst.* **2024**, *27*, 2179–2189. [CrossRef]

https://doi.org/10.3390/d18020099

268. Walsh, J.C.; Gibson, M.R.; Simmonds, J.S.; Mayfield, H.J.; Bracey, C.; Melton, C.B.; Reside, A.E.; Maron, M. Effectiveness conservation interventions for Australian woodland birds: A systematic review. *Biol. Conserv.* **2023**, *282*, 11003. [CrossRef]
269. Miller, M.R. Descartes on animals revisited. *J. Philos. Res.* **2013**, *38*, 89–114. [CrossRef]
270. Tinbergen, N. *The Herring Gull's World: A Study of the Social Behaviour of Birds*; Collins: London, UK, 1953.
271. Tinbergen, N. *Social Behaviour in Animals, with Special Reference to Vertebrates*; Psych. Press: London, UK, 1964.
272. Wright, P.; Caryl, P.G.; Vowles, D.M. (Eds.) *Neural and Endocrine Aspects of Behaviour in Birds*; Elsevier Scientific Publishi Company: Amsterdam, The Netherlands; Oxford, UK; New York, NY, USA, 1975.
273. Puig, M.V.; Rose, J.; Schmidt, R.; Freund, N. Dopamine modulation of learning and memory in the prefrontal cortex: Insig from studies in primates, rodents, and birds. *Front. Neural Circuits* **2014**, *8*, 93. [CrossRef]
274. Dietl, M.M.; Palacios, J.M. Neurotransmitter receptors in the avian brain. I. Dopamine receptors. *Brain Res.* **1988**, *439*, 354–3 [CrossRef]
275. Pimm, S.; Raven, P.; Peterson, A.; Şekercioğlu, Ç.H.; Ehrlich, P.R. Human impacts on the rates of recent, present, and future bi extinctions. *Proc. Natl. Acad. Sci. USA* **2006**, *103*, 10941–10946. [CrossRef] [PubMed]
276. Taylor-Brown, A.; Booth, R.; Gillett, A.; Mealy, E.; Ogbourne, M.; Polkinghorne, A.; Conroy, G.C. The impact of human activit on Australian wildlife. *PLoS ONE* **2019**, *14*, e0206958. [CrossRef]
277. Feldman-Hall, O.; Glimcher, P.; Baker, A.L.; Nyu Prospec Collaboration; Phelps, E.A. The functional roles of the amygdala a prefrontal cortex in processing uncertainty. *J. Cogn. Neurosci.* **2019**, *31*, 1742–1754. [CrossRef]
278. LeDoux, J.E. The Amygdala: Contributions to fear and stress. In *Seminars in Neuroscience*; Academic Press: New York, NY, US 1994; Volume 6, pp. 231–237.
279. Agnvall, B.; Katajamaa, R.; Altimiras, J.; Jensen, P. Is domestication driven by reduced fear of humans? Boldness, metabolism a serotonin levels in divergently selected red junglefowl (Gallus gallus). *Biol. Lett.* **2015**, *11*, 20150509. [CrossRef]
280. Tinbergen, N. On aims and methods of ethology. *Z. Für Tierpsychol.* **1963**, *20*, 410–433. [CrossRef]
281. Hinde, A. The biological significance of the territories of birds. *IBIS* **1956**, *98*, 340–369. [CrossRef]
282. Redondo, T. Avian nest defence: Theoretical models and evidence. *Behaviour* **1989**, *1111*, 161–195. [CrossRef]
283. Oberski, I.M.; Wilson, J.D. Territoriality and site-related dominance: On two related concepts in avian social organization. *Etholo* **1991**, *87*, 225–236. [CrossRef]
284. Briffa, M. Agonistic signals: Integrating analysis of functions and mechanisms. In *Animal Signaling and Function: An Integrat Approach*; Irschick, D.J., Briffa, M., Podos, J., Eds.; Wiley Online Library: Hoboken, NJ, USA, 2014; Chapter 6, pp. 141–1 [CrossRef]
285. Brandl, H.B.; Farine, D.R. Stress in the social environment: Behavioural and social consequences of stress transmission in bi flocks. *Proc. B* **2024**, *291*, 20241961. [CrossRef]
286. Lima, S.L. Nonlethal effects in the ecology of predator-prey interactions. *Bioscience* **1998**, *48*, 25–34. [CrossRef]
287. Lin, H.Y.; Song, G.; Lei, F.; Li, D.; Qu, Y. Avian corticosteroid-binding globulin: Biological function and regulatory mechanisms physiological stress responses. *Front. Zool.* **2021**, *18*, 22. [CrossRef] [PubMed]
288. Beattie, U.K.; Ysrael, M.C.; Lok, S.E.; Romero, L.M. The effect of a combined fast and chronic stress on body mass, blo metabolites, corticosterone, and behavior in house sparrows (*Passer domesticus*). *Yale J. Biol. Med.* **2022**, *95*, 19. [PubMed]
289. Cyr, N.E.; Romero, L.M. Chronic stress in free-living European starlings reduces corticosterone concentrations and reproducti success. *Gen. Comp. Endocrinol.* **2007**, *151*, 82–89. [CrossRef] [PubMed]
290. Wada, H.; Coutts, V. Detrimental or beneficial? Untangling the literature on developmental stress studies in birds. *J. Exp. B* **2021**, *224*, jeb227363. [CrossRef]
291. Exnerová, A.; Kang, C.; Rowland, H.M.; Kikuchi, D.W. Evolution of multiple prey defences: From predator cognition community ecology. *J. Evol. Biol.* **2023**, *36*, 961–966. [CrossRef]
292. Clinchy, M.; Schulkin, J.; Zanette, L.Y.; Sheriff, M.J.; McGowan, P.O.; Boonstra, R. The neurological ecology of fear: Insigh neuroscientists and ecologists have to offer one another. *Front. Behav. Neurosci.* **2011**, *5*, 21. [CrossRef]
293. Friedman, M.J.; Keane, T.M.; Resick, P.A. (Eds.) *Handbook of PTSD: Science and Practice*; Guilford Press: New York, NY, USA, 20
294. Yenkosky, J.P.; Bradshaw, G.A.; McCarthy, E. Post-Traumatic Stress Disorder among parrots in captivity: Treatment consideratio In *Proceedings of the Association of Avian Veterinarians*; Association of Avian Veterinarians: Watkinsville, GA, USA, 2010; pp. 17–
295. Blumstein, D.T. Developing an evolutionary ecology of fear: How life history and natural history traits affect disturban tolerance in birds. *Anim. Behav.* **2006**, *71*, 389–399. [CrossRef]
296. Clinchy, M.; Sheriff, M.J.; Zanette, L.Y. Predator-induced stress and the ecology of fear. *Funct. Ecol.* **2013**, *27*, 56–65. [CrossRef
297. Clinchy, M.; Zanette, L.Y.; Roberts, D.; Suraci, J.P.; Buesching, C.D.; Newman, C.; Macdonald, D.W. Fear of the human "sup predator" far exceeds the fear of large carnivores in a model mesocarnivore. *Behav. Ecol.* **2016**, *27*, 1826–1832. [CrossRef]
298. Zanette, L.Y.; Clinchy, M. Ecology of fear. *Curr. Biol.* **2019**, *29*, R309–R313. [CrossRef]
299. Young, C. Agonistic Behavior. In *Encyclopedia of Animal Cognition and Behavior*; Springer International Publishing: Cha Switzerland, 2022; pp. 143–148.

https://doi.org/10.3390/d180200

Blackburn, G.; Ashton, B.J.; Ridley, A.R. Evidence that multiple anthropogenic stressors cumulatively affect foraging and vigilance in an urban-living bird. *Anim. Behav.* **2024**, *211*, 81–90. [CrossRef]
Caro, T.M. *Antipredator Defenses in Birds and Mammals*; University of Chicago Press: Chicago, IL, USA, 2005.
Willcox, K. Using Cognition to Assess the Welfare of Wild Birds. Master's Thesis, School of Life Sciences, University of Lincoln, Lincoln, UK, 2021.
Schmidt, K.; Kuijper, D.P. A "death trap" in the landscape of fear. *Mammal Res.* **2015**, *60*, 275–284. [CrossRef]
Laundré, J.W.; Hernández, L.; Ripple, W.J. The landscape of fear: Ecological implications of being afraid. *Open Ecol. J.* **2010**, *3*, 1–7. [CrossRef]
Pike, G.H.; Christy, M.; Major, R.E. Territoriality in Honeyeaters: Reviewing the concept and evaluating available information. *Aust. J. Zool.* **1996**, *44*, 297–317. [CrossRef]
Teckentrup, L.; Kramer-Schadt, S.; Jeltsch, F. The risk of ignoring fear: Underestimating the effects of habitat loss and fragmentation on biodiversity. *Landsc. Ecol.* **2019**, *34*, 2851–2868. [CrossRef]
Harding, E.J.; Paul, E.S.; Mendl, M. Cognitive bias and affective state. *Nature* **2004**, *427*, 312. [CrossRef] [PubMed]
Bateson, M. Optimistic and pessimistic biases: A primer for behavioural ecologists. *Curr. Opin. Behav. Sci.* **2016**, *12*, 115–121. [CrossRef]
Siewert, V.; Kaiser, S.; Sachser, N.; Richter, S.H. Optimism and pessimism: A concept for behavioural ecology. *Biol. Rev.* **2025**, *100*, 1152–1162. [CrossRef]
Koolhaas, J.M.; Korte, S.M.; De Boer, S.F.; Van Der Vegt, B.J.; Van Reenen, C.G.; Hopster, H.; De Jong, I.C.; Ruis, M.A.W.; Blokhuis, H.J. Coping styles in animals: Current status in behavior and stress-physiology. *Neurosci. Biobehav. Rev.* **1999**, *23*, 925–935. [CrossRef]
Kaplan, G. Vigilance. Encyclopaedic essays. In *Encyclopaedia of Animal Behaviour and Animal Cognition*; Vonk, J., Shackelford, T.K., Eds.; Springer International Publishing AG/Nature: Cham, Switzerland, 2022.
Beauchamp, G. A comparative analysis of vigilance in birds. *Evol. Ecol.* **2010**, *24*, 1267–1276. [CrossRef]
Elgar, M.A. The establishment of foraging flocks in house sparrows: Risk of predation and daily temperature. *Behav. Ecol. Sociobiol.* **1986**, *19*, 433–438. [CrossRef]
Beschta, R.L.; Ripple, W.J. Riparian vegetation recovery in Yellowstone: The first two decades after wolf reintroduction. *Biol. Conserv.* **2016**, *198*, 93–103. [CrossRef]
Valone, T.J.; Lima, S.L. Carrying food items to cover for consumption: The behavior of ten bird species feeding under the risk of predation. *Oecologia* **1987**, *71*, 286–294. [CrossRef]
Phillips, L.J.J.; Barati, A.; McDonald, P.G. New culling technique provides localised reductions in interspecific aggression and density of Noisy Miners (*Manorina melanocephala*) without recolonisation. *Emu-Austral Ornithol.* **2025**, *125*, 2–13. [CrossRef]
Gaynor, K.M.; Brown, J.S.; Middleton, A.D.; Power, M.E.; Brashares, J.S. Landscapes of fear: Spatial patterns of risk perception and response. *Trends Ecol. Evol.* **2019**, *34*, 355–368. [CrossRef] [PubMed]
Şekercioğlu, Ç.H.; Sutherland, W.J.; Buechley, E.R.; Li, B.V.; Ocampo-Peñuela, N.; Mahamued, B.A. Avian biodiversity collapse in the Anthropocene: Drivers and consequences. *Front. Ecol. Evol.* **2023**, *11*, 1202621. [CrossRef]
Armstrong, E.A. *Bird Display and Behavior*; Dover Publications: New York, NY, USA, 1965.
Igic, B.; Magrath, R. Turning the tables: A tiny bird uses alarm calls and mimicry to deceive its nest predator. *Biol. Lett.* **2025**, *21*, 20240710. [CrossRef] [PubMed]
Lumsden, H.G. The display of the capercaillie. *Br. Birds* **1961**, *54*, 257–272.
Fielding, M. The Cascading Effects of Top Scavenger Loss on Bird Communities: A Case Study on a Shifting Strait. Ph.D. Thesis, University of Tasmania, Hobart, Australia, 2024. [CrossRef]
Natsukawa, H.; Sergio, F.J. Top predators as biodiversity indicators: A meta-analysis. *Ecol. Lett.* **2022**, *25*, 2062–2075. [CrossRef]
O'Connell, J.F.; Allen, J.; Williams, M.A.J.; Williams, A.N.; Turney, C.S.M.; Spooner, N.A.; Kamminga, J.; Brown, G.; Cooper, A. When did Homo sapiens first reach Southeast Asia and Sahul? *Proc. Natl. Acad. Sci. USA* **2018**, *115*, 8482–8490. [CrossRef]
Hindwood, K.A. John Gould in Australia. *Emu-Austral Ornith.* **1938**, *38*, 95–118. [CrossRef]
Carpenter, G.; Black, A. John Gould in South Australia and a reappraisal of his type locality,' The Belts of the Murray'. *South Austral. Ornithologist* **2015**, *15*, 41.
Maciag, T.; Remisiewicz, M. Climate Change Impact on the Populations of Goldcrest *Regulus regulus* and Firecrest *regulus ignicapilla* Migrating Through the Southern Baltic Coast. *Sustainability* **2025**, *17*, 1243. [CrossRef]
Bennett, A.F.; Holland, G.J.; Haslem, A.; Stewart, A.; Radford, J.Q.; Clarke, R.H. Restoration promotes recovery of woodland birds in agricultural environments: A comparison of 'revegetation' and 'remnant' landscapes. *J. Appl. Ecol.* **2022**, *59*, 1334–1346. [CrossRef]
Maron, M. Threshold effect of eucalypt density on an aggressive avian competitor. *Biol. Conserv.* **2007**, *136*, 100–107. [CrossRef]
Marion, L. Territorial feeding and colonial breeding are not mutually exclusive: The case of the Grey Heron (*Ardea cinerea*). *J. Anim. Ecol.* **1989**, *58*, 693–710. [CrossRef]

https://doi.org/10.3390/d18020099

331. Hegner, R.E.; Emlen, S.T.; Demong, N.J. Spatial organization of the white-fronted bee-eater. *Nature* **1982**, *298*, 264–266. [CrossR
332. Reyer, H.U. Flexible helper structure as an ecological adaptation in the pied kingfisher (*Ceryle rudis rudis*). *Behav. Ecol. Sociol* **1980**, *6*, 219–227. [CrossRef]
333. Trouwborst, A.; McCormack, P.C.; Camacho, E.M. Domestic cats and their impacts on biodiversity: A blind spot in the applicat of nature conservation law. *People Nat.* **2020**, *2*, 235–250. [CrossRef]
334. Chace, J.F.; Walsh, J.J. Urban effects on native avifauna: A review. *Landsc. Urban Plan.* **2006**, *74*, 46–69. [CrossRef]
335. Matthews, T.J.; Triantis, K.A.; Wayman, J.P.; Martin, T.E.; Hume, J.P.; Cardoso, P.; Faurby, S.; Mendenhall, C.D.; Dufour, P.; Ri F.; et al. The global loss of avian functional and phylogenetic diversity from anthropogenic extinctions. *Science* **2024**, *386*, 55- [CrossRef]

https://doi.org/10.3390/d18020(

MDPI AG
Grosspeteranlage 5
4052 Basel
Switzerland
Tel.: +41 61 683 77 34

Diversity Editorial Office
E-mail: diversity@mdpi.com
www.mdpi.com/journal/diversity

www.ingramcontent.com/pod-product-compliance
Lightning Source LLC
LaVergne TN
LVHW070459061126
851858LV00011B/265